APPETITE AND
BODY WEIGHT

—————

INTEGRATIVE SYSTEMS AND THE DEVELOPMENT
OF ANTI-OBESITY DRUGS

APPETITE AND BODY WEIGHT

Integrative Systems and the Development of Anti-Obesity Drugs

Edited by

TIM C. KIRKHAM

and

STEVEN J. COOPER

University of Liverpool, School of Psychology, Liverpool, UK

ELSEVIER

AMSTERDAM • BOSTON • HEIDELBERG • LONDON
NEW YORK • OXFORD • PARIS • SAN DIEGO
SAN FRANCISCO • SINGAPORE • SYDNEY • TOKYO

Academic Press is an imprint of Elsevier

Academic Press is an imprint of Elsevier
30 Corporate Drive, Suite 400, Burlington, MA 01803, USA
525 B Street, Suite 1900, San Diego, California 92101-4495, USA
84 Theobald's Road, London WC1X 8RR, UK

This book is printed on acid-free paper. ∞

Library of Congress Cataloging-in-Publication Data
Application submitted

British Library Cataloguing-in-Publication Data
A catalogue record for this book is available from the British Library.

ISBN 13: 978-0-12-370633-1
ISBN 10: 0-12-370633-5

For information on all Academic Press publications
visit our Web site at www.books.elsevier.com

Transferred to Digital Print 2008
Printed and bound by CPI Antony Rowe, Eastbourne

Contents

Contributors   vii

1. Introduction and Overview   1
Steven J. Cooper and Tim C. Kirkham

2. Cortical Systems Involved in Appetite and Food Consumption   5
Morten L. Kringelbach

I.   Introduction   5
II.   Food Motivation   6
III. Cortical Representations of Sensory Inputs   8
IV.   Conclusion   20
Acknowledgments   20
References   20

3. The Nucleus Accumbens Shell as a Model of Integrative Subcortical Forebrain Systems Regulating Food Intake   27
Thomas R. Stratford

I.   Introduction   28
II.   The Regulation of Feeding Behavior by the Nucleus Accumbens Shell   29
III. Afferent Projections to the AcbSh That May Be Involved in Regulating Food Intake   38
IV.   Downstream Components of the Functional AcbSh Feeding Circuit   43
V.   The AcbSh Feeding Circuit As a Potential Site for Pathology and Therapeutic Intervention in the Treatment of Eating Disorders   53
References   55

4. Hypothalamic Neuropeptides and Feeding Regulation   67
Bernard Beck

I.   Introduction   68
II.   Neuropeptide Y   69

III.   The Orexins   75
IV.   Melanin-Concentrating Hormone   78
V.   Conclusion   81
Acknowledgments   81
References   81

5. **Brainstem-Hypothalamic Neuropeptides and the Regulation of Feeding   99**
Simon M. Luckman

I.   The Control of Feeding   100
II.   Neural Integration by the Dorsal Vagal Complex   103
III.   Considerations in Assessing the Role of "Neuropeptides"   105
IV.   "Humoral" Interactions with the Dorsal Vagal Complex   107
V.   Intrinsic Peptidergic Neurons of the Dorsal Vagal Complex   114
VI.   Descending Peptidergic Regulation of Brainstem Feeding Circuits   119
VII.   Summary and Potential for Drug Development   122
References   123

6. **The Gut-Brain Axis in the Control of Eating   143**
Thomas A. Lutz and Nori Geary

I.   Introduction   144
II.   Gastric Mechanoreception   145
III.   Intestinal Cholecystokinin (CCK)   146
IV.   Amylin   152
V.   Ghrelin   155
VI.   Potentials and Problems of Gut–Brain Axis Signals in the Treatment of Obesity   158
References   160

7. **Integration of Peripheral Adiposity Signals and Psychological Controls of Appetite   167**
Dianne Figlewicz Lattemann, Nicole M. Sanders, Amy MacDonald Naleid and Alfred J. Sipols

I.   Introduction and Overview   168
II.   Mesolimbic Dopamine Circuitry and Energy Regulatory Signals   170
III.   Brain Opioid Systems and Energy Regulatory Signals   171
IV.   Endocannabinoids and Energy Regulatory Signals   175
V.   LHA Circuitry and Energy Regulatory Signals   176
VI.   Other CNS Sites: Target for Future Studies?   178
VII.   Human and Clinical Studies: At the Forefront of Our Knowledge   179
VIII. Concluding Remarks   183
Acknowledgments   183
References   183

8. **Brain Reward Systems for Food Incentives and Hedonics in Normal Appetite and Eating Disorders   191**
Kent C. Berridge

I.   Introduction   192
II.   Possible Roles of Brain Reward Systems in Eating Disorders   192
III.   Understanding Brain Reward Systems for Food "Liking" and "Wanting"   193

IV. "Wanting" Without "Liking" 201
V. A Brief History of Appetite: Food Incentives, Not Hunger Drives 208
VI. Connecting Brain Reward and Regulatory Systems 210
VII. Conclusion 210
Acknowledgments 211
References 211

9. **Pharmacology of Food, Taste and Learned Flavor Preferences 217**
Steven J. Cooper

I. Introduction 217
II. Pharmacology of Food Preference 219
III. Pharmacology of Unlearned Taste Preference and Reactivity 224
IV. Pharmacology of Learned Flavor Preference 232
V. Concluding Remarks 236
Acknowledgments 238
References 238

10. **The Role of Palatability in Control of Human Appetite: Implications for Understanding and Treating Obesity 247**
Martin R. Yeomans

I. Introduction 248
II. Assessing the Effects of Palatability on Appetite 248
III. Palatability and the Control of Normal Appetite 250
IV. Palatability and Obesity 262
V. Conclusion 264
References 264

11. **Learned Influences on Appetite, Food Choice, and Intake: Evidence in Human Beings 271**
E.L. Gibson and J.M. Brunstrom

I. Introduction 272
II. Innate Influences on Human Eating 273
III. Types of Learning 274
IV. The Learned Appetite for Energy 277
V. Learned Modulation of Appetite and Meal Size by Associated States 278
VI. Nutrient-Specific Learned Appetites 283
VII. Flavor-Flavor Learning 286
VIII. Awareness and Dietary Learning 290
IX. Summary 293
References 294

12. **Gene Environment Interactions and the Origin of the Modern Obesity Epidemic: A Novel "Nonadaptive Drift" Scenario 301**
John R. Speakman

I. Introduction 302
II. Evidence Supporting the Famine Hypothesis 303

III.   Evidence Against the Famine Hypothesis   304
IV.   The Challenge Facing Evolutionary Scenarios for the Genetic Predisposition to Obesity in Modern Societies   306
V.   An Alternative Model for the Evolution of the Genetic Basis for Obesity   306
VI.   Time Trends in Obesity Prevalence: The Interaction of Physiology and Social Factors   314
VII.   Implications   316
Acknowledgments   318
References   318

13. Preclinical Developments in Antiobesity Drugs   323
Steven P. Vickers and Sharon C. Cheetham

I.   Introduction   324
II.   CNS Targets for Novel Antiobesity Drugs   325
III.   Peripheral Targets for Novel Antiobesity Drugs   330
IV.   Conclusion   332
References   333

14. Clinical Investigations of Antiobesity Drugs   337
John Wilding

I.   Introduction   338
II.   A Brief History of Obesity Pharmacotherapy   338
III.   Currently Available Drugs Licensed for Obesity Treatment   339
IV.   Potential New Targets for Obesity Drug Development   342
V.   Preclinical Testing   345
VI.   Regulatory Requirements for Clinical Development   346
References   352

Index   357

Contributors

Numbers in parenthesis indicate the chapter to which the author has contributed

Bernard Beck (4) INSERM U.724, Neurocal Group—Faculté de Médecine, B.P. 184—Avenue de la Forêt de Haye, 54500 vandŒuvre les Nancy, France

Kent C. Berridge (8) The University of Michigan, Department of Psychology (Biopsychology Program), 525 E. University, Ann Arbor, MI, 48109, USA

J.M. Brunstrom (11) Bristol University, Department of Experimental Psychology, 8 Woodland Road, Bristol, BS8 1TN, United Kingdom

Sharon C. Cheetham (13) RenaSci Consultancy Ltd, BioCity, Nottingham, Pennyfoot Street, Nottingham, NG1 1GF, United Kingdom

Steven J. Cooper (1, 9) University of Liverpool, School of Psychology, Liverpool, L69 7ZA, United Kingdom

Dianne Figlewicz Lattemann (7) VA Puget Sound Health Care System, Metabolism/Endocrinology (151), 1660 So. Columbian Way, Seattle, WA, 98108, USA

Nori Geary (6) Swiss Federal Institute of Technology Zurich (ETHZ), Institute of Animal Sciences, Schorenstrasse 16, 8603 Schwerzenbach, Switzerland

E.L. Gibson (11) Roehampton University, Whitelands College, School of Human and Life Sciences, West Hill, Holybourne Avenue, London, SW15 3SN, United Kingdom

Tim C. Kirkham (1) University of Liverpool, School of Psychology, Liverpool, L69 7ZA, United Kingdom

Morten L. Kringelbach (2) University of Oxford, Department of Physiology, Anatomy and Genetics, Parks Road, Oxford, OX1 3PT, United Kingdom

Simon M. Luckman (5) University of Manchester, School of Biological Sciences, Oxford Road, Manchester, M13 3PT, United Kingdom

Thomas A. Lutz (6) University of Zurich, Vetsuisse Faculty, Institute of Veterinary Physiology, Winterthurerstrasse 260, CH-8057 Zurich, Switzerland

Amy MacDonald Naleid (7) University of Washington, Department of Psychiatry and Behavioral Sciences, Box 356560, Seattle, WA, 98195-6560, USA

Nicole M. Sanders (7) VA Puget Sound Health Care System, Metabolism/Endocrinology (151), 1660 So. Columbian Way, Seattle, WA, 98108, USA

Alfred J. Sipols (7) University of Latvia, Institute for Experimental and Clinical Medicine, Department of Medicine, Riga LV-1004, Latvia

John R. Speakman (12) University of Aberdeen, School of Biological Sciences, Aberdeen, AB24 2TZ, United Kingdom

Thomas R. Stratford (3) University of Illinois at Chicago, Department of Psychology, 1007 West Harrison Street, Chicago, IL, 60607, USA

Steven P. Vickers (13) RenaSci Consultancy Ltd, BioCity, Nottingham, Pennyfoot Street, Nottingham, NG1 1GF, United Kingdom

John Wilding (14) University Hospital Aintree, Clinical Sciences Centre, Department of Medicine, Diabetes and Endocrinology Clinical Research Group, Longmoor Lane, Liverpool, L9 7AL, United Kingdom

Martin R. Yeomans (10) University of Sussex, Department of Psychology, Falmer, Brighton, BN1 9QH, United Kingdom

1

Introduction and Overview

STEVEN J. COOPER
School of Psychology, University of Liverpool
TIM C. KIRKHAM
School of Psychology, University of Liverpool

It is generally accepted that the average temperatures of the atmosphere, the landmasses, and the seas are perceptibly rising. The trajectory of the rising temperatures is such that informed predictions envisage a range of disastrous consequences that will affect everyone's lives. Yet we also face another potential calamity, and that is the global rise in obesity. How many people will be affected in 10, 20, or 50 years' time is very difficult to predict. However, the associated health care costs alone will be prohibitively high, and there will be overwhelming economic and social costs too. Obesity is a global problem that demands our urgent attention. Modern methods of food production, distribution, and retailing have led to surplus levels of high-quality food production, which in turn require high levels of consumer demand and consumption to be sustained economically.

This volume highlights some of the basic scientific issues and recent progress in the investigation of the causes of obesity, and more specifically the normal regulation of appetite and body weight. It does not deal directly with clinical issues. Rather, to understand why body weight increases to the "point of obesity," we have to understand the central and peripheral physiological mechanisms that normally underlie the

control of food intake, and the processes that monitor fuel storage and utilization—operating both in the short term and the long-term. Alongside this traditionally mechanistic form of inquiry, we must also understand better the psychological factors which operate to drive food consumption in excess of apparent need—and to appreciate that people are not simple homeostatic machines, but are adaptive, emotional organisms whose behavior is influenced by experience and context. Moreover, we should improve our understanding of the gene–environment interactions that help determine susceptibility to obesity. Out of these basic scientific inquiries there is a hope that new therapeutic measures will emerge. If effective, in the short term and the long term, and if safe, basic science advances may provide clinicians with effective therapeutic interventions. A likely avenue is the development of new drug treatments that will assist in achieving clinically significant reductions in body weight.

We have organized this volume to reflect each of these themes. First, we offer an integrated overview of the systems in the central nervous system, the endocrine system, peripheral nervous system, and gastrointestinal system that together operate to determine the control of food intake and long-term body weight regulation. We cannot pretend to be comprehensive, but we hope to provide enough information to give a genuine sense of the integrated activities of multiple systems engaged in determining these outcomes. Second, we consider in some detail a number of psychological approaches that emphasize the motivational, emotional, and learning factors which determine food choice and food consumption. Of course, we should strive for a consolidated integration of the physiological and psychological approaches, to take us beyond the more common mechanistic models of energy homeostasis. Some progress has been made, and we have highlighted some of the best examples in contemporary research where the psychological and neuroscientific approaches are brought together. Third, we deal with obesity. The question of gene–environment interactions that may be pushing us toward obesity is considered. Then, we focus on the possible new developments in the pharmacotherapy of obesity. Preclinical investigations of new drug candidates are discussed, together with an examination of the requirements for satisfactory evaluation of candidate drugs in human clinical trials. This volume does not strike an unduly optimistic note. There is no doubting the urgency of the clinical, social, and economic problems associated with rapidly escalating levels of obesity. However, in our view, it is imperative to accelerate basic research on the interrelations between the multiple systems implicated in determining appetite, food consumption, and body weight control. Too narrow a focus, too simplistic a model—at any level, will not yield sufficient knowledge to understand the nature of the problem, let alone solve it. Our intention, in this volume, is to encourage an integrated approach to basic research, from which more effective therapies may emerge.

We have reversed the traditional approach in functional neuroscience of beginning at the periphery and working toward the center, gradually ascending from the lower brainstem to the cerebral cortex. That approach is good for analytical, experimental purposes, but it is less useful when we wish to emphasize the integration of individual elements distributed throughout the central and peripheral systems, and their sensitivity to environmental or experiential factors. Integration requires top-down controls. We begin, then, with Kringelbach's contribution which develops a model of cortical function. In the human brain, the orbito-frontal cortex serves as a key, central node that links sensory and hedonic systems to mechanisms determining appetite and food consumption. Stratford's chapter then takes us to the nucleus accumbens shell, which also appears to be a key nodal structure, where many modulatory influences are integrated, and from which downstream projections strongly influence hypothalamic functions. Beck picks up the hypothalamic theme, and reviews the rich brew of orexigenic neuropeptides located there, which help determine food intake. Dysregulation in these neuropeptide systems may underlie at least some varieties of obesity. Luckman follows the trail downwards, and examines the integrative role of the brainstem in relation to information from higher brain centers and the gut. He draws attention to the neuropeptide complexity of one such structure, the dorsal vagal complex. Lutz and Geary establish further the importance of the gut–brain axis, by describing the peptide signals that relay information to the brain, through a variety of endocrine and neural channels, about the transit and processing of nutrients through the gastrointestinal tract. In turn, these signals play a key role in coordinating the size and frequency of meals. Survival depends on getting this right. The brain is also in receipt of information about levels of energy storage, and Figlewicz-Latteman details the signals about peripheral adiposity which affect brain function, and which critically link psychological and neurophysiological processes through changes in the rewarding value of food. No one can yet provide a complete blueprint of the integration between the brain and peripheral systems, which determine food intake and long-term body weight regulation. However, we believe our authors provide a valid impression of how the blueprint will emerge, in greater detail, as research is pursued.

In terms of brain systems, reward and motivation, we enjoy the advantage of an extremely active research environment. Considerable advances, theoretical and empirical, are being made. It is an opportune time to integrate these psychological insights with the rapidly accumulating knowledge about food-intake and body weight controls. Berridge reviews his fertile distinction between "liking" systems in the brain (responsible for the hedonic evaluation of food) and the "wanting" system (underpinning incentive motivation or food craving). Critically, these two psychological domains map onto distinctive brain systems, which can be distinguished anatomically and neurochemically. Yeomans picks up the theme of palatability and applies it to issues of human appetite. He emphasizes the importance of individual experiential factors in the development of preferences for palatable foods, and argues that an overresponsivity to palatability may be key to the development of obesity and the unsuccessful nature of many antiobesity interventions. Cooper considers the question of food and taste preferences, both innate and learned. He describes pharmacological studies that suggest that key neurochemical differences may emerge for different sources of food preference. Gibson and Brunstrom provide a comprehensive assessment of learned influences on human appetite and food intake, and explore how food choice and meal size are influenced by motivational/nutritional state. Continued experience of eating foods rich in fats and energy may impair inhibitory controls in learned

appetite, contributing to the development of obesity. These chapters indicate how our investigation of the determinants of obesity must be sufficiently broad in scope to take full account of motivational and learning factors that help determine the control of appetite, and to integrate these with physiological signals generated in the gut and the brain.

Obesity is not solely a question of too much food available in the environment. Some people remain slim in the midst of plenty. Speakman considers the role of genetic endowment in the etiology of obesity, and the central question of genetic–environmental interactions. Vickers and Cheetham review the latest developments in the field of antiobesity drugs. Drugs currently licensed for use may have limited practical value, and there is therefore a considerable incentive to developing new approaches in pharmacotherapy. New ideas have emerged from the new discoveries at a basic research level, exemplified perhaps by recent developments in endocannabinoid pharmacology and biochemistry. For example, the antiobesity agent rimonabant, a cannabinoid receptor antagonist, has helped build a model of a novel neuromodulatory system which impacts on central and peripheral and behavioral and metabolic processes in appetite and weight regulation. We can be optimistic, therefore, in believing that advances at a basic research level will eventually lead to improved treatment possibilities for obese patients. Wilding completes the volume with an account of ongoing clinical investigation into antiobesity drugs. Many candidate drugs will fail, and Wilding details the conditions which acceptable future therapies must satisfy if they are to be licensed for clinical use.

It will be apparent from individual chapters, and the overlap of some of their content, that our authors are exemplars of this integrative approach to the study of appetite and body weight regulation. Of course, we have imposed to some extent an editorial segregation of brain regions, and of central and peripheral processes, to obtain authoritative accounts of essential subcomponents. But we hope that, overall, this volume captures and stimulates an integrative, interdisciplinary spirit.

Obesity is not a simple issue, to be quickly addressed. Understanding its determinants and developing effective therapeutic interventions will require very considerable efforts by research scientists and clinicians. Our aim is to provide a useful guide to current research, the knowledge so far acquired, and promising new lines of investigation before us. A more complete understanding of the bases of human appetite and food consumption will provide important information with which to understand the rising tide of human obesity, and how best to manage it to avoid severe and widespread clinical and economic consequences.

2

Cortical Systems Involved in Appetite and Food Consumption

MORTEN L. KRINGELBACH
Department of Physiology, Anatomy, and Genetics,
University of Oxford

I. Introduction
II. Food Motivation
III. Cortical Representations of Sensory Inputs
   A. The Taste System
   B. The Olfactory System
   C. Multimodal Integration
   D. Hedonic Representations
IV. Conclusion
   Acknowledgments
   References

Food intake consumption and the control of appetite rely on cortical processing in humans and other primates to a much larger degree than other mammals. This chapter describes the evidence from neurophysiology, neuropsychology, and neuroimaging. Four main computational principles are proposed: (1) motivation-independent processing of identity and intensity; (2) formation of learning-dependent multimodal sensory representations; (3) reward representations using mechanisms including selective satiation, and (4) representations of hedonic experience, monitoring/learning, or direct behavioral change. A model incorporating these computational principles is proposed for the orbitofrontal cortex, which is one of the most important nodes linking sensory and hedonic systems involved in appetite and food consumption in the human brain.

I. INTRODUCTION

Human food intake relies on a complex hierarchy of cortical processing which include obtaining stable sensory information, evaluation for desirability, and choosing the appropriate behavior. Part of this processing is linked to basic homeostatic regulation, which has been elucidated in great details in animal models with mammals, including humans sharing many subcortical circuits and molecules (such as leptin and ghrelin) as outlined elsewhere in this book and in other recent reviews (Saper et al., 2002).

However, human food intake is not only regulated by homeostatic processes as illustrated by our easy overindulgence on sweet foods beyond homeostatic needs and the epidemic proportions of obesity, which has become a major health problem (Kohn and Booth, 2003). Instead the regulation of human food intake relies on the interaction between homeostatic regulation and hedonic pleasure. This complex subcortical and cortical processing involves higher order processing, including learning, memory, planning, and prediction, and gives rise to conscious experience of not only the sensory properties of the food (such as the identity, intensity, temperature, fat content, and viscosity) but also the valence elicited by the food (including, most importantly, the pleasure experienced) (Kringelbach, 2005).

This chapter reviews the evidence linking cortical regions of the human brain to aspects of human food intake including appetite and food consumption. This evidence was obtained through neurophysiological findings in other primates and then further elaborated in human neuroimaging and neuropsychology studies.

The emphasis on cortical processing in humans and other primates relies on the direct projection from the rostral part of the nucleus of the solitary tract, which is different from the pontine taste area and associated subcortical projections found in rodents (Norgren, 1984; Pritchard et al., 1986). It is proposed here that the human brain utilizes at least four main computational principles underlying food intake. First, the identity of a food is established in primary cortical areas which contain neural activity that encode the sensory (taste, smell, somatosensory, auditory, and visual) aspects of the food. These cortical representations are to a large extent invariant and thus independent of current motivational state (such as hunger or thirst). Second, this multitude of unimodal sensory information is subsequently combined to form multimodal representations in polysensory cortical regions (including the orbitofrontal cortex). These multimodal representations use slowly acting learning mechanisms to associate taste, smell, and other sensory food information. Third, reward representations are computed in further cortical regions by taking into account the desirability of a food, the current motivational state, and previous learned associations. These reward representations include selective satiation mechanisms that selectively suppress further food intake of previously ingested foods while other foods can still be readily ingested. Fourth, these reward representations may either result directly in behavioral change or in stable monitoring, learning, and memory processing. This cortical processing may also lead to subjective hedonic experience of the food, which is correlated with neural activity in circumscribed regions of the orbitofrontal cortex. This link between food intake and hedonic experience may still yield important insights into the core of subjective experience and is thus a promising avenue for further research with wide-ranging consequences.

II. FOOD MOTIVATION

Food motivation (and motivation in general) is closely related to emotion and is generally defined in opposition to cognition as that which moves us in some way, as implied by the common Latin root of both words (*movere*, to move). It is only recently that the fields of motivation and emotion have tended to go different ways, and have in the past been considered together (Gray, 1975; Papez, 1937; Weiskrantz, 1968). One often-used functional distinction between motivation and emotion is that motivated behaviors are elicited by rewards (and punishers) related to internal homeostatic states associated with, say, hunger and thirst, while emotional states and behaviors are elicited by the great majority of rewards and punishers that are external stimuli and not associated with these internal need states (Rolls, 1999).

The field of emotion research has advanced tremendously (Kringelbach, 2004a), and may provide conceptualizations that could benefit the field of motivation and motivational hedonics. One important insight from emotion research is to divide the concept of emotion into two parts: the *emotional state* that can be measured through physiological changes such as autonomic and endocrine responses, and *feelings*, seen as the subjective experience of emotion. The emotional state relies on basic implicit brain mechanisms, which are rarely available for immediate conscious introspection. We do, however, appraise our emotional state on a regular basis and this conscious appraisal must partly rely on interfacing our language system with some introspection of our current emotional state. One could argue that a very similar division could be made for motivational processes, where the motivational state relies on basic brain mechanisms that can subsequently be appraised or evaluated as hedonic experience.

Historically, early drive theories of motivation proposed that hedonic behavior is controlled by need states (Hull, 1951). But these theories do not, for example, explain why people still continue to eat when sated. This led to theories of incentive motivation where hedonic behavior is mostly determined by the incentive value of a stimulus or its capacity to function as a reward (Bindra, 1978). Need states such as hunger are still important but only work indirectly on the stimulus' incentive value. The principle of modulation of the hedonic value of a consummatory sensory stimulus by homeostatic factors was labeled *alliesthesia* (from *allios*, changed, and *esthesia*, sensation) (Cabanac, 1971). A useful distinction has been proposed between two aspects of reward: hedonic impact and incentive salience, where the former refers to the liking or pleasure related to the reward, and the latter to the wanting or desire for the reward (Berridge, 1996; Berridge and Robinson, 1998).

Food intake is a precisely controlled act that can potentially be fatal if the wrong decision is taken to swallow toxins, microorganisms, or nonfood objects on the basis of erroneously determining the sensory properties of the food. Humans have therefore developed elaborate food behaviors which are aimed at balancing conservative risk-minimizing and life-preserving strategies, with occasional novelty seeking in the hope of discovering new sources of nutrients (Rozin, 2001).

Food intake must provide the right balance of carbohydrates, fats, amino acids, vitamins, and minerals (apart from sodium) to sustain life. The neural mechanisms regulating food intake are complicated and must like any regulatory system include at least four features: system variables, detectors for the system variables, set points for these system variables, and correctional mechanisms. A simple regulatory feedback system operates best with immediate changes, and it becomes significantly more complex when the feedback is not immediate. In the case of controlling food intake, there are significant delays in system changes caused by relatively slow metabolic processes, and therefore the regulatory neural systems controlling food intake must include sophisticated mechanisms to learn to predict in advance when a meal should be initiated and terminated. Many of the basic components and principles of food intake have been elucidated in great detail and have been described in reviews elsewhere (LeMagnen, 1985; Woods and Stricker, 1999).

Most research in this area has been carried out in nonhuman animals, which has not helped us understand the strong hedonic component of human food intake which is directly linked to appetite. It is clear that much of complex human behavior related to food intake must be linked to neural activity in the cerebral cortex integrating the complex multitude of stimuli and situational variables. Important examples of such complex behavior include the decrease of the rated pleasantness of sweet tastes when subjects are sated relative to when they are hungry (negative gustatory alliesthesia) (Cabanac, 1971) and satiation signals that selectively suppress further food intake of previously ingested foods while other foods can still be readily ingested (sensory-specific satiety) (Rolls et al., 1981).

III. CORTICAL REPRESENTATIONS OF SENSORY INPUTS

All of the classic five senses (vision, hearing, smell, taste, and touch) are involved in the regulation of food intake. However, in addition to these sensory systems there are also other sensory receptors such as those in the digestive tract which are sensitive to gastric distension, or those in the circulatory system that are sensitive to changes in blood pressure or carbon dioxide gas in the blood.

Our sensory systems help identify and evaluate potential food sources. This can happen even from a distance using olfactory, visual, and (to some extent) auditory systems. Olfactory evaluation relies on receptors in the nose that are used to identify volatile airborne molecules. The sense of smell has limited use at longer distances and has to be aided by other more precise distance and directional senses such as the visual system (Gottfried and Dolan, 2003). Even at close range visual influences can override the olfactory system, such as demonstrated by experiments manipulating the olfactory perception of wine by artificially coloring a white wine red (Morrot et al., 2001).

Foremost, of course, the sensing of food occurs when a food is grasped and delivered to the mouth. This includes taste, smell, and somatosensory (such as temperature, viscosity, pungency, and irritation) input primarily from our oral and nasal cavity (but also in rare cases from the eyes, such as in the case of irritation excreted from, e.g., onions). This sensory input is essential in making the vital decision of whether to swallow or reject a potentially poisonous food. Such decisions are so important that mammals have brainstem reflexes (stereotypical for each basic taste) that are based on rudimentary analyses of the chemical composition, and which are not altered even by the loss of all neural tissue above the level of the midbrain (Grill and Norgren, 1978).

An important computational principle of sensory processing in the brain is that the primary sensory cortices receive sensory information from receptor cells and process this information to form neural representations of the identity of the stimulus (see Fig. 1). It has been shown in other primates that this representation of stimulus identity remains to a large degree constant and is not modulated by motivational state (Rolls, 1999). This principle is important in forming stable and accurate representations of the world, and this unimodal information is then integrated further on in the hierarchy of cortical processing in the secondary and tertiary sensory cortices. Some of these multimodal representations can be modulated by motivational state which depends on hunger, thirst, and gastric distension but also on learned associations with pleasure states. This can then influence behavior both internally (via the hypothalamus and brainstem structures) and externally (via the motor systems).

A. The Taste System

Taste is sensed by taste receptor cells arranged in taste buds which are primarily found on the tongue but also on other areas in the oral cavity such as the soft palate, the pharynx, the larynx, and the epiglottis

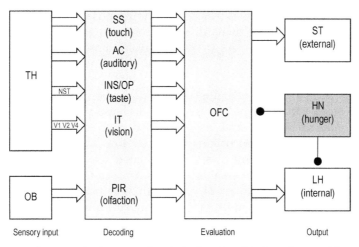

FIGURE 1 Schematic diagram of information flow linked to the orbitofrontal cortex. The orbitofrontal cortex receives input from all the sensory modalities: gustatory, olfactory, somatosensory, auditory, and visual. This information is then represented and made available for pattern association between primary (e.g., taste) and secondary (e.g., visual) reinforcers. The reward value of this representation can be modulated by hunger neurons (HN). The output from the orbitofrontal cortex to both striatum (external) and lateral hypothalamus (internal) can then lead to behavior. Abbreviations utilized: TH, thalamus; OB, olfactory bulb; NST, nucleus of the solitary tract; V1, V2, V4, primary and secondary visual areas; SS, somatosensory cortex (3,1,2); AC, auditory cortex; INS/OP, insula cortex/frontal operculum; IT, inferior temporal visual cortex; PIR, piriform cortex; OFC, orbitofrontal cortex; HN, hunger neurons; ST, striatum; LH, lateral hypothalamus.

(Scott and Plata-Salaman, 1999). Taste receptor cells are constantly being renewed and have a turnover period of between 7 to 10 days. As a convenient way of organizing the taste system, most researchers agree that there are four different taste qualities with specific receptor types: sweet (glucose, sucrose), salty (NaCl), sour (HCl, citric acid), and bitter (quinine). Some have also argued for the inclusion of the amino acid taste *umami* (of which an exemplar is monosodium glutamate) that corresponds to what is sometimes described as the taste of protein. The argument for including umami as a basic taste has received support from the recent discovery of specific receptors for glutamate in lingual tissue with taste buds (Chaudhari et al., 2000).

1. Central Taste Pathways

The taste information is relayed from the taste buds to the cortex by way of the nucleus of the solitary tract of the medulla and the parvicellular part of the ventral posterior medial (VPMpc) nucleus of the thalamus. The three cranial nerves involved are the facial nerve (VII) which conveys information from the anterior two thirds of the tongue, the glossopharyngeal nerve (IX) which innervates the posterior third of the tongue, and the vagus nerve (X) which conveys information from the remaining taste buds from other areas in the mouth (Norgren, 1990). It should be noted that these cranial nerves also carry information about touch, temperature, and pain sensitivity on the tongue. These projections are primarily ipsilateral in higher primates (Pritchard et al., 1989) but other evidence suggests that crossed and bilateral projections may exist (Norgren, 1984, 1990).

In primates, gustatory information is relayed to cortical areas via the VPMpc nucleus in the thalamus. In rats a further nonthalamic gustatory system synapses on the parabrachial taste nuclei before projecting directly to the amygdala and the hypothalamus (Norgren, 1984; Norgren and

Leonard, 1973), but this pathway has not been found in primates (Pritchard, 1991).

2. Primary Taste Cortex

The primary taste cortex is the next gustatory relay station from the VPMpc and is found in the frontal operculum and the dorsal part of the anterior insula (Bornstein, 1940a,b; Pritchard et al., 1986; Scott et al., 1986; Sudakov et al., 1971). Another smaller area of the anterior primary somatosensory cortex (area 3b) also receives taste information from the VPMpc (Norgren, 1990). Neurophysiological investigations in primates found that 4% of neurons in the primary taste cortex showed broadly tuned responses to different tastes and that responses of these neurons were more specific than neurons in the nucleus of the solitary tract (Ogawa et al., 1989; Scott et al., 1986). Other sensory information related to feeding such as temperature and somatosensory input from the mouth is processed by other neurons in the primary taste cortex (Norgren, 1990).

3. Higher Order Taste Cortices

Following the general principle of sensory processing, the responses of neurons in the primary taste cortex in primates are not influenced by motivational state and thus maintain stable neural properties for correctly decoding taste input (Yaxley et al., 1988). The primary taste cortex projects to the secondary taste cortex found in the adjoining caudolateral orbitofrontal cortex which does not receive projections from the VPMpc of the thalamus and is therefore defined as the secondary taste cortex (Baylis et al., 1994). Neurons in this region are more finely tuned than those found in the primary taste cortex and respond to a single taste, whereas other neurons in more anterior and medial orbitofrontal cortex are more broadly tuned and some neurons have multimodal responses to both gustatory and olfactory (Rolls and Baylis, 1994). It has

been proposed that these multimodal neurons may encode the flavor of food.

Furthermore, it has been shown that the taste representation carried by neurons in the orbitofrontal cortex can be influenced by motivational states such that neurons will decrease their response to a taste when a food with this taste is being eaten to satiety but still respond to other foods not eaten to satiety—an effect known as sensory-specific satiety (Rolls, 1999).

Another important brain area receiving gustatory information is the lateral hypothalamus, where neurons were found to be modulated by hunger (Burton et al., 1976). Neurons responding to tastes have also been found in the amygdala, which may be part of a system involved in learning associations between primary taste reinforcers and other arbitrary stimuli (Sanghera et al., 1979). Such a system would be important for reliably selecting food for ingestion.

Further brain areas receiving taste information include the ventral striatum (including the nucleus accumbens and the olfactory tubercle), which could interface reward signals with initiation of action in output motor systems (Williams et al., 1993).

4. Human Lesion Studies

The literature on patients with gustatory symptoms is rather sparse. One of the first studies found ageusia in patients with bullet wounds to the parietal operculum and thus established this site as the most probable for the human primary taste cortex (Bornstein, 1940a,b). This was extended to the insula following the observations by Penfield and Faulk (1955) of a patient with an aura of unpleasant taste stemming from a tumor in the insula. Further support came from cortical stimulation of the insula in epileptics, which was found to elicit taste perception. In a more recent study, gustatory auras were found in three patients with magnetic resonance imaging evidence of lesions to the insula (Cascino and Karnes, 1990).

Published case studies of gustation in patients with surgical lesions are even more rare. Only two studies have been published to date and both studies concern detection and recognition thresholds in patients with resection of the temporal lobe for intractable epilepsy (Henkin et al., 1977; Small et al., 1997b). Both studies found normal detection thresholds in the patients but differed in their results on how recognition thresholds were affected by the lesions. Small and colleagues (1997b) found a general impairment in recognizing citric acid in their group of patients with right surgical resection of the temporal lobe, while Henkin and colleagues (1977) found a deficit for recognizing citric acid in their patient group with *left* resection. This discrepancy is not easily resolved and more data are clearly needed. Yet, these experiments show that the anterior temporal lobe is involved in the recognition of taste quality. The observed deficits in patients in recognizing taste are likely to be due to the disconnection of the amygdala, orbitofrontal cortex, and interacting brain areas in temporal lobectomy.

5. Neuroimaging Studies of Taste

The functional neuroimaging of taste in humans has been delayed due to a number of methodological challenges. Taste experiments naturally involve some tongue, mouth, and swallowing related movement, which have been seen as problematic by some investigators. Early investigators used manual application and removal of taste stimuli from the tongue of the subject, which created more problems than it solved, given that it still involved mouth movement and would not necessarily elicit the best taste response.

Most investigators are therefore using tubes that deliver the taste directly to the oral cavity and the taste is subsequently swallowed. A control stimulus applied in exactly the same fashion can then serve to act as control for movement. It is important that the control stimulus is not water because water is a primary reinforcer that animals will work for. Thus neurons in primary and secondary gustatory cortex have been found to respond to water when thirsty (Rolls et al., 1990). Instead, the best option currently available is to use a tasteless control substance containing the main ionic components of saliva.

i. Motivation-Independent Taste Representations The early neuroimaging studies were affected by the methodological challenges mentioned but have in general corroborated the findings from human lesion studies and the neurophysiological findings in primates. The primary gustatory area in humans has been found to be located in the anterior insula/frontal operculum (Frey and Petrides, 1999; Kinomura et al., 1994; O'Doherty et al., 2001b; Small et al., 1997a, 1999).

The largest functional magnetic resonance imaging (fMRI) study of taste processing to date used 40 data sets from 38 right-handed subjects (13 women and 25 men, of which two subjects participated in two experiments) in four taste investigations that used (1) identical delivery of the taste stimuli, (2) the same control procedure in which a tasteless solution was delivered after every taste stimulus, and (3) event-related interleaved designs (Kringelbach et al., 2004). A total of eight unimodal and six multimodal taste stimuli (oral stimuli that produce typically taste, olfactory, and somatosensory stimulation) ranging from pleasant to unpleasant were used in the four experiments. The results of the main analysis included both unimodal and multimodal taste stimuli, which was then confirmed in a separate analysis using only unimodal taste stimuli.

Stringent random effects analysis of taste activity across the 40 data sets revealed three cortical activation foci to the main effects of taste in the human brain (which were corrected for multiple comparisons). Bilateral activation of the anterior insular/frontal opercular cortex was found with a slightly stronger response on the right side. The locations in standard brain space in Montreal Neurological Institute (MNI) coordinates: $[x,y,z: 38,20,-4]$ and $[x,y,z: -32,22,0]$ are the likely bilateral sites of the primary taste cortices. This slight asymmetry in bilateral taste processing fits with an early meta-analysis of gustatory responses gathered from neuroimaging studies suggesting that the preponderance of activity peaks to taste fall in the right hemisphere (Small et al., 1999).

Taste stimuli also produced activity in the medial caudal orbitofrontal cortex. The location in MNI coordinates: $[x,y,z: 6,22, -16]$ is likely to coincide with the secondary taste cortex, which fits well with subsequent neurophysiological recordings in medial parts of the macaque orbitofrontal cortex (Pritchard et al., 2005).

In addition, activity was also found in the left dorsolateral prefrontal cortex in the posterior part of the middle frontal gyrus (Brodmann Area 46): $[x,y,z: -42,26,36]$. This could aid higher cognitive processes in guiding complex motivational and emotional behavior. The finding is consistent with neurophysiological recordings in the dorsolateral prefrontal and orbitofrontal cortices in nonhuman primates in a reward preference task (Wallis and Miller, 2003). This experiment demonstrated that neurons in the dorsolateral prefrontal cortex encode both the reward amount and the monkeys' forthcoming response, while neurons in the orbitofrontal cortex more often encode the reward amount alone. Furthermore, the reward selectivity arose more rapidly in the orbitofrontal cortex than the dorsolateral prefrontal cortex, which is consistent with the idea that reward information enters the prefrontal cortex via the orbitofrontal cortex, where it is passed to the dorsolateral prefrontal cortex and used to control behavior. The finding is also consistent with a monkey PET (positron emission tomography) study on olfactory and gustatory processing that found activity of the inferior frontal gyrus (Kobayashi et al., 2002).

Furthermore, at a lower statistical threshold, taste-related activity was found in the anterior cingulate cortex: $[x,y,z: 8,18,50; p < 0.001$, uncorrected for multiple comparisons]. Quite a few other studies have found taste-related activity in the cingulate cortex (De Araujo et al., 2003a,b,c; Francis et al., 1999; Kringelbach et al., 2003; O'Doherty et al., 2001b, 2002; Small et al., 2003; Zald et al., 1998, 2002). Regions of the anterior cingulate cortex thus may contain taste-related activity which can help the role of this region in executive control.

ii. Reward-Dependent Taste Representations In contrast to these motivation-independent representations of reinforcer identity, one neuroimaging study was able to dissociate the brain regions responding to taste intensity and taste affective valence (Small et al., 2003). It was found that the cerebellum, pons, middle insula, and amygdala responded to intensity regardless of valence, while valence-specific responses were observed in the orbitofrontal cortex with the right caudolateral orbitofrontal cortex responding preferentially to pleasant compared to unpleasant taste, irrespective of intensity.

Another study used neuroimaging to show that subjective ratings of taste pleasantness (but not intensity) correlate with activity in the orbitofrontal cortex and in the anterior cingulate cortex (De Araujo et al., 2003b). Moreover, in this study investigating the effects of thirst and subsequent replenishment, it was found that the orbitofrontal cortex and a region of mid-insula was correlated with subjective pleasantness ratings of water across the whole experiment (De Araujo et al., 2003b).

Further evidence of neural correlates of subjective experience of pure taste was found in an experiment investigating true taste synergism, which is the phenomenon whereby the intensity of a taste is dramatically enhanced by adding minute doses of another taste. The results of this neuroimaging experiment showed that the strong subjective enhancement of umami taste that occurs when 0.005 M inosine 5′-monophosphate is added to 0.5 M monosodium glutamate (compared to both delivered separately) was correlated with increased activity in an anterior part of the orbitofrontal cortex (see Fig. 2b) (De Araujo et al., 2003a).

B. The Olfactory System

Smell is sensed by olfactory receptors placed in the upper part of the nasal cavity. There are about 100 million bipolar olfactory cells in humans with lifelong continuous replacement every 4–8 weeks from stem cells in the sensory epithelium. Each bipolar olfactory cell bears up to a thousand hairs (cilia), which are the points of contact for aromatic molecules. The cilia must be moistened continuously by mucous glands roughly every 10 min to avoid desiccation. The mucus consists of a water base with dissolved mucopolysaccharides, salts, and proteins, including odorant binding proteins, enzymes, and antibodies (which help protect the brain from odorant-borne illnesses).

The specific olfactory receptor proteins are genetically specified by about 1000 different genes (Buck and Axel, 1991), which correspond to the largest family of genes found in humans and underlie the profound importance of smell in humans. The resulting number of unique odors which can be discriminated is likely to be on the order of around half a million odors (Mori et al., 1998).

The main olfactory system relies on these receptors for the conscious perception of odors but also relies on the trigeminal system mediating somatosensory information such as irritation, tickling, burning, and cooling of odors. In addition, other animals have a vomeronasal system which plays a major role in the perception of pheromones that are extremely important in controlling reproduction and related sexual behavior including aggression. It remains highly controversial whether humans have a vomeronasal olfactory system (VNO, also called Jacobson's organ) (Monti-Bloch et al., 1998), although the presence of such a system (operating nonconsciously) has been proposed to mediate pheromones synchronizing the menstrual cycle of cohabiting women (Stern and McClintock, 1998). Although nearly all humans have paired VNO-like structures at the base of the septum and paired VNO ducts near the posterior aspect of the external nares, the human VNO appears to lose receptor cells and associated neural elements in the second trimester of pregnancy (Smith and Bhatnagar, 2000). The human VNO thus appears to lack basic elements needed for a functioning VNO and is therefore most likely a nonfunctioning unit (Bhatnagar and Meisami, 1998).

1. Primary Olfactory Cortex

Olfactory receptor neurons run through the bony cribiform plate to the ipsilateral olfactory bulb forming the first cranial nerve. The input layer in each olfactory bulb contains several thousand glomeruli (Royet et al., 1988). Within a glomerulus, several thousand primary olfactory axons converge and terminate on the dendrites of a few hundred second-order olfactory relay neurons. The projections of these neurons form the olfactory tract.

The axons in the lateral olfactory tract run ipsilateral through the olfactory peduncle synapsing on the anterior olfactory nucleus, the piriform cortex, the periamygdaloid cortex, the anterior cortical nucleus of the amygdala, and the anteromedial part of the entorhinal cortex. These cortical areas receiving direct projections from the lateral olfactory tract by definition constitute the primary olfactory cortices. Olfaction is unique among the senses in that the sensory information is not relayed via the thalamus (Carmichael et al., 1994). In higher primates (including humans) the primary olfactory cortex is situated near the border of the dorsomedial anterior temporal lobes and the caudolateral part of orbitofrontal cortex. In the human brain the piriform cortex is also located anatomically at this junction of the temporal and frontal lobes based on its cytoarchitectonic definition (Eslinger et al., 1982).

The projections from the olfactory bulb are all primarily ipsilateral although there are contralateral connections running through the anterior commissure via the anterior olfactory nucleus and the anterior piriform cortex.

2. Higher Order Olfactory Cortices

Numerous brain regions receive projections from the primary olfactory cortex including the orbitofrontal cortex, medial thalamus, ventral striatum and pallidum, agranular insular cortex, hippocampus, and medial and lateral hypothalamus. One of the most important olfactory pathways converges on the orbitofrontal cortex, and experiments in nonhuman primates have shown olfactory areas in the posterior orbitofrontal cortex (Takagi, 1986; Tanabe et al., 1975a,b) and in the medial central posterior orbitofrontal cortex (part of area 13) (Carmichael et al., 1994; Takagi, 1986; Yarita et al., 1980). These regions of the orbitofrontal cortex then, by definition, constitute the secondary olfactory cortices (Rolls and Baylis, 1994) and have onward connections to the secondary taste cortex (Carmichael et al., 1994; Carmichael and Price, 1994).

Single-neuron activity in the orbitofrontal cortex was recorded during an olfactory discrimination task, where behavioral responses following eight odors were rewarded with sucrose while responses following two other odors were punished with saline (Rolls and Baylis, 1994). It was found that 3.1% of the 1580 neurons recorded had olfactory responses and that 2.2% responded differentially to the different odors in the task. The findings showed that a 65% subset of olfactory representation within the orbitofrontal cortex reflects the odor irrespective of its reward

association and that the remaining 35% reflects the taste associations of the odor.

Further neurophysiological investigations of orbitofrontal cortex neurons with olfactory responses were carried out in a series of experiments using an olfactory discrimination task where the taste reward contingencies of two odors were reversed once they had been successfully acquired (Rolls et al., 1996). Of the 28 odor-responsive neurons analyzed it was found that 68% of these neurons modified their responses following changes in taste reward associations, with 25% of the neurons reversing their responses and task-reversal extinction of differential neuronal responses seen in 43% of the neurons. The remaining 32% of the neurons did not change their responses. This modification of odor-taste reward associations was found to be much slower and inflexible than visual-taste associations which usually take place with one-trial learning (Thorpe et al., 1983). This relative inflexibility of olfactory responses could be important for forming stable odor-taste associations needed for the formation and perception of flavors.

3. Human Lesion Studies

The bulk of patients studied for olfactory processing are patients who have undergone surgery for intractable epilepsy. Almost all patients therefore have unilateral selective surgical excisions to the temporal lobes, while only a few cases have lesions in the frontal lobes. Most olfactory studies have investigated detection thresholds, odor identification, quality discrimination, and odor memory in such patients. Across these studies there is great variation with regards to stimulation type (mono- or birhinal), stimuli type (e.g., odors are presented from bottles, on cotton wands, or on paper labels) and number of stimuli. With such variation it is hard to draw firm conclusions but all studies have shown that detection thresholds have been shown to be unimpaired following lesions of the anterior temporal lobe (Henkin et al., 1977; Jones-Gotman and Zatorre, 1988). This would suggest that odor detection is computed earlier in the olfactory pathways or perhaps in a parallel pathway (Zatorre et al., 2000). Mild deficits with discrimination of odor quality have been found in patients with temporal lobe excision although they appear to be confined to the ipsilateral nostril (Zatorre and Jones-Gotman, 1991). The temporal lobes also appear to be involved in odor memory, odor identification (Jones-Gotman et al., 1994, 1997), and affective responses and intensity judgments (Rouby et al., 1997). The relatively mild deficits following temporal lobe excisions should be contrasted with the odor deficits in patients with lesions to the orbitofrontal cortex, which cause severe impairments to all aspects of odor processing except odor detection (Jones-Gotman and Zatorre, 1993). This fits well with the neurophysiological data from primates where the frontal and in particular the orbitofrontal cortex appear to play a more important role in olfactory processing (Rolls and Baylis, 1994).

The evidence for hemispheric specialization in odor processing in patients is not very clear. Some studies find a right hemispheric dominance in patients with right-sided lesions (Eskenazi et al., 1983; Jones-Gotman and Zatorre, 1993), whereas other studies find olfactory deficits following left hemisphere lesions (Henkin et al., 1977; Jones-Gotman et al., 1994). There does, however, seem to be more patient studies implicating the right hemisphere in olfaction although the numbers are too small to be significant. It would also be interesting to look at the performance of patients with bilateral surgical lesions.

4. Neuroimaging Studies of Olfaction

Relatively few neuroimaging studies have investigated olfaction. Most of those have dealt with the identification of the human brain areas involved in odor processing in general. Specifically brain activ-

ity in the piriform cortex, which is part of the primary olfactory cortices, has proven quite difficult to detect. Lately neuroimaging studies have started to explore brain activity related to hedonic processes.

Some of the main difficulties with imaging olfactory processes are that they involve brain structures from which it is quite hard to get a good signal with fMRI because their proximity to air- and bone-tissue gives rise to susceptibility artefacts and geometric distortion (Wilson et al., 2002). Such problems are virtually absent when using PET, but instead other problems arise with imaging brain regions such as the piriform cortex exhibiting fast habituation, which is problematic given the temporal resolution of PET.

i. Motivation-Independent Odor Representations Similar to taste stimuli, pure olfactory stimuli activate dissociable brain areas for motivation-independent representations of reinforcer identity and hedonic representations. Neuroimaging studies have found representations of olfactory identity in primary olfactory cortices (Anderson et al., 2003; Gottfried et al., 2002a; O'Doherty et al., 2000; Rolls et al., 2003a; Royet et al., 2000, 2001; Zald and Pardo, 1997), which are distinct from hedonic representations in other brain areas.

Zatorre and colleagues (1992) were among the first to present PET data of odor processing in normal volunteer subjects. Two conditions were used which both involved birhinally sniffing for seven seconds from a cotton wand which was either odorless in the control condition or one of eight different odorants. Activation was seen bilaterally in a region at the junction of the temporal and frontal lobes in what probably corresponds to primary olfactory cortex. Further activity was also seen in the right orbitofrontal cortex, prompting the authors to claim a hemispheric asymmetry in olfactory processing. It is interesting to note, however, that activity just below the chosen threshold was also seen in left orbitofrontal cortex (Zatorre et al., 2000). The asymmetry argument for olfactory is weak as demonstrated in a large meta-analysis of neuroimaging studies (Kringelbach and Rolls, 2004). While some studies found similar asymmetric activity in the orbitofrontal cortex (Dade et al., 1998; Small et al., 1997a; Sobel et al., 1998), other studies have found activation exclusively in the left orbitofrontal cortex (Francis et al., 1999; Gottfried et al., 2002a; Rolls et al., 2003a; Yousem et al., 1997; Zald and Pardo, 1997).

ii. Reward-Dependent Odor Representations Several neuroimaging studies have found dissociable encoding of olfactory stimuli with the intensity encoded in the amygdala and nearby regions, and the pleasantness correlated with activity in the orbitofrontal cortex and anterior cingulate cortex (Anderson et al., 2003; Gottfried et al., 2002b; Rolls et al., 2003a). This is consistent with studies that have found that hedonic judgments activate the orbitofrontal cortex (Royet et al., 2001) and that the unpleasantness of aversive odors correlates with activity in the orbitofrontal cortex (Zald and Pardo, 1997). Furthermore, it has been found that the orbitofrontal cortex represents the sensory-specific decrease of smell (O'Doherty et al., 2000), which is clear evidence that the reward value of olfactory stimuli is represented in the orbitofrontal cortex.

Other recent strong evidence for the role of the orbitofrontal cortex in the representation of the reward value of olfactory stimuli comes from an appetitive conditioning neuroimaging experiment which measured the brain activity related to two arbitrary visual stimuli both before and after olfactory devaluation (Gottfried et al., 2003). In the amygdala and the orbitofrontal cortex, responses evoked by a predictive target stimulus were decreased after devaluation, whereas responses to the non-devalued stimulus were maintained. It would thus appear that differential activity in the amygdala and the orbitofrontal cortex encodes the current value of reward representations accessible to predictive cues. It should also be noted that the affect and intensity judgments of odor in a paired discrimination task activates the orbitofrontal cortex (Zatorre et al., 2000).

C. Multimodal Integration

In addition to multimodal information from taste and smell, decisions about food intake also integrate, for example, somatosensory information, which is sensed by receptors in the oral and nasal cavity. This sensory information includes temperature, viscosity, fat content, pungency, and irritation and is mediated by a large variety of neural systems. This integrated information is processed and made available for the crucial decision of ingestion or rejection of a potentially poisonous food (although, as mentioned before, simple brainstem mechanisms also exist).

Food consumption relies on swallowing, which in itself is a complex physiological process that involves numerous central processes including sensorimotor integration, reflexive and voluntary motor activity, autonomic regulation, and salivatory processes. Emphasis has been placed on the brainstem control of swallowing (Jean, 1984, 2001) but recently noninvasive neuroimaging techniques has revealed some of the cortical areas involved in reflexive and voluntary swallowing (Hamdy et al., 1999a,b; Martin et al., 2001; Martin and Sessle, 1993; Zald and Pardo, 1999). Dysphagia is caused by a variety of neurological disorders with, for example, up to one-third of patients suffering unilateral hemispheric stroke (Barer, 1989; Gordon et al., 1987), which may reflect the widespread network involved in swallowing.

Furthermore, once a decision regarding ingestion or rejection has been taken it is important that this choice is integrated in the cortical processes governing learning, memory, planning, and prediction. It is also important that further information from the autonomic system about internal states such as hunger, thirst, nausea, and gastric distension are integrated in these processes.

Differences have been reported in olfactory experience depending on whether a smell reaches the nasal cavity through the nose (orthonasal) or mouth via the posterior nares of the nasopharynx (retronasal) (Pierce and Halpern, 1996), which is likely to be related to differences in somatosensory influences (e.g., mastication). Consequently, several neuroimaging studies have found differences in cortical activity patterns between ortho- and retronasal olfaction (Cerf-Ducastel and Murphy, 2001; De Araujo et al., 2003c; Small et al., 2005).

As mentioned previously, an important principle of cross-modal integration between taste and smell is the formation of odor-taste associations, which are required to be more stable than, say, visual-taste associations to facilitate the formation and perception of flavors which guide food intake (Rolls et al., 1996). Within the orbitofrontal cortex there are direct projections from secondary olfactory cortices to the secondary taste cortices (Carmichael et al., 1994; Carmichael and Price, 1994). Neurophysiological recordings have found that odor-taste reward associations are much slower and inflexible (Rolls et al., 1996) than visual-taste associations which usually take place with one-trial learning (Thorpe et al., 1983).

D. Hedonic Representations

The evidence from neuroimaging studies of pure taste and smell cited previously shows that the orbitofrontal cortex is consistently correlated with the subjective pleasantness ratings of the stimuli. Therefore, it is to be expected that studies using multimodal combinations of taste and smell should find correlations between pleasantness and activity in these brain regions.

Compelling evidence that this is indeed the case comes from a sensory-specific satiety neuroimaging study which has shown that a region of the left orbitofrontal cortex showed not only a sensory-specific decrease in the reward value to the whole food eaten to satiety (and not to the whole food not eaten), but also a correlation with pleasantness ratings (see Fig. 2a) (Kringelbach et al., 2003). This result strongly indicates that the reward value of the taste, olfactory, and somatosensory components of a whole food are represented in the orbitofrontal cortex, and that the subjective pleasantness of food thus might be represented here.

Further evidence comes from a study investigating the nonspecific satiation effects of chocolate (with both olfactory and gustatory components), which found a correlation between the decrease in pleasantness and activity in the orbitofrontal cortex (Small et al., 2001). Another multimodal study investigating the link between olfaction and vision found activity in the anterior orbitofrontal cortex for semantically congruent trials (Gottfried and Dolan, 2003).

Other multimodal neuroimaging studies have investigated the interaction between taste and smell, and found significant activity in more posterior parts of the orbitofrontal cortex and nearby agranular insula for the combination of taste and smell (De Araujo et al., 2003c; Small et al., 1997a). When investigating the synergistic enhancement of a matched taste and retronasal smell it was again found that a region of the orbitofrontal cortex was significantly activated (see Fig. 2c) (De Araujo et al., 2003c). This region was located very near to the region of the orbitofrontal cortex activated by the synergistic combinations of umami (De Araujo et al., 2003a).

Further neuroimaging studies have investigated the hedonic systems involved in other sensory modalities and, consistent with the results obtained with food relevant stimuli, patterns of activity were found that

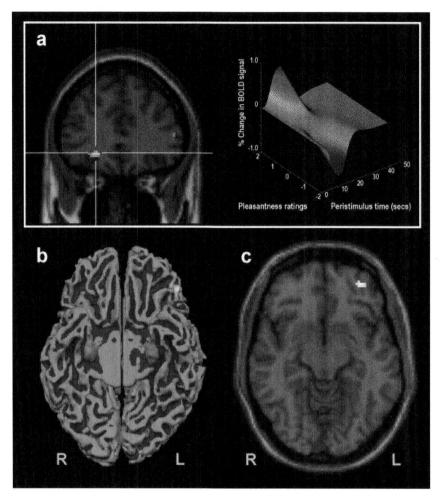

FIGURE 2 Hedonic experience. (a) A neuroimaging study using selective satiation found that mid-anterior parts of the orbitofrontal cortex are correlated with the subjects' subjective pleasantness ratings of the foods throughout the experiment (Kringelbach et al., 2003). On the right is shown a plot of the magnitude of the fitted hemodynamic response from a representative single subject against the subjective pleasantness ratings (on a scale from −2 to +2) and peristimulus time in seconds. (b) Additional evidence for the role of the orbitofrontal cortex in subjective experience comes from another neuroimaging experiment investigating the supra-additive effects of combining the umami tastants monosodium glutamate and inosine monophosphate (De Araujo et al., 2003a). The figure shows the region of mid-anterior orbitofrontal cortex showing synergistic effects (rendered on the ventral surface of human cortical areas with the cerebellum removed). The perceived synergy is unlikely to be expressed in the taste receptors themselves and the activity in the orbitofrontal cortex may thus reflect the subjective enhancement of umami taste which must be closely linked to subjective experience. (c) Adding strawberry odor to a sucrose taste solution makes the combination significantly more pleasant than the sum of each of the individual components. The supralinear effects reflecting the subjective enhancement were found to significantly activate a lateral region of the left anterior orbitofrontal cortex, which is remarkably similar to that found in the other experiments (De Araujo et al., 2003c). (Color Plate)

link hedonic processing to activity in the orbitofrontal cortex. In a study of thermal stimulation it was found that the perceived thermal intensity was correlated with activity in the insula and orbitofrontal cortices (Craig, 2002; Craig et al., 2000). In another study investigating the effects of touch to the hand it was found that dissociable regions of the anterior cingulate and orbitofrontal cortex were activated by painful and pleasant (but not neutral) touch (Rolls et al., 2003b). Another pain study found that the lateral orbitofrontal cortex and the anterior cingulate cortex are mediating the placebo effect in placebo responders (Petrovic et al., 2002). Studies investigating the effects of auditory stimulation and music have found that activity in the orbitofrontal cortex correlates with the negative dissonance (pleasantness) of musical chords (Blood et al., 1999) and that intensely pleasurable responses (chills) elicited by music are correlated with activity in the orbitofrontal cortex, ventral striatum, cingulate, and insula cortex (Blood and Zatorre, 2001).

Even more abstract reinforcers such as monetary reward (O'Doherty et al., 2001a; Thut et al., 1997) and punishment (O'Doherty et al., 2001a) have been found to activate the orbitofrontal cortex, and there is evidence to suggest that dissociable regions of the medial and lateral orbitofrontal cortex correlate with the amount of monetary gain and loss, respectively (O'Doherty et al., 2001a). More generally, it has been demonstrated by another neuroimaging study that the lateral orbitofrontal cortex and a region of the anterior cingulate cortex are together responsible for supporting general reversal learning in the human brain (Kringelbach and Rolls, 2003). Further compelling evidence from drug studies have found responses to cocaine in the orbitofrontal cortex, ventral striatum, and other reward-related brain structures (e.g., Breiter et al., 1997). A correlation was also found between a reliable index of the rush of i.v. methamphetamine in drug-naïve subjects (mind-racing) and activity in the orbitofrontal cortex (Völlm et al., 2004).

The preceding evidence implicates the orbitofrontal cortex in a wide variety of tasks. A large meta-analysis has demonstrated medio-lateral and anterior-posterior functional distinctions in the orbitofrontal cortex (Kringelbach, 2002; Kringelbach and Rolls, 2004). Activity in the medial orbitofrontal cortex has been shown to

relate to the monitoring of the reward value of many different reinforcers (involving, of course, mechanisms for learning and memory), whereas lateral orbitofrontal cortex activity has been proposed to be related to the evaluation of reinforcers which leads to a change in ongoing behavior. Furthermore, it has also been demonstrated that there is a posterior-anterior distinction with more complex or abstract reinforcers (such as monetary gain and loss) being represented more anteriorly in the orbitofrontal cortex than simpler reinforcers such as taste or pain (Kringelbach, 2002; Kringelbach and Rolls, 2004).

1. Lesion Studies and Reward

Neuroimaging studies are essentially only correlative measures of behavior, and it is therefore important to briefly consider the evidence from lesions in both humans and higher primates to assess the validity of the findings. In humans, damage to the orbitofrontal cortex causes major changes in motivation, emotion, personality, behavior, and social conduct. A classic case of orbitofrontal damage is that of Phineas Gage, whose medial frontal lobes were penetrated by a metal rod (Harlow, 1848). Miraculously Gage survived but his personality and emotional processing were changed completely (although care should be taken since our sources are severely limited) (Macmillan, 2000). In more recent cases of orbitofrontal cortex damage, patients often show lack of affect, social inappropriateness, and irresponsibility (Anderson et al., 1999; Hornak et al., 2003; Rolls et al., 1994). It has been shown that patients are impaired at correctly identifying social signals including, for example, face and voice expressions (Hornak et al., 1996, 2003). Interestingly, the recent experiments on human patients with surgical lesions to the orbitofrontal cortex have found functional heterogeneity in that bilateral lesions to the lateral orbitofrontal cortex—but not unilateral or lesions to medial parts of the orbitofrontal cortex—produce significant impairments in reversal learning (Hornak et al., 2004).

Furthermore, in the context of the evidence for a role of the orbitofrontal cortex in motivational hedonics, it is interesting to note that frontotemporal dementia (a progressive neurodegenerative disorder attacking the frontal lobes) does not only produce major and pervasive behavioral changes in personality and social conduct resembling those produced by orbitofrontal lesions. The condition is also associated with profound changes in eating habits with escalating desire for sweet food coupled with reduced satiety, which is often followed by enormous weight gain (Rahman et al., 1999).

Lesion studies in nonhuman primates support the hypothesis that reward value is represented in the orbitofrontal cortex. In devaluation paradigms monkeys with lesions to the orbitofrontal cortex were able to respond normally to associations between food and conditioners but fail to modify their behavior to the cues when probing these representations by reducing the incentive value of the food (Butter et al., 1963). Other studies found that lesions to the orbitofrontal cortex alter food preferences in monkeys (Baylis and Gaffan, 1991), while another lesion study used unilateral crossed lesions to show that the orbitofrontal cortex and the amygdala are important for the alteration of stimulus-reward associations (Baxter et al., 2000). Even in rats, it has been demonstrated that lesions to the orbitofrontal cortex blocks the ability to adapt behavior by accessing representational information about the incentive value of the associated food reinforcement (Gallagher et al., 1999). When making comparisons between rats and higher primates it is important to note that many brain areas have undergone considerable development. As mentioned previously, the elaboration of some of these brain areas has been so extensive in primates that even ancient systems such as the taste system appear to have been rewired to

place more emphasis on cortical processing in areas such as the orbitofrontal cortex.

IV. CONCLUSION

Food intake is essential to sustain life, and the sensory systems of taste and smell are among the most fundamental building blocks of the brain's natural reward systems (Kelley and Berridge, 2002). The special importance of food in human life is underlined by the predominance of food symbols and metaphors in human expressions across cultures (Lévi-Strauss, 1964) and the elaborate social constructions regarding purity and taboo of foods (Douglas, 1966). Food intake and food choice constitute a fundamental and frequent part of human life and have played a major role in the cultural evolution of nonfood systems such as ritual, religion, and social exchange as well as in the advancement of technology, development of cities, illnesses, and warfare through agriculture and domestication (Diamond, 1999).

It has been proposed that humans' higher cognitive functions may have evolved to support the required cognitive processing involved in the sophisticated foraging necessary for the sustained food intake needed for omnivores such as humans (Kringelbach, 2004b).

The evidence reviewed here suggests that food consumption and the control of appetite rely on cortical processing in humans and other primates. Four main computational principles have been proposed: (1) motivation-independent processing of identity and intensity; (2) formation of learning-dependent multimodal sensory representations; (3) reward representations using mechanisms including selective satiation, and (4) representations of hedonic experience, monitoring/learning, or direct behavioral change.

This processing relies to a large extent on the orbitofrontal cortex. In line with earlier proposals (Kringelbach, 2004b, 2005) a possible model is proposed which implements these computational principles for the interaction between sensory and hedonic systems in the human brain (see Fig. 3). Sensory information about primary (e.g., taste and smell) and secondary (e.g., visual) reinforcers is carried from the periphery to the primary sensory cortices (e.g., anterior insula/frontal operculum for taste and pyriform cortex for smell), where stimulus identity is decoded into stable representations (Small et al., 1999; Zatorre et al., 1992). This information is then conveyed for further multimodal integration in brain structures such as the posterior parts of the orbitofrontal cortex (De Araujo et al., 2003c; Small et al., 1997a). The reward value of the reinforcer is assigned (e.g., in more anterior parts of the orbitofrontal cortex) (Gottfried et al., 2003; Small et al., 2003) from where it can then be used for influencing subsequent behavior [e.g., lateral parts of the anterior orbitofrontal cortex, anterior cingulate cortex (Kringelbach and Rolls, 2003), and dorsolateral prefrontal cortex (Kringelbach et al., 2004; Wallis and Miller, 2003)], monitored as part of learning and memory mechanisms (e.g., in medial parts of the anterior orbitofrontal cortex) (De Araujo et al., 2003b; Gottfried and Dolan, 2003), and made available for subjective hedonic experience (e.g., mid-anterior orbitofrontal cortex) (Kringelbach et al., 2003). The reward value, and thus also the subjective hedonic experience of a reinforcer, can be modulated by hunger and other internal states (Gottfried et al., 2003; Kringelbach et al., 2003; O'Doherty et al., 2000), while the identity representation is remarkably stable and not subject to modulation (De Araujo et al., 2003b; Rolls et al., 2003a).

This model of the orbitofrontal cortex is obviously simplified but begins to address the basic principles of how food consumption and appetite are controlled in the human brain. More research into the cortical mechanisms of food intake is likely not only to further elucidate the workings of this network but may perhaps also prove

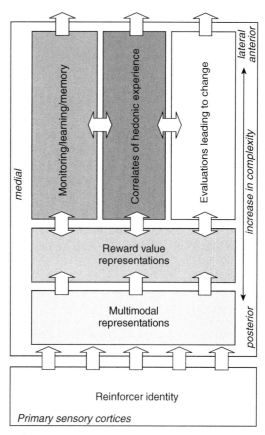

FIGURE 3 Proposed model for the interaction between sensory and hedonic systems in the human brain using as an example one hemisphere of the orbitofrontal cortex. Information is flowing from the bottom to the top of the figure. Sensory information about primary (e.g., taste, smell, touch, and pain) and secondary (e.g., visual) reinforcers is sent from the periphery to the primary sensory cortices (e.g., anterior insula/frontal operculum for taste), where stimulus identity is decoded into stable cortical representations. This information is then conveyed for further multimodal integration in brain structures in the posterior parts of the orbitofrontal cortex. The reward value of the reinforcer is assigned in more anterior parts of the orbitofrontal cortex from where it can then be used for influencing subsequent behavior (in lateral parts of the anterior orbitofrontal cortex from where it is sent to anterior cingulate cortex and dorsolateral prefrontal cortex), stored for learning (medial parts of the anterior orbitofrontal cortex), and made available for subjective hedonic experience (mid-anterior orbitofrontal cortex). The reward value and thus also the subjective hedonic experience of a reinforcer can be modulated by hunger and other internal states, while the identity representations in primary sensory cortices are remarkably stable and not subjected to modulation. It should be noted that there is, of course, important reciprocal information flowing between the higher level regions of the orbitofrontal cortex.

important to larger questions such as the neural correlates of subjective experience.

Acknowledgments

This research has been supported by the Danish Research Agency, the Wellcome Trust, and the MRC (to FMRIB, where most of the neuroimaging was carried out). The author would like to thank Caroline Andrews, Annie Cattrell, Phil Cowen, Ivan De Araujo, Peter Hobden, Julia Hornak, John O'Doherty, Edmund Rolls, Birgit Völlm, and James Wilson for their collaboration on some of the research reviewed here.

References

Anderson, A. K., Christoff, K., Stappen, I., Panitz, D., Ghahremani, D. G., Glover, G., Gabrieli, J. D., and Sobel, N. (2003). Dissociated neural representations of intensity and valence in human olfaction. *Nature Neurosci.* **6**, 196–202.

Anderson, S. W., Bechara, A., Damasio, H., Tranel, D., and Damasio, A. R. (1999). Impairment of social and moral behavior related to early damage in human prefrontal cortex. *Nature Neurosci.* **2**, 1032–1037.

Barer, D. H. (1989). The natural history and functional consequences of dysphagia after hemispheric stroke. *J. Neurol. Neurosurg. Psychiatry* **52**, 236–241.

Baxter, M. G., Parker, A., Lindner, C. C., Izquierdo, A. D., and Murray, E. A. (2000). Control of response selection by reinforcer value requires interaction of amygdala and orbital prefrontal cortex. *J. Neurosci.* **20**, 4311–4319.

Baylis, L. L., and Gaffan, D. (1991). Amygdaloctomy and ventromedial prefrontal ablation produce similar deficits in food choice and in simple object discrimination learning for an unseen reward. *Exp. Brain Res.* **86**, 617–622.

Baylis, L. L., Rolls, E. T., and Baylis, G. C. (1994). Afferent connections of the orbitofrontal cortex taste area of the primate. *Neuroscience* **64**, 801–812.

Berridge, K. C. (1996). Food reward: brain substrates of wanting and liking. *Neurosci. Biobehav. Rev.* **20**, 1–25.

Berridge, K. C., and Robinson, T. E. (1998). What is the role of dopamine in reward: hedonic impact, reward learning, or incentive salience? *Brain Res. Brain Res. Rev.* **28**, 309–369.

Bhatnagar, K. P., and Meisami, E. (1998). Vomeronasal organ in bats and primates: extremes of structural variability and its phylogenetic implications. *Microsc. Res. Tech.* **43**, 465–475.

Bindra, D. (1978). How adaptive behavior is produced: A perceptual-motivational alternative to response-reinforcement. *Behavioral Brain Sci.* **1**, 41–91.

Blood, A. J., and Zatorre, R. J. (2001). Intensely pleasurable responses to music correlate with activity in brain regions implicated in reward and emotion. *Proc. Natl. Acad. Sci. U.S.A.* **98**, 11818–11823.

Blood, A. J., Zatorre, R. J., Bermudez, P., and Evans, A. C. (1999). Emotional responses to pleasant and unpleasant music correlate with activity in paralimbic brain regions. *Nature Neurosci.* **2**, 382–387.

Bornstein, W. S. (1940a). Cortical representation of taste in man and monkey. I. Functional and anatomical relations of taste, olfaction and somatic sensibility. *Yale J. Biol. Med.* **12**, 719–736.

Bornstein, W. S. (1940b). Cortical representation of taste in man and monkey. II. The localization of the cortical taste area in man, a method of measuring impairment of taste in man. *Yale J. Biol. Med.* **13**, 133–156.

Breiter, H. C., Gollub, R. L., Weisskoff, R. M., Kennedy, D. N., Makris, N., Berke, J. D., Goodman, J. M., Kantor, H. L., Gastfriend, D. R., Riorden, J. P., Mathew, R. T., Rosen, B. R., and Hyman, S. E. (1997). Acute effects of cocaine on human brain activity and emotion. *Neuron* **19**, 591–611.

Buck, L., and Axel, R. (1991). A novel multigene family may encode odorant receptors: a molecular basis for odor recognition. *Cell* **65**, 175–187.

Burton, M. J., Rolls, E. T., and Mora, F. (1976). Effects of hunger on the responses of neurones in the lateral hypothalamus to the sight and taste of food. *Exp. Neurol.* **51**, 668–677.

Butter, C. M., Mishkin, M., and Rosvold, H. E. (1963). Conditioning and extinction of a food-rewarded response after selective ablations of frontal cortex in rhesus monkeys. *Exp. Neurol.* **7**, 65–75.

Cabanac, M. (1971). Physiological role of pleasure. *Science* **173**, 1103–1107.

Carmichael, S. T., and Price, J. L. (1994). Architectonic subdivision of the orbital and medial prefrontal cortex in the macaque monkey. *J. Comp. Neurol.* **346**, 366–402.

Carmichael, S. T., Clugnet, M.-C., and Price, J. L. (1994). Central olfactory connections in the macaque monkey. *J. Comp. Neurol.* **346**, 403–434.

Cascino, G., and Karnes, W. (1990). Gustatory and second sensory seizures associated with lesions in the insular cortex seen on magnetic resonance imaging. *J. Epilepsy* **3**, 185–187.

Cerf-Ducastel, B., and Murphy, C. (2001). fMRI activation in response to odorants orally delivered in aqueous solutions. *Chem. Senses* **26**, 625–637.

Chaudhari, N., Landin, A. M., and Roper, S. D. (2000). A metabotropic glutamate receptor variant functions as a taste receptor. *Nature Neurosci.* **3**, 113–119.

Craig, A. D. (2002). Opinion: How do you feel? Interoception: the sense of the physiological condition of the body. *Nature Rev. Neurosci.* **3**, 655–666.

Craig, A. D., Chen, K., Bandy, D., and Reiman, E. M. (2000). Thermosensory activation of insular cortex. *Nature Neurosci.* **3**, 184–190.

Dade, L. A., Jones-Gotman, M., Zatorre, R., and Evans, A. C. (1998). Human brain function during odor encoding and recognition: A PET activation study. *In* "Olfaction and Taste XII: An International Symposium" (C. Murphy, ed.), (*Ann. N.Y. Acad. Sci*), Vol. 855, pp. 572–575. New York Academy of Sciences, New York.

De Araujo, I. E. T., Kringelbach, M. L., Rolls, E. T., and Hobden, P. (2003a). The representation of umami taste in the human brain. *J. Neurophysiol.* **90**, 313–319.

De Araujo, I. E. T., Kringelbach, M. L., Rolls, E. T., and McGlone, F. (2003b). Human cortical responses to water in the mouth, and the effects of thirst. *J. Neurophysiol.* **90**, 1865–1876.

De Araujo, I. E. T., Rolls, E. T., Kringelbach, M. L., McGlone, F., and Phillips, N. (2003c). Taste-olfactory convergence, and the representation of the pleasantness of flavour, in the human brain. *Eur. J. Neurosci.* **18**, 2059–2068.

Diamond, J. M. (1999). "Guns, Germs, and Steel: The Fates of Human Societies." Norton, New York.

Douglas, M. (1966). "Purity and Danger: An Analysis of Concepts of Pollution and Taboo." Routledge & Kegan Paul, London.

Eskenazi, B., Cain, W., Novelly, R., and Friend, K. (1983). Olfactory functioning in temporal lobectomy patients. *Neuropsychologia* **21**, 365–374.

Eslinger, P. J., Damasio, A. D., and Van Hoesen, G. W. (1982). Olfactory dysfunction in man: anatomical and behavioural aspects. *Brain Cogn.* **1**, 259–285.

Francis, S., Rolls, E. T., Bowtell, R., McGlone, F., O'Doherty, J., Browning, A., Clare, S., and Smith, E. (1999). The representation of pleasant touch in the brain and its relationship with taste and olfactory areas. *Neuroreport* **10**, 453–459.

Frey, S., and Petrides, M. (1999). Re-examination of the human taste region: a positron emission tomography study. *Eur. J. Neurosci.* **11**, 2985–2988.

Gallagher, M., McMahan, R. W., and Schoenbaum, G. (1999). Orbitofrontal cortex and representation of incentive value in associative learning. *J. Neurosci.* **19**, 6610–6614.

Gordon, C., Hewer, R. L., and Wade, D. T. (1987). Dysphagia in acute stroke. *Br. Med. J. (Clin. Res. Ed.)* **295**, 411–414.

Gottfried, J. A., and Dolan, R. J. (2003). The nose smells what the eye sees: crossmodal visual facilitation of human olfactory perception. *Neuron* **39**, 375–386.

Gottfried, J. A., Deichmann, R., Winston, J. S., and Dolan, R. J. (2002a). Functional heterogeneity in human olfactory cortex: an event-related functional magnetic resonance imaging study. *J. Neurosci.* **22**, 10819–10828.

Gottfried, J. A., O'Doherty, J., and Dolan, R. J. (2002b). Appetitive and aversive olfactory learning in humans studied using event-related functional magnetic resonance imaging. *J. Neurosci.* **22**, 10829–10837.

Gottfried, J. A., O'Doherty, J., and Dolan, R. J. (2003). Encoding predictive reward value in human amygdala and orbitofrontal cortex. *Science* **301**, 1104–1107.

Gray, J. A. (1975). "Elements of a Two-Process Theory of Learning." Academic Press, London.

Grill, H. J., and Norgren, R. (1978). The taste reactivity test. II. Mimetic responses to gustatory stimuli in chronic thalmic and chronic decerebrate rats. *Brain Res.* **143**, 281–297.

Hamdy, S., Mikulis, D. J., Crawley, A., Xue, S., Lau, H., Henry, S., and Diamant, N. E. (1999a). Cortical activation during human volitional swallowing: an event-related fMRI study. *Am. J. Physiol.* **277**, G219–G225.

Hamdy, S., Rothwell, J. C., Brooks, D. J., Bailey, D., Aziz, Q., and Thompson, D. G. (1999b). Identification of the cerebral loci processing human swallowing with H2(15)O PET activation. *J. Neurophysiol.* **81**, 1917–1926.

Harlow, J. M. (1848). Passage of an iron rod through the head. *Boston Med. Surg. J.* **39**, 389–393.

Henkin, R., Comiter, H., Fedio, P., and O'Doherty, D. (1977). Defects in taste and smell recognition following temporal lobectomy. *Trans. Am. Neurol. Assoc.* **102**, 146–150.

Hornak, J., Rolls, E. T., and Wade, D. (1996). Face and voice expression identification in patients with emotional and behavioural changes following ventral frontal lobe damage. *Neuropsychologia* **34**, 247–261.

Hornak, J., Bramham, J., Rolls, E. T., Morris, R. G., O'Doherty, J., Bullock, P. R., and Polkey, C. E. (2003). Changes in emotion after circumscribed surgical lesions of the orbitofrontal and cingulate cortices. *Brain* **126**, 1671–1712.

Hornak, J., O'Doherty, J., Bramham, J., Rolls, E. T., Morris, R. G., Bullock, P. R., and Polkey, C. E. (2004). Reward-related reversal learning after surgical excisions in orbitofrontal and dorsolateral prefrontal cortex in humans. *J. Cogn. Neurosci.* **16**, 463–478.

Hull, C. L. (1951). *Essentials of Behavior*. Yale University Press, New Haven, CT.

Jean, A. (1984). Brainstem organization of the swallowing network. *Brain Behav. Evol.* **25**, 109–116.

Jean, A. (2001). Brainstem control of swallowing: neuronal network and cellular mechanisms. *Physiol. Rev.* **81**, 929–969.

Jones-Gotman, M., and Zatorre, R. J. (1988). Olfactory identification deficits in patients with focal cerebral excision. *Neuropsychologia* **26**, 387–400.

Jones-Gotman, M., and Zatorre, R. J. (1993). Odor recognition memory in humans: Role of right temporal and orbitofrontal regions. *Brain Cogn.* **22**, 182–198.

Jones-Gotman, M., Zatorre, R. J., Olivier, A., Andermann, F., Cendes, F., Staunton, H., McMackin, D., Siegel, A. M., and Wieser, H. G. (1994). Odor memory in patients with excision from medial or lateral temporal lobe structures. *Soc. Neurosci. Abstr.* **20**, 358.

Jones-Gotman, M., Zatorre, R. J., Cendes, F., Olivier, A., Andermann, F., McMackin, D., Staunton, H., Siegel, A. M., and Wieser, H. G. (1997). Contribution of medial versus lateral temporal-lobe structures to human odour identification. *Brain* **120**, 1845–1856.

Kelley, A. E., and Berridge, K. C. (2002). The neuroscience of natural rewards: relevance to addictive drugs. *J. Neurosci.* **22**, 3306–3311.

Kinomura, S., Kawashima, R., Yamada, K., Ono, S., Itoh, M., Yoshioka, S., Yamaguchi, T., Matsui, H., Miyazawa, H., Itoh, H., Goto, R., Fujiwara, T., Satoh, K., and Fukuda, H. (1994). Functional anatomy of taste perception in the human brain studied with positron emission tomography. *Brain Res.* **659**, 263–266.

Kobayashi, M., Sasabe, T., Takeda, M., Kondo, Y., Yoshikubo, S., Imamura, K., Onoe, H., Sawada, T., and Watanabe, Y. (2002). Functional anatomy of chemical senses in the alert monkey revealed by positron emission tomography. *Eur. J. Neurosci.* **16**, 975–980.

Kohn, M., and Booth, M. (2003). The worldwide epidemic of obesity in adolescents. *Adolesc. Med.* **14**, 1–9.

Kringelbach, M. L. (2002). D.Phil. Thesis: The Functional Neuroanatomy of Emotion. University of Oxford, Oxford.

Kringelbach, M. L. (2004a). Emotion. *In* "The Oxford Companion to the Mind" (R. L. Gregory, ed.), 2nd Ed. (pp. 287–290). Ed. R. L. Gregory. Oxford University Press, Oxford.

Kringelbach, M. L. (2004b). Food for thought: hedonic experience beyond homeostasis in the human brain. *Neuroscience* **126**, 807–819.

Kringelbach, M. L. (2005). The orbitofrontal cortex: linking reward to hedonic experience. *Nature Rev. Neurosci.* **6**, 691–702.

Kringelbach, M. L., and Rolls, E. T. (2003). Neural correlates of rapid context-dependent reversal learning in a simple model of human social interaction. *Neuroimage* **20**, 1371–1383.

Kringelbach, M. L., and Rolls, E. T. (2004). The functional neuroanatomy of the human orbitofrontal cortex: evidence from neuroimaging and neuropsychology. *Prog. Neurobiol.* **72**, 341–372.

Kringelbach, M. L., O'Doherty, J., Rolls, E. T., and Andrews, C. (2003). Activation of the human orbitofrontal cortex to a liquid food stimulus is correlated with its subjective pleasantness. *Cereb. Cortex* **13**, 1064–1071.

Kringelbach, M. L., De Araujo, I. E. T., and Rolls, E. T. (2004). Taste-related activity in the human dorsolateral prefrontal cortex. *Neuroimage* **21**, 781–788.

LeMagnen, J. (1985). "Hunger." Cambridge University Press, London.

Lévi-Strauss, C. (1964). "Le Cru et Le Cuit. [1969] [The Raw and the Cooked: Introduction to a Science of Mythology. Jonathan Cape, London]." Librairie Plon, Paris.

Macmillan, M. (2000). "An Odd Kind of Fame: Stories of Phineas Gage." MIT Press, Cambridge, MA.

Martin, R. E., and Sessle, B. J. (1993). The role of the cerebral cortex in swallowing. *Dysphagia* **8**, 195–202.

Martin, R. E., Goodyear, B. G., Gati, J. S., and Menon, R. S. (2001). Cerebral cortical representation of automatic and volitional swallowing in humans. *J. Neurophysiol.* **85**, 938–950.

Monti-Bloch, L., Jennings-White, C., and Berliner, D. L. (1998). The human vomeronasal system. A review. *Ann. N. Y. Acad. Sci.* **855**, 373–389.

Mori, K., Nagao, H., and Sasaki, Y. F. (1998). Computation of molecular information in mammalian olfactory systems. *Network* **9**, R79–R102.

Morrot, G., Brochet, F., and Dubourdieu, D. (2001). The color of odors. *Brain Lang.* **79**, 309–320.

Norgren, R. (1984). Central neural mechanisms of taste. *In* "Handbook of Physiology—The Nervous System III. Sensory Processes 1", (J. M. Brookhart, V. B. Mountcastle, and I. Darien-Smith, eds.) pp. 1087–1128. American Physiological Society, Washington, D.C.

Norgren, R. (1990). Gustatory system. *In* "The Human Nervous System" (G. Paxinos, ed.), pp. 845–861. Academic Press, New York.

Norgren, R., and Leonard, C. M. (1973). Ascending central gustatory pathways. *J. Comp. Neurol.* **150**, 217–237.

O'Doherty, J., Rolls, E. T., Francis, S., Bowtell, R., McGlone, F., Kobal, G., Renner, B., and Ahne, G. (2000). Sensory-specific satiety-related olfactory activation of the human orbitofrontal cortex. *Neuroreport* **11**, 893–897.

O'Doherty, J., Kringelbach, M. L., Rolls, E. T., Hornak, J., and Andrews, C. (2001a). Abstract reward and punishment representations in the human orbitofrontal cortex. *Nat. Neurosci.* **4**, 95–102.

O'Doherty, J., Rolls, E. T., Francis, S., Bowtell, R., and McGlone, F. (2001b). Representation of pleasant and aversive taste in the human brain. *J. Neurophysiol.* **85**, 1315–1321.

O'Doherty, J. P., Deichmann, R., Critchley, H. D., and Dolan, R. J. (2002). Neural responses during anticipation of a primary taste reward. *Neuron* **33**, 815–826.

Ogawa, H., Ito, S. L., and Nomura, T. (1989). Oral cavity representation at the frontal operculum of macaque monkeys. *Neurosci. Res.* **6**, 283–298.

Papez, J. W. (1937). A proposed mechanism for emotion. *Arch. Neurol. Psychiatry* **38**, 725–743.

Penfield, W., and Faulk, M. E. (1955). The insula. Further observations on its function. *Brain* **78**, 445–470.

Petrovic, P., Kalso, E., Petersson, K. M., and Ingvar, M. (2002). Placebo and opioid analgesia—imaging a shared neuronal network. *Science* **295**, 1737–1740.

Pierce, J., and Halpern, B. P. (1996). Orthonasal and retronasal odorant identification based upon vapor phase input from common substances. *Chem. Senses* **21**, 529–543.

Pritchard, T. C. (1991). The primate gustatory system. *In* "Smell and Taste in Health and Disease" (T. Getchell, R. L. Doty, L. M. Bartoshuk, and J. Snow, eds.), Raven Press, New York.

Pritchard, T. C., Hamilton, R. B., Morse, J. R., and Norgren, R. (1986). Projections of thalamic gustatory and lingual areas in the monkey, *Macaca fascicularis*. *J. Comp. Neurol.* **244**, 213–228.

Pritchard, T. C., Hamilton, R. B., and Norgren, R. (1989). Neural coding of gustatory information in the thalamus of *Macaca mulatta*. *J. Neurophysiol.* **61**, 1–14.

Pritchard, T. C., Edwards, E. M., Smith, C. A., Hilgert, K. G., Gavlick, A. M., Maryniak, T. D., Schwartz, G. J., and Scott, T. R. (2005). Gustatory neural responses in the medial orbitofrontal cortex of the old world monkey. *J. Neurosci.* **25**, 6047–6056.

Rahman, S., Sahakian, B. J., Hodges, J. R., Rogers, R. D., and Robbins, T. W. (1999). Specific cognitive deficits in mild frontal variant frontotemporal dementia. *Brain* **122**, 1469–1493.

Rolls, B. J., Rolls, E. T., Rowe, E. A., and Sweeney, K. (1981). Sensory specific satiety in man. *Physiol. Behav.* **27**, 137–142.

Rolls, E. T. (1999). "The Brain and Emotion." Oxford University Press, Oxford.

Rolls, E. T., and Baylis, L. L. (1994). Gustatory, olfactory, and visual convergence within the primate orbitofrontal cortex. *J. Neurosci.* **14**, 5437–5452.

Rolls, E. T., Yaxley, S., and Sienkiewicz, Z. J. (1990). Gustatory responses of single neurons in the caudolateral orbitofrontal cortex of the macaque monkey. *J. Neurophysiol.* **64**, 1055–1066.

Rolls, E. T., Hornak, J., Wade, D., and McGrath, J. (1994). Emotion-related learning in patients with social and emotional changes associated with frontal lobe damage. *J. Neurol. Neurosurg. Psychiatry* **57**, 1518–1524.

Rolls, E. T., Critchley, H. D., Mason, R., and Wakeman, E. A. (1996). Orbitofrontal cortex neurons: role in olfactory and visual association learning. *J. Neurophysiol.* **75**, 1970–1981.

Rolls, E. T., Kringelbach, M. L., and De Araujo, I. E. T. (2003a). Different representations of pleasant and unpleasant odours in the human brain. *Eur. J. Neurosci.* **18**, 695–703.

Rolls, E. T., O'Doherty, J., Kringelbach, M. L., Francis, S., Bowtell, R., and McGlone, F. (2003b). Representations of pleasant and painful touch in the human orbitofrontal and cingulate cortices. *Cereb. Cortex* **13**, 308–317.

Rouby, C., Zatorre, R. J., Jones-Gotman, M., and Forster, L. (1997). Dissociation of odor intensity and pleasantness judgments: effect of temporal lobe lesions. *Chem. Senses* **22**, 780.

Royet, J. P., Souchier, C., Jourdan, F., and Ploye, H. (1988). Morphometric study of the glomerular population in the mouse olfactory bulb: numerical density and size distribution along the rostrocaudal axis. *J. Comp. Neurol.* **270**, 559–568.

Royet, J. P., Zald, D., Versace, R., Costes, N., Lavenne, F., Koenig, O., and Gervais, R. (2000). Emotional responses to pleasant and unpleasant olfactory, visual, and auditory stimuli: a positron emission tomography study. *J. Neurosci.* **20**, 7752–7759.

Royet, J. P., Hudry, J., Zald, D. H., Godinot, D., Gregoire, M. C., Lavenne, F., Costes, N., and Holley, A. (2001). Functional neuroanatomy of different olfactory judgments. *Neuroimage* **13**, 506–519.

Rozin, P. (2001). Food preference. *In* "International Encyclopedia of the Social & Behavioral Sciences" (N. J. Smelser and P. B. Baltes, eds.), pp. 5719–5722. Elsevier, Amsterdam.

Sanghera, M. K., Rolls, E. T., and Roper-Hall, A. (1979). Visual responses of neurons in the dorsolateral amygdala of the alert monkey. *Exp. Neurol.* **63**, 610–626.

Saper, C. B., Chou, T. C., and Elmquist, J. K. (2002). The need to feed: homeostatic and hedonic control of eating. *Neuron* **36**, 199–211.

Scott, T. R., and Plata-Salaman, C. R. (1999). Taste in the monkey cortex. *Physiol. Behav.* **67**, 489–511.

Scott, T. R., Yaxley, S., Sienkiewicz, Z. J., and Rolls, E. T. (1986). Gustatory responses in the frontal opercular cortex of the alert cynomolgus monkey. *J. Neurophysiol.* **56**, 876–890.

Small, D. M., Jones-Gotman, M., Zatorre, R. J., Petrides, M., and Evans, A. C. (1997a). Flavor processing: more than the sum of its parts. *Neuroreport* **8**, 3913–3917.

Small, D. M., Jones-Gotman, M., Zatorre, R. J., Petrides, M., and Evans, A. C. (1997b). A role for the right anterior temporal lobe in taste quality recognition. *J. Neurosci.* **17**, 5136–5142.

Small, D. M., Zald, D. H., Jones-Gotman, M., Zatorre, R. J., Pardo, J. V., Frey, S., and Petrides, M. (1999). Human cortical gustatory areas: a review of functional neuroimaging data. *Neuroreport* **10**, 7–14.

Small, D. M., Zatorre, R. J., Dagher, A., Evans, A. C., and Jones-Gotman, M. (2001). Changes in brain activity related to eating chocolate: from pleasure to aversion. *Brain* **124**, 1720–1733.

Small, D. M., Gregory, M. D., Mak, Y. E., Gitelman, D., Mesulam, M. M., and Parrish, T. (2003). Dissociation of neural representation of intensity and affective valuation in human gustation. *Neuron* **39**, 701–711.

Small, D. M., Gerber, J. C., Mak, Y. E., and Hummel, T. (2005). Differential neural responses evoked by orthonasal versus retronasal odorant perception in humans. *Neuron* **47**, 593–605.

Smith, T. D., and Bhatnagar, K. P. (2000). The human vomeronasal organ: Part II. Prenatal development. *J. Anat.* **197**, 421–436.

Sobel, N., Prabhakaran, V., Desmond, J. E., Glover, G. H., Goode, R. L., Sullivan, E. V., and Gabrieli, J. D. E. (1998). Sniffing and smelling: separate subsystems in the human olfactory cortex. *Nature* **392**, 282–286.

Stern, K., and McClintock, M. K. (1998). Regulation of ovulation by human pheromones. *Nature* **392**, 177–179.

Sudakov, K., MacLean, P. D., Reeves, A., and Marino, R. (1971). Unit study of exteroceptive inputs to claustrocortex in awake, sitting, squirrel monkey. *Brain Res.* **28**, 19–34.

Takagi, S. F. (1986). Studies on the olfactory nervous system of the old world monkey. *Prog. Neurobiol.* **27**, 195–250.

Tanabe, T., Iino, M., and Takagi, S. F. (1975a). Discrimination of odors in olfactory bulb, pyriform-amygdaloid areas, and orbitofrontal cortex of the monkey. *J. Neurophysiol.* **38**, 1284–1296.

Tanabe, T., Yarita, H., Iino, M., Ooshima, Y., and Takagi, S. F. (1975b). An olfactory projection area in orbitofrontal cortex of the monkey. *J. Neurophysiol.* **38**, 1269–1283.

Thorpe, S. J., Rolls, E. T., and Maddison, S. (1983). The orbitofrontal cortex: neuronal activity in the behaving monkey. *Exp. Brain Res.* **49**, 93–115.

Thut, G., Schultz, W., Roelcke, U., Nienhusmeier, M., Missimer, J., Maguire, R. P., and Leenders, K. L. (1997). Activation of the human brain by monetary reward. *Neuroreport* **8**, 1225–1228.

Völlm, B. A., De Araujo, I. E. T., Cowen, P. J., Rolls, E. T., Kringelbach, M. L., Smith, K. A., Jezzard, P., Heal, R. J., and Matthews, P. M. (2004). Methamphetamine activates reward circuitry in drug naïve human subjects. *Neuropsychopharmacology* in press.

Wallis, J. D., and Miller, E. K. (2003). Neuronal activity in primate dorsolateral and orbital prefrontal cortex during performance of a reward preference task. *Eur. J. Neurosci.* **18**, 2069–2081.

Weiskrantz, L. (1968). Emotion. *In* "Analysis of Behavioural Change" (L. Weiskrantz, ed.), pp. 50–90. Harper & Row, New York and London.

Williams, G. V., Rolls, E. T., Leonard, C. M., and Stern, C. (1993). Neuronal responses in the ventral striatum

of the behaving macaque. *Behav. Brain Res.* **55**, 243–252.

Wilson, J., Jenkinson, M., De Araujo, I. E. T., Kringelbach, M. L., Rolls, E. T., and Jezzard, P. (2002). Fast, fully automated global and local magnetic field optimization for fMRI of the human brain. *Neuroimage* **17**, 967–976.

Woods, S. C., and Stricker, E. M. (1999). Central control of food intake. *In* "Fundamental Neuroscience" (M. J. Zigmond, F. E. Bloom, S. C. Landis, J. L. Roberts, and L. R. Squire, eds.), volume Chap. 41, pp. 1091–1109. Academic Press, San Diego, CA.

Yarita, H., Iino, M., Tanabe, T., Kogure, S., and Takagi, S. F. (1980). A transthalamic olfactory pathway to the orbitofrontal cortex in the monkey. *J. Neurophysiol.* **43**, 69–85.

Yaxley, S., Rolls, E. T., and Sienkiewicz, Z. J. (1988). The responsiveness of neurons in the insular gustatory cortex of the macaque monkey is independent of hunger. *Physiol. Behav.* **42**, 223–229.

Yousem, D. M., Williams, S. C. R., Howard, R. O., Andrew, C., Simmons, A., Allin, M., Geckle, R. J., Suskind, D., Bullmore, E. T., Brammer, M. J., and Doty, R. L. (1997). Functional MR imaging during odor stimulation: preliminary data. *Radiology* **204**, 833–838.

Zald, D. H., and Pardo, J. V. (1997). Emotion, olfaction, and the human amygdala: amygdala activation during aversive olfactory stimulation. *Proc. Natl. Acad. Sci. U.S.A.* **94**, 4119–4124.

Zald, D. H., and Pardo, J. V. (1999). The functional neuroanatomy of voluntary swallowing. *Ann. Neurol.* **46**, 281–286.

Zald, D. H., Lee, J. T., Fluegel, K. W., and Pardo, J. V. (1998). Aversive gustatory stimulation activates limbic circuits in humans. *Brain* **121**, 1143–1154.

Zald, D. H., Hagen, M. C., and Pardo, J. V. (2002). Neural correlates of tasting concentrated quinine and sugar solutions. *J. Neurophysiol.* **87**, 1068–1075.

Zatorre, R. J., and Jones-Gotman, M. (1991). Human olfactory discrimination after unilateral frontal or temporal lobectomy. *Brain* **114**, 71–84.

Zatorre, R. J., Jones-Gotman, M., Evans, A. C., and Meyer, E. (1992). Functional localization and lateralization of human olfactory cortex. *Nature* **360**, 339–340.

Zatorre, R. J., Jones-Gotman, M., and Rouby, C. (2000). Neural mechanisms involved in odor pleasantness and intensity judgments. *Neuroreport* **11**, 2711–2716.

The Nucleus Accumbens Shell as a Model of Integrative Subcortical Forebrain Systems Regulating Food Intake

THOMAS R. STRATFORD

Department of Psychology, University of Illinois at Chicago

I. Introduction
II. The Regulation of Feeding Behavior by the Nucleus Accumbens Shell
   A. Excitatory and Inhibitory Amino Acid Circuits in the AcbSh Regulate Food Intake
   B. AcbSh-Mediated Changes in Feeding Behavior Are Behaviorally Specific
   C. Neuropeptides Systems in the AcbSh Are Involved in the Control of Feeding Behavior
   D. The AcbSh Feeding Circuit Is Likely to Regulate Normal Homeostatic Feeding Behavior
III. Afferent Projections to the AcbSh That May Be Involved in Regulating Food Intake
   A. Excitatory Amino Acid Inputs
   B. Additional Inputs
   C. The Role of Reentrant Projections in the AcbSh Circuit
IV. Downstream Components of the Functional AcbSh Feeding Circuit
   A. Classes of Evidence
   B. Identification of Brain Regions Activated by Inhibition of the AcbSh
   C. Role of the Ventral Pallidum in the Expression of AcbSh-Mediated Feeding
   D. Role of the Hypothalamus in the Expression of AcbSh-Mediated Feeding
V. The AcbSh Feeding Circuit As a Potential Site for Pathology and Therapeutic Intervention in the Treatment of Eating Disorders
   References

This chapter presents the nucleus accumbens shell (AcbSh) feeding circuit as an example of how integrative processes in the subcortical forebrain contribute to the regulation of food intake. A persuasive body of evidence has been assembled over the past decade indicating that AcbSh circuits containing GABA, glutamate, and a number of neuropeptides play a specific and fundamental role in controlling food intake. The AcbSh receives an exceptionally wide variety of information from afferent fiber systems originating in a number of cortical and subcortical brain regions. Processed multimodal sensory data are integrated with information regarding energy balance, taste, reward, and previously formed associations and results in an output signal that potently regulates neural activity in feeding-related circuits involving the lateral, arcuate, and paraventricular hypothalamic nuclei. The AcbSh appears to mediate activity in the lateral hypothalamus both through direct projections and by altering the firing rate of neurons in the interposed medial ventral pallidum. The magnitude and behavioral specificity of AcbSh-mediated feeding suggest that pathology of the circuit might be implicated in the etiology of some eating disorders. Furthermore, amino acid and neuropeptide receptors in the AcbSh may be promising targets for pharmacological therapies designed to treat such disorders.

I. INTRODUCTION

Feeding in mammals is a complex behavior, the successful expression of which involves the coordinated integration of multimodal signals from numerous brain regions and the periphery. While endocrine and paracrine systems are used to convey information concerning energy stores and physiological state, all relevant information is eventually transduced to neural signals. Many of the factors involved in the control of feeding behavior are the result of integrative processes in the forebrain. While the isolated hindbrain can sustain feeding to satiation when food is placed directly in the mouth, the forebrain initiates and controls the appetitive, or motivational, aspects of feeding. It is only more recently that a sufficient body of data has accumulated to allow us to begin to understand the specific roles played by discrete cell populations, to assess the importance of their relative contributions to the control of normal and pathological feeding, and to investigate

how these functional populations work together to control food intake.

Several subcortical forebrain regions have been demonstrated to play a role in the control of food intake, although the precise behavioral contribution and neurocircuitry through which these changes are effected often remain to be elucidated. Substantive changes in feeding behavior can be elicited by manipulations of the amygdala, nucleus accumbens shell, ventral pallidum, ventral tegmental area, lateral and medial hypothalamic areas, and a diffusely distributed set of opioid receptor-bearing neurons in the ventral forebrain. Data, much of it gathered over the past 10 years, suggest that many of these brain regions work in an integrated fashion to form a functional circuit subserving the control of food intake. One of the primary components of this circuit, and perhaps a useful conceptual focal point, is the nucleus accumbens shell (AcbSh), which appears to be a primary point of integration of feeding-related information in the subcortical rostral forebrain. The nature of inputs to the AcbSh indicates that this region is positioned to integrate information concerning taste and its current incentive value, previous associations made between environmental stimuli and the act of feeding, and information about the current state of energy balance, cardiovascular tone, and other visceral functions. Furthermore, recurrent collaterals from AcbSh projection neurons and reentrant processes from the ventral pallidum, lateral hypothalamus, and other downstream components provide constant feedback on neuronal activity within the system itself. The result of this neural integration in the AcbSh is a largely inhibitory output signal that suppresses food intake through its powerful effects on hypothalamic circuits. In fact, the AcbSh appears to be unique among extrahypothalamic brain regions in the magnitude and specificity of its influence on feeding behavior. Because of this, much of the work on the forebrain mechanisms mediating food intake has centered on defining the role of the AcbSh. The substantial body of data that has resulted from these efforts suggests that it may be most profitable to center the discussion of the extrahypothalamic forebrain feeding circuitry on the AcbSh.

Of course, the preceding list is not exhaustive and other subcortical forebrain regions are likely to mediate some aspects of feeding behavior. Furthermore, although each of these brain regions is believed to contain circuits that subserve multiple functions, we will focus on their demonstrated roles in the control of food intake. The goal of this chapter is not to present a complete review of all forebrain feeding systems, or of the complete behavioral role of each system, but rather to detail our current best approximation of a system that seems an exceptionally promising target for potential pharmacological therapies aimed at altering food intake.

It would be prudent to add a brief caveat about the terms used to delineate brain regions. While traditional anatomical nomenclature tends to convey the impression that brain nuclei are specific, unitary entities, it should be remembered that each contains heterogeneous populations of cells and are believed to subserve multiple functions. Nevertheless, it is becoming increasingly clear that certain populations of cells within each of these anatomically defined structures work in concert to specifically mediate food intake. Therefore, given the current limitations of our knowledge, we necessarily use such terms as general anatomical markers, but with the understanding that, in the current context, we are referring to specific, but as yet undefined, subsets of functionally related cells located in the vicinity.

II. THE REGULATION OF FEEDING BEHAVIOR BY THE NUCLEUS ACCUMBENS SHELL

It has been recognized since the 1970s that pharmacological and surgical manipu-

lations of the neural elements within the nucleus accumbens can alter feeding behavior (Lorens et al., 1970; Koob et al., 1978). Although most of the early behavioral investigations involving the accumbens considered the region as a unitary structure, both anatomically and functionally, insight into the neural substrates responsible for the behavioral changes was greatly facilitated by anatomical advances in the late 1980s. During that time, emerging detail about differential patterns of connectivity and neurochemical profiles in the basal forebrain led to the identification of anatomical subdivisions within the Acb proper, most notably the delineation of a central "core" and an adjacent "shell" subregion, composed of a band of tissue surrounding the ventral and medial aspects of the pericommissural core (Zaborszky et al., 1985; Heimer et al., 1997). From the perspective of extrapolating data obtained in rats to human conditions, it is important to note that the core and shell subdivisions are also present in primates, including humans (Ikemoto et al., 1995; Meredith et al., 1996).

The identification of the anatomical core–shell distinction subsequently led to the initiation of studies designed to evaluate the specific behavioral functions subserved by each region. Subsequently, a convergent body of data has emerged indicating that both inhibitory and excitatory amino acid systems located in a select region of the AcbSh play a specific and fundamental role in the control of food intake. The data gathered to date support the hypothesis that pharmacological manipulations resulting in a net inhibition of activity in medium spiny projection neurons located in the medial AcbSh result in a behaviorally specific increase in food intake. Moreover, the AcbSh appears to be one component of a feeding circuit that includes the ventral pallidum and that effects changes in feeding behavior by altering activity in the lateral, arcuate, and paraventricular hypothalamic regions.

A. Excitatory and Inhibitory Amino Acid Circuits in the AcbSh Regulate Food Intake

Chemical inhibition of neurons in the medial AcbSh elicits behaviorally specific increases in short-term food intake in satiated rats. Fast synaptic activity in the AcbSh is mediated principally by glutamate and GABA, and these neurotransmitter systems appear to be the primary mechanisms responsible for controlling this behavior. Blocking the AMPA- and kainate-preferring subsets of ionotropic glutamate receptors in the AcbSh with local injections of 6,7-dinitroquinoxaline-2,3-dione (DNQX) elicits robust feeding in satiated rats (Maldonado-Irizarry et al., 1995; Stratford et al., 1998; Reynolds and Berridge, 2002). Even larger increases in short-term food intake are elicited by activating GABA$_A$ or GABA$_B$ receptors with local administration of the selective agonists muscimol (Stratford and Kelley, 1997, 1999; Reynolds and Berridge, 2001; Znamensky et al., 2001; Stratford, 2005) or baclofen (Stratford and Kelley, 1997; Ward et al., 2000; Znamensky et al., 2001), suggesting that the acute inhibition of neural activity in the AcbSh is the relevant event that initiates feeding. After either treatment, rats show a dose-dependent increase in food intake. Feeding effects are characterized by a rapid onset, with a corresponding dose-dependent decrease in eating latency. Feeding elicited by inhibiting activity in the AcbSh is characterized by the magnitude of the effect; muscimol-elicited feeding results in mean intakes of dry chow that are among the highest reported in the literature. Although AcbSh inhibition does not alter water intake, it does increase intake of palatable, nutritive sucrose solutions, demonstrating that the elicited motor response is situationally appropriate and is not restricted to the ingestion of solid food (Stratford et al., 1998; Basso and Kelley, 1999).

Importantly, the elicitation of GABA-mediated feeding in the AcbSh is not restricted to injections of exogenous GABA mimetics. Microinjections of γ-vinyl-GABA, a selective inhibitor of the metabolic enzyme GABA transaminase, also elicits intense, dose-related feeding of the same magnitude (Stratford and Kelley, 1997). By blocking the metabolism of GABA, this pharmacological intervention increases local levels of endogenous GABA at terminals (Halonen et al., 1991). The fact that feeding is induced by inhibiting AcbSh GABA metabolism indicates that there is a tonic release of GABA in the region and that GABA tone in the AcbSh is maintained, in part, by GABA transaminase in satiated rats. These facts constitute strong evidence that endogenous GABA in the AcbSh plays a role in the normal regulation of food intake.

Feeding-relevant cells constitute a subpopulation of neurons that are distributed heterogeneously within the AcbSh proper. Microinjection mapping studies have revealed that the most sensitive basal forebrain site for eliciting feeding through manipulations of local amino acid systems is the medial aspect of the AcbSh (relative to any prime axis) (Kelley and Swanson, 1997; Stratford and Kelley, 1997; Basso and Kelley, 1999; Reynolds and Berridge, 2001). Although various relativistic qualifiers have been used to identify the precise level from which feeding can be elicited, the relevant neural elements in the AcbSh appear to be located primarily in the middle third of the rostrocaudal extent of the medial AcbSh as traditionally delineated.

As might be expected from this model, the converse treatment, activating AcbSh neurons with injections of low doses of AMPA, suppresses deprivation-induced feeding and intake of palatable sucrose solutions (Stratford et al., 1998). Like the elicitation of feeding by glutamate antagonists and GABA agonists, this suppression appears to be behaviorally specific since intake of both solid and liquid food was reduced by a dose of AMPA that did not alter locomotor activity or, more importantly, water intake in dehydrated rats (Stratford et al., 1998). While it is clear that lower doses of AMPA suppress intake under a number of conditions during shorter tests, injections of AMPA at doses 25 to 50 times higher than that which suppresses deprivation- or palatability-induced feeding have been reported to increase food intake in satiated rats (Echo et al., 2001). Interestingly, feeding under these conditions is characterized by a very slow onset (>1 h), with maximal intakes about half those seen after intra-AcbSh DNQX, muscimol, or baclofen. These results could be accounted for by diffusion of effective concentrations of the higher dose of AMPA to surrounding regions. However, a perhaps more parsimonious explanation, suggested by the greatly increased latency to feed, is the possibility that the feeding may be the consequence of a type of cellular rebound effect resulting from transmitter depletion or neuronal fatigue that leaves AcbSh projection cells essentially in a suppressed state of activity. Regardless of the underlying cause, such results underscore the complexity of studying behaviors elicited by the disruption of local transmitter systems.

To date, no significant differences have been noted between the behavioral effects of inhibiting the AcbSh with DNQX or GABA agonists. It is believed they exert their effects through actions on the same local circuits in the AcbSh and result in a similar disruption of activity in medium spiny projection neurons.

B. AcbSh-Mediated Changes in Feeding Behavior Are Behaviorally Specific

One of the most striking aspects of AcbSh-mediated feeding given its intensity is the behavioral specificity of the effect. The common perception that the accumbens acts primarily as a general mediator of reward makes it tempting to speculate that the influence of the AcbSh on feeding is simply one manifestation of a more general influence on motivational or reward processes. However, it is reasonable to expect that manipulating a "reward system" would have a more general effect on behavior. In fact, manipulations of medial AcbSh amino acid systems very specifically affect food intake. Water intake, gnawing behavior, and locomotor activity are not changed by inhibiting neurons in the medial AcbSh (Stratford et al., 1998; Ward et al., 2000). Particularly telling is the fact that not only is prandial drinking not increased, but AcbSh inhibition does not increase water intake even when the rats are dehydrated and predisposed to drink. In water-deprived rats, blocking AcbSh glutamate receptors neither facilitates nor suppresses water intake despite the fact that the rats engage in an intense ingestive act (Stratford et al., 1998). However, AcbSh-mediated feeding does not take precedence over all other behaviors, as dehydrated, but food-satiated, rats will drink to replete their water stores before feeding—which they then do to excess. This demonstrates that despite being driven to feed, the rats continue to prioritize their ingestive behaviors. It is only after an uninterrupted period of drinking water that the differential feeding response is elicited.

All of these facts suggest that the changes in feeding behavior are not simply the result of a nonspecific behavioral activation of the animal. Rather, a select group of neurons in the medial AcbSh appears to be a fundamental component of a neural system that is capable of profoundly, but very specifically, altering feeding behavior. These results stand in contrast to other brain systems proposed to control feeding behavior that also exert powerful effects on activity, water intake, and gnawing. Such general increases in ingestive-related behaviors can be elicited by electrical stimulation of the lateral hypothalamus (Roberts, 1969, 1980; Valenstein et al., 1970) or the midbrain (Waldbillig, 1975), by injection of opioid agonists into the ventral tegmental area (VTA) (Badiani et al., 1995), and by inhibition of cells in the median raphe nucleus (Klitenick and Wirtshafter, 1989; Wirtshafter et al., 1993; Wirtshafter, 2001). Thus, the amino acid systems of the AcbSh, rather than being general modulators of behavior, are specifically involved in feeding. In fact, the specificity of the elicited feeding effect suggests that these particular microcircuits in the AcbSh may be involved exclusively in the control of food intake.

C. Neuropeptide and Endocannabinoid Systems in the AcbSh Are Involved in the Control of Feeding Behavior

Given the demonstrated relevance of glutamatergic afferent input to the AcbSh and the GABAergic phenotype of most AcbSh interneurons and projection neurons, GABA and glutamate appear to have the most basic and global effects on cells in the Acb. Thus, much of the early research centered on the role of AcbSh amino acid circuits in the control of food intake. However, research into the AcbSh feeding system has demonstrated that a number of endogenous neuroactive compounds can alter food intake when administered in the AcbSh, reinforcing the idea that AcbSh neural circuits integrate varied information from numerous sources that serve as parameters in the neural calculations that modulate AcbSh output and which are ultimately manifested as modified feeding behavior.

1. Opioids

Defining the functional role of opioid systems in the accumbens and immediate vicinity has been the subject of numerous investigations. Opioid receptors are distributed broadly throughout the ventral striatum and there is a rich literature detailing their role in the control of food intake (Bodnar and Klein, 2005). It has been known since the 1980s that activation of μ or δ, but not κ, opioid receptors in the Acb increase food intake, with manipulations of μ receptors resulting in the largest and most

consistent changes in intake (Majeed et al., 1986; Mucha and Iversen, 1986). Additionally, intra-Acb injections of μ or κ receptor antagonists potently suppress deprivation-induced feeding (Bodnar et al., 1995).

Acb μ receptor activation preferentially increases intake of preferred diets (Evans and Vaccarino, 1990; Zhang and Kelley, 1997) and the prevailing hypothesis is that opioid peptides are likely to be preferentially involved in mediating the hedonic evaluation of ingesta (Cooper and Kirkham, 1993; Berridge, 1996; Yeomans and Gray, 2002), although it is likely that they subserve other feeding-related functions as well (Glass et al., 1999). As such, the system is considered in detail in Chapter 8 of the present volume.

While feeding can be elicited through manipulations of amino acid systems or opioid systems in the AcbSh, there are several lines of evidence that suggest the two systems are fundamentally discrete. For instance, the areas of maximal sensitivity to manipulations of amino acid and opioid systems are not identical. In contrast to the relatively circumscribed region of the AcbSh that is involved in amino acid modulated feeding, a much wider distribution of feeding-relevant opioid receptors exists in the ventral forebrain. Feeding can be elicited with injections of μ receptor agonists throughout the striatum and surrounding area (Bakshi and Kelley, 1993b). Within the Acb, the ventral and lateral aspects of the nucleus are most sensitive to μ receptor stimulation, with a much smaller response being elicited from injections in the medial AcbSh (Zhang and Kelley, 2000). Furthermore, within the AcbSh proper, morphine-induced feeding is elicited most efficaciously by injections into the caudal AcbSh (Pecina and Berridge, 2000), an area from which muscimol injections elicit defensive reactions and not feeding (Reynolds and Berridge, 2001). Thus, while feeding can be elicited by activation of opioid agonists into the AcbSh, this appears to be due to the activation of a portion of a system that is not limited to that region, but which extends throughout much of the ventral striatum and to some extent probably even the dorsal striatum.

Additionally, feeding behavior itself is expressed differently after the two treatments. As compared to muscimol-induced feeding, there is a considerably longer latency to begin feeding after activation of ventral striatal μ opioid receptors and the feeding episode is longer in duration, but less intense (Bakshi and Kelley, 1993a). Also, locomotor activity during the feeding test is increased significantly by μ receptor activation, but not by muscimol. Arguably, these data may suggest that while intra-AcbSh GABA agonists can directly invoke the expression of feeding sequences, opioid receptor agonists might only modify intrameal processes, related to food palatability, that determine the persistence of ingestion after a meal has been initiated. Additionally, AcbSh opioid agonists increase intake of saccharine and ethanol solutions (Zhang and Kelley, 2002), unlike AcbSh muscimol which suppresses intake of these compounds (Stratford and Gosnell, 2000). When high carbohydrate and high fat diets are presented together, activation of opioid receptors in the Acb results in a preferential increase in fat intake (Zhang et al., 1998), whereas AcbSh muscimol preferentially increases carbohydrate intake (Basso and Kelley, 1999).

While the qualitative differences between the actions of opioid agonists or GABA agonists suggest that two separate systems may mediate feeding behavior in the medial AcbSh, an interesting series of studies reported by Bodnar and colleagues demonstrates an important, though incompletely understood, relationship between the two. Blocking μ, δ, or κ receptors in the AcbSh suppresses muscimol-induced feeding, while DAMGO-induced feeding is suppressed by blocking postsynaptic $GABA_A$ receptors, but is increased by blocking the primarily presynaptic $GABA_B$ receptors (Ragnauth et al., 2000; Znamensky et al., 2001; Khaimova et al., 2004). While it remains uncertain to what extent opioids and amino acids share a common neural substrate, this work suggests that both opioid and amino acid feeding in the medial AcbSh involves a complex arrangement of pre- and postsynaptic receptors on afferent terminals, interneurons, and projection neurons. Glutamate antagonists block activity at all synapses regardless of the functional cell assembly to which they belong. Unless glutamatergic fibers only terminate on a feeding-related assembly, these treatments will compromise multiple functional systems. Similarly, the often employed GABA agonist muscimol inhibits activity in most neurons expressing the widely distributed $GABA_A$ receptor. If the opioid hedonic system and the amino acid circuits are separate assemblies that overlap spatially in the medial AcbSh, then it is probable that these treatments are affecting activity in both. Unraveling this complicated relationship will be crucial to exploiting the system therapeutically.

2. Cocaine- and Amphetamine-Regulated Transcript Peptide

Cocaine- and amphetamine-regulated transcript (CART) peptide is an anorectic peptide synthesized in neurons located in a variety of brain regions, including the AcbSh (Yang et al., 2005). CART levels are particularly high in the region of the AcbSh known to be most sensitive for eliciting muscimol-induced feeding. Injections of CART peptide into the AcbSh suppress spontaneous food intake, as well as that induced by food deprivation or intra-AcbSh muscimol, and do so without affecting locomotor activity, suggesting a specific effect on feeding (Yang et al., 2005).

The majority of CART peptide containing cells in the Acb appear to be GABAergic medium spiny projection neurons that also contain substance P and are more numerous in neurons located in the rostral half of the Acb than the caudal half, which maps well with the feeding related region of the AcbSh (Smith et al., 1997b; Hubert and Kuhar, 2005). However, some fibers from AcbSh CART cells also appear to terminate locally (Yang et al., 2005), although whether these are interneurons or recurrent collaterals from projection neurons has yet to be established.

CART peptide is found in moderate to high levels in many AcbSh afferent and efferent sites as well, including the ventral pallidum (VP), subiculum, basolateral amygdala (BLA), infralimbic and prelimbic cortices, paraventricular thalamic nucleus (PVT), ventral tegmental area (VTA), nucleus of the solitary tract (NTS), perifornical lateral hypothalamus (LH), and the closely aligned arcuate (Arc) and paraventricular (PVN) hypothalamic nuclei (Koylu et al., 1997, 1998; Smith et al., 1997b). CART synthesis in the hypothalamus is regulated in part by levels of circulating leptin (Kristensen et al., 1998). Thus, CART peptide is well positioned to modulate the overall tone of activity in many components of the AcbSh feeding circuit based on long-term adiposity signals. Food deprivation decreases CART mRNA in the AcbSh, perifornical LH, Arc, and PVN in a time-dependent manner demonstrating that CART peptide synthesis in each of these structures is intimately related to energy stores (Kristensen et al., 1998; Yang et al., 2005). Furthermore, over 50% of the CART neurons in the Arc and perifornical LH project directly to the AcbSh (Yang et al., 2005), providing a direct route through which the AcbSh may receive leptin-relayed information concerning current stores of adipose tissue.

3. Melanin-Concentrating Hormone

Melanin-concentrating hormone (MCH) is an orexigenic peptide synthesized in hypothalamic cells (Bittencourt et al., 1992). The AcbSh contains MCH fibers and high levels of MCH receptor mRNA (Saito et al., 2001). Acb projection neurons express MCH_1 receptors and food intake is increased by intra-AcbSh MCH and suppressed by application of a specific MCH_1 receptor antagonist (Georgescu et al., 2005). MCH inhibits GABAergic neurons in the LH, both postsynaptically and by suppressing glutamate release through actions on presynaptic terminals (Gao and van den Pol, 2001). Similar actions on neurons in the AcbSh may provide the cellular mechanism through which intra-AcbSh MCH effects increases in food intake. MCH-synthesizing cells are found exclusively in the vicinity of the LH, therefore, the presence of MCH fibers in the AcbSh indicates that MCH neurons contribute to the direct projection the AcbSh receives from the LH. MCH neurons are activated by circulating insulin (Bahjaoui-Bouhaddi et al., 1994b) and under some conditions involving food deprivation (Herve and Fellmann, 1997), providing yet another possible mechanism through which direct information on current energy balance can be transmitted to the AcbSh.

4. Orexin

The orexin peptides, so termed because of the initial belief that their primary physiological role was to control feeding behavior, are synthesized exclusively by neurons in the vicinity of the LH (Sakurai et al., 1998). Neurons in the AcbSh express orexin receptors (Trivedi et al., 1998; Marcus et al., 2001; Cluderay et al., 2002) and application of orexin peptides increase GABA-evoked currents in isolated Acb neurons while decreasing NMDA-evoked currents (Martin et al., 2002). This dual inhibitory effect stands in contrast to the increased neuronal activity orexin elicits in the LH (de Lecea et al., 1998) and indicates that the peptide induces cellular effects in both regions appropriate for the initiation of feeding behavior. Although intra-AcbSh injections of orexin initially were reported to have no effect on food intake (Baldo and Kelley, 2001), a more recent study found that the peptide potently increased feeding, an effect that was suppressed by coadministration of a selective orexin receptor blocker (Thorpe and Kotz, 2005). Further work will be required to resolve this discrepancy, however, this result makes it appear likely that orexin-synthesizing neurons in the LH contribute to the AcbSh control of feeding behavior.

5. Nociceptin

The AcbSh contains moderately high levels of the opioid receptor-like 1 (ORL1) receptor (Anton et al., 1996) and intra-AcbSh administration of nociceptin, the endogenous ligand for the ORL1 receptor, dose-dependently increases food intake without altering locomotor activity (Stratford et al., 1997). As with manipulations of the amino acid systems, there is a very short latency to begin feeding and most of the ingestion occurs during the initial 30 to 45 min of the test. Nociceptin has been demonstrated to inhibit neurons expressing the opioid receptor-like 1 (ORL1) receptor through effects on the same population of potassium channels targeted by baclofen (Vaughan and Christie, 1996). Interestingly, the nociceptin receptor is located on CART-synthesizing neurons in the hypothalamus, and local application of nociceptin inhibits CART release in hypothalamic cells (Bewick et al., 2005). If the same holds true for neurons in the AcbSh, it may provide the mechanism through which nociceptin increases food intake in that region (Stratford et al., 1997).

6. Amylin

Relatively high levels of receptors for the anorectic peptide amylin are located in the caudal AcbSh (Beaumont et al., 1993; Christopoulos et al., 1995) and injections of the peptide into this region are reported to suppress food intake in deprived rats (Baldo and Kelley, 2001). However, intra-AcbSh amylin also potently suppresses water intake and locomotor activity, suggesting that the peptide does not act in the AcbSh to control feeding behavior specifically. In fact, despite the existence of the caudal AcbSh amylin receptors, the effects

on ingestion seem largely due to leakage into the overlying lateral ventricle, as amylin has no effect on food or water intake when cannulae placements avoid breaching that structure (Baldo and Kelley, 2001).

7. Endocannabinoids

Although investigations have begun only more recently, a promising avenue of research involves the role of AcbSh endocannabinoids in the control of feeding. Endocannabinoids are important regulators of appetite and body weight that are being investigated intensively as potential therapeutic targets for the treatment of obesity and eating disorders (Kirkham, 2003, 2005). The nucleus accumbens contains a moderately high density of the CB1 (cannabinoid receptors 1) that are more abundant in the shell than in the core (Herkenham et al., 1991; Tsou et al., 1998; Pickel et al., 2004, Matyas et al., 2006). While a majority of CB1 receptors in the Acb appear to be located on excitatory presynaptic terminals that synapse on GABAergic neurons, smaller postsynaptic populations are also found on dendrites and soma in the region (Robbe et al., 2001; Pickel et al., 2004, 2006) [although there is evidence that a substantial population of CB1 receptors in the Acb is located on GABAergic terminals (Matyas et al., 2006)]. Stimulation of excitatory afferents from the prefrontal cortex leads to depolarization-bound release of postsynaptic endocannabinoids in the Acb which act at presynaptic CB1 receptors to potently suppressing stimulation-elicited glutamate release, thereby inhibiting Acb projection neurons (Robbe et al., 2001; Pistis et al., 2002). This activation of presynaptic CB1 receptors yields a long-term synaptic depression in the Acb, suggesting that the endocannabinoid-coded retrograde signal is part of a negative feedback loop that modulates the strength of glutamatergic synapses during periods of sustained cortical activity, and that manipulation of AcbSh CB1 receptors may result in relatively long-lasting changes in AcbSh glutamate transmission, a fact that may indicate important therapeutic potential.

While CB receptors are found in several brain regions related to food intake, given the intense feeding observed in rats after disruption of glutamate transmission in the AcbSh and the fact that activation of CB receptors down-regulates activity in the same system, it is tempting to speculate that a suppression of glutamate transmission elicited by the effects of exogenous cannabinoid administration on CB1 receptors in the AcbSh may well underlie the overeating (munchies) that characterizes marijuana self-administration in humans.

In keeping with this ability to suppress glutamate transmission in the AcbSh, activation of AcbSh CB1 receptors with the endocannabinoid 2-AG results in a potent and dose-dependent increase in food intake that is suppressed by pretreatment with a CB1 receptor antagonist (Kirkham et al., 2002). In line with the actions of AcbSh muscimol, cannabinoid-induced feeding also is associated with a reduction in eating latency (Williams and Kirkham, 2002).

Other data support the idea that endocannabinoids in the AcbSh play a role in the mediation of food intake. CB1 receptors in the Acb are down-regulated in rats that become overweight through consumption of a palatable diet (Harrold et al., 2002). More importantly, fasting increases levels of the endogenous cannabinoids anandamide and 2-AG by 200 to 300% in the limbic forebrain, an area encompassing the AcbSh (Kirkham et al., 2002). Levels were not changed in rats consuming a palatable diet, suggesting that endocannabinoid levels in the region are related to metabolic state or the motivation to eat associated with fasting, but not to the act of feeding or the hedonic qualities of the food. Interestingly, mice that lack fatty acid amide hydrolase, the enzyme responsible for the metabolism of the endogenous cannabinoid anandamide, show reduced CART levels in the AcbSh, a condition that should facilitate feeding (Osei-Hyiaman et al., 2005). A similar reduction in CART levels is seen in several hypothalamic areas as well, suggesting that changes in CART activity in the AcbSh or hypothalamus may be one of the mechanisms through which endocannabinoids effect changes in feeding behavior.

8. *Cholecystokinin*

The presence of high levels of the satiety peptide cholecystokinin (CCK) is one of the defining characteristics of the AcbSh (Zaborszky et al., 1985). CCK is a potent suppressor of food intake; however, its specific role in the AcbSh has yet to be investigated. Most AcbSh CCK originates in the VTA and is coexpressed in DA fibers (Lanca et al., 1998). Although the cellular effects of CCK are somewhat complicated in that the peptide acts to directly suppress both excitatory and inhibitory postsynaptic currents in the Acb, the overall effect of CCK administration appears to be a potent excitation of the medium spiny projection neurons (Kombian et al., 2004, 2005). This effect is reflected in the robust, dose-dependent increase in Fos expression in the AcbSh elicited by intracerebroventricular injections of CCK (Li et al., 1995). Behaviorally, Acb CCK appears to modulate intake of some diets. While intra-Acb injections of CCK do not suppress intake of a highly palatable diet in mildly deprived animals (Blevins et al., 2000), they do inhibit sucrose intake in rats with high baseline intakes, but not in those with low baselines (Sills and Vaccarino, 1991). In turn, blockade of Acb CCK receptors increased sucrose intake in the low baseline intake rats, but actually decreased it in rats with high baselines (Sills and Vaccarino, 1996). These intriguing results suggest the issue is deserving of further study.

The number of endogenous neuromodulators that affect feeding behavior when injected into the AcbSh support the notion of the AcbSh as an important focal point for the convergence and integration of a wide variety of feeding-related information. Furthermore, they provide an exceptionally large number of avenues that have the potential to be exploited therapeutically.

D. The AcbSh Feeding Circuit Is Likely to Regulate Normal Homeostatic Feeding Behavior

While the eating stimulated by AcbSh manipulations possesses all the hallmarks of species-typical feeding sequences, an important question concerns whether the AcbSh plays a role in the normal, physiological control of food intake. Though difficult to answer definitively, evidence from a number of diverse studies indicates that it does.

- *The AcbSh has the capacity to induce feeding in satiated animals or suppress intake in animals that are predisposed to eat.* Not only can feeding be induced in satiated rats by inhibiting neurons in the AcbSh, but activating AcbSh neurons with AMPA suppresses deprivation- and palatability-induced feeding (Stratford et al., 1998), indicating that the expression of these normal behaviors results from altered AcbSh activity. Furthermore, the magnitude and bidirectionality of the feeding responses indicate that changes in AcbSh activity are capable of accounting for most, if not all, changes in normal feeding behavior. While not sufficient to confirm a role for the AcbSh in normal feeding, such "bidirectionality" would be expected in any structure posited to gate feeding behavior.
- *Feeding induced by deprivation or AcbSh muscimol are suppressed by the same manipulations.* Both deprivation-induced feeding and AcbSh-mediated feeding are suppressed by blocking glutamate transmission in the perifornical LH (Stanley et al., 1996; Stratford and Kelley, 1999) or GABA activity in the AcbSh (Kandov et al., 2006), data consistent with the hypothesis that, at some level, a common neural substrate underlies the expression of these behaviors.
- *Feeding alters glutamate levels in the AcbSh.* Extracellular glutamate levels in the AcbSh are reduced during feeding and increased when food is replaced by an inedible object (Rada et al., 1997; Saul'skaya and Mikhailova, 2002), further evidence that endogenous glutamate levels in the AcbSh are tightly correlated with feeding behavior and that a reduction in AcbSh activity is involved in initiating feeding behavior.
- *Food deprivation alters peptide levels in the AcbSh.* Acute food deprivation decreases AcbSh levels of mRNA encoding the peptide CART, which is released locally and inhibits feeding when administered into the AcbSh (Yang et al., 2005), data consistent with a decrease in AcbSh CART activity being involved in the initiation of deprivation-induced feeding. Furthermore, fasting increases levels of the endogenous cannabinoids anandamide and 2-AG in the limbic forebrain, an area encompassing the AcbSh (Kirkham et al., 2002).
- *Feeding in satiated animals or suppression of food intake in feeding animals is induced by manipulating the activity of endogenous neuroactive compounds in the AcbSh.* It is important to remember that intense feeding behavior can be elicited by increasing local GABA levels (Stratford and Kelley, 1997), or by blocking the activity of normally released glutamate in the AcbSh (Maldonado-Irizarry et al., 1995; Stratford et al., 1998; Echo et al., 2001; Reynolds and Berridge, 2003). Significantly, both manipulations increase food intake simply by altering the activity of tonically released endogenous neurotransmitters, suggesting that the normal physiological role of GABA and glutamate in the AcbSh involves the routine control of food intake. Similarly,

more recent studies have shown that intra-AcbSh injections of the endo-cannabinoid 2-AG or the endogenous orexigenic hypothalamic peptides MCH or orexin induce intense feeding, while administration of the endogenous anorectic peptide CART or blocking the activity of endogenous MCH in the AcbSh significantly suppresses feeding (Georgescu et al., 2005; Thorpe and Kotz, 2005; Yang et al., 2005).

- *AcbSh neurons are necessary for the long-term control of body weight.* Perhaps most importantly, bilateral excitotoxic lesions of the ventromedial AcbSh result in a significant postoperative increase in weight gain as compared to unlesioned rats (Maldonado-Irizarry and Kelley, 1995), indicating that neurons in the region play an important inhibitory role in the control of normal feeding.

Taken together, these data comprise a convergent body of evidence suggesting that the AcbSh participates in the normal, physiological control of food intake. However, it is important to emphasize that even if the AcbSh were ultimately shown not to be critically involved in the routine control of feeding, the magnitude and specificity of the effects described earlier in this chapter raises the very real possibility that pathology of the neural circuitry controlled by the AcbSh might underlie the etiology of some eating disorders and that a thorough understanding of the neural circuitry involved is likely to identify novel avenues that will aid in the development of efficacious pharmacological therapies for treating those disorders.

III. AFFERENT PROJECTIONS TO THE AcbSh THAT MAY BE INVOLVED IN REGULATING FOOD INTAKE

Although little is known about precisely which Acb inputs are involved in control-

ling food intake, the AcbSh is the terminus for afferent fibers from a number of cortical and subcortical regions that have been implicated in the control of ingestive behavior. These inputs, through changes in firing rate or number of active projections, inform the AcbSh that a certain set of conditions exist under which it would be beneficial for the animal to feed. Much of the input to the system arises in neocortical, hippocampal, and thalamic regions, suggesting that some of the information provided to the AcbSh has already been extensively processed. However, the circuit also receives input from brain regions monitoring taste, energy stores, and arousal level. Whether the feeding-related cell assemblies in the AcbSh receive this information has yet to be determined, although the functional similarities between the AcbSh and some of the regions are suggestive.

A. Excitatory Amino Acid Inputs

As noted earlier, avid feeding is elicited in satiated rats by acutely blocking non-NMDA ionotropic glutamate receptors or by activating $GABA_A$ or $GABA_B$ receptors in the AcbSh. While GABA in the AcbSh arises from interneurons, recurrent collaterals of projection neurons, and afferent terminals, all glutamate originates extrinsically. This indicates the existence of one or more tonically active glutamatergic afferent projections to the AcbSh that act specifically to inhibit feeding behavior. These feeding-relevant, excitatory projections to the AcbSh have yet to be identified; however, presumed glutamatergic fibers in the AcbSh originate primarily in the ventromedial prefrontal cortex, midline thalamic complex, basolateral amygdala, and ventral subiculum (Christie et al., 1987; Fuller et al., 1987), making these the most likely candidates.

1. Medial Prefrontal Cortex

The medial prefrontal cortex (PFCtx) sends a dense projection to the AcbSh that originates primarily in the infralimbic cortex, with additional fibers arriving from the ventral prelimbic and dorsal peduncular cortices (Beckstead, 1979; Sesack et al., 1989; Hurley et al., 1991; Brog et al., 1993; Vertes, 2004). The ventromedial PFCtx also sends projections to several feeding-related brain regions that, in turn, innervate the AcbSh; potentially permitting the region to modify AcbSh activity through short multisynaptic pathways. Both infralimbic and prelimbic cortices project heavily to midline thalamic nuclei, including the paratenial, mediodorsal, and paraventricular thalamic nuclei, of which the latter sends a dense projection to the AcbSh (Brog et al., 1993). The infralimbic cortex projects strongly to the LH and VTA, although fibers from the prelimbic cortex reach these areas as well. In addition, the infralimbic cortex innervates the parabrachial nucleus and NTS, while the prelimbic cortex sends fibers to the BLA and dorsal and median raphe nuclei (Beckstead, 1979; Sesack et al., 1989; Hurley et al., 1991; Vertes, 2004).

The infralimbic cortex, which provides most of the cortical input to the medial AcbSh, has been strongly implicated in the control of cardiovascular function, and in coordinating adaptive autonomic and behavioral responses to stress or anticipated tasks (Neafsey, 1990; Frysztak and Neafsey, 1991, 1994; Verberne and Owens, 1998; Amat et al., 2005). At least some of these effects appear to be mediated through the direct infralimbic projection to the NTS (Owens et al., 1999). The medial PFCtx also plays a critical role in more complex processes, such as the ability of rats to alter behavioral response strategies under changing conditions (Ragozzino et al., 1999, 2003), in generating situation-appropriate preparatory motor responses in anticipation of a stimulus (Risterucci et al., 2003), and in the process by which animals make appetitive cost–benefit analyses (Walton et al., 2002).

Although activity in the infralimbic cortex is not altered by food deprivation per se, more than 85% of infralimbic neurons alter their firing rate in response to food presentation (Valdes et al., 2006). However, it is important to note that even very large lesions of the PFCtx, which necessarily remove a substantial proportion of AcbSh glutamate, do not alter spontaneous feeding (Kolb, 1974; Recabarren et al., 2005; Valdes et al., 2006). These results stand in contrast to the strong feeding elicited by blocking glutamate receptors in a circumscribed region of the AcbSh. Such lesions do reduce locomotor activity associated with food presentation (Valdes et al., 2006) and manipulations of the PFCtx have robust effects on gastric motility (Hurley-Gius and Neafsey, 1986), facts that suggest these neurons may be involved in controlling feeding-related preparatory autonomic and behavioral responses rather than in the initiation of feeding itself. Of course, it also is possible that when AcbSh-mediated feeding is initiated, these adaptive responses are triggered by information diverted through the ventropallidal-mediodorsal thalamic-infralimbic cortex pathway and are effected through infralimbic connections with the NTS or other brainstem nuclei.

2. Basolateral Amygdala

The caudal BLA sends a moderately dense projection to the caudal AcbSh (Phillipson and Griffiths, 1985; Brog et al., 1993; Wright et al., 1996) and these fibers have been shown to terminate in the vicinity of AcbSh neurons projecting directly to the LH (Kirouac and Ganguly, 1995). While there is little evidence that the BLA is important for the expression of spontaneous feeding, the nucleus is involved in controlling some behavioral responses that affect specific aspects of food intake. For example, lesions of the BLA disrupt formation of conditioned taste aversions, reduce neophobia for palatable food, and suppress intake of palatable food elicited by activation of μ opioid receptors in the ventral striatum (Rollins et al., 2001; Will et al.,

2004). Chronic, electrical stimulation of the BLA has been reported to increase weight gain (Loscher et al., 2003); however, this relatively nonspecific technique is likely to have affected perikarya or fibers in the posterodorsal medial amygdala as well, an area known to be involved in the control of spontaneous feeding and body weight (Rollins and King, 2000; King et al., 2003). The BLA plays a critical role in cue-potentiated increases in food intake (Petrovich et al., 2002), although more recent work suggests that this effect is not mediated through the BLA-AcbSh projection, but rather through the connection with the LH (Holland and Petrovich, 2005).

Intra-BLA injections of muscimol do not potentiate the submaximal feeding elicited by administration of a low dose of muscimol into the AcbSh (Baldo et al., 2005), suggesting that reduced activity in this excitatory input is not involved in the initiation of AcbSh-mediated feeding. More important, although the projection from the BLA to the caudal AcbSh is almost certainly glutamatergic, neither chronic lesions nor transient inhibition of the BLA induce increases in feeding or weight gain as would be expected when removing a major excitatory input to the AcbSh (Rollins et al., 2001; Baldo et al., 2005). However, the BLA also projects directly to the LH (Petrovich et al., 2001). Glutamate transmission in the LH is necessary for the expression of AcbSh-mediated feeding (Stratford and Kelley, 1999) and coincident removal of an important excitatory input to this region could conceivably block feeding that would have been elicited by removal of the BLA input to the AcbSh.

3. Paraventricular Thalamic Nucleus

One of the largest projections to the AcbSh arises in the paraventricular thalamic (PVT) nucleus, a member of the midline and intralaminar thalamic complex (Zaborszky et al., 1985; Carlsen and Heimer, 1986; Berendse and Groenewegen, 1990; Brog et al., 1993; Otake and Nakamura,

1998). Some neural elements in the PVT appear to play a critical role in the control of autonomic responses to stress (Bhatnagar et al., 2000). However, this AcbSh afferent may also act as a potential regulator of AcbSh-mediated feeding since lesions of the posterior PVT increase food intake and body weight—although the effect is modest compared to blockade of AcbSh glutamate receptors (Bhatnagar and Dallman, 1999). The PVT has also been implicated in the control of entrained food anticipatory behavior which is likely to be mediated, in part, through its strong reciprocal connections with the medial PFCtx, a brain region showing similar functionality (Mendoza et al., 2005; Pereira de Vasconcelos et al., 2006). Furthermore, the PVT is richly innervated by various peptides and monoamines with important roles in the control of food intake. Of particular interest is the fact that PVT cells projecting to the AcbSh are located primarily in the dorsal PVT, a region heavily innervated by orexin fibers from the LH, α-melanocyte-stimulating hormone fibers from the Arc, CART fibers that likely arise from the Arc and/or LH, and CCK fibers from the dorsomedial hypothalamic and dorsal raphe nuclei (Guy et al., 1981; Freedman and Cassell, 1994; Kirouac et al., 2005; Otake, 2005; Parsons et al., 2006). In fact, a more recent study showed that CART and orexin fibers appear to terminate directly on PVN neurons that project to the AcbSh, and are sometimes apposed to one another on the processes of these cells—further evidence that the hypothalamus can tightly control activity in the PVT-AcbSh projection (Parsons et al., 2006). Additional input to the PVT arises primarily from neurons in the infralimbic and prelimbic cortices, amygdala, suprachiasmatic nucleus, median raphe nucleus, parabrachial nucleus, locus coeruleus, and NTS (Chen and Su, 1990; Risold et al., 1997; Krout and Loewy, 2000). Thus, the PVT is positioned to receive information on energy balance, visceral condition, and arousal level, and furthermore, can integrate that information with important circadian information received from the suprachiasmatic nucleus before passing the resulting signal directly to the AcbSh.

In addition to sending a large monosynaptic projection to the AcbSh, the PVT may also be able to influence AcbSh activity through short multisynaptic projection involving the infralimbic cortex, BLA, and ventral subiculum, or may mediate feeding behavior independently of the AcbSh through its direct projections to the central amygdaloid nucleus, dorsomedial hypothalamic nucleus, or ventromedial hypothalamic nucleus (Moga et al., 1995). A moderate proportion (15–20%) of PVT neurons that project to the AcbSh also send collateral fibers to either the ventromedial PFCtx or central amygdaloid nucleus, which may provide a mechanism for coordinating activity between the AcbSh and these regions under certain conditions (Freedman and Cassell, 1994; Bubser and Deutch, 1998).

Although blocking glutamate transmission in the AcbSh elicits intense feeding, disrupting the major excitatory inputs to the AcbSh individually has limited effects on spontaneous food intake. This suggests that the elicitation of feeding may require the simultaneous reduction of activity in each of these afferents; further evidence, perhaps, of the integrative role of the AcbSh. The AcbSh is likely to control multiple behaviors through an integration of the actions of multiple excitatory inputs. In fact, the characteristics of membrane potentials in Acb projection neurons suggest that the near simultaneous activation of excitatory inputs from multiple brain regions may be necessary to change Acb output, particularly for some functional cell assemblies in the AcbSh (Pennartz et al., 1992; O'Donnell and Grace, 1993, 1995; O'Donnell et al., 1999). An intricate pattern of excitatory afferent terminals is present in the AcbSh, and individual projection neurons have been shown to receive inputs from multiple brain regions. Afferent terminals

arising from the ventromedial PFCtx and BLA overlap in the AcbSh, while the PVT innervates a separate population of cells (Wright and Groenewegen, 1995). Similarly, fibers from the ventral subiculum terminate on the same AcbSh neurons as do fibers from the PFCtx and BLA (French and Totterdell, 2002, 2003). The convergence of inputs from the PFCtx, BLA, and ventral subiculum suggest that together they may contribute to the control of a common physiological or behavioral response, whereas the PVT fibers are likely to subserve a different function. While it is premature to attempt to assign a definitive functional role to any of the excitatory afferents to the AcbSh, the behavioral data that have been collected—and the pattern of AcbSh innervation—suggest that the PVT may be the most relevant excitatory input in terms of basic control of food intake. In turn, the medial PFCtx, BLA, and hippocampal input may be preferentially involved in other functions which may or may not involve gating feeding behavior.

B. Additional Inputs

1. Ventral Pallidum

Neurons throughout the rostrocaudal extent of the subcommissural ventral pallidum (VPm) contribute to a primarily GABAergic projection to the AcbSh (Brog et al., 1993; Groenewegen et al., 1993; Churchill and Kalivas, 1994). These fibers originate in neurons located in the terminal field of AcbSh afferents and therefore appear to represent a reentrant segment of an accumbens–ventral pallidal feedback circuit. The effect of activating this pathway on activity in AcbSh projection neurons is unknown and depends on whether the VPm fibers terminate primarily on AcbSh interneurons or on the medium spiny projection neurons themselves.

2. Hypothalamus

The AcbSh receives a substantial peptidergic innervation from different hypothalamic regions, including orexin- and MCH-containing fibers originating in the LH- and CART-containing fibers from the LH and Arc (Peyron et al., 1998; Yang et al., 2005). Each of these feeding-related peptides is regulated, either directly or indirectly, by circulating levels of leptin and ghrelin, and therefore are in a position to modify AcbSh activity based on the current status of energy balance in the animal. As noted before, orexin fibers from the LH and CART fibers from the LH and/or Arc are found in close apposition on the processes of PVT neurons projecting to the AcbSh, providing an additional bisynaptic route through which feeding-related hypothalamic neurons can exert a strong influence over AcbSh activity.

The AcbSh also receives input from several nuclei in the caudal midbrain and hindbrain. Although the absolute number of cells contributing fibers to these pathways is much smaller than those in the corticostriatal or thalamostriatal pathways, they are located in brain regions with documented roles in the control of food intake and may well provide afferent information important for the appropriate expression of AcbSh-mediated feeding behavior.

3. Dorsal and Median Raphe Nuclei

Separate populations of dopaminergic and serotonergic neurons in the dorsal raphe nucleus innervate the medial Acb, as do serotonergic and other neurochemically uncharacterized cells from the median raphe nucleus (Stratford and Wirtshafter, 1990; Vertes, 1991; Brog et al., 1993; Compan et al., 1996). Serotonergic elements in the dorsal raphe are involved in controlling feeding (Bendotti et al., 1986; Fletcher and Davies, 1990) and inhibition of neurons in the median raphe elicits a particularly powerful behavioral activation, of which intense feeding is a prime component (Klitenick and Wirtshafter, 1989; Wirtshafter et al., 1993; Wirtshafter, 2001). Unlike AcbSh feeding, manipulations of the dorsal or median raphe also increase water intake,

gnawing, and general locomotor activity, suggesting that they may play more of a role in general arousal than in the specific control of food intake. Still, it is possible that the feeding component of the arousal is related to raphe-mediated changes in AcbSh activity.

Neurons in the dorsal raphe accumulate circulating leptin, and both dorsal and median raphe neurons express functional orexin receptors (Brown et al., 2001; Fernandez-Galaz et al., 2002) and receive projections from LH orexin neurons (Date et al., 1999). Consequently, activity in these nuclei may be regulated to some degree by the LH and correlated with available energy stores. Thus, the rostral raphe nuclei may be one of the sites through which the LH effects changes in arousal level and information related to current behavioral state may make its way to the AcbSh through fibers originating in the rostral raphe nuclei.

4. Nucleus of the Solitary Tract

There is a substantial noradrenergic innervation of the mid-caudal AcbSh that originates primarily in the caudal NTS, with a smaller contribution from the locus coeruleus (Berridge et al., 1997; Delfs et al., 1998). Neurons in the dorsal, medial, and commissural subnuclei of the caudal NTS, containing cholecystokinin, neurotensin, substance P, and dopamine, also contribute fibers to the projection (Li et al., 1990; Wang et al., 1992). Thus, the NTS is positioned to provide information to the AcbSh concerning the state of the gastrointestinal tract, the activity of peripherally acting gut satiety hormones, and other visceral conditions arriving along vagal afferents.

Few, if any, fibers terminating in the AcbSh originate in the rostral NTS where gustatory signals are processed. However, some taste information may be transmitted through the projections to the AcbSh that arise from the parabrachial nucleus (Brog et al., 1993), an important component of the afferent taste pathway (Di Lorenzo and Monroe, 1997).

C. The Role of Reentrant Projections in the AcbSh Circuit

It is interesting to note that the AcbSh receives reentrant projections from virtually all of the brain regions to which it projects. In addition to bringing in new information related to the shared function of intra- and extra-AcbSh loci, these fibers may also provide important feedback concerning the state of activity in the downstream site. This may allow the AcbSh to regulate its output in response to changes in downstream activity, regardless of the causes underlying those changes. Such a self-regulating system may permit the AcbSh to maintain its influence on downstream circuit components even when they are perturbed by input from other systems. Additionally, feeding programs must be sensitive to instantaneous contextual changes that alter the emotional valence of food or that require a response to nonfood aspects of the environment.

IV. DOWNSTREAM COMPONENTS OF THE FUNCTIONAL AcbSh FEEDING CIRCUIT

A. Classes of Evidence

Behavior results from physiological changes in the brain and techniques evaluating changes in gene expression, glucose utilization, or electrophysiological responses can be used to establish that one brain region has the ability to exert a physiological effect on another. However, to determine whether that observed physiological effect is related to the expression of a particular behavior requires further data. The functional circuitry through which the AcbSh effects behavioral changes has been investigated by examining: (1) stimulus bound changes in the expression of the immediate-early gene c-*fos*, (2) anatomical connections, and (3) behavioral effects of disrupting neural activity in suspected downstream effector sites. Each provides a

different class of information which, by itself, is simply suggestive. Taken together, however, a profile emerges that can be strongly indicative of a behaviorally functional relationship.

1. *Immediate-Early Gene Studies*

Changes in immediate-early gene expression act as a marker for brain regions in which neuronal activity is altered as a result of inhibiting the AcbSh. These studies have used the immunohistochemical localization of the c-*fos* gene product Fos to label cells activated by injections of muscimol into the AcbSh. Fos is synthesized rapidly in acutely activated neurons and is used routinely as a marker for increased neuronal activity (Dragunow and Faull, 1989; Sheng and Greenberg, 1990). Thus, an increase in Fos expression in response to inhibiting AcbSh cells can be used to identify brain regions that demonstrate a physiological, and perhaps behaviorally functional, relationship to that particular stimulus.

One of the principal strengths of the Fos technique is that it permits the transsynaptic mapping of activated brain regions and the simultaneous demonstration of multiple nodes in a functional circuit. It is also capable of demonstrating functionally related cell populations, regardless of whether they are activated through neural connections or as the result of secondarily elicited physiological or behavioral changes.

There is an important caveat that must be taken into consideration when interpreting any Fos results. It should be remembered that the use of Fos to map functional circuits is by no means complete, in that not all activated neurons express Fos and brain regions in which neuronal activity is inhibited are not generally identified. Therefore, it is almost certain that any functional circuit delineated by Fos expression is incomplete, and is likely to contain unidentified interposed brain regions that either are activated and do not express Fos or are normally inhibited by the eliciting treatment. Furthermore, mapping increases in Fos synthesis alone does not allow us to identify definitively the specific path taken by projection neurons between any brain structures in the circuit, although some of this missing information can be inferred from previously described anatomy and neuropharmacology.

2. Anatomical Connections

Hodological studies in which the terminal fields of AcbSh projection neurons are mapped can provide information on which brain regions are potential components of a functional circuit. The existence of direct monosynaptic pathways from the AcbSh to other brain regions provides evidence of a functional relationship between the AcbSh and neural elements in the terminal fields, but does not tell us under what conditions information is transmitted. Therefore, anatomical studies alone do not provide direct data on the physiological or behavioral relevance of the pathway. This consideration is particularly important when investigating heterogeneous areas where projections may arise from functionally discrete subpopulations of neurons.

Of course, information transfer between brain regions is not limited to monosynaptic pathways. However, there is always a danger in invoking multisynaptic pathways as primary effector routes because of the likelihood that a certain amount of information is altered at each intervening node in the circuit. Therefore, while the existence of a multisynaptic connection demonstrates that the capacity for passing relevant information between regions, the greater the number of intervening synapses, the greater is the potential for integration of secondary information and dilution of the originally encoded information. Also, without evidence of direct synaptic contact between the relevant afferent and efferent elements in each node, the chance of segregation of information in unidentified subpopulations is always

present. The AcbSh itself is an example of a region in which adjacent clusters of cells do not always form functional assemblies and the heterogeneous nature of the AcbSh is reflected in the complicated patterns of afferent innervation (Groenewegen et al., 1999).

3. Behavioral Effects of Disrupting Neural Activity in Suspected Downstream Effector Sites

The third category of studies are those evaluating whether disruption of neural activity in suspected downstream components of the functional circuit affect AcbSh-mediated feeding. While increased Fos expression indicates a physiological relationship exists, and the presence of a strong projection to an activated region provides a ready mechanism through which these changes may be effected, we rely on a demonstrated alteration of behavior to determine whether two brain regions are part of a behaviorally functional circuit.

Of course, by itself, the suppression of AcbSh-mediated feeding by transient or permanent lesions of another part of the brain only indicates that normal activity in the lesioned region permits the expression of the elicited behavior. Even if a brain region is found to be essential for the expression of the elicited behavior, it does not necessarily indicate that the region is a node in the AcbSh feeding circuit. Such effects could possibly reflect some "nonspecific" consequence of the lesions, such as malaise, sedation, hormonal changes, or taste or oral motor disturbances, rather than a specific interruption in the circuitry through which the AcbSh influences feeding. Alternatively, both regions could influence a common effector without actually being functionally related to one another. Many brain sites may play such "permissive" roles in the control of food intake, in that while not directly involved in the initiation or maintenance of eating, feeding can only proceed when they are in certain activational states. It takes a convergence of anatomical, physiological, and behavioral data to tentatively assign a site as a component of a functional system.

We do not yet have each of these classes of data for the putative components of the AcbSh feeding circuit. Nor do we have essential data on AcbSh-mediated electrophysiological responses or *in vivo* neurotransmitter release. Nevertheless, the following summary of the available data indicates the broad extent of current knowledge already derived through the use of the different strategies.

B. Identification of Brain Regions Activated by Inhibition of the AcbSh

Perhaps the best current overview of the functional circuits through which the AcbSh effects changes in feeding behavior comes from localization of Fos expression after AcbSh inhibition. The vast majority of projection neurons in the AcbSh are GABAergic (Meredith et al., 1993) and it follows that inhibiting the firing rate of these cells should result in a disinhibition of postsynaptic neurons located in the terminal fields of these projections. Of course, the AcbSh is a functionally heterogeneous region with multiple cell assemblies projecting to different brain regions and a subset of the activated regions are undoubtedly involved in the control of behaviors other than feeding. Of the several brain regions activated by acute AcbSh inhibition that play a well documented role in the control of feeding behavior, perhaps the most interesting are hypothalamic sites. Bilateral AcbSh injections of muscimol greatly increased the number of cells expressing Fos throughout the entire rostrocaudal axis of the LH, with the largest increase seen in the perifornical region (Stratford and Kelley, 1999). The treatment also increases Fos expression in the arcuate and paraventricular hypothalamic nuclei (Stratford and Kelley, 1999; Zheng et al., 2003; Stratford, 2005), brain regions which, together with the LH, form an intricately

interconnected network that plays a primary role in the control of ingestive behavior (Bernardis and Bellinger, 1996; Schwartz et al., 2000; Williams et al., 2001). These results suggest that activity within this circuit is mediated in part by inputs arising in the AcbSh, which is capable of influencing the LH both through direct projections and through relays in the ventral pallidum. Extrahypothalamic feeding sites that are activated by AcbSh inhibition include the medial ventral pallidum, ventral tegmental area, central amygdaloid nucleus, and nucleus of the solitary tract (Stratford and Kelley, 1999).

These findings demonstrate that a physiological relationship exists between the AcbSh and these feeding-related structures and provides a mechanism through which a behaviorally functional circuit may act. While consistent with the idea that the AcbSh elicits feeding through a neurally-mediated influence on the activity of hypothalamic neurons, changes in Fos expression do need to be interpreted with caution. When combined with data from anatomical tract-tracing studies, mapping Fos allows for inferences to be made about the pathways through which one brain region may alter activity in another. Ultimately, however, the technique only demonstrates whether a temporal relationship exists between the eliciting stimulus and the change in gene expression. However, when drugs are administered systemically, or gain access to both sides of the brain through intracerebroventricular or bilateral intracerebral microinjections, we cannot determine whether changes in Fos are the result of alterations in the activity of neural circuits linking two brain regions, or are secondary to stimulus-bound changes in autonomic activity, hormone levels, arousal, stress, or other systemically or behaviorally mediated factors. This is a particularly important issue when investigating brain regions such as the LH and PVN, in which Fos expression can be induced by a variety of stressors (Chastrette et al., 1991; Silveira et al., 1993; Smith et al., 1997a; Briski and Gillen, 2001) and by changes in circulating levels of glucose or a number of hormones (Bonaz et al., 1993; Chen et al., 1993; Bahjaoui-Bouhaddi et al., 1994a; Niimi et al., 1995; Roberts et al., 1995; Turton et al., 1996; Elmquist et al., 1998; Moriguchi et al., 1999; Cai et al., 2001; Lawrence et al., 2002). Therefore, the most recent effort to clarify the mechanisms through which the AcbSh activates down-stream structures involved an examination of Fos synthesis throughout the brain after *unilateral* injections of muscimol into the AcbSh (Stratford, 2005).

While bilateral labeling of structures may be the result of indirect neuronal acti-vation induced by changes in interoceptive cues, behavioral state, or circulating levels of orexigenic or satiety compounds, the presence of unilateral Fos expression is strong evidence that neurons in the labeled structure are activated principally by the actions of muscimol on AcbSh neurons and are nodes in an uninterrupted AcbSh neural circuit. Unilaterally inhibiting the AcbSh by injecting muscimol into one AcbSh and the saline vehicle into the contralateral AcbSh only increases Fos expression in the LH *ipsi-lateral* to the drug injection. This ipsilateral increase is as large as that seen after bilat-eral muscimol injections and, again, occurs throughout the rostrocaudal extent of the LH, with the largest increase in Fos expres-sion seen in the perifornical region of the LH. The unilateral nature of the Fos expres-sion reinforces the idea that neurons in the LH are activated as part of a functional neural circuit and not in response to an AcbSh-mediated change in behavior or in autonomic variables or circulating levels of glucose or other feeding-related com-pounds. These findings are, of course, con-sistent with the concept that the AcbSh controls food intake by directly altering the activity of LH neurons.

Unilateral muscimol injections also re-sulted in a primarily ipsilateral activation of the VPm, Arc, and PVN, demonstrating that these structures are also part of an uninterrupted, lateralized neural circuit controlled by neurons in the AcbSh, sup-porting the idea that they are also nodes in a functional circuit controlling the expres-sion of AcbSh-mediated feeding. The AcbSh does not project directly to the Arc or PVN, but potentially can influence them through its actions on the LH, which is heavily inter-connected with the medial hypothalamus. The demonstration that unilateral injections of muscimol into the AcbSh are sufficient to induce robust feeding strongly suggests that the structures showing unilateral changes in gene expression are likely to be involved in the expression of the behavior (Stratford, 2005).

Another important conclusion can be drawn from the observation that unilateral inhibition of the AcbSh increases Fos to a much greater degree in hypothalamic sites ipsilateral to the drug injection than it does on the contralateral side. Neurons in the PVN respond to a wide variety of systemic changes. They are activated by certain orex-igenic conditions such as hypoglycemia (Niimi et al., 1995; Cai et al., 2001) and intracerebroventricular injections of ghrelin (Lawrence et al., 2002), but also by in-creased levels of hormones that reduce food intake (Bonaz et al., 1993; Chen et al., 1993; Turton et al., 1996; Elmquist et al., 1998). PVN neurons are also activated in response to a variety of noxious or anxio-genic stimuli (Chastrette et al., 1991; Silveira et al., 1993). However, in contrast to the effects of unilateral AcbSh inhibition, all of these conditions induce Fos equally on both sides of the brain. This surprising result suggests that the PVN Fos response to AcbSh muscimol must be mediated largely through uncrossed neural pathways origi-nating in the AcbSh, and allows us to exclude most secondary systemic changes in the rat as primary factors in the activa-tion of downstream component nuclei.

A final important advantage to the uni-lateral approach is that it can be used to eliminate from consideration potential pathways through which the AcbSh may alter activity in downstream components, in that bilaterally activated structures are unlikely to be interposed in the unilateral circuit. Given the general anatomical symmetry and laterality of fiber projections and the circuits they form, once the initially unilateral signal is transmitted through bifurcating projections to homomorphic structures on both sides of the brain, it is unlikely that those bilateral signals will subsequently revert to strictly unilateral signals. Thus, we can conclude that the bilaterally activated NTS and central amygdaloid nucleus are not interposed in the lateralized circuit composed, in part, of the AcbSh, VPm, LH, Arc, and PVN.

C. Role of the Ventral Pallidum in the Expression of AcbSh-Mediated Feeding

The medial subcommissural ventral pallidum is a basal forebrain structure anatomically interposed between the AcbSh and the LH that has a demonstrated role in the control of food intake. Furthermore, current data suggest that the VPm is a component of the functional AcbSh–LH feeding circuit. Although much less is known about the contribution of the ventral pallidum to the control of food intake, it has been established that disrupting transmission at GABAergic synapses in the VP with local injections of the GABA$_A$ receptor blocker bicuculline elicits robust, dose-dependent feeding in satiated rats (Stratford et al., 1999; Smith and Berridge, 2005). The VPm is the terminus for many GABAergic fibers arising from neurons in the AcbSh (Nauta et al., 1978; Churchill et al., 1990; Zahm and Heimer, 1990; Heimer et al., 1991) and inhibition of AcbSh neurons by local injections of muscimol should reduce GABA transmission in the VPm, a condition mimicked by intra-VP bicuculline. The relevant GABA$_A$ receptors appear to be distributed in a relatively homogeneous fashion, as bicuculline injections throughout the rostrocaudal extent of the VPm are effective at

increasing food intake (Smith and Berridge, 2005).

Inhibiting the AcbSh and disinhibiting the VP do not elicit identical behavioral responses, however. Unlike intra-AcbSh muscimol, blocking GABA$_A$ receptors in the VP with bicuculline also yields a potent dose-dependent increase in locomotor activity even when food is present (Austin and Kalivas, 1990; van den Bos and Cools, 1991; Stratford et al., 1999; Smith and Berridge, 2005). Although further studies on the behavioral specificity of this effect need to be performed, the fact that water intake is not affected and that increases in feeding and locomotion are dissociated at low doses of bicuculline suggest that disinhibition of cells in the VPm does not result in a nonspecific behavioral activation, but rather may selectively increase both food intake and locomotor activity. Currently, it is not known whether the same population of neurons is responsible for the observed changes in feeding behavior and locomotor activity and it remains to be seen whether these behaviors can be dissociated by other manipulations of the VPm.

From a feeding behavior perspective, however, it should be noted that even when locomotor activity is at its highest, the overall behavior of the rats remain strongly oriented toward food (Stratford et al., 1999; Smith and Berridge, 2005), suggesting that locomotor activity does not displace the feeding response. As locomotor activity increases with higher doses of bicuculline, the rats often carry food pellets in their mouths as they run around the perimeter of the cage and are reluctant to relinquish the food even when they are not actively feeding. During the periods of maximal hyperactivity, rats carrying a food pellet will often stop and attempt to place a second food pellet in the mouth. This pellet carrying activity is very similar to portions of the behavioral pattern exhibited during hoarding behavior. The VPm sends substantial projections to the mediodorsal thalamic nucleus (MD) (Haber et al., 1985;

Zahm et al., 1996; O'Donnell et al., 1997) and the LH (Haber et al., 1985; Groenewegen et al., 1993), two brain regions that have been implicated in the control of hoarding behavior (Blundell and Herberg, 1973; Mogenson and Wu, 1988). It is possible that under the appropriate environmental conditions the VP output signals contribute to the coordination of food-directed locomotor patterns involved in hoarding and that a fragmented form of this behavior is elicited by administration of high doses of bicuculline into the VP that result in changes in activity in the MD, LH, or both. It is interesting to note that lesions of the PFCtx, which is a component of the VPm–MD–PFCtx–AcbSh loop, also result in permanent disruption of hoarding behavior (Kolb, 1974).

Both the μ opioid receptor agonist morphine and the GABA$_A$ antagonist bicuculline have similar effects on evoked activity in the VP (Chrobak and Napier, 1993) and, therefore, it is not surprising that feeding is also elicited by activating μ receptors in the region (Smith and Berridge, 2005). Activation of VP μ receptors also increases locomotor activity, although the pellet-carrying behavior seen after intra-VP bicuculline injections was not observed. Interestingly, while injections of the μ receptor agonist DAMGO into the posterior VP increase food intake, μ receptor activation in the rostral VP actually suppresses intake (Smith and Berridge, 2005). These rostrocaudal intake differences are also reflected in altered hedonic responses, where DAMGO injections into the posterior VP increase positive taste reactions to intraoral sucrose, while those in the rostral VP decrease such responses. The fact that the bicuculline and DAMGO ingestive responses do not map equivalently throughout the VP, that they have the same effect on feeding in the caudal VP, but actually have opposite effects in the rostral VP, suggests that the two drugs are likely to be acting on different populations of neurons within the VP.

The largest efferent projection of the AcbSh terminates in the subcommissural VP with most of these fibers originating in the subpopulation of GABAergic projection neurons of the AcbSh that coexpress dynorphin and substance P (Zahm and Heimer, 1990; Heimer et al., 1991; Napier et al., 1995; Usuda et al., 1998; Zahm et al., 1999; Zhou et al., 2003). Substance P generally increases spontaneous activity in VP neurons, however, at low levels it appears to act preferentially at presynaptic receptors to suppress the release of glutamate and, thus, may result in an overall inhibition of activity in the region (Mitrovic and Napier, 1998). The corelease of GABA and substance P in the VPm may help explain the rather complex effects on the activity of VP neurons elicited by AcbSh stimulation. While approximately half of the responding neurons show a brief excitation or a more sustained inhibition, the other half show more complex bi- or triphasic responses, which may indicate the existence of separate functional populations (Chrobak and Napier, 1993). However, blocking GABA$_A$ receptors in those terminal fields does reliably suppress the inhibitory effects of AcbSh stimulation on VP neurons (Chrobak and Napier, 1993), suggesting that bicuculline acts in the VP to disinhibit at least some population of neurons receiving GABAergic AcbSh fibers. Coupled with the behavioral effects observed after intra-VP bicuculline, these data support the idea that disinhibiting a population of cells in the VPm is the relevant event for inducing feeding.

Other inputs to the VP include a dopaminergic projection from the VTA and glutamatergic projections from the medial PFCtx and BLA (Fuller et al., 1987; Klitenick et al., 1992), all again duplicating inputs from those regions to the AcbSh and suggestive of a shared functionality among them.

At present, nothing definite is known about the efferent pathways through which the VP effects changes in food intake.

Neurons in the region of the VP that receive input from the AcbSh send a large projection back to the AcbSh, in addition to sending fibers to the ventral prelimbic and infralimbic cortices, BLA, VTA, LH, and median raphe nucleus (Haber et al., 1985; Zahm and Heimer, 1990; Groenewegen et al., 1993; Zahm et al., 1996). A massive GABAergic projection to the mediodorsal thalamic nucleus arises from ventral pallidal cells contacted by AcbSh fibers (O'Donnell et al., 1997). The MD has strong reciprocal connections with ventromedial PFCtx, and it is possible that these structures are components of a multisynaptic feedback loop in which information from the AcbSh passes through the VPm to the MD, and back to the ventromedial PFCtx to modify activity in the AcbSh. Such a signal could act to regulate AcbSh activity based on the current level of AcbSh output integrated with additional information brought into the circuit at each node.

The presence of the dense, monosynaptic projection from the VPm to the LH (Haber et al., 1985; Groenewegen et al., 1993) leads naturally to the hypothesis that, like the AcbSh, the VPm mediates feeding through its actions on LH neurons. Like that from the AcbSh, the VPm projection terminates in the lateral perifornical region of the LH, at which feeding can be elicited most effectively by activation of glutamate receptors (Duva et al., 2005). With the exception of the substantial projection to the MD, the efferent connections of the VPm are very similar to those of the AcbSh, providing further evidence that the two regions share a close functional relationship (Groenewegen et al., 1993; Zahm et al., 1996).

The hypothesis that the VPm is an intervening node in the AcbSh–LH circuit more recently gained additional support with the observation that fiber-sparing excitotoxic lesions of the VPm attenuate intra-AcbSh muscimol hyperphagia by more than 70% (Stratford and Wirtshafter, unpublished observations). Although these lesions did not completely block the elicited feeding, it currently is not known whether the remaining feeding is related to incomplete lesions or perhaps indicates a continued influence of the AcbSh on LH activity mediated through the direct AcbSh–LH projection. Regardless, combined with the anatomical, physiological, and behavioral data obtained previously, the finding that neurons in the VPm are essential for the full expression of AcbSh-mediated feeding provides compelling evidence that the AcbSh mediates feeding behavior, in part, through an action on VPm neurons and that this structure is an essential component in the functional AcbSh–LH feeding circuit.

D. Role of the Hypothalamus in the Expression of AcbSh-Mediated Feeding

1. Lateral Hypothalamus

The LH is a neurochemically heterogeneous region that plays a critical role in the control of food intake (Bernardis and Bellinger, 1996) as well as in a number of other ingestive-related processes, including thermogenesis (Lupien et al., 1986; Cerri and Morrison, 2005), salivation (Schallert et al., 1978), gastric acid secretion and gastric motility (Glavcheva et al., 1972; Shiraishi, 1988), predatory attack behavior (Nikulina, 1991; Siegel and Schubert, 1995), and water intake in response to regulatory challenges (Stricker, 1976; Stratford and Wirtshafter, 2000). It is well-established that electrical or chemical stimulation of LH neurons induces intense feeding in satiated animals (Delgado and Anand, 1953; Valenstein et al., 1970; Stanley and Leibowitz, 1984; Stanley et al., 1993a). The lateral perifornical region of the tuberal LH appears to be a critical zone for the mediation of food intake, and glutamate transmission in particular has been shown to play an important role in the expression of feeding (Stanley et al., 1993b, 1996; Khan et al., 2004). Injections of glutamate or NMDA into the LH stimulate intense feeding which is suppressed by blocking local NMDA receptors (Stanley et al., 1993a). More importantly, blocking NMDA receptors in the perifornical LH potently suppresses feeding induced by food deprivation (Stanley et al., 1996), suggesting that this system plays a role in the normal, physiological control of food intake. Further support for this hypothesis comes from the observation that glutamate is released in the LH when rats feed, with levels rising during the first third of a meal followed by a decrease to below baseline levels (Rada et al., 1997, 2003), a time-course suggesting that LH glutamate is preferentially involved in the initiation, rather than the maintenance, of feeding.

It has been recognized for some time that the medial Acb is the only striatal region that sends a direct projection to the lateral hypothalamus (Williams et al., 1977; Heimer et al., 1991; Zahm and Brog, 1992). While anterograde tracers demonstrate that the fibers originating in the AcbSh are found throughout the rostrocaudal extent of the LH (Heimer et al., 1991; Zahm et al., 1999), data obtained using retrograde tract-tracers suggest that the AcbSh projection terminates preferentially in the lateral perifornical region of the mid-rostral LH (Duva et al., 2005), the region most sensitive to glutamate-elicited feeding. Importantly, blocking NMDA receptors in this particular region of the LH with local administration of D(−)-2-amino-5-phosphonopentanoic acid (AP-5) potently suppresses AcbSh-mediated feeding, demonstrating that the expression of feeding behavior elicited by inhibition of neurons in the AcbSh depends on the NMDA-mediated activation of LH neurons (Stratford and Kelley, 1999). Because the same treatment suppresses deprivation-induced food intake (Stanley et al., 1996), this result also is consistent with the hypothesis that inhibition of cells in the AcbSh and food deprivation induce feeding through activation of a common neural pathway involving cells in the perifornical LH. The VPm also projects to the glutamate-sensitive portion of the LH (Duva et al., 2005) and while the AcbSh and

VPm provide just two of the many inputs to this region, it is noteworthy that manipulations of these LH afferents alone are sufficient to elicit a full feeding response.

There is evidence that a GABAergic component is also involved in the control of food intake by the LH. GABA receptors in the region play at least a permissive role in mediating food intake under some conditions as injections of muscimol into the perifornical LH suppress both spontaneous and AcbSh-mediated feeding (Kelly et al., 1979; Maldonado-Irizarry et al., 1995). However, the precise conditions under which LH GABA acts to alter feeding behavior is not yet clear. It is known that while extracellular GABA levels in the LH remain stable during insulin-induced hypoglycemia or 2-deoxyglucose glucoprivation if the animals are not allowed to eat, they decrease by over 20% during a 2-deoxyglucose-elicited meal (Beverly et al., 1995, 2001). In contrast, when feeding is elicited by a period of food deprivation, extracellular GABA levels begin increasing at the start of the meal and continue rising until they peak when feeding is terminated (Rada et al., 2003). These data suggest that an increase in LH GABA transmission may be involved in terminating feeding elicited by deprivation, but not by glucoprivation. Interestingly, a more recent study reported that disinhibiting AcbSh projection neurons with local injections of GABA receptor blockers, a treatment that should increase GABA release in terminal fields in the perifornical LH, suppresses deprivation-induced food intake, but not 2-deoxyglucose-induced feeding (Kandov et al., 2006). Taken together, these intriguing findings strongly suggest that AcbSh-mediated GABA release in the LH is preferentially involved in the control of deprivation-induced feeding and not in feeding elicited by acute glucoprivation or hypoglycemia.

AcbSh muscimol injections greatly increase Fos synthesis in perifornical LH neurons (Stratford and Kelley, 1999) and a subpopulation of these activated cells contain orexin (Zheng et al., 2003; Baldo et al., 2004). GABA receptors are located on lateral hypothalamic orexin neurons and can directly mediate their activity, providing support for the idea that the GABAergic projection from the AcbSh may play a role in tonically inhibiting those cells (Backberg et al., 2003; Xie et al., 2006). This circumscribed population of orexin-containing cells projects diffusely throughout the neuraxis and has been implicated in the control of sleep, arousal level, and feeding behavior (Willie et al., 2001; Sakurai, 2005). Orexin fibers innervate the AcbSh directly, as well as several brain regions that send direct projections to the AcbSh, including the infralimbic and prelimbic cortices, VTA, BLA, dorsal PVT, Arc, dorsal and median raphe nuclei, and NTS (Date et al., 1999; Fadel and Deutch, 2002), providing multiple short polysynaptic pathways through which LH orexin neurons may mediate activity in the AcbSh. Activation of orexin receptors in the LH increases Fos expression in many of the same regions as does intra-AcbSh muscimol, including, the central amygdaloid nucleus, Arc, PVN, NTS, and the AcbSh itself (Mullett et al., 2000). Because the AcbSh does not project directly to any of these sites, it is possible that the increased Fos expression seen in these brain regions after inhibition of the AcbSh is a direct result of the AcbSh-mediated activation of LH orexin neurons. It is important to note, however, that the neurochemical phenotype of the vast majority of activated LH cells remains unknown and other neurotransmitter systems undoubtedly participate in the expression of AcbSh-mediated feeding.

While AcbSh inhibition does not induce Fos in MCH-containing neurons in the LH (Zheng et al., 2003), this result needs to be interpreted cautiously, given the demonstration that intra-AcbSh MCH increases food intake (Georgescu et al., 2005) and that MCH neurons are activated by another orexigenic treatment, systemic injections of

insulin, but do not express Fos under those conditions (Bahjaoui-Bouhaddi et al., 1994b). Therefore, a potential role for MCH neurons in the control of AcbSh-mediated feeding cannot be dismissed based on the Fos results. Taken together, these data suggest that activation of the LH-AcbSh MCH projection are likely to be involved in increasing food intake under some conditions, such as in response to insulin-induced hypoglycemia.

It is clear that the AcbSh mediates activity in the LH, however, the exact mechanism through which it does so has not yet been determined. Although the AcbSh sends a monosynaptic, GABAergic projection to the LH, there are data suggesting that this may not be the only pathway through which AcbSh inhibition mediates food intake. In particular, blocking GABA receptors in the terminal fields of these fibers, a treatment that should mimic the local effects of AcbSh inhibition, does not increase food intake in satiated rats (Stratford and Kelley, 1999). The issue is complicated though, as injections of bicuculline or picrotoxin into the LH have been reported to elicit modest feeding under some conditions (Kelly et al., 1977; Tsujii and Bray, 1991). Further studies will be required to determine under what, if any, conditions blocking GABA transmission in the perifornical LH will consistently elicit feeding. However, it is important to note that under the same test conditions in which manipulations of the AcbSh elicit such large increases in food intake, injections of GABA antagonists into the specific region of the LH in which the largest increases in Fos expression were observed and in which disrupting glutamate transmission potently suppressed AcbSh-mediated feeding, did not increase food intake.

The fact that blocking GABA receptors in the perifornical LH had no effect on food intake would appear to indicate that GABAergic projections terminating in the region are not critically important for the expression of feeding behavior, regardless of whether they originate in the AcbSh, VP, or elsewhere. Of course, several alternative explanations are possible. It is possible that simultaneous blockade of GABA$_A$ and GABA$_B$ receptors in the perifornical LH is required to consistently elicit feeding. Also, because AcbSh efferents target a select subset of LH neurons, inhibiting the cell bodies at the source of the projection is likely to result in a more complete and specific disinhibition of relevant neurons in the terminal field. The lack of feeding could conceivably be the result of local injections of GABA receptor blockers not affecting a sufficient number of GABA receptors to initiate feeding or alternatively, of antagonist actions on additional cell populations mediating opposing behavioral effects that do not receive AcbSh fibers and would not normally be disinhibited by AcbSh inhibition.

As noted before, the tuberal level of the perifornical LH appears to be the critical region for mediating food intake, and blocking NMDA receptors in this specific region of the LH suppresses both deprivation-induced and AcbSh-mediated feeding. However, inhibiting cells in the AcbSh activates neurons throughout the entire rostrocaudal extent of the LH and the magnitude of the suppression induced by perifornical AP-5 administration is surprising given the small percentage of AcbSh-activated LH neurons that must be affected directly by the injections. The relationship between the widely distributed Fos expression seen after AcbSh muscimol and the more circumscribed region sensitive to manipulations of the glutamate system has yet to be investigated. Currently it is not known whether inhibition of LH neurons outside the perifornical region have any effect on AcbSh-mediated feeding.

2. Medial Hypothalamus

AcbSh inhibition also activates neurons in the medial hypothalamic arcuate and paraventricular nuclei (Stratford and Kelley, 1999; Zheng et al., 2003; Stratford,

2005). Modulation of Arc activity by the AcbSh is particularly interesting as the Arc contains one population of neurons producing the orexigenic peptides neuropeptide Y (NPY) and agouti-related protein (AGRP) and a separate population synthesizing the anorexic peptides proopiomelanocortin (POMC) and CART protein (Elias et al., 1998; Hahn et al., 1998). Activity in these neurons is regulated in part by levels of circulating factors such as insulin, the adipocyte-derived peptide leptin, and the gastric peptide ghrelin, which work together to signal the availability of energy stores and comprise the backbone of the system through which changes in available energy stores regulate feeding behavior (Ahima et al., 2000; Schwartz et al., 2000). Importantly, intra-AcbSh muscimol exerts multiple orexigenic effects on cells in the Arc, increasing Fos expression in a subset of neurons containing neuropeptide Y/AgRP, while decreasing Fos in POMC/CART-containing neurons (Zheng et al., 2003). Thus, AcbSh mediated feeding appears to involve the activation of neurons that closely monitor levels of circulating factors signaling energy availability. Although it is not yet clear whether such circulating peripheral metabolic indicators play a permissive role in AcbSh feeding, the fact that unilateral injections of muscimol into the AcbSh increase Fos expression primarily in the ipsilateral hypothalamus indicates that circulating insulin, leptin, or ghrelin are not directly responsible for the activation of hypothalamic neurons (Stratford, 2005). Both NPY/AGRP and POMC/CART neurons in the Arc project to the PVN (Bai et al., 1985; Legradi and Lechan, 1998; Fekete et al., 2004), suggesting that these pathways may be involved in the control of PVN activity by the AcbSh.

Although the effects of Arc or PVN lesions on AcbSh hyperphagia have not yet been evaluated, further evidence that AcbSh-mediated feeding involves changes in medial hypothalamic activity comes from the demonstration that activation of

NPY receptors near the third ventricle is essential for the expression of AcbSh-mediated food intake (Stratford and Wirtshafter, 2004). Central NPY is an important regulator of feeding behavior, and studies suggest that food intake is mediated through the Y1 and Y5 receptor subtypes (Inui, 1999; Duhault et al., 2000), both of which are found in significant numbers in the Arc and PVN (Parker and Herzog, 1999; Wolak et al., 2003). Interestingly, AcbSh-mediated feeding is potently suppressed by injections of either NPY Y1 or Y5 receptor blockers into the third ventricle (Stratford and Wirtshafter, 2004). While the exact location of the relevant NPY receptors has yet to be determined, these results provide compelling evidence that neurons perfused by the third ventricle and expressing Y1 and Y5 receptors are functional components of a feeding-related neural circuit controlled by the AcbSh.

The precise sequence in which information from the AcbSh flows in the hypothalamus is not yet known; however, the LH, Arc, and PVN are heavily interconnected and work together to modify feeding behavior. Although neither the AcbSh nor the closely aligned VP send a substantial projection to the medial hypothalamus, neurons in both regions project to the perifornical LH which, in turn, projects to the PVN and is reciprocally connected with the Arc. These data suggest the interesting possibility that the AcbSh and/or VPm may mediate activity in these medial hypothalamus through a bisynaptic pathway involving the orexin-containing LH neurons that are activated by AcbSh muscimol. Arcuate NPY neurons receive projections from LH orexin neurons (Horvath et al., 1999), express functional orexin receptors (Backberg et al., 2002), and increase Fos expression in response to intracerebroventricular or intra-LH administration of orexin (Mullett et al., 2000; Yamanaka et al., 2000). Furthermore, like AcbSh-mediated feeding, orexin increases food intake through actions on a NPY system (Jain

et al., 2000; Yamanaka et al., 2000). Together, these data support the hypothesis that AcbSh-mediated feeding is effected, at least in part, through an activation of LH orexin neurons which, in turn, activate NPY neurons in the Arc.

The fact that the AcbSh and VPm directly innervate the perifornical LH, but not the Arc and PVN suggest that AcbSh-mediated LH activation is not downstream of medial hypothalamic activation. However, information from the Arc can influence LH activity. Perifornical LH orexin and MCH cells are innervated by Arc NPY and POMC neurons and activity in these projections is regulated to some degree by circulating levels of leptin (Broberger et al., 1998; Elias et al., 1999). Therefore, the possibility remains that AcbSh-mediated activation of LH neurons depends on those cells receiving an appropriate excitatory signal from Arc neurons monitoring circulating leptin or other feeding-related factors.

Finally, it is important to remember that the specific efferent pathways through which the hypothalamus effects changes in feeding behavior are virtually unknown. Determining both the precise pathways through which information from the AcbSh enters and exits the hypothalamus and the sequence in which hypothalamic nuclei are activated is of critical importance in understanding the mechanisms through which the AcbSh mediates feeding behavior and how the AcbSh feeding circuit is related to hypothalamic systems controlling feeding under other circumstances.

V. THE AcbSh FEEDING CIRCUIT AS A POTENTIAL SITE FOR PATHOLOGY AND THERAPEUTIC INTERVENTION IN THE TREATMENT OF EATING DISORDERS

The AcbSh is a region with the potential to integrate processed multimodal sensory data with information regarding energy balance, taste, reward, previously formed associations, and current levels of activity in pallidal and hypothalamic circuits. Whether it actually does so remains to be determined. What is known is that, regardless of the factors that mediate activity in the feeding-related cell assemblies in the AcbSh, the output signal of those neurons, at least part of which passes through the medial ventral pallidum, plays an important role in determining the activation state of feeding-related circuits in the lateral and medial hypothalamic areas. In essence, it appears that integrative physical properties in the AcbSh result in a "decision" being made on whether the animal should be feeding at any given time and the AcbSh communicates this "advice" to the hypothalamus, which integrates that input with numerous others to determine whether the animal will eat. It seems reasonable to assume that the AcbSh is an important regulator of feeding behavior, since inhibiting activity in the structure can override the collection of internal and external cues that suppress feeding in the nondeprived rat, as well as the potent orosensory and postingestive satiety signals that are present after ingesting such large quantities of food.

Although the exact role of the AcbSh is not known, it is possible that the integration of such diverse information in the AcbSh is used to help determine the appropriateness of engaging in feeding behavior in a particular environmental situation. Feeding behavior, while vital to the ultimate survival of the animal, can often be postponed without adverse results. The expression of situationally appropriate adaptive behaviors depends on a constant assessment of external and internal information and feeding is but one of the appetitive behaviors an organism needs to place correctly in a response hierarchy in order to survive. The ability of the AcbSh to integrate a wide variety of exteroceptive and interoceptive stimuli and to potently mediate activity in hypothalamic feeding circuits would

appear to make it an ideal candidate for such a role.

Although the topic has yet to be investigated, it is plausible that pathology of the AcbSh or VPm may be implicated in the etiology of human obesity or some eating disorders. For example, there is evidence that, at least in rats, some AcbSh projections develop postnatally making them susceptible to damage by perinatal stress, malnutrition, or exposure to environmental toxins (Zahm et al., 2001). Damage to AcbSh circuits during this formative period could result in long-term disruption of normal eating patterns. Interestingly, the tonic inhibition of the LH by the AcbSh circuit indicates that pathology disrupting activity in the circuit will likely result in a condition in which the animal overeats when it would normally be satiated. While less adaptive than a properly functioning system, such pathologies may present little problems when food supplies are limited. The resulting disorders may become seriously maladaptive only when food is exceptionally plentiful and, therefore, might be readily propagated in populations with limited access to food.

While a therapy would ideally be able to target the specific pathology underlying an eating disorder, a more general approach may also be useful. There is much to recommend the AcbSh circuit as a target for therapeutic intervention even in cases in which the circuit remains intact. Regardless of exactly what information is being integrated in the AcbSh, the fact that AcbSh output acts powerfully on hypothalamic cells that appear to be specifically involved in the control of feeding behavior suggests that AcbSh circuits are a worthy focus for potential pharmacological interventions. The multiple neurotransmitter systems and rich variety of receptors in the AcbSh that already have been demonstrated to control food intake may each provide a conduit through which AcbSh output could be altered. The key to developing efficacious, systemically administered pharmacological therapies might lie in identifying a unique structural feature, or combination of features, that would provide a means of targeting the specific subpopulation of AcbSh neurons comprising the feeding-related functional assembly. Such advances are likely to come through the identification of differentially localized membrane receptor subunits, vesicular transport systems, or unique combinations of receptors and the development of drugs that specifically target those features.

Over the past decade, a substantial body of evidence has been gathered strongly supporting an important, possibly critical, role for the AcbSh feeding circuit in the control of normal, homeostatic food intake. Furthermore, the work serves to underscore the importance of investigating the AcbSh feeding circuit and what such studies may reveal about the functional organization of the neural mechanisms involved in the regulation of normal, homeostatic feeding in mammals, particularly as regards the extrahypothalamic control of hypothalamic circuits. While the importance of the AcbSh feeding circuit has become evident, there is a vast amount of work that remains to be done in order to obtain a detailed anatomical, neurochemical, and behavioral characterization of the system. The results of these continuing investigations will allow us to understand more clearly the nature of the neural mechanisms involved in the regulation of feeding behavior and may provide information critical to the effort to discover efficacious treatments to ameliorate the suffering caused by dysregulation of feeding in humans.

References

Ahima, R. S., Saper, C. B., Flier, J. S., and Elmquist, J. K. (2000). Leptin regulation of neuroendocrine systems. *Front. Neuroendocrinol.* **21**, 263–307.

Amat, J., Baratta, M. V., Paul, E., Bland, S. T., Watkins, L. R., and Maier, S. F. (2005). Medial prefrontal cortex determines how stressor controllability affects behavior and dorsal raphe nucleus. *Nat. Neurosci.* **8**, 365–371.

Anton, B., Fein, J., To, T., Li, X., Silberstein, L., and Evans, C. J. (1996). Immunohistochemical localization of ORL-1 in the central nervous system of the rat. *J. Comp. Neurol.* **368**, 229–251.

Austin, M. C., and Kalivas, P. W. (1990). Enkephalinergic and GABAergic modulation of motor activity in the ventral pallidum. *J. Pharmacol. Exp. Ther.* **252**, 1370–1377.

Backberg, M., Hervieu, G., Wilson, S., and Meister, B. (2002). Orexin receptor-1 (OX-R1) immunoreactivity in chemically identified neurons of the hypothalamus: focus on orexin targets involved in control of food and water intake. *Eur. J. Neurosci.* **15**, 315–328.

Backberg, M., Collin, M., Ovesjo, M. L., and Meister, B. (2003). Chemical coding of GABA(B) receptor-immunoreactive neurones in hypothalamic regions regulating body weight. *J. Neuroendocrinol.* **15**, 1–14.

Badiani, A., Leone, P., Noel, M. B., and Stewart, J. (1995). Ventral tegmental area opioid mechanisms and modulation of ingestive behavior. *Brain Res.* **670**, 264–276.

Bahjaoui-Bouhaddi, M., Fellmann, D., and Bugnon, C. (1994a). Induction of Fos-immunoreactivity in prolactin-like containing neurons of the rat lateral hypothalamus after insulin treatment. *Neurosci. Lett.* **168**, 11–15.

Bahjaoui-Bouhaddi, M., Fellmann, D., Griffond, B., and Bugnon, C. (1994b). Insulin treatment stimulates the rat melanin-concentrating hormone-producing neurons. *Neuropeptides* **27**, 251–258.

Bai, F. L., Yamano, M., Shiotani, Y., Emson, P. C., Smith, A. D., Powell, J. F., and Tohyama, M. (1985). An arcuato-paraventricular and -dorsomedial hypothalamic neuropeptide Y-containing system which lacks noradrenaline in the rat. *Brain Res.* **331**, 172–175.

Bakshi, V. P., and Kelley, A. E. (1993a). Feeding induced by opioid stimulation of the ventral striatum: role of opiate receptor subtypes. *J. Pharmacol. Exp. Ther.* **265**, 1253–1260.

Bakshi, V. P., and Kelley, A. E. (1993b). Striatal regulation of morphine-induced hyperphagia: an anatomical mapping study. *Psychopharmacology (Berlin)* **111**, 207–214.

Baldo, B. A., and Kelley, A. E. (2001). Amylin infusion into rat nucleus accumbens potently depresses motor activity and ingestive behavior. *Am. J. Physiol.* **281**, R1232–R1242.

Baldo, B. A., Gual-Bonilla, L., Sijapati, K., Daniel, R. A., Landry, C. F., and Kelley, A. E. (2004). Activation of a subpopulation of orexin/hypocretin-containing hypothalamic neurons by GABAA receptor-mediated inhibition of the nucleus accumbens shell, but not by exposure to a novel environment. *Eur. J. Neurosci.* **19**, 376–386.

Baldo, B. A., Alsene, K. M., Negron, A., and Kelley, A. E. (2005). Hyperphagia induced by GABA$_A$ receptor-mediated inhibition of the nucleus accumbens shell: Dependence on intact neural output from the central amygdaloid region. *Behav. Neurosci.* **119**, 1195–1206.

Basso, A. M., and Kelley, A. E. (1999). Feeding induced by GABA(A) receptor stimulation within the nucleus accumbens shell: regional mapping and characterization of macronutrient and taste preference. *Behav. Neurosci.* **113**, 324–336.

Beaumont, K., Kenney, M. A., Young, A. A., and Rink, T. J. (1993). High affinity amylin binding sites in rat brain. *Mol. Pharmacol.* **44**, 493–497.

Beckstead, R. M. (1979). An autoradiographic examination of corticocortical and subcortical projections of the mediodorsal-projection (prefrontal) cortex in the rat. *J. Comp. Neurol.* **184**, 43–62.

Bendotti, C., Garattini, S., and Samanin, R. (1986). Hyperphagia caused by muscimol injection in the nucleus raphe dorsalis of rats: its control by 5-hydroxytryptamine in the nucleus accumbens. *J. Pharm. Pharmacol.* **38**, 541–543.

Berendse, H. W., and Groenewegen, H. J. (1990). Organization of the thalamostriatal projections in the rat, with special emphasis on the ventral striatum. *J. Comp. Neurol.* **299**, 187–228.

Bernardis, L. L., and Bellinger, L. L. (1996). The lateral hypothalamic area revisited: ingestive behavior. *Neurosci. Biobehav. Rev.* **20**, 189–287.

Berridge, C. W., Stratford, T. R., Foote, S. L., and Kelley, A. E. (1997). Distribution of dopamine beta-hydroxylase-like immunoreactive fibers within the shell subregion of the nucleus accumbens. *Synapse* **27**, 230–241.

Berridge, K. C. (1996). Food reward: brain substrates of wanting and liking. *Neurosci. Biobehav. Rev.* **20**, 1–25.

Beverly, J. L., Beverly, M. F., and Meguid, M. M. (1995). Alterations in extracellular GABA in the ventral hypothalamus of rats in response to acute glucoprivation. *Am. J. Physiol.* **269**, R1174–R1178.

Beverly, J. L., De Vries, M. G., Bouman, S. D., and Arseneau, L. M. (2001). Noradrenergic and GABAergic systems in the medial hypothalamus are activated during hypoglycemia. *Am. J. Physiol.* **280**, R563–R569.

Bewick, G. A., Dhillo, W. S., Darch, S. J., Murphy, K. G., Gardiner, J. V., Jethwa, P. H., Kong, W. M., Ghatei, M. A., and Bloom, S. R. (2005). Hypothalamic cocaine- and amphetamine-regulated transcript (CART) and agouti-related protein (AgRP) neurons coexpress the NOP1 receptor and nociceptin alters CART and AgRP release. *Endocrinology* **146**, 3526–3534.

Bhatnagar, S., and Dallman, M. F. (1999). The paraventricular nucleus of the thalamus alters rhythms in core temperature and energy balance in a state-dependent manner. *Brain Res.* **851**, 66–75.

Bhatnagar, S., Viau, V., Chu, A., Soriano, L., Meijer, O. C., and Dallman, M. F. (2000). A cholecystokinin-mediated pathway to the paraventricular thalamus is recruited in chronically stressed rats and regulates hypothalamic-pituitary-adrenal function. *J. Neurosci.* **20**, 5564–5573.

Bittencourt, J. C., Presse, F., Arias, C., Peto, C., Vaughan, J., Nahon, J.-L., Vale, W., and Sawchenko, P. E. (1992). The melanin-concentrating hormone system of the rat brain: an immuno- and hybridization histochemical characterization. *J. Comp. Neurol.* **319**, 218–245.

Blevins, J. E., Stanley, B. G., and Reidelberger, R. D. (2000). Brain regions where cholecystokinin suppresses feeding in rats. *Brain Res.* **860**, 1–10.

Blundell, J. E., and Herberg, L. J. (1973). Effectiveness of lateral hypothalamic stimulation, arousal, and food deprivation in the initiation of hoarding behaviour in naive rats. *Physiol. Behav.* **10**, 763–767.

Bodnar, R. J., and Klein, G. E. (2005). Endogenous opiates and behavior: 2004. *Peptides* **26**, 2629–2711.

Bodnar, R. J., Glass, M. J., Ragnauth, A., and Cooper, M. L. (1995). General, μ and κ opioid antagonists in the nucleus accumbens alter food intake under deprivation, glucoprivic and palatable conditions. *Brain Res.* **700**, 205–212.

Bonaz, B., De Giorgio, R., and Taché, Y. (1993). Peripheral bombesin induces c-*fos* protein in the rat brain. *Brain Res.* **600**, 353–357.

Briski, K., and Gillen, E. (2001). Differential distribution of Fos expression within the male rat preoptic area and hypothalamus in response to physical vs. psychological stress. *Brain Res. Bull.* **55**, 401–408.

Broberger, C., De Lecea, L., Sutcliffe, J. G., and Hokfelt, T. (1998). Hypocretin/orexin- and melanin-concentrating hormone-expressing cells form distinct populations in the rodent lateral hypothalamus: relationship to the neuropeptide Y and agouti gene-related protein systems. *J. Comp. Neurol.* **402**, 460–474.

Brog, J. S., Salyapongse, A., Deutch, A. Y., and Zahm, D. S. (1993). The patterns of afferent innervation of the core and shell in the "accumbens" part of the rat ventral striatum: immunohistochemical detection of retrogradely transported Fluoro-gold. *J. Comp. Neurol.* **338**, 255–278.

Brown, R. E., Sergeeva, O., Eriksson, K. S., and Haas, H. L. (2001). Orexin A excites serotonergic neurons in the dorsal raphe nucleus of the rat. *Neuropharmacology* **40**, 457–459.

Bubser, M., and Deutch, A. Y. (1998). Thalamic paraventricular nucleus neurons collateralize to innervate the prefrontal cortex and nucleus accumbens. *Brain Res.* **787**, 304–310.

Cai, X. J., Evans, M. L., Lister, C. A., Leslie, R. A., Arch, J. R., Wilson, S., and Williams, G. (2001). Hypoglycemia activates orexin neurons and selectively increases hypothalamic orexin-B levels: responses inhibited by feeding and possibly mediated by the nucleus of the solitary tract. *Diabetes* **50**, 105–112.

Carlsen, J., and Heimer, L. (1986). The projection from the parataenial thalamic nucleus, as demonstrated by the phaseolus vulgaris-leucoagglutinin (PHA-L) method, identifies a subterritorial organization of the ventral striatum. *Brain Res.* **374**, 375–379.

Cerri, M., and Morrison, S. F. (2005). Activation of lateral hypothalamic neurons stimulates brown adipose tissue thermogenesis. *Neuroscience* **135**, 627–638.

Chastrette, N., Pfaff, D. W., and Gibbs, R. B. (1991). Effects of daytime and nighttime stress on Fos-like immunoreactivity in the paraventricular nucleus of the hypothalamus, the habenula, and the posterior paraventricular nucleus of the thalamus. *Brain Res.* **563**, 339–344.

Chen, D.-Y., Deutsch, J. A., Gonzalez, M. F., and Gu, Y. (1993). The induction and suppression of c-*fos* expression in the rat brain by cholecystokinin and its antagonist L364, 718. *Neurosci. Lett.* **149**, 91–94.

Chen, S., and Su, H. S. (1990). Afferent connections of the thalamic paraventricular and parataenial nuclei in the rat—a retrograde tracing study with iontophoretic application of Fluoro-Gold. *Brain Res.* **522**, 1–6.

Christie, M. J., Summers, R. J., Stephenson, J. A., Cook, C. J., and Beart, P. M. (1987). Excitatory amino acid projections to the nucleus accumbens septi in the rat: a retrograde transport study utilizing D[^{3}H]aspartate and [^{3}H]GABA. *Neuroscience* **22**, 425–439.

Christopoulos, G., Paxinos, G., Huang, X. F., Beaumont, K., Toga, A. W., and Sexton, P. M. (1995). Comparative distribution of receptors for amylin and the related peptides calcitonin gene related peptide and calcitonin in rat and monkey brain. *Can. J. Physiol. Pharmacol.* **73**, 1037–1041.

Chrobak, J. J., and Napier, T. C. (1993). Opioid and GABA modulation of accumbens-evoked ventral pallidal activity. *J. Neural Transm.* **93**, 123–143.

Churchill, L., and Kalivas, P. W. (1994). A topographically organized gamma-aminobutyric acid projection from the ventral pallidum to the nucleus accumbens in the rat. *J. Comp. Neurol.* **345**, 579–595.

Churchill, L., Dilts, R. P., and Kalivas, P. W. (1990). Changes in gamma-aminobutyric acid, mu-opioid and neurotensin receptors in the accumbens-pallidal projection after discrete quinolinic acid lesions in the nucleus accumbens. *Brain Res.* **511**, 41–54.

Cluderay, J. E., Harrison, D. C., and Hervieu, G. J. (2002). Protein distribution of the orexin-2 receptor in the rat central nervous system. *Regul. Pept.* **104**, 131–144.

Compan, V., Daszuta, A., Salin, P., Sebben, M., Bockaert, J., and Dumuis, A. (1996). Lesion study of the distribution of serotonin 5-HT4 receptors in rat basal ganglia and hippocampus. *Eur. J. Neurosci.* **8**, 2591–2598.

Cooper, S. J., and Kirkham, T. C. (1993). Opioid mechanisms in the control of food consumption and taste preferences. *In* "Handbook of Experimental Pharmacology, Volume 104/II: Opioids II" (A. Herz, ed.), pp. 239–262. Springer-Verlag, Berlin.

Date, Y., Ueta, Y., Yamashita, H., Yamaguchi, H., Matsukura, S., Kangawa, K., Sakurai, T., Yanagisawa, M., and Nakazato, M. (1999). Orexins, orexigenic hypothalamic peptides, interact with autonomic, neuroendocrine and neuroregulatory systems. *Proc. Nat. Acad. Sci. U.S.A.* **96**, 748–753.

de Lecea, L., Kilduff, T. S., Peyron, C., Gao, X., Foye, P. E., Danielson, P. E., Fukuhara, C., Battenberg, E. L., Gautvik, V. T., Bartlett II, F. S., Frankel, W. N., van den Pol, A. N., Bloom, F. E., Gautvik, K. M., and Sutcliffe, J. G. (1998). The hypocretins: hypothalamus-specific peptides with neuroexcitatory activity. *Proc. Natl. Acad. Sci. U.S.A.* **95**, 322–327.

Delfs, J. M., Zhu, Y., Druhan, J. P., and Aston-Jones, G. S. (1998). Origin of noradrenergic afferents to the shell subregion of the nucleus accumbens: anterograde and retrograde tract-tracing studies in the rat. *Brain Res.* **806**, 127–140.

Delgado, J., and Anand, B. K. (1953). Increases in food intake induced by electrical stimulation of the lateral hypothalamus. *Am. J. Physiol.* **172**, 162–168.

Di Lorenzo, P. M., and Monroe, S. (1997). Transfer of information about taste from the nucleus of the solitary tract to the parabrachial nucleus of the pons. *Brain Res.* **763**, 167–181.

Dragunow, M., and Faull, R. (1989). The use of c-fos as a metabolic marker in neuronal pathway tracing. *J. Neurosci. Methods* **29**, 261–265.

Duhault, J., Boulanger, M., Chamorro, S., Boutin, J. A., Della Zuana, O., Douillet, E., Fauchere, J. L., Feletou, M., Germain, M., Husson, B., Vega, A. M., Renard, P., and Tisserand, F. (2000). Food intake regulation in rodents: Y5 or Y1 NPY receptors or both? *Can. J. Physiol. Pharmacol.* **78**, 173–185.

Duva, M. A., Tomkins, E. M., Moranda, L. M., Kaplan, R., Sukhaseum, A., and Stanley, B. G. (2005). Origins of lateral hypothalamic afferents associated with *N*-methyl-*d*-aspartic acid-elicited eating studied using reverse microdialysis of NMDA and Fluorogold. *Neurosci. Res.* **52**, 95–106.

Echo, J. A., Lamonte, N., Christian, G., Znamensky, V., Ackerman, T. F., and Bodnar, R. J. (2001). Excitatory amino acid receptor subtype agonists induce feeding in the nucleus accumbens shell in rats: opioid antagonist actions and interactions with mu-opioid agonists. *Brain Res.* **921**, 86–97.

Elias, C. F., Lee, C., Kelly, J., Aschkenasi, C., Ahima, R. S., Couceyro, P. R., Kuhar, M. J., Saper, C. B., and Elmquist, J. K. (1998). Leptin activates hypothalamic CART neurons projecting to the spinal cord. *Neuron* **21**, 1375–1385.

Elias, C. F., Aschkenasi, C., Lee, C., Kelly, J., Ahima, R. S., Bjorbaek, C., Flier, J. S., Saper, C. B., and Elmquist, J. K. (1999). Leptin differentially regulates NPY and POMC neurons projecting to the lateral hypothalamic area. *Neuron* **23**, 775–786.

Elmquist, J. K., Ahima, R. S., Elias, C. F., Flier, J. S., and Saper, C. B. (1998). Leptin activates distinct projections from the dorsomedial and ventromedial hypothalamic nuclei. *Proc. Natl. Acad. Sci. U.S.A.* **95**, 741–746.

Evans, K. R., and Vaccarino, F. J. (1990). Amphetamine- and morphine-induced feeding: evidence for involvement of reward mechanisms. *Neurosci. Biobehav. Rev.* **14**, 9–22.

Fadel, J., and Deutch, A. Y. (2002). Anatomical substrates of orexin-dopamine interactions: lateral hypothalamic projections to the ventral tegmental area. *Neuroscience* **111**, 379–387.

Fekete, C., Wittmann, G., Liposits, Z., and Lechan, R. M. (2004). Origin of cocaine- and amphetamine-regulated transcript (CART)-immunoreactive innervation of the hypothalamic paraventricular nucleus. *J. Comp. Neurol.* **469**, 340–350.

Fernandez-Galaz, M. C., Diano, S., Horvath, T. L., and Garcia-Segura, L. M. (2002). Leptin uptake by serotonergic neurones of the dorsal raphe. *J. Neuroendocrinol.* **14**, 429–434.

Fletcher, P. J., and Davies, M. (1990). Dorsal raphe microinjection of 5-HT and indirect 5-HT agonists induces feeding in rats. *Eur. J. Pharmacol.* **184**, 265–271.

Freedman, L. J., and Cassell, M. D. (1994). Relationship of thalamic basal forebrain projection neurons to the peptidergic innervation of the midline thalamus. *J. Comp. Neurol.* **348**, 321–342.

French, S. J., and Totterdell, S. (2002). Hippocampal and prefrontal cortical inputs monosynaptically converge with individual projection neurons of the nucleus accumbens. *J. Comp. Neurol.* **446**, 151–165.

French, S. J., and Totterdell, S. (2003). Individual nucleus accumbens-projection neurons receive both basolateral amygdala and ventral subicular afferents in rats. *Neuroscience* **119**, 19–31.

Frysztak, R. J., and Neafsey, E. J. (1991). The effect of medial frontal cortex lesions on respiration, "freezing," and ultrasonic vocalizations during conditioned emotional responses in rats. *Cereb. Cortex* **1**, 418–425.

Frysztak, R. J., and Neafsey, E. J. (1994). The effect of medial frontal cortex lesions on cardiovascular conditioned emotional responses in the rat. *Brain Res.* **643**, 181–193.

Fuller, T. A., Russchen, F. T., and Price, J. L. (1987). Sources of presumptive glutamergic/aspartergic afferents to the rat ventral striatopallidal region. *J. Comp. Neurol.* **258**, 317–338.

Gao, X. B., and van den Pol, A. N. (2001). Melanin concentrating hormone depresses synaptic activity of glutamate and GABA neurons from rat lateral hypothalamus. *J. Physiol.* **533**, 237–252.

Georgescu, D., Sears, R. M., Hommel, J. D., Barrot, M., Bolanos, C. A., Marsh, D. J., Bednarek, M. A., Bibb, J. A., Maratos-Flier, E., Nestler, E. J., and DiLeone, R. J. (2005). The hypothalamic neuropeptide melanin-concentrating hormone acts in the nucleus accumbens to modulate feeding behavior and forced-swim performance. *J. Neurosci.* **25**, 2933–2940.

Glass, M. J., Billington, C. J., and Levine, A. S. (1999). Opioids and food intake: distributed functional neural pathways? *Neuropeptides* **33**, 360–368.

Glavcheva, L., Manchanda, S. K., Box, B., and Stevenson, J. A. (1972). Gastric motor activity during feeding induced by stimulation of the lateral hypothalamus in the rat. *Can. J. Physiol. Pharmacol.* **50**, 1091–1098.

Groenewegen, H. J., Berendse, H. W., and Haber, S. N. (1993). Organization of the output of the ventral striatopallidal system in the rat: ventral pallidal efferents. *Neuroscience* **57**, 113–142.

Groenewegen, H. J., Wright, C. I., Beijer, A. V., and Voorn, P. (1999). Convergence and segregation of ventral striatal inputs and outputs. *Ann. N.Y. Acad. Sci.* **877**, 49–63.

Guy, J., Vaudry, H., and Pelletier, G. (1981). Differential projections of two immunoreactive alpha-melanocyte stimulating hormone (alpha-MSH) neuronal systems in the rat brain. *Brain Res.* **220**, 199–202.

Haber, S. N., Groenewegen, H. J., Grove, E. A., and Nauta, W. J. (1985). Efferent connections of the ventral pallidum: evidence of a dual striatopallidofugal pathway. *J. Comp. Neurol.* **235**, 322–335.

Hahn, T. M., Breininger, J. F., Baskin, D. G., and Schwartz, M. W. (1998). Coexpression of Agrp and NPY in fasting-activated hypothalamic neurons. *Nat. Neurosci* **1**, 271–272.

Halonen, T., Pitkanen, A., Saano, V., and Riekkinen, P. J. (1991). Effects of vigabatrin (γ-vinyl GABA) on neurotransmission-related amino acids and on GABA and benzodiazepine receptor binding in rats. *Epilepsia* **32**, 242–249.

Harrold, J. A., Elliott, J. C., King, P. J., Widdowson, P. S., and Williams, G. (2002). Down-regulation of cannabinoid-1 (CB-1) receptors in specific extrahypothalamic regions of rats with dietary obesity: a role for endogenous cannabinoids in driving appetite for palatable food? *Brain Res.* **952**, 232–238.

Heimer, L., Zahm, D. S., Churchill, L., Kalivas, P. W., and Wohltmann, C. (1991). Specificity in the projection patterns of accumbal core and shell in the rat. *Neuroscience* **41**, 89–125.

Heimer, L., Alheid, G. F., de Olmos, J. S., Groenewegen, H. J., Haber, S. N., Harlan, R. E., and Zahm, D. S. (1997). The accumbens: beyond the core-shell dichotomy. *J. Neuropsychiatry Clin. Neurosci.* **9**, 354–381.

Herkenham, M., Lynn, A. B., de Costa, B. R., and Richfield, E. K. (1991). Neuronal localization of cannabinoid receptors in the basal ganglia of the rat. *Brain Res.* **547**, 267–274.

Herve, C., and Fellmann, D. (1997). Changes in rat melanin-concentrating hormone and dynorphin messenger ribonucleic acids induced by food deprivation. *Neuropeptides* **31**, 237–242.

Holland, P. C., and Petrovich, G. D. (2005). A neural systems analysis of the potentiation of feeding by conditioned stimuli. *Physiol. Behav.* **86**, 747–761.

Horvath, T. L., Diano, S., and van den Pol, A. N. (1999). Synaptic interaction between hypocretin (orexin) and neuropeptide Y cells in the rodent and primate hypothalamus: a novel circuit implicated in metabolic and endocrine regulations. *J. Neurosci.* **19**, 1072–1087.

Hubert, G. W., and Kuhar, M. J. (2005). Colocalization of CART with substance P but not enkephalin in the rat nucleus accumbens. *Brain Res.* **1050**, 8–14.

Hurley-Gius, K. M., and Neafsey, E. J. (1986). The medial frontal cortex and gastric motility: microstimulation results and their possible significance for the overall pattern of organization of rat frontal and parietal cortex. *Brain Res.* **365**, 241–248.

Hurley, K. M., Herbert, H., Moga, M. M., and Saper, C. B. (1991). Efferent projections of the infralimbic cortex of the rat. *J. Comp. Neurol.* **308**, 249–276.

Ikemoto, K., Satoh, K., Maeda, T., and Fibiger, H. C. (1995). Neurochemical heterogeneity of the primate nucleus accumbens. *Exp. Brain Res.* **104**, 177–190.

Inui, A. (1999). Neuropeptide Y feeding receptors: are multiple subtypes involved? *Trends Pharmacol. Sci.* **20**, 43–46.

Jain, M. R., Horvath, T. L., Kalra, P. S., and Kalra, S. P. (2000). Evidence that NPY Y1 receptors are involved in stimulation of feeding by orexins (hypocretins) in sated rats. *Regul. Pept.* **87**, 19–24.

Kandov, Y., Israel, Y., Kest, A., Dostova, I., Verasammy, J., Bernal, S. Y., Kasselman, L., and Bodnar, R. J. (2006). GABA receptor subtype antagonists in the nucleus accumbens shell and ventral tegmental area differentially alter feeding responses induced by deprivation, glucoprivation and lipoprivation in rats. *Brain Res.* in press.

Kelley, A. E., and Swanson, C. J. (1997). Feeding induced by blockade of AMPA and kainate receptors within the ventral striatum: a microinfusion mapping study. *Behav. Brain Res.* **89**, 107–113.

Kelly, J., Alheid, G. F., Newberg, A., and Grossman, S. P. (1977). GABA stimulation and blockade in the hypothalamus and midbrain: effects on feeding and locomotor activity. *Pharmacol. Biochem. Behav.* **7**, 537–541.

Kelly, J., Rothstein, J., and Grossman, S. P. (1979). GABA and hypothalamic feeding systems. I. Topographic analysis of the effects of microinjections of muscimol. *Physiol. Behav.* **23**, 1123–1134.

Khaimova, E., Kandov, Y., Israel, Y., Cataldo, G., Hadjimarkou, M. M., and Bodnar, R. J. (2004). Opioid receptor subtype antagonists differentially alter GABA agonist-induced feeding elicited from either the nucleus accumbens shell or ventral tegmental area regions in rats. *Brain Res.* **1026**, 284–294.

Khan, A. M., Cheung, H. H., Gillard, E. R., Palarca, J. A., Welsbie, D. S., Gurd, J. W., and Stanley, B. G. (2004). Lateral hypothalamic signaling mechanisms underlying feeding stimulation: differential contributions of Src family tyrosine kinases to feeding triggered either by NMDA injection or by food deprivation. *J. Neurosci.* **24**, 10603–10615.

King, B. M., Cook, J. T., Rossiter, K. N., and Rollins, B. L. (2003). Obesity-inducing amygdala lesions: examination of anterograde degeneration and retrograde transport. *Am. J. Physiol.* **284**, R965–R982.

Kirkham, T. C. (2003). Endogenous cannabinoids: a new target in the treatment of obesity. *Am. J. Physiol.* **284**, R343–R344.

Kirkham, T. C. (2005). Endocannabinoids in the regulation of appetite and body weight. *Behav. Pharmacol.* **16**, 297–313.

Kirkham, T. C., Williams, C. M., Fezza, F., and Di Marzo, V. (2002). Endocannabinoid levels in rat limbic forebrain and hypothalamus in relation to fasting, feeding and satiation: stimulation of eating by 2-arachidonoyl glycerol. *Br. J. Pharmacol.* **136**, 550–557.

Kirouac, G. J., and Ganguly, P. K. (1995). Topographical organization in the nucleus accumbens of afferents from the basolateral amygdala and efferents to the lateral hypothalamus. *Neuroscience* **67**, 625–630.

Kirouac, G. J., Parsons, M. P., and Li, S. (2005). Orexin (hypocretin) innervation of the paraventricular nucleus of the thalamus. *Brain Res.* **1059**, 179–188.

Klitenick, M. A., and Wirtshafter, D. (1989). Elicitation of feeding, drinking, and gnawing following microinjections of muscimol into the median raphe nucleus of rats. *Behav. Neural Biol.* **51**, 436–441.

Klitenick, M. A., Deutch, A. Y., Churchill, L., and Kalivas, P. W. (1992). Topography and functional role of dopaminergic projections from the ventral mesencephalic tegmentum to the ventral pallidum. *Neuroscience* **50**, 371–386.

Kolb, B. (1974). Prefrontal lesions alter eating and hoarding behavior in rats. *Physiol. Behav.* **12**, 507–511.

Kombian, S. B., Ananthalakshmi, K. V., Parvathy, S. S., and Matowe, W. C. (2004). Cholecystokinin activates CCKB receptors to excite cells and depress EPSCs in the rat rostral nucleus accumbens in vitro. *J. Physiol.* **555**, 71–84.

Kombian, S. B., Ananthalakshmi, K. V., Parvathy, S. S., and Matowe, W. C. (2005). Cholecystokinin inhibits evoked inhibitory postsynaptic currents in the rat nucleus accumbens indirectly through gamma-aminobutyric acid and gamma-aminobutyric acid type B receptors. *J. Neurosci. Res.* **79**, 412–420.

Koob, G. F., Riley, S. J., Smith, S. C., and Robbins, T. W. (1978). Effects of 6-hydroxydopamine lesions of the nucleus accumbens septi and olfactory tubercle on feeding, locomotor activity, and amphetamine anorexia in the rat. *J. Comp. Physiol. Psychol.* **92**, 917–927.

Koylu, E. O., Couceyro, P. R., Lambert, P. D., Ling, N. C., DeSouza, E. B., and Kuhar, M. J. (1997). Immunohistochemical localization of novel CART peptides in rat hypothalamus, pituitary and adrenal gland. *J. Neuroendocrinol.* **9**, 823–833.

Koylu, E. O., Couceyro, P. R., Lambert, P. D., and Kuhar, M. J. (1998). Cocaine- and amphetamine-regulated transcript peptide immunohistochemical localization in the rat brain. *J. Comp. Neurol.* **391**, 115–132.

Kristensen, P., Judge, M. E., Thim, L., Ribel, U., Christjansen, K. N., Wulff, B. S., Clausen, J. T., Jensen, P. B., Madsen, O. D., Vrang, N., Larsen, P. J., and Hastrup, S. (1998). Hypothalamic CART is a new anorectic peptide regulated by leptin. *Nature* **393**, 72–76.

Krout, K. E., and Loewy, A. D. (2000). Parabrachial nucleus projections to midline and intralaminar thalamic nuclei of the rat. *J. Comp. Neurol.* **428**, 475–494.

Lanca, A. J., De Cabo, C., Arifuzzaman, A. I., and Vaccarino, F. J. (1998). Cholecystokinergic innervation of nucleus accumbens subregions. *Peptides* **19**, 859–868.

Lawrence, C. B., Snape, A. C., Baudoin, F. M., and Luckman, S. M. (2002). Acute central ghrelin and GH secretagogues induce feeding and activate brain appetite centers. *Endocrinology* **143**, 155–162.

Legradi, G., and Lechan, R. M. (1998). The arcuate nucleus is the major source for neuropeptide Y-innervation of thyrotropin-releasing hormone neurons in the hypothalamic paraventricular nucleus. *Endocrinology* **139**, 3262–3270.

Li, X., Chadi, G., and Fuxe, K. (1995). Cholecystokinin octapeptide and the D2 antagonist raclopride induce Fos-like immunoreactivity in the shell part of the rat nucleus accumbens via different mechanisms. *Brain Res.* **684**, 225–229.

Li, Y. Q., Rao, Z. R., and Shi, J. W. (1990). Substance P-like immunoreactive neurons in the nucleus tractus solitarii of the rat send their axons to the nucleus accumbens. *Neurosci. Lett.* **120**, 194–196.

Lorens, S. A., Sorensen, J. P., and Harvey, J. A. (1970). Lesions in the nuclei accumbens septi of the rat: behavioral and neurochemical effects. *J. Comp. Physiol. Psychol.* **73**, 284–290.

Loscher, W., Brandt, C., and Ebert, U. (2003). Excessive weight gain in rats over extended kindling of the basolateral amygdala. *Neuroreport* **14**, 1829–1832.

Lupien, J. R., Tokunaga, K., Kemnitz, J. W., Groos, E., and Bray, G. A. (1986). Lateral hypothalamic lesions and fenfluramine increase thermogenesis in brown adipose tissue. *Physiol. Behav.* **38**, 15–20.

Majeed, N. H., Przewlocka, B., Wedzony, K., and Przewlocki, R. (1986). Stimulation of food intake following opioid microinjection into the nucleus accumbens septi in rats. *Peptides* **7**, 711–716.

Maldonado-Irizarry, C. S., and Kelley, A. E. (1995). Excitotoxic lesions of the core and shell subregions of the nucleus accumbens differentially disrupt body weight regulation and motor activity in rat. *Brain Res. Bull.* **38**, 551–559.

Maldonado-Irizarry, C. S., Swanson, C. J., and Kelley, A. E. (1995). Glutamate receptors in the nucleus accumbens shell control feeding behavior via the lateral hypothalamus. *J. Neurosci.* **15**, 6779–6788.

Marcus, J. N., Aschkenasi, C. J., Lee, C. E., Chemelli, R. M., Saper, C. B., Yanagisawa, M., and Elmquist, J. K. (2001). Differential expression of orexin receptors 1 and 2 in the rat brain. *J. Comp. Neurol.* **435**, 6–25.

Martin, G., Fabre, V., Siggins, G. R., and de Lecea, L. (2002). Interaction of the hypocretins with neurotransmitters in the nucleus accumbens. *Regul. Pept.* **104**, 111–117.

Matyas, F., Yanovsky, Y., Mackie, K., Kelsch, W., Misgeld, U., and Freund, T. F. (2006). Subcellular localization of type 1 cannabinoid receptors in the rat basal ganglia. *Neuroscience* **137**, 337–361.

Mendoza, J., Angeles-Castellanos, M., and Escobar, C. (2005). A daily palatable meal without food deprivation entrains the suprachiasmatic nucleus of rats. *Eur. J. Neurosci.* **22**, 2855–2862.

Meredith, G. E., Pennartz, C. M., and Groenewegen, H. J. (1993). The cellular framework for chemical signaling in the nucleus accumbens. *Prog. Brain Res.* **99**, 3–24.

Meredith, G. E., Pattiselanno, A., Groenewegen, H. J., and Haber, S. N. (1996). Shell and core in monkey and human nucleus accumbens identified with antibodies to calbindin-D28k. *J. Comp. Neurol.* **365**, 628–639.

Mitrovic, I., and Napier, T. C. (1998). Substance P attenuates and DAMGO potentiates amygdala glutamatergic neurotransmission within the ventral pallidum. *Brain Res.* **792**, 193–206.

Moga, M. M., Weis, R. P., and Moore, R. Y. (1995). Efferent projections of the paraventricular thalamic nucleus in the rat. *J. Comp. Neurol.* **359**, 221–238.

Mogenson, G. J., and Wu, M. (1988). Disruption of food hoarding by injections of procaine into mediodorsal thalamus, GABA into subpallidal region and haloperidol into accumbens. *Brain Res. Bull.* **20**, 247–251.

Moriguchi, T., Sakurai, T., Nambu, T., Yanagisawa, M., and Goto, K. (1999). Neurons containing orexin in the lateral hypothalamic area of the adult rat brain are activated by insulin-induced acute hypoglycemia. *Neurosci. Lett.* **264**, 101–104.

Mucha, R. F., and Iversen, S. D. (1986). Increased food intake after opioid microinjections into nucleus accumbens and ventral tegmental area of rat. *Brain Res.* **397**, 214–224.

Mullett, M. A., Billington, C. J., Levine, A. S., and Kotz, C. M. (2000). Hypocretin I in the lateral hypothalamus activates key feeding-regulatory brain sites. *Neuroreport* **11**, 103–108.

Napier, T. C., Mitrovic, I., Churchill, L., Klitenick, M. A., Lu, X. Y., and Kalivas, P. W. (1995). Substance P in the ventral pallidum: projection from the ventral striatum, and electrophysiological and behavioral consequences of pallidal substance P. *Neuroscience* **69**, 59–70.

Nauta, W. J., Smith, G. P., Faull, R. L., and Domesick, V. B. (1978). Efferent connections and nigral afferents of the nucleus accumbens septi in the rat. *Neuroscience* **3**, 385–401.

Neafsey, E. J. (1990). Prefrontal cortical control of the autonomic nervous system: anatomical and physiological observations. *Prog. Brain Res.* **85**, 147–165; discussion 165–166.

Niimi, M., Sato, M., Tamaki, M., Wada, Y., Takahara, J., and Kawanishi, K. (1995). Induction of Fos protein in the rat hypothalamus elicited by insulin-induced hypoglycemia. *Neurosci. Res.* **23**, 361–364.

Nikulina, E. M. (1991). Neural control of predatory aggression in wild and domesticated animals. *Neurosci. Biobehav. Rev.* **15**, 545–547.

O'Donnell, P., and Grace, A. A. (1993). Physiological and morphological properties of accumbens core and shell neurons recorded in vitro. *Synapse* **13**, 135–160.

O'Donnell, P., and Grace, A. A. (1995). Synaptic interactions among excitatory afferents to nucleus accumbens neurons: hippocampal gating of prefrontal cortical input. *J. Neurosci.* **15**, 3622–3639.

O'Donnell, P., Lavin, A., Enquist, L. W., Grace, A. A., and Card, J. P. (1997). Interconnected parallel circuits between rat nucleus accumbens and thalamus revealed by retrograde transsynaptic transport of pseudorabies virus. *J. Neurosci.* **17**, 2143–2167.

O'Donnell, P., Greene, J., Pabello, N., Lewis, B. L., and Grace, A. A. (1999). Modulation of cell firing in the nucleus accumbens. *Ann. N.Y. Acad. Sci.* **877**, 157–175.

Osei-Hyiaman, D., Depetrillo, M., Harvey-White, J., Bannon, A. W., Cravatt, B. F., Kuhar, M. J., Mackie, K., Palkovits, M., and Kunos, G. (2005). Cocaine- and amphetamine-related transcript peptide is involved in the orexigenic effect of endogenous anandamide. *Neuroendocrinology* **81**, 273–282.

Otake, K. (2005). Cholecystokinin and substance P immunoreactive projections to the paraventricular thalamic nucleus in the rat. *Neurosci. Res.* **51**, 383–394.

Otake, K., and Nakamura, Y. (1998). Single midline thalamic neurons projecting to both the ventral striatum and the prefrontal cortex in the rat. *Neuroscience* **86**, 635–649.

Owens, N. C., Sartor, D. M., and Verberne, A. J. (1999). Medial prefrontal cortex depressor response: role of the solitary tract nucleus in the rat. *Neuroscience* **89**, 1331–1346.

Parker, R. M., and Herzog, H. (1999). Regional distribution of Y-receptor subtype mRNAs in rat brain. *Eur. J. Neurosci.* **11**, 1431–1448.

Parsons, M. P., Li, S., and Kirouac, G. J. (2006). The paraventricular nucleus of the thalamus as an interface between the orexin and CART peptides and the shell of the nucleus accumbens. *Synapse* **59**, 480–490.

Pecina, S., and Berridge, K. C. (2000). Opioid site in nucleus accumbens shell mediates eating and hedonic "liking" for food: map based on microinjection Fos plumes. *Brain Res.* **863**, 71–86.

Pennartz, C. M., Dolleman-Van der Weel, M. J., and Lopes da Silva, F. H. (1992). Differential membrane properties and dopamine effects in the shell and core of the rat nucleus accumbens studied in vitro. *Neurosci. Lett.* **136**, 109–112.

Pereira de Vasconcelos, A., Bartol-Munier, I., Feillet, C. A., Gourmelen, S., Pevet, P., and Challet, E. (2006). Modifications of local cerebral glucose utilization during circadian food-anticipatory activity. *Neuroscience*.

Petrovich, G. D., Canteras, N. S., and Swanson, L. W. (2001). Combinatorial amygdalar inputs to hippocampal domains and hypothalamic behavior systems. *Brain Res. Brain Res. Rev.* **38**, 247–289.

Petrovich, G. D., Setlow, B., Holland, P. C., and Gallagher, M. (2002). Amygdalo-hypothalamic circuit allows learned cues to override satiety and promote eating. *J. Neurosci.* **22**, 8748–8753.

Peyron, C., Tighe, D. K., van den Pol, A. N., de Lecea, L., Heller, H. C., Sutcliffe, J. G., and Kilduff, T. S. (1998). Neurons containing hypocretin (orexin) project to multiple neuronal systems. *J. Neurosci.* **18**, 9996–10015.

Phillipson, O. T., and Griffiths, A. C. (1985). The topographic order of inputs to nucleus accumbens in the rat. *Neuroscience* **16**, 275–296.

Pickel, V. M., Chan, J., Kash, T. L., Rodriguez, J. J., and MacKie, K. (2004). Compartment-specific localization of cannabinoid 1 (CB1) and micro-opioid receptors in rat nucleus accumbens. *Neuroscience* **127**, 101–112.

Pickel, V. M., Chan, J., Kearn, C. S., and Mackie, K. (2006). Targeting dopamine D2 and cannabinoid-1 (CB1) receptors in rat nucleus accumbens. *J. Comp. Neurol.* **495**, 299–313.

Pistis, M., Muntoni, A. L., Pillolla, G., and Gessa, G. L. (2002). Cannabinoids inhibit excitatory inputs to neurons in the shell of the nucleus accumbens: an in vivo electrophysiological study. *Eur. J. Neurosci.* **15**, 1795–1802.

Rada, P., Tucci, S., Murzi, E., and Hernandez, L. (1997). Extracellular glutamate increases in the lateral hypothalamus and decreases in the nucleus accumbens during feeding. *Brain Res.* **768**, 338–340.

Rada, P., Mendialdua, A., Hernandez, L., and Hoebel, B. G. (2003). Extracellular glutamate increases in the lateral hypothalamus during meal initiation, and GABA peaks during satiation: microdialysis measurements every 30 s. *Behav. Neurosci.* **117**, 222–227.

Ragnauth, A., Moroz, M., and Bodnar, R. J. (2000). Multiple opioid receptors mediate feeding elicited by mu and delta opioid receptor subtype agonists in the nucleus accumbens shell in rats. *Brain Res.* **876**, 76–87.

Ragozzino, M. E., Wilcox, C., Raso, M., and Kesner, R. P. (1999). Involvement of rodent prefrontal cortex subregions in strategy switching. *Behav. Neurosci.* **113**, 32–41.

Ragozzino, M. E., Kim, J., Hassert, D., Minniti, N., and Kiang, C. (2003). The contribution of the rat prelimbic-infralimbic areas to different forms of task switching. *Behav. Neurosci.* **117**, 1054–1065.

Recabarren, M. P., Valdes, J. L., Farias, P., Seron-Ferre, M., and Torrealba, F. (2005). Differential effects of infralimbic cortical lesions on temperature and locomotor activity responses to feeding in rats. *Neuroscience* **134**, 1413–1422.

Reynolds, S. M., and Berridge, K. C. (2001). Fear and feeding in the nucleus accumbens shell: rostrocaudal segregation of GABA-elicited defensive behavior versus eating behavior. *J. Neurosci.* **21**, 3261–3270.

Reynolds, S. M., and Berridge, K. C. (2002). Positive and negative motivation in nucleus accumbens shell: bivalent rostrocaudal gradients for GABA-elicited eating, taste "liking"/"disliking" reactions, place preference/avoidance, and fear. *J. Neurosci.* **22**, 7308–7320.

Reynolds, S. M., and Berridge, K. C. (2003). Glutamate motivational ensembles in nucleus accumbens: rostrocaudal shell gradients of fear and feeding. *Eur. J. Neurosci.* **17**, 2187–2200.

Risold, P. Y., Thompson, R. H., and Swanson, L. W. (1997). The structural organization of connections between hypothalamus and cerebral cortex. *Brain Res. Brain Res. Rev.* **24**, 197–254.

Risterucci, C., Terramorsi, D., Nieoullon, A., and Amalric, M. (2003). Excitotoxic lesions of the prelimbic-infralimbic areas of the rodent prefrontal cortex disrupt motor preparatory processes. *Eur. J. Neurosci.* **17**, 1498–1508.

Robbe, D., Alonso, G., Duchamp, F., Bockaert, J., and Manzoni, O. J. (2001). Localization and mechanisms of action of cannabinoid receptors at the glutamatergic synapses of the mouse nucleus accumbens. *J. Neurosci.* **21**, 109–116.

Roberts, K. A., Krebs, L. T., Kramar, E. A., Shaffer, M. J., Harding, J. W., and Wright, J. W. (1995). Autoradiographic identification of brain angiotensin IV binding sites and differential c-Fos expression following intracerebroventricular injection of angiotensin II and IV in rats. *Brain Res.* **682**, 13–21.

Roberts, W. W. (1969). Are hypothalamic motivational mechanisms functionally and anatomically specific? *Brain Behav. Evol.* **2**, 317–342.

Roberts, W. W. (1980). [^{14}C]deoxyglucose mapping of first-order projections activated by stimulation of lateral hypothalamic sites eliciting gnawing, eating, and drinking in rats. *J. Comp. Neurol.* **194**, 617–638.

Rollins, B. L., and King, B. M. (2000). Amygdala-lesion obesity: what is the role of the various amygdaloid nuclei? *Am. J. Physiol.* **279**, R1348–R1356.

Rollins, B. L., Stines, S. G., McGuire, H. B., and King, B. M. (2001). Effects of amygdala lesions on body

weight, conditioned taste aversion, and neophobia. *Physiol. Behav.* **72**, 735–742.

Saito, Y., Cheng, M., Leslie, F. M., and Civelli, O. (2001). Expression of the melanin-concentrating hormone (MCH) receptor mRNA in the rat brain. *J. Comp. Neurol.* **435**, 26–40.

Sakurai, T. (2005). Roles of orexin/hypocretin in regulation of sleep/wakefulness and energy homeostasis. *Sleep Med. Rev.* **9**, 231–241.

Sakurai, T., Amemiya, A., Ishii, M., Matsuzaki, I., Chemelli, R. M., Tanaka, H., Williams, S. C., Richardson, J. A., Kozlowski, G. P., Wilson, S., Arch, J. R., Buckingham, R. E., Haynes, A. C., Carr, S. A., Annan, R. S., McNulty, D. E., Liu, W. S., Terrett, J. A., Elshourbagy, N. A., Bergsma, D. J., and Yanagisawa, M. (1998). Orexins and orexin receptors: a family of hypothalamic neuropeptides and G protein-coupled receptors that regulate feeding behavior. *Cell* **92**, 573–585.

Saul'skaya, N. B., and Mikhailova, M. O. (2002). Feeding-induced decrease in extracellular glutamate level in the rat nucleus accumbens: dependence on glutamate uptake. *Neuroscience* **112**, 791–801.

Schallert, T., Leach, L. R., and Braun, J. J. (1978). Saliva hypersecretion during aphagia following lateral hypothalamic lesions. *Physiol. Behav.* **21**, 461–463.

Schwartz, M. W., Woods, S. C., Porte, D., Jr., Seeley, R. J., and Baskin, D. G. (2000). Central nervous system control of food intake. *Nature* **404**, 661–671.

Sesack, S. R., Deutch, A. Y., Roth, R. H., and Bunney, B. S. (1989). Topographical organization of the efferent projections of the medial prefrontal cortex in the rat: an anterograde tract-tracing study with phaseolus vulgaris leucoagglutinin. *J. Comp. Neurol.* **290**, 213–242.

Sheng, M., and Greenberg, M. E. (1990). The regulation and function of c-fos and other immediate early genes in the nervous system. *Neuron* **4**, 477–485.

Shiraishi, T. (1988). Hypothalamic control of gastric acid secretion. *Brain Res. Bull.* **20**, 791–797.

Siegel, A., and Schubert, K. (1995). Neurotransmitters regulating feline aggressive behavior. *Rev. Neurosci.* **6**, 47–61.

Sills, T. L., and Vaccarino, F. J. (1991). Facilitation and inhibition of feeding by a single dose of amphetamine: relationship to baseline intake and accumbens cholecystokinin. *Psychopharmacology (Berlin)* **105**, 329–334.

Sills, T. L., and Vaccarino, F. J. (1996). Individual differences in the feeding response to CCKB antagonists: role of the nucleus accumbens. *Peptides* **17**, 593–599.

Silveira, M. C., Sandner, G., and Graeff, F. G. (1993). Induction of Fos immunoreactivity in the brain by exposure to the elevated plus-maze. *Behav. Brain Res.* **56**, 115–118.

Smith, K. S., and Berridge, K. C. (2005). The ventral pallidum and hedonic reward: neurochemical maps of sucrose "liking" and food intake. *J. Neurosci.* **25**, 8637–8649.

Smith, W. J., Stewart, J., and Pfaus, J. G. (1997a). Tail pinch induces fos immunoreactivity within several regions of the male rat brain: effects of age. *Physiol. Behav.* **61**, 717–723.

Smith, Y., Koylu, E. O., Couceyro, P., and Kuhar, M. J. (1997b). Ultrastructural localization of CART (cocaine- and amphetamine-regulated transcript) peptides in the nucleus accumbens of monkeys. *Synapse* **27**, 90–94.

Stanley, B. G., and Leibowitz, S. F. (1984). Neuropeptide Y: stimulation of feeding and drinking by injection into the paraventricular nucleus. *Life Sci.* **35**, 2635–2642.

Stanley, B. G., Ha, L. H., Spears, L. C., and Dee, M. G. (1993a). Lateral hypothalamic injections of glutamate, kainic acid, *d,l*-α-amino3-hydroxy-5-methylisoxazole propionic acid or *N*-methyl-*d*-aspartic acid rapidly elicit intense transient eating in rats. *Brain Res.* **613**, 88–95.

Stanley, B. G., Willett, V. L., Donias, H. W., Ha, L. H., and Spears, L. C. (1993b). The lateral hypothalamus: a primary site mediating excitatory amino acid-elicited eating. *Brain Res.* **630**, 41–49.

Stanley, B. G., Willett, V. L., Donias, H. W., Dee, M. G., and Duva, M. A. (1996). Lateral hypothalamic NMDA receptors and glutamate as physiological mediators of eating and weight control. *Am. J. Physiol.* **270**, R443–R449.

Stratford, T. R. (2005). Activation of feeding-related neural circuitry after unilateral injections of muscimol into the nucleus accumbens shell. *Brain Res.* **1048**, 241–250.

Stratford, T. R., and Gosnell, B. A. (2000). Activation of GABA$_A$ receptors in the nucleus accumbens shell increases intake of a sweetened ethanol solution, but suppresses intake of unsweetened ethanol. *Soc. Neurosci. Abstr.* **26**, 990.

Stratford, T. R., and Kelley, A. E. (1997). GABA in the nucleus accumbens shell participates in the central regulation of feeding behavior. *J. Neurosci.* **17**, 4434–4440.

Stratford, T. R., and Kelley, A. E. (1999). Evidence of a functional relationship between the nucleus accumbens shell and lateral hypothalamus subserving the control of feeding behavior. *J. Neurosci.* **19**, 11040–11048.

Stratford, T. R., and Wirtshafter, D. (1990). Ascending dopaminergic projections from the dorsal raphe nucleus in the rat. *Brain Res.* **511**, 173–176.

Stratford, T. R., and Wirtshafter, D. (2000). Forebrain lesions differentially affect drinking elicited by dipsogenic challenges and injections of muscimol into the median raphe nucleus. *Behav. Neurosci.* **114**, 760–771.

Stratford, T. R., and Wirtshafter, D. (2004). NPY mediates the feeding elicited by muscimol injections into

the nucleus accumbens shell. *Neuroreport* **15**, 2673–2676.

Stratford, T. R., Holahan, M. R., and Kelley, A. E. (1997). Microinjections of nociceptin into the nucleus accumbens shell or ventromedial hypothalamic nucleus increase food intake in the rat. *Neuroreport* **8**, 423–426.

Stratford, T. R., Swanson, C. J., and Kelley, A. E. (1998). Specific changes in food intake elicited by blockade or activation of glutamate receptors in the nucleus accumbens shell. *Behav. Brain Res.* **93**, 43–50.

Stratford, T. R., Kelley, A. E., and Simansky, K. J. (1999). Blockade of GABA$_A$ receptors in the medial ventral pallidum elicits feeding in satiated rats. *Brain Res.* **825**, 199–203.

Stricker, E. M. (1976). Drinking by rats after lateral hypothalamic lesions: a new look at the lateral hypothalamic syndrome. *J. Comp. Physiol. Psychol.* **90**, 127–143.

Thorpe, A. J., and Kotz, C. M. (2005). Orexin A in the nucleus accumbens stimulates feeding and locomotor activity. *Brain Res.* **1050**, 156–162.

Trivedi, P., Yu, H., MacNeil, D. J., Van der Ploeg, L. H., and Guan, X. M. (1998). Distribution of orexin receptor mRNA in the rat brain. *FEBS Lett.* **438**, 71–75.

Tsou, K., Brown, S., Sanudo-Pena, M. C., Mackie, K., and Walker, J. M. (1998). Immunohistochemical distribution of cannabinoid CB1 receptors in the rat central nervous system. *Neuroscience* **83**, 393–411.

Tsujii, S., and Bray, G. A. (1991). GABA-related feeding control in genetically obese rats. *Brain Res.* **540**, 48–54.

Turton, M. D., Oshea, D., Gunn, I., Beak, S. A., Edwards, C. M. B., Meeran, K., Choi, S. J., Taylor, G. M., Heath, M. M., Lambert, P. D., Wilding, J. P. H., Smith, D. M., Ghatei, M. A., Herbert, J., and Bloom, S. R. (1996). A role for glucagon-like peptide-1 in the central regulation of feeding. *Nature* **379**, 69–72.

Usuda, I., Tanaka, K., and Chiba, T. (1998). Efferent projections of the nucleus accumbens in the rat with special reference to subdivision of the nucleus: biotinylated dextran amine study. *Brain Res.* **797**, 73–93.

Valdes, J. L., Maldonado, P., Recabarren, M., Fuentes, R., and Torrealba, F. (2006). The infralimbic cortical area commands the behavioral and vegetative arousal during appetitive behavior in the rat. *Eur. J. Neurosci.* **23**, 1352–1364.

Valenstein, E. S., Cox, V. C., and Kakolewski, J. W. (1970). Reexamination of the role of the hypothalamus in motivation. *Psychol. Rev.* **77**, 16–31.

van den Bos, R., and Cools, A. R. (1991). Motor activity and the GABA$_A$-receptor in the ventral pallidum/substantia innominata complex. *Neurosci. Lett.* **124**, 246–250.

Vaughan, C. W., and Christie, M. J. (1996). Increase by the ORL(1) receptor (opioid receptor-like(1)) ligand, nociceptin, of inwardly rectifying K conductance in dorsal raphe nucleus neurones. *Br. J. Pharmacol.* **117**, 1609–1611.

Verberne, A. J., and Owens, N. C. (1998). Cortical modulation of the cardiovascular system. *Prog. Neurobiol.* **54**, 149–168.

Vertes, R. P. (1991). A PHA-L analysis of ascending projections of the dorsal raphe nucleus in the rat. *J. Comp. Neurol.* **313**, 643–668.

Vertes, R. P. (2004). Differential projections of the infralimbic and prelimbic cortex in the rat. *Synapse* **51**, 32–58.

Waldbillig, R. J. (1975). Attack, eating, drinking, and gnawing elicited by electrical stimulation of rat mesencephalon and pons. *J. Comp. Physiol. Psychol.* **89**, 200–212.

Walton, M. E., Bannerman, D. M., and Rushworth, M. F. (2002). The role of rat medial frontal cortex in effort-based decision making. *J. Neurosci.* **22**, 10996–11003.

Wang, Z.-J., Rao, Z.-R., and Shi, J.-W. (1992). Tyrosine hydroxylase-, neurotensin-, or cholecystokinin-containing neurons in the nucleus tractus solitarii send projection fibers to the nucleus accumbens in the rat. *Brain Res.* **578**, 347–350.

Ward, B. O., Somerville, E. M., and Clifton, P. G. (2000). Intraaccumbens baclofen selectively enhances feeding behavior in the rat. *Physiol. Behav.* **68**, 463–468.

Will, M. J., Franzblau, E. B., and Kelley, A. E. (2004). The amygdala is critical for opioid-mediated binge eating of fat. *Neuroreport* **15**, 1857–1860.

Williams, C. M., and Kirkham, T. C. (2002). Observational analysis of feeding induced by Delta9-THC and anandamide. *Physiol. Behav.* **76**, 241–250.

Williams, D. J., Crossman, A. R., and Slater, P. (1977). The efferent projections of the nucleus accumbens in the rat. *Brain Res.* **130**, 217–227.

Williams, G., Bing, C., Cai, X. J., Harrold, J. A., King, P. J., and Liu, X. H. (2001). The hypothalamus and the control of energy homeostasis: different circuits, different purposes. *Physiol. Behav.* **74**, 683–701.

Willie, J. T., Chemelli, R. M., Sinton, C. M., and Yanagisawa, M. (2001). To eat or to sleep? Orexin in the regulation of feeding and wakefulness. *Annu. Rev. Neurosci.* **24**, 429–458.

Wirtshafter, D. (2001). The control of ingestive behavior by the median raphe nucleus. *Appetite* **36**, 99–105.

Wirtshafter, D., Stratford, T. R., and Pitzer, M. R. (1993). Studies on the behavioral activation produced by stimulation of GABA$_B$ receptors in the median raphe nucleus. *Behav. Brain Res.* **59**, 83–93.

Wolak, M. L., DeJoseph, M. R., Cator, A. D., Mokashi, A. S., Brownfield, M. S., and Urban, J. H. (2003). Comparative distribution of neuropeptide Y Y1 and Y5 receptors in the rat brain by using immunohistochemistry. *J. Comp. Neurol.* **464**, 285–311.

Wright, C. I., and Groenewegen, H. J. (1995). Patterns of convergence and segregation in the medial nucleus accumbens of the rat: relationships of pre-

frontal cortical, midline thalamic, and basal amygdaloid afferents. *J. Comp. Neurol.* **361**, 383–403.

Wright, C. I., Beijer, A. V., and Groenewegen, H. J. (1996). Basal amygdaloid complex afferents to the rat nucleus accumbens are compartmentally organized. *J. Neurosci.* **16**, 1877–1893.

Xie, X., Crowder, T. L., Yamanaka, A., Morairty, S. R., Lewinter, R. D., Sakurai, T., and Kilduff, T. S. (2006). GABA$_B$ receptor-mediated modulation of hypocretin/orexin neurones in mouse hypothalamus. *J. Physiol.*

Yamanaka, A., Kunii, K., Nambu, T., Tsujino, N., Sakai, A., Matsuzaki, I., Miwa, Y., Goto, K., and Sakurai, T. (2000). Orexin-induced food intake involves neuropeptide Y pathway. *Brain Res.* **859**, 404–409.

Yang, S. C., Shieh, K. R., and Li, H. Y. (2005). Cocaine- and amphetamine-regulated transcript in the nucleus accumbens participates in the regulation of feeding behavior in rats. *Neuroscience.*

Yeomans, M. R., and Gray, R. W. (2002). Opioid peptides and the control of human ingestive behaviour. *Neurosci. Biobehav. Rev.* **26**, 713–728.

Zaborszky, L., Alheid, G. F., Beinfeld, M. C., Eiden, L. E., Heimer, L., and Palkovits, M. (1985). Cholecystokinin innervation of the ventral striatum: a morphological and radioimmunological study. *Neuroscience* **14**, 427–453.

Zahm, D. S., and Heimer, L. (1990). Two transpallidal pathways originating in the rat nucleus accumbens. *J. Comp. Neurol.* **302**, 437–446.

Zahm, D. S., and Brog, J. S. (1992). On the significance of subterritories in the "accumbens" part of the rat ventral striatum. *Neuroscience* **50**, 751–767.

Zahm, D. S., Williams, E., and Wohltmann, C. (1996). Ventral striatopallidothalamic projection: IV. Relative involvements of neurochemically distinct subterritories in the ventral pallidum and adjacent parts of the rostroventral forebrain. *J. Comp. Neurol.* **364**, 340–362.

Zahm, D. S., Jensen, S. L., Williams, E. S., and Martin, J. R. (1999). Direct comparison of projections from the central amygdaloid region and nucleus accumbens shell. *Eur. J. Neurosci.* **11**, 1119–1126.

Zahm, D. S., Williams, E. A., Latimer, M. P., and Winn, P. (2001). Ventral mesopontine projections of the caudomedial shell of the nucleus accumbens and extended amygdala in the rat: double dissociation by organization and development. *J. Comp. Neurol.* **436**, 111–125.

Zhang, M., and Kelley, A. E. (1997). Opiate agonists microinjected into the nucleus accumbens enhance sucrose drinking in rats. *Psychopharmacology (Berlin)* **132**, 350–360.

Zhang, M., and Kelley, A. E. (2000). Enhanced intake of high-fat food following striatal mu-opioid stimulation: microinjection mapping and fos expression. *Neuroscience* **99**, 267–277.

Zhang, M., and Kelley, A. E. (2002). Intake of saccharin, salt, and ethanol solutions is increased by infusion of a mu opioid agonist into the nucleus accumbens. *Psychopharmacology* (*Berlin*) **159**, 415–423.

Zhang, M., Gosnell, B. A., and Kelley, A. E. (1998). Intake of high-fat food is selectively enhanced by mu opioid receptor stimulation within the nucleus accumbens. *J. Pharmacol. Exp. Ther.* **285**, 908–914.

Zheng, H., Corkern, M., Stoyanova, I., Patterson, L. M., Tian, R., and Berthoud, H. R. (2003). Peptides that regulate food intake: appetite-inducing accumbens manipulation activates hypothalamic orexin neurons and inhibits POMC neurons. *Am. J. Physiol.* **284**, R1436–R1444.

Zhou, L., Furuta, T., and Kaneko, T. (2003). Chemical organization of projection neurons in the rat accumbens nucleus and olfactory tubercle. *Neuroscience* **120**, 783–798.

Znamensky, V., Echo, J. A., Lamonte, N., Christian, G., Ragnauth, A., and Bodnar, R. J. (2001). gamma-Aminobutyric acid receptor subtype antagonists differentially alter opioid-induced feeding in the shell region of the nucleus accumbens in rats. *Brain Res.* **906**, 84–91.

4

Hypothalamic Neuropeptides and Feeding Regulation

BERNARD BECK
Neurocal Group—Faculté de Médecine, INSERM U. 724

I. Introduction
II. Neuropeptide Y
   A. Neuropeptide Y Is a Potent Orexigenic Peptide
   B. NPY, Macronutrient Intake, and Dietary Preferences
   C. NPY, Feeding State, and Hormonal and Metabolic Control
   D. NPY and Obesity
   E. NPY and Antiobesity Treatments
III. The Orexins
   A. Orexins Stimulate Food Intake
   B. Orexins–NPY Interactions
   C. Factors of Regulation of the OX System
   D. Interactions with Other Neuromodulators
   E. Orexins and Obesity
   F. Orexins and Antiobesity Treatment
IV. Melanin-Concentrating Hormone
   A. MCH Is an Orexigenic Peptide
   B. Factors Affecting Hypothalamic MCH
   C. Interactions with Other Feeding-Related Peptides
   D. MCH and Obesity
   E. MCH and Antiobesity Treatment
V. Conclusion
   Acknowledgments
   References

This chapter focuses on three neuropeptides, neuropeptide Y, melanin-concentrating hormone, and orexins, that stimulate food intake. Their variations in different feeding states and interactions with hormones such as leptin are described.

All three neuropeptidergic systems are dysregulated in a variety of animal models of obesity, including obesity linked to defective leptin signaling and diet-induced obesity. The specific receptors through which these hypothalamic neuropeptides regulate food intake and body weight constitute potential targets for antiobesity drugs.

I. INTRODUCTION

Feeding behavior is a very complex behavior that is regulated by many pathways involving both the central nervous system and the periphery. Feeding has been studied intensively during the last 50 years, in association with the growing prevalence in almost all parts of the world of obesity and its adverse metabolic disorders. Special attention has centered on the role of the hypothalamus, where multiple pathways form regulatory networks to integrate all parameters linked to energy intake and expenditure. These networks incorporate several interconnected hypothalamic nuclei including the ventromedial nucleus (VMN) and lateral hypothalamus (LH), previously considered as respective satiety and feeding "centers"; and also the arcuate (Arc), dorsomedial (DMN), and paraventricular (PVN) nuclei. These areas are connected with the hindbrain (nucleus of the solitary tract, area postrema, etc.) and, in the forebrain, to the limbic system, allowing integration of additional, nonenergetic information.

Numerous neuromodulators are present in these areas. They include classic neurotransmitters such as serotonin, GABA (γ-aminobutyric acid), or dopamine (Maitre, 1997; Meguid et al., 2000), molecules derived from fatty acids like endocannabinoids (DiMarzo and Matias, 2005) or neuropeptides. The neuropeptides have been well-studied during the last 20 years. They are very numerous and can be divided into stimulatory (orexigenic) or inhibitory (anorexigenic) peptides (Fig. 1). Their effects on food intake are very variable in intensity and they often function in a

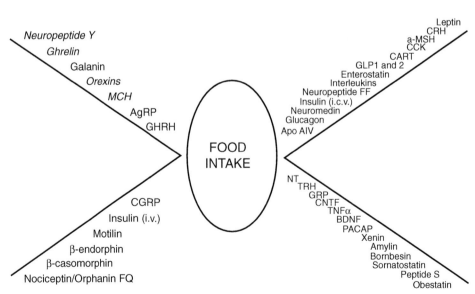

FIGURE 1 Stimulatory peptides (placed around the >0 sign; left part of the figure) and inhibitory peptides (placed around the <0 sign; right part of the figure) involved in food intake regulation. AgRP: agouti-related peptide; CART: cocaine- and amphetamine-related peptide; GHRH: growth hormone-releasing hormone; MCH: melanin-concentrating hormone; TRH: thyrotropin-releasing hormone; α-MSH: alpha-melanocyte-stimulating hormone; CNTF: ciliary neurotrophic factor; BDNF: brain-derived neurotrophic factor; CGRP: calcitonin gene-related peptide; PACAP: pituitary adenylate cyclase-activated peptide; GLP: glucagon-like peptide; NT: neurotensin; TNFα: tumor necrosis factor alpha; Apo AIV: apolipoprotein AIV; CRH: corticotropin-releasing hormone; CCK: cholecystokinin.

synergistic or antagonistic manner. This chapter will focus on three of these peptides: neuropeptide Y (NPY), which is the most potent stimulator of food intake; orexins (OX), and melanin-concentrating hormone (MCH). These three peptides share the common characteristic of being intensively studied in pharmacology departments and pharmaceutical companies in order to develop drugs against obesity. Other peptides are important in the regulation of feeding but have yet to receive the same level of attention in drug research in this domain. Information about these peptides can be found in other chapters, and in recent reviews on cholecystokinin (Rehfeld, 2004), ghrelin (Kojima et al., 2004), galanin (Crawley, 1999), corticotropin-releasing hormone (Richard et al., 2002), or melanocortins (Cone, 2005).

II. NEUROPEPTIDE Y

Neuropeptide Y is one of the most abundant peptides found in the brain, although it is also present in the peripheral nervous system (Allen et al., 1987; Everitt et al., 1984; Zukowska et al., 2003). That its amino acid composition has been well-conserved during evolution (Larhammar, 1996) suggests that NPY is particularly important for the regulation of basic physiological functions such as feeding behavior.

In the brain, NPY is almost ubiquitous and is found in abundance in the hypothalamus (Chronwall et al., 1985). It is synthesized by a large population of neurons in the arcuate nucleus (Arc) (Chronwall, 1985; Beck, 2005), and is colocalized with another orexigenic peptide, agouti-related-peptide (AGRP) (Hahn et al., 1998). Arcuate NPY neurons project to the paraventricular nucleus (PVN) but also to the dorsomedial nucleus (DMN) and the median preoptic area (Bai et al., 1985; Kerkerian and Pelle). The PVN also receives projections from NPY-synthesizing neurons

present in the brainstem (Sahu et al., 1988a) particularly in the A1 and C1 areas (Sawchenko et al., 1985).

A. Neuropeptide Y Is a Potent Orexigenic Peptide

When NPY is injected in the rat brain either in ventricles or in different hypothalamic nuclei, it induces robust feeding (Clark et al., 1984; Levine and Morley, 1984; Stanley and Leibowitz, 1984), at any time during the dark–light cycle (Tempel and Leibowitz, 1990; Stanley and Thomas, 1993). This orexigenic effect is not observed after peripheral infusion. Central stimulatory effects are also observed in other distinct species, such as mouse, rabbit, and snake (Parrott et al., 1986; Kuenzel et al., 1987; Morley et al., 1987b; Pau et al., 1988; Miner et al., 1989; Morris and Crews, 1990). Food intake is increased severalfold by NPY, with the effect lasting several hours and significant increases still observed after 24 h. NPY stimulates the appetitive phase of feeding, which consists of the searching and acquiring food (Seeley et al., 1995; Ammar et al., 2000; Benoit et al., 2005; Day et al., 2005). While NPY reduces the latency to eat (Clark et al., 1984; Stanley and Leibowitz, 1985; Morley et al., 1987b; Corp et al., 1990), the absence of NPY in NPY knockout mice is associated with a pronounced delay in the initiation of feeding (Sindelar et al., 2005). NPY augmentation of the motivation to eat (Flood and Morley, 1991) is evidenced by the finding that NPY-injected animals will eat more even when they have to work to get food, tolerate electric shocks to the tongue to drink milk, and can consume large quantities of quinine-adulterated milk (Flood and Morley, 1991; Jewett et al., 1992, 1995). NPY may also play a role in mediating food anticipatory responses. Rats conditioned with daily NPY injections and a restricted feeding schedule eat more than rats conditioned with saline after saline injection in the third ventricle (Drazen et al., 2005).

NPY may also play a role in the consummatory phase of feeding behavior, modulating the processes that maintain ingestion. The effects of NPY on consummatory behavior have been tested in animals receiving intraoral infusion of sugar solutions. Contrasting data (stimulation, inhibition, or no change) have been reported (Seeley et al., 1995; Ammar et al., 2000; Benoit et al., 2005; Day et al., 2005). These discrepancies might be related to the training of the rats to habituate them to the ingestion of a sweet solution through an implanted intraoral cannula. Indeed, it has been reported that naïve untrained rats ingest significantly more intraoral sucrose solution after NPY injection (Benoit et al., 2005). Additionally, NPY delays satiety and therefore augments meal size, time spent eating and meal duration—irrespective of the type of food provided (Leibowitz and Alexander, 1991; Stricker-Krongrad et al., 1994; Lynch et al., 1994).

The orexigenic effects of NPY can be diminished or suppressed by the immunoneutralization of the NPY system through central infusion of NPY antibodies or antisense oligodeoxynucleotides (Shibasaki et al., 1993; Walter et al., 1994; Dube et al., 1994; Hulsey et al., 1995). When antisense to NPY is expressed in the Arc through gene therapy, treated animals release 50% less NPY, gain less weight, and eat less than the controls for a long (50 days) period of time (Gardiner et al., 2005).

The perifornical area, the paraventricular nuclei, and ventromedial nuclei (VMN) are the most sensitive areas in which NPY injection induces feeding (Stanley et al., 1985a, 1993; Morley et al., 1987a; Jolicoeur et al., 1995). The connections between hindbrain and hypothalamus also play a role in the NPY effects. Food intake is stimulated after NPY injection in the fourth ventricle (Corp et al., 1990; Steinman et al., 1994) and there is an augmentation of the early gene c-Fos in the PVN after hindbrain injection, demonstrating the activation of this nucleus (Xu et al., 1995).

The nyctohemeral variations of hypothalamic NPY are closely associated with feeding patterns. Light-to-dark transition is a triggering signal for food intake in the rat, and NPY content peaks 1 h before dark onset in the PVN and decreases 1 h after the lights are off (Jhanwar-Uniyal et al., 1990). There is also a dusk rise in NPY mRNA in the mediobasal hypothalamus, including the arcuate nucleus (Akabayashi et al., 1994). In the NPY knockout mice, intake during the first 4 h of the dark period is reduced by one-third (Sindelar et al., 2005).

B. NPY, Macronutrient Intake, and Dietary Preferences

In addition to increasing the quantity of ingested food, NPY also modifies food choice. In a food-choice paradigm, acute NPY injection preferentially increases the intake of carbohydrates when rats can choose between either pure macronutrient sources or two nutritionally complete diets, a high carbohydrate (HC) diet and a high fat (HF) diet (Stanley et al., 1985b, 1989; Welch et al., 1994). When NPY is continuously infused into the lateral ventricle for 2 weeks from osmotic minipumps, both HF intake and HC intake are stimulated. The effect on HF intake is, however, much shorter than that on HC intake (Beck et al., 1992a).

More generally, diet palatability, and particularly sweetness, facilitates NPY hyperphagia. Sucrose-containing diets, and sweet caloric or noncaloric solutions are susceptible to its orexigenic effects, and to preference formation after pairing with NPY, suggesting that orosensory mechanisms could be involved in the peptide's action (Lynch et al., 1993; Glass et al., 1997). A link between NPY and carbohydrate intake is further attested by several lines of evidence. First, when injected in the paraventricular nucleus, NPY increases the respiratory quotient (Menendez et al., 1990; Currie and Coscina, 1995), indicating an increased utilization of carbohydrates as an energy source. Second, the hypothalamic content and expression of NPY are dependent on the carbohydrate content of the diets. Thus, NPY concentration in the parvocellular part of the PVN is reduced after 2 weeks of ingestion of a HC diet than after the ingestion of a HF diet (Beck et al., 1990d). In fact, a significant negative correlation exists between NPY concentrations and the carbohydrate-to-fat ratio (Beck et al., 1992b). In the longer term, when rats are fed from weaning for 2 months with an obesogenic diet (e.g., a high starch diet together with a 25% sucrose solution to drink), the diet-induced hyperphagia is associated with a significant decrease in NPY content in the Arc (Beck et al., 1994a). These data are consistent with the existence of counterregulatory mechanisms mediated to some extent by NPY to limit energy consumption. Accordingly, NPY knockout mice eat less of a palatable diet than wild-type mice (Sindelar et al., 2005).

The sensitivity of the NPY system to macronutrient content, and particularly to carbohydrates, is emphasized by the changes to its functioning in the offspring of dams fed a HC or HF diet during the differentiation and maturation of the system, for example, during gestation and the early postdelivery period (Kagotani et al., 1989; Bouret et al., 2004). Offspring of HC dams are characterized by changes in sensitivity to NPY injection and dietary preferences and an increased peptide release after a glucoprivic challenge (Kozak et al., 1998, 2000, 2005).

Finally, rats with marked dietary preferences are characterized by a specific neuropeptidergic profile. Carbohydrate-preferring rats are characterized by lower basal NPY content in the parvocellular part of the PVN (Beck et al., 2001b), perhaps reflecting increased peptide release. Indeed, at the beginning of the dark period when carbohydrates are preferentially ingested (Tempel et al., 1989) there is a peak in NPY release (Stricker-Krongrad et al., 1997).

C. NPY, Feeding State, and Hormonal and Metabolic Control

Neuropeptide Y interacts with several peripheral hormones and central peptides to induce food intake, either through receptors present in the Arc and PVN, direct intra-arcuate connections between different neuronal populations (Beck, 2005), or through vagal connections arriving in the hindbrain. Many of these factors are sensors of the available energy necessary for metabolism. A negative energy balance, induced by food deprivation or restriction, is the strongest stimulus to activate the NPY system. Fasting overnight or longer induces an increase in NPY content and expression in the Arc and of NPY content in the PVN (Sahu et al., 1988b; Calza et al., 1989; White and Kershaw, 1990; Sanacora et al., 1990; Brady et al., 1990; Beck et al., 1990e; O'Shea and Gundlach, 1991; Chua et al., 1991a; Marks et al., 1992; Schwartz et al., 1993; Lewis et al., 1993; Davies and Marks, 1994). Chronic food restriction increases NPY expression in the hindbrain in addition to the hypothalamic areas (Chua et al., 1991a; McShane et al., 1993; White et al., 1994; Kim et al., 1998; McShane et al., 1999; Ishizaki et al., 2003). Refeeding rapidly returns NPY levels to normal values. In vivo, NPY release measured by the push–pull technique is increased in fasted rats in the PVN and food ingestion induces a diminution in NPY release (Kalra et al., 1991; Stricker-Krongrad et al., 1993; Lambert et al., 1994). Additionally, NPY release stimulated by hyperosmotic KCl is associated with feeding episodes (Stricker-Krongrad et al., 1993).

Glucose availability in the brain is an important factor in NPY regulation. This is mediated through NPY-containing, glucose-sensitive neurons in the Arc (Muroya et al., 1999). Glucoprivation induced by injection of the metabolic blocker, 2-deoxyglucose (2DG) induces feeding and activates hypothalamic NPY neurons (Minami et al., 1995). NPY expression and content in the Arc are augmented

(Akabayashi et al., 1993; Giraudo et al., 1998; Sergeyev et al., 2000). A similar augmentation is also noted in the hindbrain (Li and Ritter, 2004). By contrast, NPY-stimulated food intake is much less marked when a glucose solution is intravenously injected prior to NPY (Rowland, 1988). Arc NPY expression during the first hour of the dark period is diminished if glucose is injected i.p. 3 h before (Chang et al., 2005). Furthermore, the feeding response is much larger in normal mice than in mice lacking NPY after 2DG injection or insulin-induced hypoglycemia (Sindelar et al., 2004). Immunoneutralization of NPY in the PVN attenuates the 2DG-induced feeding (He et al., 1998).

Glucocorticoids, insulin, and leptin are other factors that are modified by fasting. Leptin, insulin, and glucocorticoid receptors are present on NPY neurons in the Arc (van Houten et al., 1980; Ceccatelli et al., 1989; Mercer et al., 1996a). The loop existing between leptin and NPY has been described in recent reviews (Porte et al., 2002; Kalra and Kalra, 2003; Sahu, 2003). Absence of insulin, as in streptozotocin-induced diabetes, is associated with hyperphagia and overactivation of the NPY system, whereas insulin treatment of these diabetic rats normalizes hypothalamic NPY and decreases food intake. These interactions with insulin have been largely described (Frankish et al., 1995; Baskin et al., 1999). There are variable results concerning the relation between NPY and glucocorticoids (see Asensio et al. (2004) for review).

NPY is also linked to the hedonic/reward systems like the endogenous opioid and cannabinoid systems, probably in relation to the functional modulation by macronutrients described before, as well as the stimulation of eating motivation. More recent studies have shown that NPY release from rat hypothalamic explants is stimulated by anandamide, a cannabinoid agonist and inhibited by AM 251, a cannabinoid receptor antagonist (Gamber

et al., 2005). NPY-induced feeding in rats is effectively decreased by pretreatment with different (μ, κ, and δ) opioid receptor antagonists (Israel et al., 2005). Finally, the NPY system interacts with the melanocortin system through the presence of both peptides in the Arc [review in (Beck, 2005)]. The precise relationships between the two systems are detailed in recent interesting reviews (Seeley et al., 2004; Cone, 2005).

The biological actions of NPY are mediated by six receptors, termed Y1 to Y6. These receptors probably derive from three Y receptor genes leading to the Y1, Y2, and Y5 subfamilies (Larhammar and Salaneck, 2004). The different subtypes have been characterized by their relative affinity for NPY fragments of different size and amino acid composition (Leibowitz and Alexander, 1991; Widdowson, 1993). All three subfamilies appear to play a role in the regulation of feeding behavior, but only the Y1 and Y5 subtypes are involved in the orexigenic effects of NPY (see Blomqvist and Herzog (1997), Balasubramaniam (1997), and Inui (1999) for review). On the contrary, chronic activation of Y2 receptors leads to hypophagia and transient weight loss (Henry et al., 2005).

D. NPY and Obesity

The Zucker *fa/fa* rat and the *ob/ob* mouse are among the most widely used rodent models for the study of obesity (Bray, 1977; Argiles, 1989; Robinson et al., 2000). Zucker (*fa/fa*) rats become obese due to a missense mutation in the leptin receptor, which diminishes but does not completely eliminate responsiveness to leptin (Iida et al., 1996; Phillips et al., 1996; Chua et al., 1996a,b). The mouse homologue of the Zucker rat is the *db/db* mouse (Truett et al., 1991; Chua et al., 1996a). In the obese *ob/ob* mouse, the inactivation of the leptin system is not due to a default in the leptin receptor but to the peptide itself (Zhang et al., 1994): a nonsense mutation in the coding region of the leptin gene prevents the production of a functional peptide.

In the obese Zucker rat, the central NPY system is profoundly altered, with distinct regional differences in NPY processing. For example, NPY concentrations are increased in the PVN and Arc in the obese rat (Beck et al., 1990b,c, 1993; McKibbin et al., 1991). Prepro-mRNA coding for NPY is also overexpressed in the Arc of the *fa/fa* rat (Sanacora et al., 1990; Mercer et al., 1996b). This profile suggests that the obese rat is effectively in a constant deprivation state. Leptin-receptor gene transfer into the Arc of female obese Zucker rats induces a suppression of NPY mRNA expression in this nucleus (Keen-Rhinehart et al., 2004) and supports the site specificity of the NPY variations.

Through chronic intraventricular (i.c.v.) NPY infusion in normal rats with osmotic minipumps, it is possible to reproduce the phenotype of the obese Zucker rat with its hyperphagia, disruption in nyctohemeral feeding rhythms, overweight, and adiposity (Beck et al., 1990a, 1992a), as well as its hormonal and metabolic changes such as high insulin, leptin, and corticosterone levels (McMinn et al., 1998; Raposinho et al., 2001; Baran et al., 2002). In these conditions, NPY effects subside after 9 to 10 days of infusion, reflecting a possible down regulation. This extinction of these effects is not observed in overweight rats after repeated (3 times/day) injection of NPY (Stanley et al., 1989) that allows a return to basal state between injections. Similar effects on body weight, adiposity, and respiratory quotient are observed after chronic infusion for 6 days of Y1 and Y5 agonists (Mashiko et al., 2003; Henry et al., 2005). These data therefore confirm the role of NPY in excess food intake and weight gain.

All the metabolic changes are present very early (from 3 to 5 weeks of age) in the life of the obese Zucker rats (Krief and Bazin, 1991). The NPY increases are also apparent early in life: content and expression are increased 30 days after birth when weaned rats begin to overeat (Bchini-Hooft van Huijsduijnen et al., 1993; Beck et al., 1993). Early determination of the *fa* genotype through genetic techniques (Smoller et al., 1993; Chua et al., 1996b) has shown that prepro-NPY mRNA levels in the Arc of *fa/fa* pups are increased as early as postnatal day 9 (Kowalski et al., 1999), whereas hypothalamic concentration is decreased (Ster et al., 2003). These changes are consistent with an augmented NPY release. During this early period, NPY could already contribute to the fat disposition through its metabolic and antithermogenic effects (Godbole et al., 1978). The changes to NPY are persistent and are still evident at 40 weeks of age (Bchini-Hooft van Huijsduijnen et al., 1993; Jhanwar-Uniyal and Chua, 1993; Sanacora et al., 1992). The high NPY mRNA expression detected in obese rats is further increased by food deprivation or restriction (Sanacora et al., 1990; Berelowitz et al., 1992; Pesonen et al., 1992; Korner et al., 2001). In adult obese rats *in vivo* NPY release in the PVN, measured by the push–pull technique, is significantly enhanced throughout the different periods of the light–dark cycle (Dryden et al., 1995; Stricker-Krongrad et al., 1997). Moreover, at the light-to-dark transition, the obese rat does not show the peak of NPY release typically observed in its lean counterpart (Stricker-Krongrad et al., 1997). Thus, NPY appears to be released in a exacerbated and anarchic manner in the obese rat.

Overproduction of NPY in the Zucker obese rat leads to down regulation of its receptors. Hypothalamic receptor density of, and binding to Y5 receptors is reduced in obese rats (McCarthy et al., 1991; Widdowson, 1997). Measurement of NPY receptor expression by RT-PCR confirms that both Y1 and Y5 receptors are down regulated (Beck et al., 2001a). This down regulation of NPY receptors consequently leads to a reduced sensitivity of the obese rats to the actions of exogenous, intraventricular NPY injection (McCarthy et al., 1991;

Stricker-Krongrad et al., 1994). The decrease in NPY binding and receptors is, however, not sufficient to normalize food intake.

A similar NPY profile is found in the *db/db* mice (Chua et al., 1991a,b; Schwartz et al., 1996; Yamamoto et al., 2000a; Bates et al., 2003) and in the obese *ob/ob* mouse (Wilding et al., 1993; Schwartz et al., 1996; Mercer et al., 1997; Jang and Romsos, 1998; Xin and Huang, 1998; Stricker-Krongrad et al., 2002). The importance of NPY in the *ob/ob* syndrome is confirmed by the attenuation of obesity in the *ob/ob* mouse knockout for NPY (Erickson et al., 1996; Marsh et al., 1998). Moreover, delivery of biologically active leptin through gene therapy in *ob/ob* mice induces a decrease of NPY expression in the Arc that is associated with body weight loss and diminished food intake (Dhillon et al., 2000).

Other genetic models of obesity such as the OLETF rat, yellow Agouti A^y mouse (a model of genetic obesity resulting from impaired melanocortin signaling) and the tubby mouse differ from the leptin-signaling deficient models as they are characterized by a decreased or normal NPY expression in the Arc instead of a marked increase, and by a significant increase in the dorsomedial nucleus. A similar profile is also described in the obese hyperphagic UCP-DTA mice (Tritos et al., 1998a). Arcuate and dorsomedial nuclei communicate in the control of food intake as NPY neurons in the Arc project their axons toward the DMH and PVN (Bai et al., 1985), and DMN neurons directly innervate the PVN (Thompson et al., 1996). NPY neurons in the DMH are also differentially regulated as they do not coexpress leptin receptors (Bi et al., 2003), contrary to NPY neurons in the Arc (Mercer et al., 1996a). The increase in the DMH might therefore be a compensatory mechanism for the Arc deficiency in order to maintain adequate NPY levels in the PVN for the feeding regulation.

Obesity can be induced by consumption of HF, HC, or high energy (HE), palatable "cafeteria" diets (Rothwell and Stock, 1979;

Louis-Sylvestre et al., 1984; Sclafani, 1987; Warwick and Schiffman, 1992). In diet-induced obese (DIO) rats, there is a decline in NPY levels after prolonged (2 months) intake of HC or HF diets (Beck et al., 1994a; Lin et al., 2000). A comparable diminution is present in rats fed a HE palatable diet (Widdowson et al., 1999; Archer et al., 2004), but in this case, it occurs earlier by 5 weeks. After 12 weeks, a decrease in NPY overflow from hypothalamic slices is observed (Hansen et al., 2004). An up regulation of Y5 receptors in adult rats fed for 6 weeks on a HE diet is consistent with a diminished NPY release (Widdowson et al., 1997).

Decreased NPY expression is also observed after longer (4 months) exposure to these diets (Hansen et al., 2004; Velkoska et al., 2005). A reduction in mRNA expression is related to the large (>200%) increase in circulating leptin levels at 8 weeks and the progressive development of leptin insensitivity with age (Qian et al., 1998). So it is evident that there is an adaptation of the NPY system when normal rats consume an energy-dense food. The diminution in hypothalamic NPY is, however, not sufficient to prevent excessive fat accretion.

Rats selected for their susceptibility to diet-induced obesity (DIO-S) are characterized by higher (>40%) NPY mRNA expression in the Arc in the preobese state (Levin and Dunn-Meynell, 1997; Levin, 1999; Gao et al., 2002), and by the absence of any regulation of this expression by 48-h fast or 50% food restriction (Levin and Dunn-Meynell, 2002b; Levin et al., 2004). These phenomena might be due to an earlier onset of reduced leptin sensitivity (Levin and Dunn-Meynell, 2002b; Levin et al., 2004) and to a decreased responsiveness of hypothalamic neurons to glucose (Levin et al., 1998). However, when obesity is well-established (as in 22-week-old rats), the NPY system again becomes responsive to diet restriction and palatability (Levin, 1999; Levin and Dunn-Meynell, 2002a), perhaps enabling these animals to defend their body weight. Similar results are obtained in selected DIO-S mice. These mice are leptin-resistant after 22 weeks on a 40% fat diet. Their NPY mRNA expression is increased and the whole NPY system is disturbed as Y5 mRNA expression is 36% higher in DIO-S mice (Huang et al., 2003). An overactive NPY system (increased peptide and receptor expression) has also been described in other rodent strains such as the Osborne-Mendel rat (Schaffhauser et al., 2002), the C57BL/6J and DBA/2J mouse (Bergen et al., 1999; Tortoriello et al., 2004). In DIO-S strains, there is a clear link between the development of obesity, an overactive NPY system, and disturbed receptor regulation. The decrease in leptin sensitivity appears to be the principal factor underlying these changes. However, other factors are likely to be involved in the short-term resistance to DIO in some strains (Bullen et al., 2004).

E. NPY and Antiobesity Treatments

The apparent importance of NPY in feeding and body weight regulation has stimulated the development of numerous agonists and antagonists for the Y1, Y2, or Y5 receptors (Balasubramaniam et al., 1994; Beck et al., 1994b; Wieland et al., 1995, 1998; Myers et al., 1995; Serradeillegal et al., 1995; Kanatani et al., 1996, 1999, 2000, 2001; Matthews et al., 1997; Morgan et al., 1998; Haynes et al., 1998; Kask et al., 1998, 2001; Wyss et al., 1998a,b; Widdowson et al., 1999; Cabrele et al., 2000; Polidori et al., 2000; Parker et al., 2000; Duhault et al., 2000; Yokosuka et al., 2001; Turnbull et al., 2002; Daniels et al., 2002; Lecklin et al., 2003; Della Zuana et al., 2004; Abbott et al., 2005). In genetic obesity, such as in Zucker rats, the reduction of receptor binding capacity can restrict the effectiveness of pharmaceutical treatments, as weak antagonists or agonists are unable to modify the feeding behavior in obese rats (Beck et al., 1994b), even if they are Y5 specific (Wyss et al., 1998a; B. Beck unpublished observations). On the other hand, DIO-S rats with their up-regulated Y1 and Y5 receptors are good

candidates for drug treatment with NPY antagonists. Because of these limitations, antagonists have mainly served as tools to understand the functioning of the NPY system.

III. THE OREXINS

The orexins (OX), or hypocretins, are involved in the regulation of the sleep–wake cycle as well as in feeding behavior (Sakurai, 2002). Indeed, transgenic OX neuron-deficient mice are characterized by narcolepsy and hypophagia (Hara et al., 2001). Orexins are synthesized in a specific population of neurons in the region of the LH (Broberger et al., 1998; deLecea et al., 1998; Sakurai et al., 1998; Sutcliffe and deLecea, 2002). Orexins exist in two forms: orexin A (or hypocretin 1) and orexin B (or hypocretin 2). Orexin B is somewhat longer than orexin A, containing 33 rather than 28 amino acids. Orexin neurons are present very early in life and their number increases until adulthood (Yamamoto et al., 2000b; van den Pol et al., 2001). They coexpress dynorphin (Chou et al., 2001) and prolactin (Qu et al., 1996). Orexin fibers are present in many brain areas and particularly those mediating food intake, stress, motivation, and arousal (Peyron et al., 1998; Date et al., 1999; Mullett et al., 2000; Baldo et al., 2003).

A. Orexins Stimulate Food Intake

Both orexin A and B stimulate food intake when injected into brain ventricles (Sakurai et al., 1998; Rodgers et al., 2000; Shiraishi et al., 2000). They are less potent than NPY, and orexin A is more potent than orexin B (Lubkin and Stricker-Krongrad, 1998; Sakurai et al., 1998; Edwards et al., 1999; Jain et al., 2000). Hyperphagic effects of orexin A are related to a delayed onset of satiety at low doses (Rodgers et al., 2000), whereas at high doses, the peptide has sedative-like effects in the initial postinjection phase. Orexins also stimulate feeding when they are injected in the fourth ventricle or directly in the dorsal vagal complex (Zheng et al., 2005). These effects might be related to the presence of OX fibers in the caudal medulla and nucleus of the solitary tract (Yang et al., 2003; Zheng et al., 2005). Direct injection of OX in the nucleus accumbens also stimulates food ingestion (Thorpe and Kotz, 2005). In addition to their effect on solid food ingestion, orexins also stimulate drinking behavior (Kunii et al., 1999), as well as food seeking, face washing, and grooming (Ida et al., 1999). They also increase spontaneous activity (Jones et al., 2001; Kiwaki et al., 2004).

Food intake is stimulated by orexins through two types of receptors OX1-R and OX2-R. These receptors are present in many brain regions (Hervieu et al., 2001; Marcus et al., 2001; Backberg et al., 2002; Cluderay et al., 2002), and orexin A has a much greater affinity for OX1-R than orexin B (Ammoun et al., 2003). In the hypothalamus, OX1-R is mostly found in the ventromedial and arcuate nuclei, whereas OX2-R is predominantly present in the PVN (Trivedi et al., 1998; Lu et al., 2000; Suzuki et al., 2002; Guan et al., 2002a). OX2-R is also detected in the brainstem (Sunter et al., 2001). Expression of both receptors increases gradually, starting at postnatal day 5 (Yamamoto et al., 2000b).

B. Orexins–NPY Interactions

Orexins are anatomically and functionally linked to the NPY system through a bidirectional pathway (Fu et al., 2004; Muroya et al., 2004). NPY neurons in the Arc project to OX neurons in the LH and, reciprocally, OX neurons project to NPY neurons (Elias et al., 1998; Horvath et al., 1999; Nambu et al., 1999). Orexin receptors are also detected on NPY neurons (Funahashi et al., 2003). Intraventricular orexin injection increases NPY mRNA expression in the Arc (Lopez et al., 2002). Food intake stimulated by orexins can be either suppressed by pretreatment with

NPY antagonists selective for the Y1 receptor (1229U91, BIBO3304) (Jain et al., 2000; Yamanaka et al., 2000; Ida et al., 2000b) and Y5 antagonists (Dube et al., 2000). The partial suppression of orexin effects observed in some experiments suggests that other modulators might be involved, including GABA. In the Arc, GABA is coexpressed in a third of the NPY neurons (Horvath et al., 1997). GABA neurons also express OX receptors, and orexins activate these neurons (Burdakov et al., 2003). The sensitivity of the OX system to GABA is further supported by the activation of OX neurons following the stimulation of feeding by a GABA agonist, muscimol (Yang et al., 2003; Zheng et al., 2003).

C. Factors of Regulation of the OX System

The orexin system is very sensitive to manipulations that suppress food intake. A small reduction of 10% induced by chronic leptin treatment induces an augmentation in OX concentrations in the lateral hypothalamus, while other peptides such as NPY are unaffected (Beck and Richy, 1999). This reactivity can be linked to a decrease in blood glucose, as OX neurons are glucosensitive (Moriguchi et al., 1999; Bayer et al., 2000), and insulin-induced hypoglycemia activates OX neurons (Griffond et al., 1999; Moriguchi et al., 1999). Some data indicate that the stimulatory effects of hypoglycemia are specific to orexin B (Cai et al., 2001). Food deprivation induces a large increase in hypothalamic preproorexin mRNA expression, peptide content, and receptors (Karteris et al., 2005), and promotes the formation of more excitatory synapses on OX neurons (Horvath and Gao, 2005). Refeeding reverses these effects. Food restriction over several weeks activates OX neurons in the LH and decreases OX2-R expression in the PVN (Kurose et al., 2002). Through its rapid adaptation to blood glucose changes, orexins can be one of the first hypothalamic factors to trigger food ingestion. It is indeed well-known that meal initiation is associated with a slight preprandial decrease in blood glucose (Louis-Sylvestre et al., 1984; Campfield and Smith, 1990). Orexins and their receptors are also present at all levels of the olfactory system (cilia, olfactory nuclei, amygdala) (Caillol et al., 2003), and these regions might incorporate another pathway for the possible adaptation of OX activity to the presence of food.

The regulation of the OX system is also dependent on glucocorticoids. Adrenalectomy induces a large decrease in OX expression (Stricker-Krongrad and Beck, 2002; Drazen et al., 2004), while dexamethasone treatment restores normal levels of expression (Stricker-Krongrad and Beck, 2002). Centrally administered orexin A can activate corticotropin-releasing hormone (CRH) neurons in the PVN and amygdala, while OX neurons are activated by stressful situations such as immobilization and cold exposure in the rat (Sakamoto et al., 2004). Under these conditions, OX mRNA expression is also increased (Ida et al., 2000a). Additionally, i.c.v. orexin injection can activate the hypothalamo-pituitary axis (HPA), in a dose-dependent manner, at all levels. CRH mRNA expression is increased in the PVN (Al Barazanji et al., 2001), and plasma ACTH and corticosterone concentrations are increased (Kuru et al., 2000; Ida et al., 2000a; Al Barazanji et al., 2001; Samson et al., 2002; Moreno et al., 2005). These effects can be blocked by pretreatment with intravenous injection of a CRH antagonist (Samson et al., 2002). In these conditions, orexin stimulates food intake and this orexigenic effect can be blocked with 1229U91, a NPY Y1 antagonist (Ida et al., 2000b). Urocortin, which also binds to CRH receptors can inhibit orexin-induced food intake (Wang and Kotz, 2002). These data provide strong indications that orexins can link stress and feeding. Another important factor influencing orexin actions on feeding is leptin. Morphological studies have shown that OX neurons project to neurons expressing leptin receptors in the arcuate and ventromedial nuclei (Funahashi et al., 2000). Leptin can also inhibit increases in the expression of OX mRNA and hypothalamic OX1-R (Lopez et al., 2000), possibly through leptin receptors present on OX neurons in the LH (Hakansson et al., 1999).

D. Interactions with Other Neuromodulators

Due to the widespread distribution of OX fibers in the brain, interactions with numerous factors have been described. Many of these relationships are with feeding inhibitory factors such as melanocortins (Guan et al., 2001), GALP (galanin-like peptide) (Takenoya et al., 2005), CCK (cholecystokinin) (Burdyga et al., 2003), GLP-1 (glucagon-like peptide 1) (Acuna-Goycolea and van den Pol, 2004), 5-HT (serotonin) (Collin et al., 2002; Muraki et al., 2004), or the cholinergic agonist, nicotine (Kane et al., 2000). Other potential interactions may occur either with peptides such as the putative orexigen neuropeptide W (Levine et al., 2005), ghrelin (Olszewski et al., 2003), opioids (Clegg et al., 2002; Sweet et al., 2004; Furudono et al., 2005), or with endocannabinoids acting via cannabinoid CB1 receptors present on OX neurons (Cota et al., 2003; Hilairet et al., 2003). All these interactions support a complex role of orexins in the regulation of food intake, but also demonstrate that the OX system is an interesting target for antiobesity drug development.

E. Orexins and Obesity

The orexins profile in the hypothalamus of the *ob/ob* mouse is very different from that of NPY. Prepro-orexin mRNA expression is decreased in the LH of the obese mouse (Yamamoto et al., 1999; Stricker-Krongrad et al., 2002), and orexin A concentrations are reduced in both the LH and PVN (Stricker-Krongrad et al., 2002). These variations reflect a true down regulation of the OX system that might constitute a counterregulatory mechanism to limit body weight gain and excessive food intake. However, the system remains sensitive to a negative energy balance in the *ob/ob* mouse, since food restriction for 2 weeks leads to its up regulation (Yamamoto et al., 2000a). In the yellow A^y mouse, OX mRNA expression is unchanged (Hanada et al., 2000).

Orexin down regulation has also been reported in the obese Zucker rat, with prepro-orexin mRNA being significantly decreased in the LH (Cai et al., 2000; Beck et al., 2001a; Yamamoto et al., 2002). However, there are some contradictory findings, with some authors detecting no difference in expression between lean and obese rats (Taheri et al., 2001; Tritos et al., 2001) or unchanged OX-A and OX-B content in the LH of obese rats (Mondal et al., 1999). These discrepancies may depend on the age of the animals (Yamamoto et al., 2002).

F. Orexins and Antiobesity Treatment

A few orexin antagonists have been developed recently (Porter et al., 2001; Smart et al., 2001; Langmead et al., 2004). One of them, SB 334867, can decrease food intake and induce weight loss (Haynes et al., 2000, 2002; Ishii et al., 2005a), reportedly by increasing satiety (Rodgers et al., 2001; Ishii et al., 2005b) in a similar way to CCK-8S, although it is unclear whether the drug's effects on feeding may be dissociated from apparent sedative effects. This compound is active in the ob/ob mouse (Haynes et al., 2002).

IV. MELANIN-CONCENTRATING HORMONE

Melanin-concentrating hormone (MCH) already has a long history as it was isolated more than 20 years ago from salmon, in which it regulates skin color (Kawauchi et al., 1983). Its production, processing, and

cell biology are well-understood (for review see Griffond and Baker, 2002). MCH is a cyclic peptide of 19 amino acids that is synthesized in the LH and zona incerta (Bittencourt et al., 1992). It is produced in a population of LH neurons that is distinct from those synthesizing the orexins (Broberger et al., 1998; Elias et al., 1998; Hakansson et al., 1999). Contrary to orexin neurons, the MCH neurons are not glucose sensitive (Grillon et al., 1997; Bayer et al., 1999, 2000). There are, however, direct anatomical connections between the two populations of neurons (Bayer et al., 2002; Guan et al., 2002b), and OX1-R are present on MCH neurons (Backberg et al., 2002). MCH neurons are present very early in life and can be detected in fetuses during the last week of gestation (Brischoux et al., 2001; Steininger et al., 2004). MCH is colocalized with an anorexigenic peptide, CART (cocaine- and amphetamine-related transcript) (Broberger, 1999; Brischoux et al., 2001; Cvetkovic et al., 2004) and acetylcholine (Chou et al., 2004). Cannabinoid CB1 receptors are also present on MCH neurons (Cota et al., 2003).

Interest in MCH was renewed when its role in feeding was discovered following central injections, and by reports that genetic manipulation of MCH systems can produce animals with distinct body weight phenotypes. This latter fact contrasts with knockout animals for other important feeding-related neuropeptides, such as NPY, in which body weight does not differ from wild-type animals (review in Beck, 2001). Mice overexpressing MCH are obese and insulin resistant (Ludwig et al., 2001), whereas those with a disruption of the MCH gene are lean and hypophagic (Shimada et al., 1998) and resistant to DIO (Kokkotou et al., 2005).

A. MCH Is an Orexigenic Peptide

There is now a consensus for the orexigenic effects of MCH injected centrally (Qu et al., 1996; Sahu, 1998b; Rossi et al., 1999;

Della Zuana et al., 2002), even if some inhibitory effects have been reported at very low doses (Presse et al., 1996). The duration of these effects is relatively short (2–4 h) as MCH is very sensitive to endoproteases and is rapidly degraded (Maulon-Feraille et al., 2002). When it is injected into brain ventricles, MCH is as potent as orexins and less potent than NPY in stimulating food intake (Edwards et al., 1999). Many areas respond to MCH injection. They include classic hypothalamic nuclei (Arc, PVN, and DMN) but not LH and VMN (Abbott et al., 2003). Extrahypothalamic areas such as the nucleus accumbens are also responsive to MCH injection (Georgescu et al., 2005). MCH stimulates both appetitive and consummatory phases of feeding behavior (Benoit et al., 2005).

Chronic MCH injections (i.c.v. 5 µg; twice a day) also increase food intake but have no impact on body weight when animals are only fed chow (Rossi et al., 1997). Tolerance develops after a few days of treatment. When the animals are fed a moderate (32%) HF diet, continuous MCH infusion (10 µg/day for 14 days) induces a sustained hyperphagia, weight gain, and increased liver and fat mass (Gomori et al., 2003; Ito et al., 2003). In addition, in these conditions, MCH induces lipogenesis in adipose tissue even in pair-fed normophagic rats. MCH not only stimulates solid food ingestion; it also stimulates water intake in the absence of food (Clegg et al., 2003). It also stimulates alcohol intake (Duncan et al., 2005) as well as sucrose intake in presence of water (Sakamaki et al., 2005).

The biological effects of MCH are mediated by at least two receptors: MCH-R1, initially named SLC-1 (Bachner et al., 1999; Chambers et al., 1999; Saito et al., 1999; Shimomura et al., 1999) and MCH-R2 that was discovered 2 years later (An et al., 2001; Mori et al., 2001; Wang et al., 2001). Both are G-protein coupled receptors (Hawes et al., 2000). The two receptors are not present in all species and MCH-R2 is not functional in rodents (Tan et al., 2002).

Food intake stimulation is mediated through the MCH-R1 receptor (Suply et al., 2001). The MCH-R1 expression and protein are widely distributed in the brain. They are detected in the cortex, in the amygdala, hippocampus, hypothalamus, and thalamus (Hervieu et al., 2000). MCH-R1 is more abundant than MCH-R2 and generally colocalized in tissues where MCH-R2 is expressed (Schlumberger et al., 2002). Deletion of the MCH-R1 gene leads to the generation of mice with normal body weight but lean (reduced fat mass) which are also hyperactive. These knockout animals are also resistant to DIO (Chen et al., 2002; Marsh et al., 2002). Chronic MCH infusions are inefficient in MCH-R1 knockout mice (Marsh et al., 2002). These data support the role of MCH-R1 in feeding.

B. Factors Affecting Hypothalamic MCH

As with the peptides described earlier, fasting and/or food restriction is the main factor modulating MCH expression and content. In almost all studies, a moderate to large increase in MCH expression is measured after 24 h of food deprivation (Presse et al., 1996; Qu et al., 1996; Herve and Fellmann, 1997; Kokkotou et al., 2001). This up regulation can be observed after longer periods of fasting, although not consistently in all studies (Herve and Fellmann, 1997; Tritos et al., 2001; Bertile et al., 2003).

As previously indicated, MCH neurons are not glucosensitive. Nonetheless, physiological concentrations of extracellular glucose in the brain dose-dependently enhance the electrical excitability of MCH neurons (Burdakov et al., 2005). Treatment with the metabolic blocker 2DG, which stimulates food intake, is associated with an increase in MCH mRNA expression (Sergeyev et al., 2000). A similar effect is noted after injection of mercaptoacetate.

Leptin is another important comodulator of the MCH system. Intraventricular leptin injection reduces both food intake and

MCH expression (Sahu, 1998a,b). The increase of MCH and MCH-R expression in fasted animals is abolished by leptin (Kokkotou et al., 2001; Tritos et al., 2001).

Finally, glucocorticoids may also be involved in the regulation of the MCH system. Adrenalectomy induces a decrease in MCH expression and adrenalectomized rats are less sensitive to the orexigenic action of MCH (Drazen et al., 2004). On the other hand, MCH stimulates the HPA axis (CRH in the PVN, ACTH, and corticosterone) (Kennedy et al., 2003). This action may be mediated in part by MCH projections to the median eminence and pituitary (Cvetkovic et al., 2003b).

C. Interactions with Other Feeding-Related Peptides

MCH has close relationships with several other neuropeptides. Intraventricular MCH increases NPY mRNA (Della Zuana et al., 2002), whereas pretreatment with NPY Y1 antagonists (BIBO 3304 or GR 231118) abolishes the stimulatory effects of MCH on food intake (Chaffer and Morris, 2002). In vitro, MCH stimulates the release of NPY from hypothalamic explants (Abbott et al., 2003). It also affects ligands of the melanocortin MC4 receptor, inhibiting the release of the anorexigen α-MSH (Cone, 2005), and stimulating that of its natural antagonist, the orexigen AGRP. The link between MCH and α-MSH is further supported by the inhibition of some biological actions of each one by the other (Ludwig et al., 1998).

A possible interaction exists with other anorectic peptides since i.c.v. pretreatment with either GLP-1 or neurotensin prevents the stimulation of food intake by MCH (Tritos et al., 1998b). On the other hand, i.p. injection of the putative satiety factor PYY$_{3-36}$ has no effect on MCH expression (Challis et al., 2003). MCH-R1 is not necessary for the orexigenic effect of ghrelin, as ghrelin is as efficient in MCHR1-knockout mice as in wild-type mice (Bjursell et al.,

2005). A functional interaction may also exist with tachykinins, as MCH neurons express neurokinin 3 receptors (Cvetkovic et al., 2003a), or with GALP since GALP neurons project on to MCH neurons (Takenoya et al., 2005) (but this latter relationship remains to be confirmed).

D. MCH and Obesity

The ob/ob mouse has a characteristic MCH profile that is comparable to that described for NPY, but opposite to that of orexins. The MCH system is indeed up regulated with an increased expression (Qu et al., 1996; Kokkotou et al., 2001; Tritos et al., 2001) and peptide content in the LH (Mondal et al., 2002). When ob/ob mice are treated with leptin, the increase in MCH is suppressed (Tritos et al., 2001). Disruption of the MCH gene in ob/ob mice attenuates the obese phenotype through a reduction in fat mass, improved glucose tolerance, and lower plasma glucocorticoids (Segal-Lieberman et al., 2003). Fasting further increases MCH expression in the ob/ob mouse (Qu et al., 1996). A similar up regulation is present in the obese Zucker rats, with a large (fivefold) increase in mRNA expression associated with a significant decrease in MCH-R1 expression (Stricker-Krongrad et al., 2001). In the yellow A^y mouse, the MCH system is also up regulated (Hanada et al., 2000; Morton et al., 2004). And in the fat/fat mouse, which has a mutation of the gene encoding carboxypeptidase E and which develops late-onset obesity, there is an increase in hypothalamic MCH compared to the fat/+ mouse (Rovere et al., 1996).

In DIO rodents, some species differences have been reported. In DIO rats, there is an increase in MCH and MCH-R1 expression, and of MCH concentrations in the Arc and PVN after 8 weeks of a high energy diet (Elliott et al., 2004). These increases disappear after 6 months of dietary treatment (Gao et al., 2002). The MCH system in DIO rat does not respond to food restriction, whereas in the DIO mouse a slight increase in hypothalamic MCH mRNA is observed (Morton et al., 2004).

E. MCH and Antiobesity Treatment

MCH receptors are interesting targets for antiobesity treatments due to the up regulation of the MCH system that is almost universally present in obesity models. Numerous antagonists of the MCH-R1 have more recently been developed (Bednarek et al., 2002; Borowsky et al., 2002; Takekawa et al., 2002; Shearman et al., 2003; Kowalski et al., 2004; Palani et al., 2005; Souers et al., 2005; Mashiko et al., 2005; review in Browning, 2004). Their potency in reducing food intake has been proved but some of them have the additional advantage of having anxiolytic actions.

V. CONCLUSION

In this chapter we have described the characteristics of three of the main neuropeptidergic systems currently being studied for their potential in antiobesity drug development. The three peptides (NPY, orexin, and MCH) are orexigenic, and two of them are peptides from the lateral hypothalamus. Studies of antagonists at their receptors are in the early phases of development in animals, and some promising results have already been published. These neuropeptides are not, however, the sole peptides that regulate feeding behavior and new ones are regularly discovered. The very latest to be described are obestatin and peptide S (Beck et al., 2005; Zhang et al., 2005). Both apparently inhibit food intake and, if their physiological roles in regulating feeding are confirmed, development of agonists might also be useful for obesity treatment.

It is likely that to fight obesity, it will be necessary to target two or three systems at the same time, as the study of knockout animals has revealed that alternative mechanisms can be invoked to compensate for the lack of any individual system, even if it has an important role. The NPY knockout mouse is a good example of this compensation, in that these animals are quite normal in terms of feeding and body weight (Beck, 2001). Of course, while we await efficient antiobesity drugs, hopefully in the near future, it is possible to normalize one's energy balance by controlling intake and energy expenditure through simple changes in eating behavior (smaller portions, low calorie food and drinks) and by augmenting energy expenditure through physical exercise. This is a valid strategy for every individual, from children to adults. Attention has also to be paid to nutritional and environmental conditions during the very early stages of life, including gestation. The neuropeptidergic systems described in this chapter are particularly sensitive to these factors during their development, and modifications to them may predispose an individual to develop earlier or more severe obesity (McMillen and Robinson, 2005). Investigation of these issues will help to limit the increase in the prevalence of obesities in our societies.

Acknowledgments

The author's research is supported by the European Commission, Quality of Life and Management of Living Resources, Key action 1 "Food, nutrition and health" programme (QLK1-2000-00515), INSERM, and the Institut Benjamin Delessert (Paris).

References

Abbott, C. R., Kennedy, A. R., Wren, A. M., Rossi, M., Murphy, K. G., Seal, L. J., Todd, J. F., Ghatei, M. A., Small, C. J., and Bloom, S. R. (2003). Identification of hypothalamic nuclei involved in the orexigenic effect of melanin-concentrating hormone. *Endocrinology* **144**, 3943–3949.

Abbott, C. R., Small, C. J., Kennedy, A. R., Neary, N. M., Sajedi, A., Ghatei, M. A., and Bloom, S. R. (2005). Blockade of the neuropeptide Y Y2 receptor with the specific antagonist BIIE0246 attenuates the effect of endogenous and exogenous peptide YY(3-36) on food intake. *Brain Res.* **1043**, 139–144.

Acuna-Goycolea, C., and van den Pol, A. (2004). Glucagon-like peptide 1 excites hypocretin/orexin neurons by direct and indirect mechanisms: Implications for viscera-mediated arousal. *J. Neurosci.* **24**, 8141–8152.

Akabayashi, A., Zaia, C. T. B. V., Silva, I., Chae, H. J., and Leibowitz, S. F. (1993). Neuropeptide-Y in the arcuate nucleus is modulated by alterations in glucose utilization. *Brain Res.* **621**, 343–348.

Akabayashi, A., Levin, N., Paez, X., Alexander, J. T., and Leibowitz, S. F. (1994). Hypothalamic neuropeptide Y and its gene expression: relation to light/dark cycle and circulating corticosterone. *Mol. Cell. Neurosci.* **5**, 210–218.

Al Barazanji, K. A., Wilson, S., Baker, J., Jessop, D. S., and Harbuz, M. S. (2001). Central orexin-A activates hypothalamic-pituitary-adrenal axis and stimulates hypothalamic corticotropin releasing factor and arginine vasopressin neurones in conscious rats. *J. Neuroendocrinol.* **13**, 421–424.

Allen, J. M., Hughes, J., and Bloom, S. R. (1987). Presence, distribution, and pharmacological effects of neuropeptide Y in mammalian gastrointestinal tract. *Dig. Dis. Sci.* **32**, 506–512.

Ammar, A. A., Sederholm, F., Saito, T. R., Scheurink, A. J. W., Johnson, A. E., and Sodersten, P. (2000). NPY-leptin: opposing effects on appetitive and consummatory ingestive behavior and sexual behavior. *Am. J. Physiol.* **278**, R1627–R1633.

Ammoun, S., Holmqvist, T., Shariatmadari, R., Oonk, H. B., Detheux, M., Parmentier, M., Akerman, K. E. O., and Kukkonen, J. P. (2003). Distinct recognition of OX1 and OX2 receptors by orexin peptides. *J. Pharmacol. Exp. Ther.* **305**, 507–514.

An, S. Z., Cutler, G., Zhao, J. J., Huang, S. G., Tian, H., Li, W. B., Liang, L. M., Rich, M., Bakleh, A., Du, J., Chen, J. L., and Dai, K. (2001). Identification and characterization of a melanin-concentrating hormone receptor. *Proc. Natl. Acad. Sci. U.S.A.* **98**, 7576–7581.

Archer, Z. A., Rayner, D. V., and Mercer, J. G. (2004). Hypothalamic gene expression is altered in underweight but obese juvenile male Sprague-Dawley rats fed a high-energy diet. *J. Nutr.* **134**, 1369–1374.

Argiles, J. M. (1989). The obese Zucker rat: a choice for fat metabolism. *Prog. Lipid Res.* **28**, 53–66.

Asensio, C., Muzzin, P., and Rohner-Jeanrenaud, F. (2004). Role of glucocorticoids in the physiopathology of excessive fat deposition and insulin resistance. *Int. J. Obes.* **28**, S45–S52.

Bachner, D., Kreienkamp, H. J., Weise, C., Buck, F., and Richter, D. (1999). Identification of melanin concentrating hormone (MCH) as the natural ligand for the orphan somatostatin-like receptor 1 (SLC-1). *FEBS Lett.* **457**, 522–524.

Backberg, M., Hervieu, G., Wilson, S., and Meister, B. (2002). Orexin receptor-1 (OX-R1) immunoreactivity in chemically identified neurons of the hypothalamus: focus on orexin targets involved in control of food and water intake. *Eur. J. Neurosci.* **15**, 315–328.

Bai, F. L., Yamano, M., Shiotani, Y., Emson, P. C., Smith, A. D., Powell, J. F., and Toyama, M. (1985). An arcuato-paraventricular and -dorsomedial hypothalamic neuropeptide Y–containing system which lacks noradrenaline in the rat. *Brain Res.* **331**, 172–175.

Balasubramaniam, A. (1997). Neuropeptide Y family of hormones: Receptor subtypes and antagonists. *Peptides* **18**, 445–457.

Balasubramaniam, A., Sheriff, S., Johnson, M. E., Prabhakaran, M., Huang, Y., Fischer, J. E., and Chance, W. T. (1994). [D-TRP(32)]Neuropeptide-Y—a competitive antagonist of NPY in rat hypothalamus. *J. Med. Chem.* **37**, 811–815.

Baldo, B. A., Daniel, R. A., Berridge, C. W., and Kelley, A. E. (2003). Overlapping distributions of orexin/hypocretin- and dopamine-beta-hydroxylase immunoreactive fibers in rat brain regions mediating arousal, motivation, and stress. *J. Comp. Neurol.* **464**, 220–237.

Baran, K., Preston, E., Wilks, D., Cooney, G. J., Kraegen, E. W., and Sainsbury, A. (2002). Chronic central melanocortin-4 receptor antagonism and central neuropeptide-Y infusion in rats produce increased adiposity by divergent pathways. *Diabetes* **51**, 152–158.

Baskin, D. G., Lattemann, D. F., Seeley, R. J., Woods, S. C., Porte, D., and Schwartz, M. W. (1999). Insulin and leptin: dual adiposity signals to the brain for the regulation of food intake and body weight. *Brain Res.* **848**, 114–123.

Bates, S. H., Stearns, W. H., Dundon, T. A., Schubert, M., Tso, A. W. K., Wang, Y. P., Banks, A. S., Lavery, H. J., Haq, A. K., Maratos-Flier, E., Neel, B. G., Schwartz, M. W., and Myers, M. G. (2003). STAT3 signalling is required for leptin regulation of energy balance but not reproduction. *Nature* **421**, 856–859.

Bayer, L., Poncet, F., Fellmann, D., and Griffond, B. (1999). Melanin-concentrating hormone expression in slice cultures of rat hypothalamus is not affected by 2-deoxyglucose. *Neurosci. Lett.* **267**, 77–80.

Bayer, L., Colard, C., Nguyen, N. U., Risold, P. Y., Fellmann, D., and Griffond, B. (2000). Alteration of the expression of the hypocretin (Orexin) gene by 2-deoxyglucose in the rat lateral hypothalamic area. *Neuroreport* **11**, 531–533.

Bayer, L., Mairet-Coello, G., Risold, P. Y., and Griffond, B. (2002). Orexin/hypocretin neurons: chemical phenotype and possible interactions with melanin-concentrating hormone neurons. *Regul. Pept.* **104**, 33–39.

Bchini-Hooft van Huijsduijnen, O., Rohner-Jeanrenaud, F., and Jeanrenaud, B. (1993). Hypothalamic neuropeptide-Y messenger ribonucleic acid levels in pre-obese and genetically obese (fa/fa) rats—potential regulation thereof by corticotropin-releasing factor. *J. Neuroendocrinol.* **5**, 381–386.

Beck, B. (2001). KO's and organisation of peptidergic feeding behavior mechanisms. *Neurosci. Biobehav. Rev.* **25**, 143–158.

Beck, B. (2005). The arcuate nucleus: its special place in the central networks that regulate feeding behavior. *In* "Nutrient and Cell Signalling" (J. Zempleni and K. Dakshinamurti, eds.), pp. 665–699. Marcel Dekker, New York.

Beck, B., and Richy, S. (1999). Hypothalamic hypocretin/orexin and neuropeptide Y: Divergent interaction with energy depletion and leptin. *Biochem. Biophys. Res. Commun.* **258**, 119–122.

Beck, B., Stricker-Krongrad, A., Nicolas, J. P., and Burlet, C. (1990a). Chronic and continuous ICV infusion of neuropeptide Y disrupts the nyctohemeral feeding patterns in rats. *Ann. N.Y. Acad. Sci.* **611**, 491–494.

Beck, B., Burlet, A., Nicolas, J. P., and Burlet, C. (1990b). Hyperphagia in obesity is associated with a central peptidergic dysregulation in rats. *J. Nutr.* **120**, 806–811.

Beck, B., Burlet, A., Nicolas, J. P., and Burlet, C. (1990c). Hypothalamic neuropeptide Y (NPY) in obese Zucker rats: implications in feeding and sexual behaviors. *Physiol. Behav.* **47**, 449–453.

Beck, B., Stricker-Krongrad, A., Burlet, A., Nicolas, J. P., and Burlet, C. (1990d). Influence of diet composition on food intake and hypothalamic neuropeptide Y (NPY) in the rat. *Neuropeptides* **17**, 197–203.

Beck, B., Jhanwar-Uniyal, M., Burlet, A., Chapleur-Chateau, M., Leibowitz, S. F., and Burlet, C. (1990e). Rapid and localized alterations of neuropeptide Y in discrete hypothalamic nuclei with feeding status. *Brain Res.* **528**, 245–249.

Beck, B., Stricker-Krongrad, A., Nicolas, J. P., and Burlet, C. (1992a). Chronic and continuous intra-cerebroventricular infusion of neuropeptide Y in Long-Evans rats mimics the feeding behavior of obese Zucker rats. *Int. J. Obes.* **16**, 295–302.

Beck, B., Stricker-Krongrad, A., Burlet, A., Nicolas, J. P., and Burlet, C. (1992b). Specific hypothalamic neuropeptide-Y variation with diet parameters in rats with food choice. *Neuroreport* **3**, 571–574.

Beck, B., Burlet, A., Bazin, R., Nicolas, J. P., and Burlet, C. (1993). Elevated neuropeptide-Y in the arcuate nucleus of young obese Zucker rats may contribute to the development of their overeating. *J. Nutr.* **123**, 1168–1172.

Beck, B., Stricker-Krongrad, A., Burlet, A., Max, J. P., Musse, N., Nicolas, J. P., and Burlet, C. (1994a). Macronutrient type independently of energy intake modulates hypothalamic neuropeptide Y in Long-Evans rats. *Brain Res. Bull.* **34**, 85–91.

Beck, B., Stricker-Krongrad, A., Musse, N., Nicolas, J. P., and Burlet, C. (1994b). Putative neuropeptide Y antagonist failed to decrease overeating in obese Zucker rats. *Neurosci. Lett.* **181**, 126–128.

Beck, B., Richy, S., Dimitrov, T., and Stricker-Krongrad, A. (2001a). Opposite regulation of hypothalamic orexin and neuropeptide Y receptors and peptide expressions in obese Zucker rats. *Biochem. Biophys. Res. Commun.* **286**, 518–523.

Beck, B., Stricker-Krongrad, A., Burlet, A., Cumin, F., and Burlet, C. (2001b). Plasma leptin and hypothalamic neuropeptide Y and galanin levels in Long-Evans rats with marked dietary preferences. *Nutr. Neurosci.* **4**, 39–50.

Beck, B., Fernette, B., and Stricker-Krongrad, A. (2005). Peptide S is a novel potent inhibitor of voluntary and fast-induced food intake in rats. *Biochem. Biophys. Res. Commun.* **332**, 859–865.

Bednarek, M. A., Hreniuk, D. L., Tan, C., Palyha, O. C., MacNeil, D. J., Van der Ploeg, L. H. Y., Howard, A. D., and Feighner, S. D. (2002). Synthesis and biological evaluation in vitro of selective, high affinity peptide antagonists of human melanin-concentrating hormone action at human melanin-concentrating hormone receptor 1. *Biochemistry* **41**, 6383–6390.

Benoit, S. C., Clegg, D. J., Woods, S. C., and Seeley, R. J. (2005). The role of previous exposure in the appetitive and consummatory effects of orexigenic neuropeptides. *Peptides* **26**, 751–757.

Berelowitz, M., Bruno, J. F., and White, J. D. (1992). Regulation of hypothalamic neuropeptide expression by peripheral metabolism. *Trends Endocrinol. Metab.* **3**, 127–133.

Bergen, H. T., Mizuno, T., Taylor, J., and Mobbs, C. V. (1999). Resistance to diet-induced obesity is associated with increased proopiomelanocortin mRNA and decreased neuropeptide Y mRNA in the hypothalamus. *Brain Res.* **851**, 198–203.

Bertile, F., Oudart, H., Criscuolo, F., LeMaho, Y., and Raclot, T. (2003). Hypothalamic gene expression in long-term fasted rats: relationship with body fat. *Biochem. Biophys. Res. Commun.* **303**, 1106–1113.

Bi, S., Robinson, B. M., and Moran, T. H. (2003). Acute food deprivation and chronic food restriction differentially affect hypothalamic NPY mRNA expression. *Am. J. Physiol.* **285**, R1030–R1036.

Bittencourt, J. C., Presse, F., Arias, C., Peto, C., Vaughan, J., Nahon, J. L., Vale, W., and Sawchenko, P. E. (1992). The melanin concentrating hormone system of the rat brain: an immuno- and hybridization histochemical characterization. *J. Comp. Neurol.* **319**, 218–245.

Bjursell, M., Egecioglu, E., Gerdin, A. K., Svensson, L., Oscarsson, J., Morgan, D., Snaith, M., Tornell, J., and Bohlooly, Y. M. (2005). Importance of melanin-concentrating hormone receptor for the acute effects of ghrelin. *Biochem. Biophys. Res. Commun.* **326**, 759–765.

Blomqvist, A. G., and Herzog, H. (1997). Y-receptor subtypes—How many more? *Trends Neurosci.* **20**, 294–298.

Borowsky, B., Durkin, M. M., Ogozalek, K., Marzabadi, M. R., DeLeon, J., Lagu, B., Heurich, R., Lichtblau, H., Shaposhnik, Z., Daniewska, I., Blackburn, T. P., Branchek, T. A., Gerald, C., Vaysse, P. J., and Forray, C. (2002). Antidepressant, anxiolytic and anorectic effects of a melanin-concentrating hormone-1 receptor antagonist. *Nat. Med.* **8**, 825–830.

Bouret, S. G., Draper, S. J., and Simerly, R. B. (2004). Formation of projection pathways from the arcuate nucleus of the hypothalamus to hypothalamic regions implicated in the neural control of feeding behavior in mice. *J. Neurosci.* **24**, 2797–2805.

Brady, L. S., Smith, M. A., Gold, P. W., and Herkenham, M. (1990). Altered expression of hypothalamic neuropeptide messenger RNAs in food-restricted and food-deprived rats. *Neuroendocrinology* **52**, 441–447.

Bray, G. A. (1977). The Zucker-fatty rat: a review. *Fed. Proc.* **36**, 146–153.

Brischoux, F., Fellmann, D., and Risold, P. Y. (2001). Ontogenetic development of the diencephalic MCH neurons: a hypothalamic "MCH area" hypothesis. *Eur. J. Neurosci.* **13**, 1733–1744.

Broberger, C. (1999). Hypothalamic cocaine- and amphetamine-regulated transcript (CART) neurons: histochemical relationship to thyrotropin-releasing hormone, melanin-concentrating hormone, orexin/hypocretin and neuropeptide Y. *Brain Res.* **848**, 101–113.

Broberger, C., DeLecea, L., Sutcliffe, J. G., and Hokfelt, T. (1998). Hypocretin/orexin- and melanin-concentrating hormone-expressing cells form distinct populations in the rodent lateral hypothalamus: relationship to the neuropeptide Y and agouti gene-related protein systems. *J. Comp. Neurol.* **402**, 460–474.

Browning, A. (2004). Recent developments in the discovery of melanin-concentrating hormone antagonists: novel antiobesity agents. *Exp. Opin. Ther. Pat.* **14**, 313–325.

Bullen, J. W., Ziotopoulou, M., Ungsunan, L., Misra, J., Alevizos, I., Kokkotou, E., Maratos-Flier, E., Stephanopoulos, G., and Mantzoros, C. S. (2004). Short-term resistance to diet-induced obesity in A/J mice is not associated with regulation of hypothalamic neuropeptides. *Am. J. Physiol.* **287**, E662–E670.

Burdakov, D., Liss, B., and Ashcroft, F. M. (2003). Orexin excites GABAergic neurons of the arcuate nucleus by activating the sodium-calcium exchanger. *J. Neurosci.* **23**, 4951–4957.

Burdakov, D., Gerasimenko, O., and Verkhratsky, A. (2005). Physiological changes in glucose differentially modulate the excitability of hypothalamic melanin-concentrating hormone and orexin neurons in situ. *J. Neurosci.* **25**, 2429–2433.

Burdyga, G., Lal, S., Spiller, D., Jiang, W., Thompson, D., Attwodd, S., Saeed, S., Grundy, D., Varro, A., Dimaline, R., and Dockray, G. J. (2003). Localization of orexin-1 receptors to vagal afferent neurons in the rat and humans. *Gastroenterology* **124**, 129–139.

Cabrele, C., Langer, M., Bader, R., Wieland, H. A., Doods, H. N., Zerbe, O., and BeckSickinger, A. G. (2000). The first selective agonist for the neuropeptide Y Y5 receptor increases food intake in rats. *J. Biol. Chem.* **275**, 36043–36048.

Cai, X. J., Lister, C. A., Buckingham, R. E., Pickavance, L., Wilding, J., Arch, J. R. S., Wilson, S., and Williams, G. (2000). Down-regulation of orexin gene expression by severe obesity in the rats: studies in Zucker fatty and Zucker diabetic fatty rats and effects of rosiglitazone. *Mol. Brain Res.* **77**, 131–137.

Cai, X. J., Evans, M. L., Lister, C. A., Leslie, R. A., Arch, J. R. S., Wilson, S., and Williams, G. (2001). Hypoglycemia activates orexin neurons and selectively increases hypothalamic orexin-B levels—Responses inhibited by feeding and possibly mediated by the nucleus of the solitary tract. *Diabetes* **50**, 105–112.

Caillol, M., Aioun, J., Baly, C., Persuy, M. A., and Salesse, R. (2003). Localization of orexins and their receptors in the rat olfactory system: possible modulation of olfactory perception by a neuropeptide synthetized centrally or locally. *Brain Res.* **960**, 48–61.

Calza, L., Giardino, L., Battistini, N., Zanni, M., Galetti, S., Protopapa, F., and Velardo, A. (1989). Increase of neuropeptide Y-like immunoreactivity in the paraventricular nucleus of fasting rats. *Neurosci. Lett.* **104**, 99–104.

Campfield, L. A., and Smith, F. J. (1990). Transient declines in blood glucose signal meal initiation. *Int. J. Obes.* **14**, 15–33.

Ceccatelli, S., Cintra, A., Hokfelt, T., Fuxe, K., Wikstrom, A., and Gustafsson, J. (1989). Coexistence of glucocorticoid receptor-like immunoreactivity with neuropeptides in the hypothalamic paraventricular nucleus. *Exp. Brain Res.* **78**, 33–42.

Chaffer, C. L., and Morris, M. J. (2002). The feeding response to melanin-concentrating hormone is attenuated by antagonism of the NPYY1-receptor in the rat. *Endocrinology* **143**, 191–197.

Challis, B. G., Pinnock, S. B., Coll, A. P., Carter, R. N., Dickson, S. L., and O'Rahilly, S. (2003). Acute effects of PYY3-36 on food intake and hypothalamic neuropeptide expression in the mouse. *Biochem. Biophys. Res. Commun.* **311**, 915–919.

Chambers, J., Ames, R. S., Bergsma, D., Muir, A., Fitzgerald, L. R., Hervieu, G., Dytko, G. M., Foley, J. J., Martin, J., Liu, W. S., Park, J., Ellis, C., Ganguly, S., Konchar, S., Cluderay, J., Leslie, R., Wilson, S., and Sarau, H. M. (1999). Melanin-concentrating hormone is the cognate ligand for the orphan G-protein-coupled receptor SLC-1. *Nature* **400**, 261–265.

Chang, G. Q., Karatayev, O., Davydova, Z., Wortley, K., and Leibowitz, S. F. (2005). Glucose injection reduces neuropeptide Y and agouti-related protein expression in the arcuate nucleus: A possible physiological role in eating behavior. *Mol. Brain Res.* **135**, 69–80.

Chen, Y. Y., Hu, C. Z., Hsu, C. K., Zhang, Q., Bi, C., Asnicar, M., Hsiung, H. M., Fox, N., Slieker, L. J., Yang, D. D., Heiman, M. L., and Shi, Y. G. (2002). Targeted disruption of the melanin-concentrating hormone receptor-1 results in hyperphagia and resistance to diet-induced obesity. *Endocrinology* **143**, 2469–2477.

Chou, T. C., Lee, C. E., Lu, J., Elmquist, J. K., Hara, J., Willie, J. T., Beuckmann, C. T., Chemelli, R. M., Sakurai, T., Yanagisawa, M., Saper, C. B., and Scammell, T. E. (2001). Orexin (Hypocretin) neurons contain dynorphin. *J. Neurosci.* **21**, U19–U24.

Chou, T. C., Rotman, S. R., and Saper, C. B. (2004). Lateral hypothalamic acetylcholinesterase-immunoreactive neurons co-express either orexin or melanin concentrating hormone. *Neurosci. Lett.* **370**, 123–126.

Chronwall, B. M. (1985). Anatomy and physiology of the neuroendocrine arcuate nucleus. *Peptides* **6 (suppl.2)**, 1–11.

Chronwall, B. M., Di Maggio, D. A., Massari, V. J., Pickel, V. M., Ruggiero, D. A., and O'Donohue, T. L. (1985). The anatomy of neuropeptide Y-containing neurons in rat brain. *Neuroscience* **15**, 1159–1181.

Chua, S. C., Leibel, R. L., and Hirsch, J. (1991a). Food deprivation and age modulate neuropeptide gene expression in the murine hypothalamus and adrenal gland. *Mol. Brain Res.* **9**, 95–101.

Chua, S. C., Brown, A. W., Kim, J. H., Hennessey, K. L., Leibel, R. L., and Hirsch, J. (1991b). Food deprivation and hypothalamic neuropeptide gene expression—effects of strain background and the diabetes mutation. *Mol. Brain Res.* **11**, 291–299.

Chua, S. C., Chung, W. K., Wupeng, X. S., Zhang, Y. Y., Liu, S. M., Tartaglia, L., and Leibel, R. L. (1996a). Phenotypes of mouse diabetes and rat fatty due to mutations in the OB (leptin) receptor. *Science* **271**, 994–996.

Chua, S. C., White, D. W., Wupeng, X. S., Liu, S. M., Okada, N., Kershaw, E. E., Chung, W. K., Powerkehoe, L., Chua, M., Tartaglia, L. A., and Leibel, R. L. (1996b). Phenotype of fatty due to Gln269Pro mutation in the leptin receptor (lepr). *Diabetes* **45**, 1141–1143.

Clark, J. T., Kalra, P. S., Crowley, W. R., and Kalra, S. P. (1984). Neuropeptide Y and human pancreatic polypeptide stimulate feeding behavior in rats. *Endocrinology* **115**, 427–429.

Clegg, D. J., Air, E. L., Woods, S. C., and Seeley, R. J. (2002). Eating elicited by orexin-A, but not melanin-concentrating hormone, is opioid mediated. *Endocrinology* **143**, 2995–3000.

Clegg, D. J., Air, E. L., Benoit, S. C., Sakai, R. S., Seeley, R. J., and Woods, S. C. (2003). Intraventricular melanin-concentrating hormone stimulates water intake independent of food intake. *Am. J. Physiol.* **284**, R494–R499.

Cluderay, J. E., Harrison, D. C., and Hervieu, G. J. (2002). Protein distribution of the orexin-2 receptor in the rat central nervous system. *Regul. Pept.* **104**, 131–144.

Collin, M., Backberg, M., Onnestam, K., and Meister, B. (2002). 5-HT1A receptor immunoreactivity in hypothalamic neurons involved in body weight control. *Neuroreport* **13**, 945–951.

Cone, R. D. (2005). Anatomy and regulation of the central melanocortin system. *Nat. Neurosci.* **8**, 571–578.

Corp, E. S., Melville, L. D., Greenberg, D., Gibbs, J., and Smith, G. P. (1990). Effect of 4th ventricular neuropeptide-Y and peptide-YY on ingestive and other behaviors. *Am. J. Physiol.* **259**, R317–R323.

Cota, D., Marsicano, G., Tschop, M., Grubler, Y., Flachskamm, C., Schubert, M., Auer, D., Yassouridis, A., ThoneReineke, C., Ortmann, S., Tomassoni, F., Cervino, C., Nisoli, E., Linthorst, A. C. E., Pasquali, R., Lutz, B., Stalla, G. K., and Pagotto, U. (2003). The endogenous cannabinoid system affects energy balance via central orexigenic drive and peripheral lipogenesis. *J. Clin. Invest.* **112**, 423–431.

Crawley, J. N. (1999). The role of galanin in feeding behavior. *Neuropeptides* **33**, 369–375.

Currie, P. J., and Coscina, D. V. (1995). Dissociated feeding and hypothermic effects of neuropeptide Y in the paraventricular and perifornical hypothalamus. *Peptides* **16**, 599–604.

Cvetkovic, V., Poncet, F., Fellmann, D., Griffond, B., and Risold, P. Y. (2003a). Diencephalic neurons producing melanin-concentrating hormone are influenced by local and multiple extra-hypothalamic tachykininergic projections through the neurokinin 3 receptor. *Neuroscience* **119**, 1113–1145.

Cvetkovic, V., Brischoux, F., Griffond, B., Bernard, G., Jacquemard, C., Fellmann, D., and Risold, P. Y. (2003b). Evidence of melanin-concentrating hormone-containing neurons supplying both cortical and neuroendocrine projections. *Neuroscience* **116**, 31–35.

Cvetkovic, V., Brischoux, F., Jacquemard, C., Fellmann, D., Griffond, B., and Risold, P. Y. (2004). Characterization of subpopulations of neurons producing melanin-concentrating hormone in the rat ventral diencephalon. *J. Neurochem.* **91**, 911–919.

Daniels, A. J., Grizzle, M. K., Wiard, R. P., Matthews, J. E., and Heyer, D. (2002). Food intake inhibition and reduction in body weight gain in lean and obese rodents treated with GW438014A, a potent and selective NPY-Y5 receptor antagonist. *Regul. Pept.* **106**, 47–54.

Date, Y., Ueta, Y., Yamashita, H., Yamaguchi, H., Matsukura, S., Kangawa, K., Sakurai, T., Yanagisawa, M., and Nakazato, M. (1999). Orexins, orexigenic hypothalamic peptides, interact with autonomic, neuroendocrine and neuroregulatory systems. *Proc. Natl. Acad. Sci. U.S.A.* **96**, 748–753.

Davies, L., and Marks, J. L. (1994). Role of hypothalamic neuropeptide Y gene expression in body weight regulation. *Am. J. Physiol.* **266**, R1687–R1691.

Day, D. E., Keen-Rinehart, E., and Bartness, T. J. (2005). Role of NPY and its receptor subtypes in foraging, food hoarding, and food intake by Siberian hamsters. *Am. J. Physiol.* **289**, R29–R36.

deLecea, L., Kilduff, T. S., Peyron, C., Gao, X., Foye, P. E., Danielson, P. E., Fukuhara, C., Battenberg,

E. L., Gautvik, V. T., Bartlett, II, F. S., Frankel, W. N., van den Pol, A. N., Bloom, F. E., Gautvik, K. M., and Sutcliffe, J. G. (1998). The hypocretins: hypothalamus-specific peptides with neuroexcitatory activity. *Proc. Natl. Acad. Sci. U.S.A.* **95**, 322–327.

Della Zuana, O., Presse, F., Ortola, C., Duhault, J., Nahon, J. L., and Levens, N. (2002). Acute and chronic administration of melanin-concentrating hormone enhances food intake and body weight in Wistar and Sprague-Dawley rats. *Int. J. Obes.* **26**, 1289–1295.

Della Zuana, O., Revereault, L., Beck-Sickinger, A., Monge, A., Caignard, D. H., Fauchere, J. L., Henlin, J. M., Audinot, V., Boutin, J. A., Chamorro, S., Feletou, M., and Levens, N. (2004). A potent and selective NPYY5 antagonist reduces food intake but not through blockade of the NPYY5 receptor. *Int. J. Obes.* **28**, 628–639.

Dhillon, H., Ge, Y. L., Minter, R. M., Prima, V., Moldawer, L. L., Muzyczka, N., Zolotukhin, S., Kalra, P. S., and Kalra, S. P. (2000). Long-term differential modulation of genes encoding orexigenic and anorexigenic peptides by leptin delivered by rAAV vector in ob/ob mice—Relationship with body weight change. *Regul. Pept.* **92**, 97–105.

DiMarzo, V., and Matias, I. (2005). Endocannabinoid control of food intake and energy balance. *Nat. Neurosci.* **8**, 585–589.

Drazen, D. L., Coolen, L. M., Strader, A. D., Wortman, M. D., Woods, S. C., and Seeley, R. J. (2004). Differential effects of adrenalectomy on melanin-concentrating hormone and orexin A. *Endocrinology* **145**, 3404–3412.

Drazen, D. L., Wortman, M. D., Seeley, R. J., and Woods, S. C. (2005). Neuropeptide Y prepares rats for scheduled feeding. *Am. J. Physiol.* **288**, R1606–R1611.

Dryden, S., Pickavance, L., Frankish, H. M., and Williams, G. (1995). Increased neuropeptide Y secretion in the hypothalamic paraventricular nucleus of obese (fa/fa) Zucker rats. *Brain Res.* **690**, 185–188.

Dube, M. G., Xu, B., Crowley, W. R., Kalra, P. S., and Kalra, S. P. (1994). Evidence that neuropeptide Y is a physiological signal for normal food intake. *Brain Res.* **646**, 341–344.

Dube, M. G., Horvath, T. L., Kalra, P. S., and Kalra, S. P. (2000). Evidence of NPY Y5 receptor involvement in food intake elicited by orexin A in sated rats. *Peptides* **21**, 1557–1560.

Duhault, J., Boulanger, M., Chamorro, S., Boutin, J. A., Zuana, O., Douillet, E., Fauchere, J. L., Feletou, M., Germain, M., Husson, B., Vega, A. M., Renard, P., and Tisserand, F. (2000). Food intake regulation in rodents: Y-5 or Y1 NPY receptors or both? *Can. J. Physiol. Pharmacol.* **78**, 173–185.

Duncan, E. A., Proulx, K., and Woods, S. C. (2005). Central administration of melanin-concentrating hormone increases alcohol and sucrose/quinine intake in rats. *Alcohol. Clin. Exp. Res.* **29**, 958–964.

Edwards, C. M. B., Abusnana, S., Sunter, D., Murphy, K. G., Ghatei, M. A., and Bloom, S. R. (1999). The effect of the orexins on food intake: comparison with neuropeptide Y, melanin-concentrating hormone and galanin. *J. Endocrinol.* **160**, R7–R12.

Elias, C. F., Saper, C. B., Maratos-Flier, E., Tritos, N. A., Lee, C., Kelly, J., Tatro, J. B., Hoffman, G. E., Ollmann, M. M., Barsh, G. S., Sakurai, T., Yanagisawa, M., and Elmquist, J. K. (1998). Chemically defined projections linking the mediobasal hypothalamus and the lateral hypothalamic area. *J. Comp. Neurol.* **402**, 442–459.

Elliott, J. C., Harrold, J. A., Brodin, P., Enquist, K., Backman, A., Bystrom, M., Lindgren, K., King, P., and Williams, G. (2004). Increases in melanin-concentrating hormone and MCH receptor levels in the hypothalamus of dietary-obese rats. *Mol. Brain Res.* **128**, 150–159.

Everitt, B. J., Hökfelt, T., Terenius, L., Tatemoto, K., Mutt, V., and Goldstein, M. (1984). Differential coexistence of neuropeptide Y (NPY)-like immunoreactivity with catecholamines in the central nervous system of the rat. *Neuroscience* **11**, 443–462.

Erickson, J. C., Hollopeter, G., and Palmiter, R. D. (1996). Attenuation of the obesity syndrome of ob/ob mice by the loss of neuropeptide Y. *Science* **274**, 1704–1707.

Flood, J. F., and Morley, J. E. (1991). Increased food intake by neuropeptide Y is due to an increased motivation to eat. *Peptides* **12**, 1329–1332.

Frankish, H. M., Dryden, S., Hopkins, D., Wang, Q., and Williams, G. (1995). Neuropeptide Y, the hypothalamus, and diabetes: insights into the central control of metabolism. *Peptides* **16**, 757–771.

Fu, L. Y., Acuna-Goycolea, C., and van den Pol, A. N. (2004). Neuropeptide Y inhibits hypocretin/orexin neurons by multiple presynaptic and postsynaptic mechanisms: tonic depression of the hypothalamic arousal system. *J. Neurosci.* **24**, 8741–8751.

Funahashi, H., Hori, T., Shimoda, Y., Mizushima, H., Ryushi, T., Katoh, S., and Shioda, S. (2000). Morphological evidence for neural interactions between leptin and orexin in the hypothalamus. *Regul. Pept.* **92**, 31–35.

Funahashi, H., Yamada, S., Kageyama, H., Takenoya, F., Guan, J. L., and Shioda, S. (2003). Co-existence of leptin- and orexin-receptors in feeding-regulating neurons in the hypothalamic arcuate nucleus—a triple labeling study. *Peptides* **24**, 687–694.

Furudono, Y., Ando, C., Kobashi, M., Yamamoto, C., and Yamamoto, T. (2005). The role of orexigenic neuropeptides in the ingestion of sweet-tasting substances in rats. *Chem. Senses* **30**(Suppl. 1), i186–i187.

Gamber, K. M., MacArthur, H., and Westfall, T. C. (2005). Cannabinoids augment the release of neuropeptide Y in the rat hypothalamus. *Neuropharmacology* **49**, 646–652.

Gao, J., Ghibaudi, L., van Heek, M., and Hwa, J. J. (2002). Characterization of diet-induced obese rats that develop persistent obesity after 6 months of high-fat followed by 1 month of low-fat diet. *Brain Res.* **936**, 87–90.

Gardiner, J. V., Kong, W. M., Ward, H., Murphy, K. G., Dhillo, W. S., and Bloom, S. R. (2005). AAV mediated expression of anti-sense neuropeptide Y cRNA in the arcuate nucleus of rats results in decreased weight gain and food intake. *Biochem. Biophys. Res. Commun.* **327**, 1088–1093.

Georgescu, D., Sears, R. M., Hommel, J. D., Barrot, M., Bolanos, C. A., Marsh, D. J., Bednarek, M. A., Bibb, J. A., Maratos-Flier, E., Nestler, E. J., and DiLeone, R. J. (2005). The hypothalamic neuropeptide melanin-concentrating hormone acts in the nucleus accumbens to modulate feeding behavior and forced-swim performance. *J. Neurosci.* **25**, 2933–2940.

Giraudo, S. Q., Kim, E. M., Grace, M. K., Billington, C. J., and Levine, A. S. (1998). Effect of peripheral 2-DG on opioid and neuropeptide Y gene expression. *Brain Res.* **792**, 136–140.

Glass, M. J., Cleary, J. P., Billington, C. J., and Levine, A. S. (1997). Role of carbohydrate type on diet selection in neuropeptide Y-stimulated rats. *Am. J. Physiol.* **42**, R2040–R2045.

Godbole, V., York, D. A., and Bloxham, D. P. (1978). Developmental changes in the fatty (fa/fa) rat: evidence for defective thermogenesis preceding the hyperlipogenesis and hyperinsulinemia. *Diabetologia* **15**, 41–44.

Gomori, A., Ishihara, A., Ito, M., Mashiko, S., Matsushita, H., Yumoto, M., Ito, M., Tanaka, T., Tokita, S., Moriya, M., Iwaasa, H., and Kanatani, A. (2003). Chronic intracerebroventricular infusion of MCH causes obesity in mice. *Am. J. Physiol.* **284**, E583–E588.

Griffond, B., and Baker, B. I. (2002). Cell and molecular cell biology of melanin-concentrating hormone. *In* "International Review of Cytology: A Survey of Cell Biology" (K. W. Jeon, ed.), Vol. 213, pp. 233–277. Academic Press, ••.

Griffond, B., Risold, P. Y., Jacquemard, C., Colard, C., and Fellmann, D. (1999). Insulin-induced hypoglycemia increases preprohypocretin (Orexin) mRNA in the rat lateral hypothalamic area. *Neurosci. Lett.* **262**, 77–80.

Grillon, S., Herve, C., Griffond, B., and Fellmann, D. (1997). Exploring the expression of the melanin-concentrating hormone messenger RNA in the rat lateral hypothalamus after goldthioglucose injection. *Neuropeptides* **31**, 131–136.

Guan, J. L., Saotome, T., Wang, Q. P., Funahashi, H., Hori, T., Tanaka, S., and Shioda, S. (2001). Orexinergic innervation of POMC-containing neurons in the rat arcuate nucleus. *Neuroreport* **12**, 547–551.

Guan, J. L., Suzuki, R., Funahashi, H., Wang, Q. P., Kageyama, H., Uehara, K., Yamada, S., Tsurugano, S., and Shioda, S. (2002a). Ultrastructural localization of orexin-1 receptor, in pre- and post-synaptic neurons in the rat arcuate nucleus. *Neurosci. Lett.* **329**, 209–212.

Guan, J. L., Uehara, K., Lu, S., Wang, Q. P., Funahashi, H., Sakurai, T., Yanagizawa, M., and Shioda, S. (2002b). Reciprocal synaptic relationships between orexin- and melanin-concentrating hormone-containing neurons in the rat lateral hypothalamus: a novel circuit implicated in feeding regulation. *Int. J. Obes.* **26**, 1523–1532.

Hahn, T. M., Breininger, J. F., Baskin, D. G., and Schwartz, M. W. (1998). Coexpression of Agrp and NPY in fasting-activated hypothalamic neurons. *Nat. Neurosci.* **1**, 271–272.

Hakansson, M. L., deLecea, L., Sutcliffe, J. G., Yanagisawa, M., and Meister, B. (1999). Leptin receptor- and STAT3-immunoreactivities in hypocretin/orexin neurones of the lateral hypothalamus. *J. Neuroendocrinol.* **11**, 653–663.

Hanada, R., Nakazato, M., Matsukura, S., Murakami, N., Yoshimatsu, H., and Sakata, T. (2000). Differential regulation of melanin-concentrating hormone and orexin genes in the agouti-related protein/melanocortin-4 receptor system. *Biochem. Biophys. Res. Commun.* **268**, 88–91.

Hansen, M. J., Jovanovska, V., and Morris, M. J. (2004). Adaptive responses in hypothalamic neuropeptide Y in the face of prolonged high-fat feeding in the rat. *J. Neurochem.* **88**, 909–916.

Hara, J., Beuckmann, C. T., Nambu, T., Willie, J. T., Chemelli, R. M., Sinton, C. M., Sugiyama, F., Yagami, K., Goto, K., Yanagisawa, M., and Sakurai, T. (2001). Genetic ablation of orexin neurons in mice results in narcolepsy, hypophagia, and obesity. *Neuron* **30**, 345–354.

Hawes, B. E., Kil, E., Green, B., ONeill, K., Fried, S., and Graziano, M. P. (2000). The melanin-concentrating hormone receptor couples to multiple G proteins to activate diverse intracellular signaling pathways. *Endocrinology* **141**, 4524–4532.

Haynes, A. C., Arch, J. R. S., Wilson, S., McClue, S., and Buckingham, R. E. (1998). Characterisation of the neuropeptide Y receptor that mediates feeding in the rat: a role for the Y5 receptor? *Regul. Pept.* **75–6**, 355–361.

Haynes, A. C., Jackson, B., Chapman, H., Tadayyon, M., Johns, A., Porter, R. A., and Arch, J. R. S. (2000). A selective orexin-1 receptor antagonist reduces food consumption in male and female rats. *Regul. Pept.* **96**, 45–51.

Haynes, A. C., Chapman, H., Taylor, C., Moore, G. B. T., Cawthorne, M. A., Tadayyon, M., Clapham, J. C., and Arch, J. R. S. (2002). Anorectic, thermogenic and anti-obesity activity of a selective orexin-1 receptor antagonist in ob/ob mice. *Regul. Pept.* **104**, 153–159.

He, B., White, B. D., Edwards, G. L., and Martin, R. J. (1998). Neuropeptide Y antibody attenuates 2-deoxy-D-glucose induced feeding in rats. *Brain Res.* **781**, 348–350.

Henry, M., Ghibaudi, L., Gao, J., and Hwa, J. J. (2005). Energy metabolic profile of mice after chronic activation of central NPY Y1, Y2, or Y5 receptors. *Obes. Res.* **13**, 36–47.

Herve, C., and Fellmann, D. (1997). Changes in rat melanin-concentrating hormone and dynorphin messenger ribonucleic acids induced by food deprivation. *Neuropeptides* **31**, 237–242.

Hervieu, G. J., Cluderay, J. E., Harrison, D., Meakin, J., Maycox, P., Nasir, S., and Leslie, R. A. (2000). The distribution of the mRNA and protein products of the melanin-concentrating hormone (MCH) receptor gene, slc-1, in the central nervous system of the rat. *Eur. J. Neurosci.* **12**, 1194–1216.

Hervieu, G. J., Cluderay, J. E., Harrison, D. C., Roberts, J. C., and Leslie, R. A. (2001). Gene expression and protein distribution of the orexin-1 receptor in the rat brain and spinal cord. *Neuroscience* **103**, 777–797.

Hilairet, S., Bouaboula, M., Carriere, D., LeFur, G., and Casellas, P. (2003). Hypersensitization of the orexin 1 receptor by the CB1 receptor—Evidence for crosstalk blocked by the specific CB1 antagonist, SR141716. *J. Biol. Chem.* **278**, 23731–23737.

Horvath, T. L., and Gao, X. B. (2005). Input organization and plasticity of hypocretin neurons: possible clues to obesity's association with insomnia. *Cell. Metab.* **1**, 279–286.

Horvath, T. L., Bechmann, I., Naftolin, F., Kalra, S. P., and Leranth, C. (1997). Heterogeneity in the neuropeptide Y-containing neurons of the rat arcuate nucleus: GABAergic and non-GABAergic subpopulations. *Brain Res.* **756**, 283–286.

Horvath, T. L., Diano, S., and van den Pol, A. N. (1999). Synaptic interaction between hypocretin (Orexin) and neuropeptide Y cells in the rodent and primate hypothalamus: a novel circuit implicated in metabolic and endocrine regulations. *J. Neurosci.* **19**, 1072–1087.

Huang, X. F., Han, M., and Storlien, L. H. (2003). The level of NPY receptor mRNA expression in diet-induced obese and resistant mice. *Mol. Brain Res.* **115**, 21–28.

Hulsey, M. G., Pless, C. M., White, B. D., and Martin, R. J. (1995). ICV administration of anti-NPY antisense oligonucleotide: effects on feeding behavior, body weight, peptide content and peptide release. *Regul. Pept.* **59**, 207–214.

Ida, T., Nakahara, K., Katayama, T., Murakami, N., and Nakazato, M. (1999). Effect of lateral cerebroventricular injection of the appetite-stimulating neuropeptide, orexin and neuropeptide Y, on the various behavioral activities of rats. *Brain Res.* **821**, 526–529.

Ida, T., Nakahara, K., Murakami, T., Hanada, R., Nakazato, M., and Murakami, N. (2000a). Possible involvement of orexin in the stress reaction in rats. *Biochem. Biophys. Res. Commun.* **270**, 318–323.

Ida, T., Nakahara, K., Kuroiwa, T., Fukui, K., Nakazato, M., Murakami, T., and Murakami, N. (2000b). Both

corticotropin releasing factor and neuropeptide Y are involved in the effect of orexin (hypocretin) on the food intake in rats. *Neurosci. Lett.* **293**, 119–122.

Iida, M., Murakami, T., Ishida, K., Mizuno, A., Kuwajima, M., and Shima, K. (1996). Substitution at codon 269 (glutamine → proline) of the leptin receptor (OB-r) cDNA is the only mutation found in the Zucker fatty (fa/fa) rat. *Biochem. Biophys. Res. Commun.* **224**, 597–604.

Inui, A. (1999). Neuropeptide Y feeding receptors: are multiple subtypes involved? *Trends Pharmacol. Sci.* **20**, 43–46.

Ishii, Y., Blundell, J. E., Halford, J. C. G., Upton, N., Porter, R., Johns, A., Jeffrey, P., Summerfield, S., and Rodgers, R. J. (2005a). Anorexia and weight loss in male rats 24 h following single dose treatment with orexin-1 receptor antagonist SB-334867. *Behav. Brain Res.* **157**, 331–341.

Ishii, Y., Blundell, J. E., Halford, J. C. G., Upton, N., Porter, R., Johns, A., and Rodgers, R. J. (2005b). Satiety enhancement by selective orexin-1 receptor antagonist SB-334867: influence of test context and profile comparison with CCK-8S. *Behav. Brain Res.* **160**, 11–24.

Ishizaki, K., Honma, S., Katsuno, Y., Abe, H., Masubuchi, S., Namihira, M., and Honma, K. (2003). Gene expression of neuropeptide Y in the nucleus of the solitary tract is activated in rats under restricted daily feeding but not under 48-h food deprivation. *Eur. J. Neurosci.* **17**, 2097–2105.

Israel, Y., Kandov, Y., Khaimova, E., Kest, A., Lewis, S. R., Pasternak, G. W., Pan, Y. X., Rossi, G. C., and Bodnar, R. J. (2005). NPY-induced feeding: pharmacological characterization using selective opioid antagonists and antisense probes in rats. *Peptides* **26**, 1167–1175.

Ito, M., Gomori, A., Ishihara, A., Oda, Z., Mashiko, S., Matsushita, H., Yumoto, M., Ito, M., Sano, H., Tokita, S., Moriya, M., Iwaasa, H., and Kanatani, A. (2003). Characterization of MCH-mediated obesity in mice. *Am. J. Physiol.* **284**, E940–E945.

Jain, M. R., Horvath, T. L., Kalra, P. S., and Kalra, S. P. (2000). Evidence that NPY Y1 receptors are involved in stimulation of feeding by orexins (hypocretins) in sated rats. *Regul. Pept.* **87**, 19–24.

Jang, M. Y., and Romsos, D. R. (1998). Neuropeptide Y and corticotropin-releasing hormone concentrations within specific hypothalamic regions of lean but not ob/ob mice respond to food-deprivation and refeeding. *J. Nutr.* **128**, 2520–2525.

Jewett, D. C., Cleary, J., Levine, A. S., Schaal, D. W., and Thompson, T. (1992). Effects of neuropeptide-Y on food-reinforced behavior in satiated rats. *Pharmacol. Biochem. Behav.* **42**, 207–212.

Jewett, D. C., Cleary, J., Levine, A. S., Schaal, D. W., and Thompson, T. (1995). Effects of neuropeptide Y, insulin, a 2-deoxyglucose, and food deprivation on food motivated behavior. *Psychopharmacology (Berlin)* **120**, 267–271.

Jhanwar-Uniyal, M., and Chua, S. C. (1993). Critical effects of aging and nutritional state on hypothalamic neuropeptide-Y and galanin gene expression in lean and genetically obese Zucker rats. *Mol. Brain Res.* **19**, 195–202.

Jhanwar-Uniyal, M., Beck, B., Burlet, C., and Leibowitz, S. F. (1990). Diurnal rhythm of neuropeptide Y-like immunoreactivity in the suprachiasmatic, arcuate and paraventricular nuclei and other hypothalamic sites. *Brain Res.* **536**, 331–334.

Jolicoeur, F. B., Bouali, S. M., Fournier, A., and St Pierre, S. (1995). Mapping of hypothalamic sites involved in the effects of NPY on body temperature and food intake. *Brain Res. Bull.* **36**, 125–129.

Jones, D. N. C., Gartlon, J., Parker, F., Taylor, S. G., Routledge, C., Hemmati, P., Munton, R. P., Ashmeade, T. E., Hatcher, J. P., Johns, A., Porter, R. A., Hagan, J. J., Hunter, A. J., and Upton, N. (2001). Effects of centrally administered orexin-1 and orexinA: a role for orexin-1 receptors in orexin-B-induced hyperactivity. *Psychopharmacology (Berlin)* **153**, 210–218.

Kagotani, Y., Hashimoto, T., Tsuruo, Y., Kawano, H., Daikoku, S., and Chihara, K. (1989). Development of the neuronal system containing neuropeptide Y in the rat hypothalamus. *Int. J. Dev. Neurosci.* **7**, 359–374.

Kalra, S. P., and Kalra, P. S. (2003). Neuropeptide Y—A physiological orexigen modulated by the feedback action of ghrelin and leptin. *Endocrine* **22**, 49–55.

Kalra, S. P., Dube, M. G., Sahu, A., Phelps, C. P., and Kalra, P. S. (1991). Neuropeptide-Y secretion increases in the paraventricular nucleus in association with increased appetite for food. *Proc. Natl. Acad. Sci. U.S.A.* **88**, 10931–10935.

Kanatani, A., Ishihara, A., Asahi, S., Tanaka, T., Ozaki, S., and Ihara, M. (1996). Potent neuropeptide YY1 receptor antagonist, 1229U91: blockade of neuropeptide Y-induced and physiological food intake. *Endocrinology* **137**, 3177–3182.

Kanatani, A., Kanno, T., Ishihara, A., Hata, M., Sakuraba, A., Tanaka, T., Tsuchiya, Y., Mase, T., Fukuroda, T., Fukami, T., and Ihara, M. (1999). The novel neuropeptide YY1 receptor antagonist J-104870: a potent feeding suppressant with oral bioavailability. *Biochem. Biophys. Res. Commun.* **266**, 88–91.

Kanatani, A., Ishihara, A., Iwaasa, H., Nakamura, K., Okamoto, O., Hidaka, M., Ito, J., Fukuroda, T., MacNeil, D. J., VanderPloeg, L. H. T., Ishii, Y., Okabe, T., Fukami, T., and Ihara, M. (2000). L-152,804: orally active and selective neuropeptide YY5 receptor antagonist. *Biochem. Biophys. Res. Commun.* **272**, 169–173.

Kanatani, A., Hata, M., Mashiko, S., Ishihara, A., Okamoto, O., Haga, Y., Ohe, T., Kanno, T., Murai, N., Ishii, Y., Fukuroda, T., Fukami, T., and Ihara, M. (2001). A typical y1 receptor regulates feeding

behaviors: effects of a potent and selective Y1 antagonist, J-115814. *Mol. Pharmacol.* **59**, 501–505.

Kane, J. K., Parker, S. L., Matta, S. G., Fu, Y., Sharp, B. M., and Li, M. D. (2000). Nicotine up-regulates expression of orexin and its receptors in rat brain. *Endocrinology* **141**, 3623–3629.

Karteris, E., Machado, R. J., Chen, J., Zervou, S., Hillhouse, E. W., and Randeva, H. S. (2005). Food deprivation differentially modulates orexin receptor expression and signaling in rat hypothalamus and adrenal cortex. *Am. J. Physiol.* **288**, E1089–E1100.

Kask, A., Rago, L., and Harro, J. (1998). Evidence for involvement of neuropeptide Y receptors in the regulation of food intake: studies with Y-1-selective antagonist BIBP3226. *Br. J. Pharmacol.* **124**, 1507–1515.

Kask, A., Vasar, E., Heidmets, L. T., Allikmets, L., and Wikberg, J. E. S. (2001). Neuropeptide YY5 receptor antagonist CGP71683A: the effects on food intake and anxiety-related behavior in the rat. *Eur. J. Pharmacol.* **414**, 215–224.

Kawauchi, H., Kawazoe, I., Tsubokawa, M., Kishida, M., and Baker, B. I. (1983). Characterization of melanin-concentrating hormone in chum salmon pituitaries. *Nature* **305**, 321–323.

Keen-Rhinehart, E., Kalra, S. P., and Kalra, P. S. (2004). Leptin-receptor gene transfer into the arcuate nucleus of female fatty Zucker rats using recombinant adeno-associated viral vectors stimulates the hypothalamo-pituitary-gonadal axis. *Biol. Reprod.* **71**, 266–272.

Kennedy, A. R., Todd, J. F., Dhillo, W. S., Seal, L. J., Ghatei, M. A., OToole, C. P., Jones, M., Witty, D., Winborne, K., Riley, G., Hervieu, G., Wilson, S., and Bloom, S. R. (2003). Effect of direct injection of melanin-concentrating hormone into the paraventricular nucleus: further evidence for a stimulatory role in the adrenal axis via SLC-1. *J. Neuroendocrinol.* **15**, 268–272.

Kerkerian, L., and Pelletier, G. (1986). Effects of monosodium L-glutamate administration on neuropeptide Y-containing neurons in the rat hypothalamus. *Brain Res.* **369**, 388–390.

Kim, E. M., Welch, C. C., Grace, M. K., Billington, C. J., and Levine, A. S. (1998). Effects of palatability-induced hyperphagia and food restriction on mRNA levels of neuropeptide-Y in the arcuate nucleus. *Brain Res.* **806**, 117–121.

Kiwaki, K., Kotz, C. M., Wang, C. F., Lanningham-Foster, L., and Levine, J. A. (2004). Orexin A (Hypocretin 1) injected into hypothalamic paraventricular nucleus and spontaneous physical activity in rats. *Am. J. Physiol.* **286**, E551–E559.

Kojima, M., Hosoda, H., and Kangawa, K. (2004). Ghrelin, a novel growth-hormone-releasing and appetite-stimulating peptide from stomach. *Best Pract. Res. Clin. Endocrinol. Metab.* **18**, 517–530.

Kokkotou, E. G., Tritos, N. A., Mastaitis, J. W., Slieker, L., and Maratos-Flier, E. (2001). Melanin-concentrating hormone receptor is a target of leptin action in the mouse brain. *Endocrinology* **142**, 680–686.

Kokkotou, E., Jeon, J. Y., Wang, X. M., Marino, F. E., Carlson, M., Trombly, D. J., and Maratos-Flier, E. (2005). Mice with MCH ablation resist diet-induced obesity through strain-specific mechanisms. *Am. J. Physiol.* **289**, R117–R124.

Korner, J., Savontaus, E., Chua, S. C., Leibel, R. L., and Wardlaw, S. L. (2001). Leptin regulation of Agrp and NPY mRNA in the rat hypothalamus. *J. Neuroendocrinol.* **13**, 959–966.

Kowalski, T. J., Houpt, T. A., Jahng, J., Okada, N., Liu, S. M., Chua, S. C., and Smith, G. P. (1999). Neuropeptide Y overexpression in the preweanling Zucker (fa/fa) rat. *Physiol. Behav.* **67**, 521–525.

Kowalski, T. J., Farley, C., Cohen-Williams, M. E., Varty, G., and Spar, B. D. (2004). Melanin-concentrating hormone-1 receptor antagonism decreases feeding by reducing meal size. *Eur. J. Pharmacol.* **497**, 41–47.

Kozak, R., Mercer, J. G., Burlet, A., Moar, K. M., Burlet, C., and Beck, B. (1998). Hypothalamic neuropeptide Y content and mRNA expression in weanling rats subjected to dietary manipulations during fetal and neonatal life. *Regul. Pept.* **75–6**, 397–402.

Kozak, R., Burlet, A., Burlet, C., and Beck, B. (2000). Dietary composition during fetal and neonatal life affects neuropeptide Y functioning in adult offspring. *Dev. Brain Res.* **125**, 75–82.

Kozak, R., Richy, S., and Beck, B. (2005). Persistent alterations in neuropeptide Y release in the paraventricular nucleus of rats subjected to dietary manipulation during early life. *Eur. J. Neurosci.* **21**, 2887–2892.

Krief, S., and Bazin, R. (1991). Genetic obesity—Is the defect in the sympathetic nervous system?—A review through developmental studies in the preobese Zucker rat. *Proc. Soc. Exp. Biol. Med.* **198**, 528–538.

Kuenzel, W. J., Douglass, L. W., and Davison, B. A. (1987). Robust feeding following central administration of neuropeptide Y or peptide YY in chicks, *Gallus domesticus. Peptides* **8**, 823–828.

Kunii, K., Yamanaka, A., Nambu, T., Matsuzaki, I., Goto, K., and Sakurai, T. (1999). Orexins/hypocretins regulate drinking behavior. *Brain Res.* **842**, 256–261.

Kurose, T., Ueta, Y., Yamamoto, Y., Serino, R., Ozaki, Y., Saito, J., Nagata, S., and Yamashita, H. (2002). Effects of restricted feeding on the activity of hypothalamic orexin (OX)-A containing neurons and OX2 receptor mRNA level in the paraventricular nucleus of rats. *Regul. Pept.* **104**, 145–151.

Kuru, M., Ueta, Y., Serino, R., Nakazato, M., Yamamoto, Y., Shibuya, I., and Yamashita, H. (2000). Centrally administered orexin/hypocretin activates HPA axis in rats. *Neuroreport* **11**, 1977–1980.

Lambert, P. D., Wilding, J. P. H., Turton, M. D., Ghatei, M. A., and Bloom, S. R. (1994). Effect of food depri-

vation and streptozotocin-induced diabetes on hypothalamic neuropeptide Y release as measured by a radioimmunoassay-linked microdialysis procedure. *Brain Res.* **656**, 135–140.

Langmead, C. J., Jerman, J. C., Brough, S. J., Scott, C., Porter, R. A., and Herdon, H. J. (2004). Characterisation of the binding of [H-3]-SB-674042, a novel non-peptide antagonist, to the human orexin-1 receptor. *Br. J. Pharmacol.* **141**, 340–346.

Larhammar, D. (1996). Evolution of neuropeptide Y, peptide YY and pancreatic polypeptide. *Regul. Pept.* **62**, 1–11.

Larhammar, D., and Salaneck, E. (2004). Molecular evolution of NPY receptor subtypes. *Neuropeptides* **38**, 141–151.

Lecklin, A., Lundell, I., Salmela, S., Mannisto, P. T., Beck-Sickinger, A. G., and Larhammar, D. (2003). Agonists for neuropeptide Y receptors Y-1 and Y-5 stimulate different phases of feeding in guinea pigs. *Br. J. Pharmacol.* **139**, 1433–1440.

Leibowitz, S. F., and Alexander, J. T. (1991). Analysis of neuropeptide Y-induced feeding—Dissociation of Y1-receptor and Y2-receptor effects on natural meal patterns. *Peptides* **12**, 1251–1260.

Levin, B. E. (1999). Arcuate NPY neurons and energy homeostasis in diet-induced obese and resistant rats. *Am. J. Physiol.* **45**, R382–R387.

Levin, B. E., and Dunn-Meynell, A. A. (1997). Dysregulation of arcuate nucleus preproneuropeptide Y mRNA in diet-induced obese rats. *Am. J. Physiol.* **41**, R1365–R1370.

Levin, B. E., and Dunn-Meynell, A. A. (2002a). Defense of body weight depends on dietary composition and palatability in rats with diet-induced obesity. *Am. J. Physiol.* **282**, R46–R54.

Levin, B. E., and Dunn-Meynell, A. A. (2002b). Reduced central leptin sensitivity in rats with diet-induced obesity. *Am. J. Physiol.* **283**, R941–R948.

Levin, B. E., Govek, E. K., and Dunn-Meynell, A. A. (1998). Reduced glucose-induced neuronal activation in the hypothalamus of diet-induced obese rats. *Brain Res.* **808**, 317–319.

Levin, B. E., Dunn-Meynell, A. A., and Banks, W. A. (2004). Obesity-prone rats have normal blood-brain barrier transport but defective central leptin signaling before obesity onset. *Am. J. Physiol.* **286**, R143–R150.

Levine, A. S., and Morley, J. E. (1984). Neuropeptide Y: a potent inducer of consummatory behavior in rats. *Peptides* **5**, 1025–1029.

Levine, A. S., Winsky-Sommerer, R., Huitron-Resendiz, S., Grace, M. K., and de Lecea, L. (2005). Injection of neuropeptide W into paraventricular nucleus of hypothalamus increases food intake. *Am. J. Physiol.* **288**, R1727–R1732.

Lewis, D. E., Shellard, L., Koeslag, D. G., Boer, D. E., McCarthy, H. D., McKibbin, P. E., Russell, J. C., and Williams, G. (1993). Intense exercise and food restriction cause similar hypothalamic neuropeptide-Y increases in rats. *Am. J. Physiol.* **264**, E279–E284.

Li, A. J., and Ritter, S. (2004). Glucoprivation increases expression of neuropeptide Y mRNA in hindbrain neurons that innervate the hypothalamus. *Eur. J. Neurosci.* **19**, 2147–2154.

Lin, S., Storlien, L. H., and Huang, X. F. (2000). Leptin receptor, NPY, POMC mRNA expression in the diet-induced obese mouse brain. *Brain Res.* **875**, 89–95.

Lopez, M., Seoane, L., Garcia, M. D., Lago, F., Casanueva, F. F., Senaris, R., and Dieguez, C. (2000). Leptin regulation of prepro-orexin and orexin receptor mRNA levels in the hypothalamus. *Biochem. Biophys. Res. Commun.* **269**, 41–45.

Lopez, M., Seoane, L. M., Garcia, M. D., Dieguez, C., and Senaris, R. (2002). Neuropeptide Y, but not agouti-related peptide or melanin-concentrating hormone, is a target peptide for orexin-A feeding actions in the rat hypothalamus. *Neuroendocrinology* **75**, 34–44.

Louis-Sylvestre, J., Giachetti, I., and Le Magnen, J. (1984). Sensory versus dietary factors in cafeteria-induced overweight. *Physiol. Behav.* **32**, 901–905.

Lu, X. Y., Bagnol, D., Burke, S., Akil, H., and Watson, S. J. (2000). Differential distribution and regulation of OX1 and OX2 orexin/hypocretin receptor messenger RNA in the brain upon fasting. *Horm. Behav.* **37**, 335–344.

Lubkin, M., and Stricker-Krongrad, A. (1998). Independent feeding and metabolic actions of orexins in mice. *Biochem. Biophys. Res. Commun.* **253**, 241–245.

Ludwig, D. S., Mountjoy, K. G., Tatro, J. B., Gillette, J. A., Frederich, R. C., Flier, J. S., and Maratos-Flier, E. (1998). Melanin-concentrating hormone: a functional melanocortin antagonist in the hypothalamus. *Am. J. Physiol.* **37**, E627–E633.

Ludwig, D. S., Tritos, N. A., Mastaitis, J. W., Kulkarni, R., Kokkotou, E., Elmquist, J., Lowell, B., Flier, J. S., and Maratos-Flier, E. (2001). Melanin-concentrating hormone overexpression in transgenic mice leads to obesity and insulin resistance. *J. Clin. Invest.* **107**, 379–386.

Lynch, W. C., Grace, M., Billington, C. J., and Levine, A. S. (1993). Effects of neuropeptide-Y on ingestion of flavored solutions in nondeprived rats. *Physiol. Behav.* **54**, 877–880.

Lynch, W. C., Hart, P., and Babcock, A. M. (1994). Neuropeptide Y attenuates satiety—evidence from a detailed analysis of patterns of ingestion. *Brain Res.* **636**, 28–34.

Maitre, M. (1997). The gamma-hydroxybutyrate signalling system in brain: organization and functional implications. *Prog. Neurobiol.* **51**, 337–361.

Marcus, J. N., Aschkenasi, C. J., Lee, C. E., Chemelli, R. M., Saper, C. B., Yanagisawa, M., and Elmquist, J. K. (2001). Differential expression of orexin receptors 1 and 2 in the rat brain. *J. Comp. Neurol.* **435**, 6–25.

Marks, J. L., Li, M., Schwartz, M., Porte, D., and Baskin, D. G. (1992). Effect of fasting on regional levels of neuropeptide-Y messenger RNA and insulin receptors in the rat hypothalamus—An autoradiographic study. *Mol. Cell. Neurosci.* **3**, 199–205.

Marsh, D. J., Hollopeter, G., Kafer, K. E., and Palmiter, R. D. (1998). Role of the Y5 neuropeptide Y receptor in feeding and obesity. *Nat. Med.* **4**, 718–721.

Marsh, D. J., Weingarth, D. T., Novi, D. E., Chen, H. Y., Trumbauer, M. E., Chen, A. S., Guan, X. M., Jiang, M. M., Feng, Y., Camacho, R. E., Shen, Z., Frazier, E. G., Yu, H., Metzger, J. M., Kuca, S. J., Shearman, L. P., Gopal-Truter, S., MacNeil, D. J., Strack, A. M., MacIntyre, D. E., VanderPloeg, L. H. T., and Qian, S. (2002). Melanin-concentrating hormone 1 receptor-deficient mice are lean, hyperactive, and hyperphagic and have altered metabolism. *Proc. Natl. Acad. Sci. U.S.A.* **99**, 3240–3245.

Mashiko, S., Ishihara, A., Iwaasa, H., Sano, H., Oda, Z., Ito, J., Yumoto, M., Okawa, M., Suzuki, J., Fukuroda, T., Jitsuoka, M., Morin, N. R., MacNeil, D. J., VanderPloeg, L. H. T., Ihara, M., Fukami, T., and Kanatani, A. (2003). Characterization of neuropeptide Y (NPY) Y5 receptor-mediated obesity in mice: chronic intracerebroventricular infusion of D-Trp(34)NPY. *Endocrinology* **144**, 1793–1801.

Mashiko, S., Ishihara, A., Gomori, A., Moriya, R., Ito, M., Iwaasa, H., Matsuda, M., Feng, Y., Shen, Z., Marsh, D. J., Bednarek, M. A., MacNeil, D. J., and Kanatani, A. (2005). Antiobesity effect of a melanin-concentrating hormone 1 receptor antagonist in diet-induced obese mice. *Endocrinology* **146**, 3080–3086.

Matthews, J. E., Jansen, M., Lyerly, D., Cox, R., Chen, W. J., Koller, K. J., and Daniels, A. J. (1997). Pharmacological characterization and selectivity of the NPY antagonist GR231118 (1229U91) for different NPY receptors. *Regul. Pept.* **72**, 113–119.

Maulon-Feraille, L., Della Zuana, O., Suply, T., Rovere-Jovene, C., Audinot, V., Levens, N., Boutin, J. A., Duhault, J., and Nahon, J. L. (2002). Appetite-boosting property of pro-melanin-concentrating hormone (131–165) (Neuropeptide-glutamic acid-isoleucine) is associated with proteolytic resistance. *J. Pharmacol. Exp. Ther.* **302**, 766–773.

McCarthy, H. D., McKibbin, P. E., Holloway, B., Mayers, R., and Williams, G. (1991). Hypothalamic neuropeptide-Y receptor characteristics and NPY-induced feeding responses in lean and obese Zucker rats. *Life Sci.* **49**, 1491–1497.

McKibbin, P. E., Cotton, S. J., McMillan, S., Holloway, B., Mayers, R., McCarthy, H. D., and Williams, G. (1991). Altered neuropeptide-Y concentrations in specific hypothalamic regions of obese (fa/fa) Zucker rats—Possible relationship to obesity and neuroendocrine disturbances. *Diabetes* **40**, 1423–1429.

McMillen, I. C., and Robinson, J. S. (2005). Developmental origins of the metabolic syndrome: prediction, plasticity, and programming. *Physiol. Rev.* **85**, 571–633.

McMinn, J. E., Seeley, R. J., Wilkinson, C. W., Havel, P. J., Woods, S. C., and Schwartz, M. W. (1998). NPY-induced overfeeding suppresses hypothalamic NPY mRNA expression: potential roles of plasma insulin and leptin. *Regul. Pept.* **75–6**, 425–431.

McShane, T. M., Petersen, S. L., McCrone, S., and Keisler, D. H. (1993). Influence of food restriction on neuropeptide-Y, proopiomelanocortin, and luteinizing hormone-releasing hormone gene expression in sheep hypothalami. *Biol. Reprod.* **49**, 831–839.

McShane, T. M., Wilson, M. E., and Wise, P. M. (1999). Effects of lifelong moderate caloric restriction on levels of neuropeptide Y, proopiomelanocortin, and galanin mRNA. *J. Gerontol. A Biol. Sci. Med. Sci.* **54**, B14–B21.

Meguid, M. M., Fetissov, S. O., Varma, M., Sato, T., Zhang, L. H., Laviano, A., and RossiFanelli, F. (2000). Hypothalamic dopamine and serotonin in the regulation of food intake. *Nutrition* **16**, 843–857.

Menendez, J. A., Mc Gregor, I. A., Healey, P. A., Atrens, D. M., and Leibowitz, S. F. (1990). Metabolic effects of neuropeptide Y injections into the paraventricular nucleus of the hypothalamus. *Brain Res.* **516**, 8–14.

Mercer, J. G., Hoggard, N., Williams, L. M., Lawrence, C. B., Hannah, L. T., Morgan, P. J., and Trayhurn, P. (1996a). Coexpression of leptin receptor and preproneuropeptide Y mRNA in arcuate nucleus of mouse hypothalamus. *J. Neuroendocrinol.* **8**, 733–735.

Mercer, J. G., Lawrence, C. B., and Atkinson, T. (1996b). Regulation of galanin gene expression in the hypothalamic paraventricular nucleus of the obese Zucker rat by manipulation of dietary macronutrients. *Mol. Brain Res.* **43**, 202–208.

Mercer, J. G., Moar, K. M., Rayner, D. V., Trayhurn, P., and Hoggard, N. (1997). Regulation of leptin receptor and NPY gene expression in hypothalamus of leptin-treated obese (ob/ob) and cold-exposed lean mice. *FEBS Lett.* **402**, 185–188.

Minami, S., Kamegai, J., Sugihara, H., Suzuki, N., Higuchi, H., and Wakabayashi, I. (1995). Central glucoprivation evoked by administration of 2-deoxy-*d*-glucose induces expression of the c-fos gene in a subpopulation of neuropeptide Y neurons in the rat hypothalamus. *Mol. Brain Res.* **33**, 305–310.

Miner, J. L., Della-Fera, M. A., Paterson, J. A., and Baile, C. A. (1989). Lateral cerebroventricular injection of neuropeptide Y stimulates feeding in the sheep. *Am. J. Physiol.* **257**, R383–R387.

Mondal, M. S., Nakazato, M., Date, Y., Murakami, N., Hanada, R., Sakata, T., and Matsukura, S. (1999). Characterization of orexin-A and orexin-B in the microdissected rat brain nuclei and their contents in two obese rat models. *Neurosci. Lett.* **273**, 45–48.

Mondal, M. S., Nakazato, M., and Matsukura, S. (2002). Characterization of orexins (Hypocretins) and melanin-concentrating hormone in genetically obese mice. *Regul. Pept.* **104**, 21–25.

Moreno, G., Perello, M., Gaillard, R. C., and Spinedi, E. (2005). Orexin A stimulates hypothalamic-pituitary-adrenal (HPA) axis function, but not food intake, in the absence of full hypothalamic NPY-ergic activity. *Endocrine* **26**, 99–106.

Morgan, D. G. A., Small, C. J., Abusnana, S., Turton, M., Gunn, I., Heath, M., Rossi, M., Goldstone, A. P., O'Shea, D., Meeran, K., Ghatei, M., Smith, D. M., and Bloom, S. (1998). The NPY Y1 receptor antagonist BIBP 3226 blocks NPY induced feeding via a non-specific mechanism. *Regul. Pept.* **75–6**, 377–382.

Mori, M., Harada, M., Terao, Y., Sugo, T., Watanabe, T., Shimomura, Y., Abe, M., Shintani, Y., Onda, H., Nishimura, O., and Fujino, M. (2001). Cloning of a novel G protein-coupled receptor, SLT, a subtype of the melanin-concentrating hormone receptor. *Biochem. Biophys. Res. Commun.* **283**, 1013–1018.

Moriguchi, T., Sakurai, T., Nambu, T., Yanagisawa, M., and Goto, K. (1999). Neurons containing orexin in the lateral hypothalamic area of the adult rat brain are activated by insulin-induced acute hypoglycemia. *Neurosci. Lett.* **264**, 101–104.

Morley, J. E., Levine, A. S., Gosnell, B. A., Kneip, J., and Grace, M. (1987a). Effect of neuropeptide Y on ingestive behaviors in the rat. *Am. J. Physiol.* **252**, R599–R609.

Morley, J. E., Hernandez, E. N., and Flood, J. F. (1987b). Neuropeptide Y increases food intake in mice. *Am. J. Physiol.* **253**, R516–R522.

Morris, Y. A., and Crews, D. (1990). The effects of exogenous neuropeptide-y on feeding and sexual behavior in the red-sided garter snake (*Thamnophis sirtalis parietalis*). *Brain Res.* **530**, 339–341.

Morton, G. J., Mystkowski, P., Matsumoto, A. M., and Schwartz, M. W. (2004). Increased hypothalamic melanin concentrating hormone gene expression during energy restriction involves a melanocortin-independent, estrogen-sensitive mechanism. *Peptides* **25**, 667–674.

Mullett, M. A., Billington, C. J., Levine, A. S., and Kotz, C. M. (2000). Hypocretin I in the lateral hypothalamus activates key feeding-regulatory brain sites. *Neuroreport* **11**, 103–108.

Muraki, Y., Yamanaka, A., Tsujino, N., Kilduff, T. S., Goto, K., and Sakurai, T. (2004). Serotonergic regulation of the orexin/hypocretin neurons through the 5-HT1A receptor. *J. Neurosci.* **24**, 7159–7166.

Muroya, S., Yada, T., Shioda, S., and Takigawa, M. (1999). Glucose-sensitive neurons in the rat arcuate nucleus contain neuropeptide Y. *Neurosci. Lett.* **264**, 113–116.

Muroya, S., Funahashi, H., Yamanaka, A., Kohno, D., Uramura, K., Nambu, T., Shibahara, M., Kuramochi, M., Takigawa, M., Yanagisawa, M., Sakurai, T., Shioda, S., and Yada, T. (2004). Orexins (Hypocretins) directly interact with neuropeptide Y, POMC and glucose-responsive neurons to regulate Ca^{2+} signaling in a reciprocal manner to leptin: orexigenic neuronal pathways in the mediobasal hypothalamus. *Eur. J. Neurosci.* **19**, 1524–1534.

Myers, R. D., Wooten, M. H., Ames, C. D., and Nyce, J. W. (1995). Anorexic action of a new potential neuropeptide Y antagonist [D-tyr(27,36), d-thr(32)]-NPY (27–36) infused into the hypothalamus of the rat. *Brain Res. Bull.* **37**, 237–245.

Nambu, T., Sakurai, T., Mizukami, K., Hosoya, Y., Yanagisawa, M., and Goto, K. (1999). Distribution of orexin neurons in the adult rat brain. *Brain Res.* **827**, 243–260.

O'Shea, R. D., and Gundlach, A. L. (1991). Preproneuropeptide-Y messenger ribonucleic acid in the hypothalamic arcuate nucleus of the rat is increased by food deprivation or dehydration. *J. Neuroendocrinol.* **3**, 11–14.

Olszewski, P. K., Li, D. H., Grace, M. K., Billington, C. J., Kotz, C. M., and Levine, A. S. (2003). Neural basis of orexigenic effects of ghrelin acting within lateral hypothalamus. *Peptides* **24**, 597–602.

Palani, A., Shapiro, S., McBriar, M. D., Clader, J. W., Greenlee, W. J., Spar, B., Kowalski, T. J., Farley, C., Cook, J., vanHeek, M., Weig, B., ONeill, K., Graziano, M., and Hawes, B. (2005). Biaryl ureas as potent and orally efficacious melanin concentrating hormone receptor 1 antagonists for the treatment of obesity. *J. Med. Chem.* **48**, 4746–4749.

Parker, E. M., Balasubramaniam, A., Guzzi, M., Mullins, D. E., Salisbury, B. G., Sheriff, S., Witten, M. B., and Hwa, J. J. (2000). [D-Trp(34)] neuropeptide Y is a potent and selective neuropeptide YY5 receptor agonist with dramatic effects on food intake. *Peptides* **21**, 393–399.

Parrott, R. F., Heavens, R. P., and Baldwin, B. A. (1986). Stimulation of feeding in the satiated pig by intracerebroventricular injection of neuropeptide Y. *Physiol. Behav.* **36**, 523–525.

Pau, M. Y. C., Pau, K. Y. F., and Spies, H. G. (1988). Characterization of central actions of neuropeptide Y on food and water intake in rabbits. *Physiol. Behav.* **44**, 797–802.

Pesonen, U., Huupponen, R., Rouru, J., and Koulu, M. (1992). Hypothalamic neuropeptide expression after food restriction in Zucker rats—evidence of persistent neuropeptide-Y gene activation. *Mol. Brain Res.* **16**, 255–260.

Peyron, C., Tighe, D. K., van den Pol, A. N., deLecea, L., Heller, H. C., Sutcliffe, J. G., and Kilduff, T. S. (1998). Neurons containing hypocretin (orexin) project to multiple neuronal systems. *J. Neurosci.* **18**, 9996–10015.

Phillips, M. S., Liu, Q. Y., Hammond, H. A., Dugan, V., Hey, P. J., Caskey, C. T., and Hess, J. F. (1996). Leptin receptor missense mutation in the fatty Zucker rat. *Nat. Genet.* **13**, 18–19.

Polidori, C., Ciccocioppo, R., Regoli, D., and Massi, M. (2000). Neuropeptide Y receptor(s) mediating feeding in the rat: characterization with antagonists. *Peptides* **21**, 29–35.

Porte, D., Baskin, D. G., and Schwartz, M. W. (2002). Leptin and insulin action in the central nervous system. *Nutr. Rev.* **60**, S20–S29.

Porter, R. A., Chan, W. N., Coulton, S., Johns, A., Hadley, M. S., Widdowson, K., Jerman, J. C., Brough, S. J., Coldwell, M., Smart, D., Jewitt, F., Jeffrey, P., and Austin, N. (2001). 1,3-Biarylureas as selective nonpeptide antagonists of the orexin-1 receptor. *Bioorg. Med. Chem. Lett.* **11**, 1907–1910.

Presse, F., Sorokovsky, I., Max, J. P., Nicolaidis, S., and Nahon, J. L. (1996). Melanin-concentrating hormone is a potent anorectic peptide regulated by food-deprivation and glucopenia in the rat. *Neuroscience* **71**, 735–745.

Qian, H., Azain, M. J., Hartzell, D. L., and Baile, C. A. (1998). Increased leptin resistance as rats grow to maturity. *Proc. Soc. Exp. Biol. Med.* **219**, 160–165.

Qu, D. Q., Ludwig, D. S., Gammeltoft, S., Piper, M., Pelleymounter, M. A., Cullen, M. J., Mathes, W. F., Przypek, J., Kanarek, R., and Maratos-Flier, E. (1996). A role for melanin-concentrating hormone in the central regulation of feeding behavior. *Nature* **380**, 243–247.

Raposinho, P. D., Pierroz, D. D., Broqua, P., White, R. B., Pedrazzini, T., and Aubert, M. L. (2001). Chronic administration of neuropeptide Y into the lateral ventricle of C57BL/6J male mice produces an obesity syndrome including hyperphagia, hyperleptinemia, insulin resistance, and hypogonadism. *Mol. Cell. Endocrinol.* **185**, 195–204.

Rehfeld, J. F. (2004). Cholecystokinin. *Best Pract. Res. Clin. Endocrinol. Metab.* **18**, 569–586.

Richard, D., Lin, Q., and Timofeeva, E. (2002). The corticotropin-releasing factor family of peptides and CRF receptors: their roles in the regulation of energy balance. *Eur. J. Pharmacol.* **440**, 189–197.

Robinson, S. W., Dinulescu, D. M., and Cone, R. D. (2000). Genetic models of obesity and energy balance in the mouse. *Annu. Rev. Genet.* **34**, 687–745.

Rodgers, R. J., Halford, J. C. G., de Souza, R. L. N., de Souza, A. L. C., Piper, D. C., Arch, J. R. S., and Blundell, J. E. (2000). Dose-response effects of orexin-A on food intake and the behavioral satiety sequence in rats. *Regul. Pept.* **96**, 71–84.

Rodgers, R. J., Halford, J. C. G., de Souza, R. L. N., de Souza, A. L. C., Piper, D. C., Arch, J. R. S., Upton, N., Porter, R. A., Johns, A., and Blundell, J. E. (2001). SB-334867, a selective orexin-1 receptor antagonist, enhances behavioral satiety and blocks the hyperphagic effect of orexin-A in rats. *Eur. J. Neurosci.* **13**, 1444–1452.

Rossi, M., Choi, S. J., O'Shea, D., Miyoshi, T., Ghatei, M. A., and Bloom, S. R. (1997). Melanin-concentrating hormone acutely stimulates feeding, but chronic administration has no effect on body weight. *Endocrinology* **138**, 351–355.

Rossi, M., Beak, S. A., Choi, S. J., Small, C. J., Morgan, D. G. A., Ghatei, M. A., Smith, D. M., and Bloom, S. R. (1999). Investigation of the feeding effects of melanin concentrating hormone on food intake—action independent of galanin and the melanocortin receptors. *Brain Res.* **846**, 164–170.

Rothwell, N., and Stock, M. (1979). Regulation of energy balance in two models of reversible obesity in the rat. *J. Comp. Physiol. Psychol.* **93**, 1024–1034.

Rovere, C., Viale, A., Nahon, J. L., and Kitabgi, P. (1996). Impaired processing of brain proneurotensin and promelanin-concentrating hormone in obese fat/fat mice. *Endocrinology* **137**, 2954–2958.

Rowland, N. E. (1988). Peripheral and central satiety factors in neuropeptide Y-induced feeding in rats. *Peptides* **9**, 989–992.

Sahu, A. (1998a). Evidence suggesting that galanin (GAL), melanin-concentrating hormone (MCH), neurotensin (NT), proopiomelanocortin (POMC) and neuropeptide Y (NPY) are targets of leptin signaling in the hypothalamus. *Endocrinology* **139**, 795–798.

Sahu, A. (1998b). Leptin decreases food intake induced by melanin-concentrating hormone (MCH), galanin (GAL) and neuropeptide Y (NPY) in the rat. *Endocrinology* **139**, 4739–4742.

Sahu, A. (2003). Leptin signaling in the hypothalamus: emphasis on energy homeostasis and leptin resistance. *Front. Neuroendocrinol.* **24**, 225–253.

Sahu, A., Kalra, S. P., Crowley, W. R., and Kalra, P. S. (1988a). Evidence that NPY-containing neurons in the brainstem project into selected hypothalamic nuclei: implication in feeding behavior. *Brain Res.* **457**, 376–378.

Sahu, A., Kalra, P. S., and Kalra, S. P. (1988b). Food deprivation and ingestion induce reciprocal changes in neuropeptide Y concentrations in the paraventricular nucleus. *Peptides* **9**, 83–86.

Saito, Y., Nothacker, H. P., Wang, Z. W., Lin, S. H. S., Leslie, F., and Civelli, O. (1999). Molecular characterization of the melanin-concentrating-hormone receptor. *Nature* **400**, 265–269.

Sakamaki, R., Uemoto, M., Inui, A., Asakawa, A., Ueno, N., Ishibashi, C., Hiron, S., Yukioka, H., Kato, A., Shinfuku, N., Kasuga, M., and Katsuura, G. (2005). Melanin-concentrating hormone enhances sucrose intake. *Int. J. Mol. Med.* **15**, 1033–1039.

Sakamoto, F., Yamada, S., and Ueta, Y. (2004). Centrally administered orexin-A activates corticotropin-releasing factor-containing neurons in the hypothalamic paraventricular nucleus and central amygdaloid nucleus of rats: possible involvement of central orexins on stress-activated central CRF neurons. *Regul. Pept.* **118**, 183–191.

Sakurai, T. (2002). Roles of orexins in regulation of feeding and wakefulness. *Neuroreport* **13**, 987–995.

Sakurai, T., Amemiya, A., Ishii, M., Matsuzaki, I., Chemelli, R. M., Tanaka, H., Williams, S. C., Richardson, J. A., Kozlowski, G. P., Wilson, S., Arch, J. R. S., Buckingham, R. E., Haynes, A. C., Carr, S. A., Annan, R. S., McNulty, D. E., Liu, W. S., Terrett, J. A., Elshourbagy, N. A., Bergsma, D. J., and Yanagisawa, M. (1998). Orexins and orexin receptors: A family of hypothalamic neuropeptides and G protein-coupled receptors that regulate feeding behavior. *Cell* **92**, 573–585.

Samson, W. K., Taylor, M. M., Follwell, M., and Ferguson, A. V. (2002). Orexin actions in hypothalamic paraventricular nucleus: physiological consequences and cellular correlates. *Regul. Pept.* **104**, 97–103.

Sanacora, G., Kershaw, M., Finkelstein, J. A., and White, J. D. (1990). Increased hypothalamic content of preproneuropeptide Y messenger ribonucleic acid in genetically obese Zucker rats and its regulation by food deprivation. *Endocrinology* **127**, 730–737.

Sanacora, G., Finkelstein, J. A., and White, J. D. (1992). Developmental aspects of differences in hypothalamic preproneuropeptide Y messenger ribonucleic acid content in lean and genetically obese Zucker rats. *J. Neuroendocrinol.* **4**, 353–357.

Sawchenko, P. E., Swanson, L. W., Grzanna, R., Howe, P. R. C., Bloom, S. R., and Polak, J. M. (1985). Colocalization of neuropeptide Y immunoreactivity in brainstem catecholaminergic neurons that project to the paraventricular nucleus of the hypothalamus. *J. Comp. Neurol.* **241**, 138–153.

Schaffhauser, A. O., Madiehe, A. M., Braymer, H. D., Bray, G. A., and York, D. A. (2002). Effects of a high-fat diet and strain on hypothalamic gene expression in rats. *Obes. Res.* **10**, 1188–1196.

Schlumberger, S. E., Talke-Messerer, C., Zumsteg, U., and Eberle, A. N. (2002). Expression of receptors for melanin-concentrating hormone (MCH) in different tissues and cell lines. *J. Recept. Signal. Transduct. Res.* **22**, 509–531.

Schwartz, M. W., Sipols, A. J., Grubin, C. E., and Baskin, D. G. (1993). Differential effect of fasting on hypothalamic expression of genes encoding neuropeptide-Y, galanin, and glutamic acid decarboxylase. *Brain Res. Bull.* **31**, 361–367.

Schwartz, M. W., Baskin, D. G., Bukowski, T. R., Kuijper, J. L., Foster, D., Lasser, G., Prunkard, D. E., Porte, D., Woods, S. C., Seeley, R. J., and Weigle, D. S. (1996). Specificity of leptin action on elevated blood glucose levels and hypothalamic neuropeptide Y gene expression in ob/ob mice. *Diabetes* **45**, 531–535.

Sclafani, A. (1987). Carbohydrate-induced hyperphagia and obesity in the rat: effects of saccharide type, form, and taste. *Neurosci. Biobehav. Rev.* **11**, 155–162.

Seeley, R. J., Payne, C. J., and Woods, S. C. (1995). Neuropeptide Y fails to increase intraoral intake in rats. *Am. J. Physiol.* **37**, R423–R427.

Seeley, R. J., Drazen, D. L., and Clegg, D. J. (2004). The critical role of the melanocortin system in the control of energy balance. *Annu. Rev. Nutr.* **24**, 133–149.

Segal-Lieberman, G., Bradley, R. L., Kokkotou, E., Carlson, M., Trombly, D. J., Wang, X. M., Bates, S., Myers, M. G., Flier, J. S., and Maratos-Flier, E. (2003). Melanin-concentrating hormone is a critical mediator of the leptin-deficient phenotype. *Proc. Natl. Acad. Sci. U.S.A.* **100**, 10085–10090.

Sergeyev, V., Broberger, C., Gorbatyuk, O., and Hokfelt, T. (2000). Effect of 2-mercaptoacetate and 2-deoxy-d-glucose administration on the expression of NPY, AGRP, POMC, MCH and hypocretin/orexin in the rat hypothalamus. *Neuroreport* **11**, 117–121.

Serradeillegal, C., Valette, G., Rouby, P. E., Pellet, A., Ourydonat, F., Brossard, G., Lespy, L., Marty, E., Neliat, G., Decointet, P., Maffrand, J. P., and Lefur, G. (1995). SR120819A, an orally-active and selective neuropeptide Y Y1 receptor antagonist. *FEBS Lett.* **362**, 192–196.

Shearman, L. P., Camacho, R. E., Stribling, D. S., Zhou, D., Bednarek, M. A., Hreniuk, D. L., Feighner, S. D., Tan, C. P., Howard, A. D., VanderPloeg, L. H. T., MacIntyre, D. E., Hickey, G. J., and Strack, A. M. (2003). Chronic MCH-1 receptor modulation alters appetite, body weight and adiposity in rats. *Eur. J. Pharmacol.* **475**, 37–47.

Shibasaki, T., Oda, T., Imaki, T., Ling, N., and Demura, H. (1993). Injection of antineuropeptide-Y gamma-globulin into the hypothalamic paraventricular nucleus decreases food intake in rats. *Brain Res.* **601**, 313–316.

Shimada, M., Tritos, N. A., Lowell, B. B., Flier, L. S., and Maratos-Flier, E. (1998). Mice lacking melanin-concentrating hormone are hypophagic and lean. *Nature* **396**, 670–674.

Shimomura, Y., Mori, M., Sugo, T., Ishibashi, Y., Abe, M., Kurokawa, T., Onda, H., Nishimura, O., Sumino, Y., and Fujino, M. (1999). Isolation and identification of melanin-concentrating hormone as the endogenous ligand of the SLC-1 receptor. *Biochem. Biophys. Res. Commun.* **261**, 622–626.

Shiraishi, T., Oomura, Y., Sasaki, K., and Wayner, M. J. (2000). Effects of leptin and orexin-A on food intake and feeding related hypothalamic neurons. *Physiol. Behav.* **71**, 251–261.

Sindelar, D. K., SteMarie, L., Miura, G. I., Palmiter, R. D., McMinn, J. E., Morton, G. J., and Schwartz, M. W. (2004). Neuropeptide Y is required for hyperphagic feeding in response to neuroglucopenia. *Endocrinology* **145**, 3363–3368.

Sindelar, D. K., Palmiter, R. D., Woods, S. C., and Schwartz, M. W. (2005). Attenuated feeding responses to circadian and palatability cues in mice lacking neuropeptide Y. *Peptides* **26**, 2597–2602.

Smart, D., Sabido-David, C., Brough, S. J., Jewitt, F., Johns, A., Porter, R. A., and Jerman, J. C. (2001). SB-334867-A: the first selective orexin-1 receptor antagonist. *Br. J. Pharmacol.* **132**, 1179–1182.

Smoller, J. W., Truett, G. E., Hirsch, J., and Leibel, R. L. (1993). A molecular genetic method for genotyping fatty (fa/fa) rats. *Am. J. Physiol.* **264**, R8–R11.

Souers, A. J., Gao, J., Wodka, D., Judd, A. S., Mulhern, M. M., Napier, J. J., Brune, M. E., Bush, E. N., Brodjian, S. J., Dayton, B. D., Shapiro, R., Hernandez, L. E., Marsh, K. C., Sham, H. L., Collins, C. A., and Kym, P. R. (2005). Synthesis and evaluation of urea-based indazoles as melanin-concentrating hormone

receptor 1 antagonists for the treatment of obesity. *Bioorg. Med. Chem. Lett.* **15**, 2752–2757.

Stanley, B. G., and Leibowitz, S. F. (1984). Neuropeptide Y: stimulation of feeding and drinking by injection into the paraventricular nucleus. *Life Sci.* **35**, 2635–2642.

Stanley, B. G., and Leibowitz, S. F. (1985). Neuropeptide Y injected in the paraventricular hypothalamus: a powerful stimulant of feeding behavior. *Proc. Natl. Acad. Sci. U.S.A.* **82**, 3940–3943.

Stanley, B. G., and Thomas, W. J. (1993). Feeding responses to perifornical hypothalamic injection of neuropeptide-Y in relation to circadian rhythms of eating behavior. *Peptides* **14**, 475–481.

Stanley, B. G., Chin, A. S., and Leibowitz, S. F. (1985a). Feeding and drinking elicited by central injection of neuropeptide Y: evidence for a hypothalamic site(s) of action. *Brain Res. Bull.* **14**, 521–524.

Stanley, B. G., Daniel, D. R., Chin, A. S., and Leibowitz, S. F. (1985b). Paraventricular nucleus injections of peptide YY and neuropeptide Y preferentially enhance carbohydrate ingestion. *Peptides* **6**, 1205–1211.

Stanley, B. G., Anderson, K. C., Grayson, M. H., and Leibowitz, S. F. (1989). Repeated hypothalamic stimulation with neuropeptide Y increases daily carbohydrate and fat intake and body weight gain in female rats. *Physiol. Behav.* **46**, 173–177.

Stanley, B. G., Magdalin, W., Seirafi, A., Thomas, W. J., and Leibowitz, S. F. (1993). The perifornical area—the major focus of (a) patchily distributed hypothalamic neuropeptide Y-sensitive feeding system(s). *Brain Res.* **604**, 304–317.

Steininger, T. L., Kilduff, T. S., Behan, M., Benca, R. M., and Landry, C. F. (2004). Comparison of hypocretin/orexin and melanin-concentrating hormone neurons and axonal projections in the embryonic and postnatal rat brain. *J. Chem. Neuroanat.* **27**, 165–181.

Steinman, J. L., Gunion, M. W., and Morley, J. E. (1994). Forebrain and hindbrain involvement of neuropeptide Y in ingestive behaviors of rats. *Pharmacol. Biochem. Behav.* **47**, 207–214.

Ster, A. M., Kowalski, T. J., Dube, M. G., Kalra, S. P., and Smith, G. P. (2003). Decreased hypothalamic concentration of neuropeptide Y correlates with onset of hyperphagia fa/fa rats on postnatal day 12. *Physiol. Behav.* **78**, 517–520.

Stricker-Krongrad, A., and Beck, B. (2002). Modulation of hypothalamic hypocretin/orexin mRNA expression by glucocorticoids. *Biochem. Biophys. Res. Commun.* **296**, 129–133.

Stricker-Krongrad, A., Barbanel, G., Beck, B., Burlet, A., Nicolas, J. P., and Burlet, C. (1993). K⁺-stimulated neuropeptide-Y release into the paraventricular nucleus and relation to feeding behavior in free-moving rats. *Neuropeptides* **24**, 307–312.

Stricker-Krongrad, A., Max, J. P., Musse, N., Nicolas, J. P., Burlet, C., and Beck, B. (1994). Increased threshold concentrations of neuropeptide Y for a stimulatory effect on food intake in obese Zucker rats—changes in the microstructure of the feeding behavior. *Brain Res.* **660**, 162–166.

Stricker-Krongrad, A., Kozak, R., Burlet, C., Nicolas, J. P., and Beck, B. (1997). Physiological regulation of hypothalamic neuropeptide Y release in lean and obese rats. *Am. J. Physiol.* **42**, R2112–R2116.

Stricker-Krongrad, A., Dimitrov, T., and Beck, B. (2001). Central and peripheral dysregulation of melanin-concentrating hormone in obese Zucker rats. *Mol. Brain Res.* **92**, 43–48.

Stricker-Krongrad, A., Richy, S., and Beck, B. (2002). Orexins/hypocretins in the ob/ob mouse: hypothalamic gene expression, peptide content and metabolic effects. *Regul. Pept.* **104**, 11–20.

Sunter, D., Morgan, I., Edwards, C. M. B., Dakin, C. L., Murphy, K. G., Gardiner, J., Taheri, S., Rayes, E., and Bloom, S. R. (2001). Orexins: effects on behavior and localisation of orexin receptor 2 messenger ribonucleic acid in the rat brainstem. *Brain Res.* **907**, 27–34.

Suply, T., Della Zuana, O., Audinot, V., Rodriguez, M., Beauverger, P., Duhault, J., Canet, E., Galizzi, J. P., Nahon, J. L., Levens, N., and Boutin, J. A. (2001). SLC-1 receptor mediates effect of melanin-concentrating hormone on feeding behavior in rat: a structure–activity study. *J. Pharmacol. Exp. Ther.* **299**, 137–146.

Sutcliffe, J. G., and deLecea, L. (2002). The hypocretins: setting the arousal threshold. *Nat. Rev. Neurosci.* **3**, 339–349.

Suzuki, R., Shimojima, H., Funahashi, H., Nakajo, S., Yamada, S., Guan, J. L., Tsurugano, S., Uehara, K., Takeyama, Y., Kikuyama, S., and Shioda, S. (2002). Orexin-1 receptor immunoreactivity in chemically identified target neurons in the rat hypothalamus. *Neurosci. Lett.* **324**, 5–8.

Sweet, D. C., Levine, A. S., and Kotz, C. M. (2004). Functional opioid pathways are necessary for hypocretin-1 (Orexin-A)-induced feeding. *Peptides* **25**, 307–314.

Taheri, S., Gardiner, J., Hafizi, S., Murphy, K., Dakin, C., Seal, L., Small, C., Ghatei, M., and Bloom, S. (2001). Orexin A immunoreactivity and preproorexin mRNA in the brain of Zucker and WKY rats. *Neuroreport* **12**, 459–464.

Takekawa, S., Asami, A., Ishihara, Y., Terauchi, J., Kato, K., Shimomura, Y., Mori, M., Murakoshi, H., Kato, K., Suzuki, N., Nishimura, O., and Fujino, M. (2002). T-226296: a novel, orally active and selective melanin-concentrating hormone receptor antagonist. *Eur. J. Pharmacol.* **438**, 129–135.

Takenoya, F., Hirayama, M., Kageyama, H., Funahashi, H., Kita, T., Matsumoto, H., Ohtaki, T., Katoh, S., Takeuchi, M., and Shioda, S. (2005). Neuronal interactions between galanin-like-peptide- and orexin- or melanin-concentrating hormone-containing neurons. *Regul. Pept.* **126**, 79–83.

Tan, C. P., Sano, H., Iwaasa, H., Pan, J., Sailer, A. W., Hreniuk, D. L., Feighner, S. D., Palyha, O. C., Pong, S. S., Figueroa, D. J., Austin, C. P., Jiang, M. M., Yu, H., Ito, J., Ito, M., Guan, X. M., MacNeil, D. J., Kanatani, A., VanderPloeg, L. H. T., and Howard, A. D. (2002). Melanin-concentrating hormone receptor subtypes 1 and 2: Species-specific gene expression. *Genomics* **79**, 785–792.

Tempel, D. L., and Leibowitz, S. F. (1990). Diurnal variations in the feeding responses to norepinephrine, neuropeptide-Y and galanin in the PVN. *Brain Res. Bull.* **25**, 821–825.

Tempel, D. L., Shor-Posner, G., Dwyer, D., and Leibowitz, S. F. (1989). Nocturnal patterns of macronutrient intake in freely feeding and food-deprived rats. *Am. J. Physiol.* **256**, R541–R548.

Thompson, R. H., Canteras, N. S., and Swanson, L. W. (1996). Organization of projections from the dorsomedial nucleus of the hypothalamus: a PHA-l study in the rat. *J. Comp. Neurol.* **376**, 143–173.

Thorpe, A. J., and Kotz, C. M. (2005). Orexin A in the nucleus accumbens stimulates feeding and locomotor activity. *Brain Res.* **1050**, 156–162.

Tortoriello, D. V., McMinn, J., and Chua, S. C. (2004). Dietary-induced obesity and hypothalamic infertility in female DBA/2J mice. *Endocrinology* **145**, 1238–1247.

Tritos, N. A., Elmquist, J. K., Mastaitis, J. W., Flier, J. S., and Maratos-Flier, E. (1998a). Characterization of expression of hypothalamic appetite-regulating peptides in obese hyperleptinemic brown adipose tissue-deficient (uncoupling protein-promoter-driven diphtheria toxin A) mice. *Endocrinology* **139**, 4634–4641.

Tritos, N. A., Vicent, D., Gillette, J., Ludwig, D. S., Flier, E. S., and Maratos-Flier, E. (1998b). Functional interactions between melanin-concentrating hormone, neuropeptide Y, and anorectic neuropeptides in the rat hypothalamus. *Diabetes* **47**, 1687–1692.

Tritos, N. A., Mastaitis, J. W., Kokkotou, E., and Maratos-Flier, E. (2001). Characterization of melanin concentrating hormone and preproorexin expression in the murine hypothalamus. *Brain Res.* **895**, 160–166.

Trivedi, P., Yu, H., MacNeil, D. J., VanderPloeg, L. H. T., and Guan, X. M. (1998). Distribution of orexin receptor mRNA in the rat brain. *FEBS Lett.* **438**, 71–75.

Truett, G. E., Bahary, N., Friedman, J. M., and Leibel, R. L. (1991). Rat obesity gene fatty (fa) maps to chromosome-5—Evidence for homology with the mouse gene diabetes (db). *Proc. Natl. Acad. Sci. U.S.A.* **88**, 7806–7809.

Turnbull, A. V., Ellershaw, L., Masters, D. J., Birtles, S., Boyer, S., Carroll, D., Clarkson, P., Loxham, S. J. G., McAulay, P., Teague, J. L., Foote, K. M., Pease, J. E., and Block, M. H. (2002). Selective antagonism of the

NPYY5 receptor does not have a major effect on feeding in rats. *Diabetes* **51**, 2441–2449.

van den Pol, A. N., Patrylo, P. R., Ghosh, P. K., and Gao, X. B. (2001). Lateral hypothalamus: Early developmental expression and response to hypocretin (Orexin). *J. Comp. Neurol.* **433**, 349–363.

van Houten, M., Posner, B., Kopriwa, B., and Brawer, J. R. (1980). Insulin binding sites localized to nerve terminals in rat median eminence and arcuate nucleus. *Science* **207**, 1081–1083.

Velkoska, E., Cole, T. J., and Morris, M. J. (2005). Early dietary intervention: long-term effects on blood pressure, brain neuropeptide Y, and adiposity markers. *Am. J. Physiol.* **288**, E1236–E1243.

Walter, M. J., Scherrer, J. F., Flood, J. F., and Morley, J. E. (1994). Effects of localized injections of neuropeptide Y antibody on motor activity and other behaviors. *Peptides* **15**, 607–613.

Wang, C. F., and Kotz, C. M. (2002). Urocortin in the lateral septal area modulates feeding induced by orexin A in the lateral hypothalamus. *Am. J. Physiol.* **283**, R358–R367.

Wang, S. K., Behan, J., ONeill, K., Weig, B., Fried, S., Laz, T., Bayne, M., Gustafson, E., and Hawes, B. E. (2001). Identification and pharmacological characterization of a novel human melanin-concentrating hormone receptor, MCH-R2. *J. Biol. Chem.* **276**, 34664–34670.

Warwick, Z. S., and Schiffman, S. S. (1992). Role of dietary fat in calorie intake and weight gain. *Neurosci. Biobehav. Rev.* **16**, 585–596.

Welch, C. C., Grace, M. K., Billington, C. J., and Levine, A. S. (1994). Preference and diet type affect macronutrient selection after morphine, NPY, norepinephrine, and deprivation. *Am. J. Physiol.* **266**, R426–R433.

White, B. D., He, B., Dean, R. G., and Martin, R. J. (1994). Low protein diets increase neuropeptide Y gene expression in the basomedial hypothalamus of rats. *J. Nutr.* **124**, 1152–1160.

White, J. D., and Kershaw, M. (1990). Increased hypothalamic neuropeptide Y expression following food deprivation. *Mol. Cell. Neurosci.* **1**, 41–48.

Widdowson, P. S. (1993). Quantitative receptor autoradiography demonstrates a differential distribution of neuropeptide-Y Y-1 and Y-2 receptor subtypes in human and rat brain. *Brain Res.* **631**, 27–38.

Widdowson, P. S. (1997). Regionally-selective downregulation of NPY receptor subtypes in the obese Zucker rat. Relationship to the Y5 "feeding" receptor. *Brain Res.* **758**, 17–25.

Widdowson, P. S., Upton, R., Henderson, L., Buckingham, R., Wilson, S., and Williams, G. (1997). Reciprocal regional changes in brain NPY receptor density during dietary restriction and dietary-induced obesity in the rat. *Brain Res.* **774**, 1–10.

Widdowson, P. S., Henderson, L., Pickavance, L., Buckingham, R., Tadayyon, M., Arch, J. R. S., and Williams, G. (1999). Hypothalamic NPY status during positive energy balance and the effects of the NPY antagonist, BW1229U91, on the consumption of highly palatable energy-rich diet. *Peptides* **20**, 367–372.

Wieland, H. A., Willim, K. D., Entzeroth, M., Wienen, W., Rudolf, K., Eberlein, W., Engel, W., and Doods, H. N. (1995). Subtype selectivity and antagonistic profile of the nonpeptide Y1 receptor antagonist BIBP 3226. *J. Pharmacol. Exp. Ther.* **275**, 143–149.

Wieland, H. A., Engel, W., Eberlein, W., Rudolf, K., and Doods, H. N. (1998). Subtype selectivity of the novel nonpeptide neuropeptide Y Y1 receptor antagonist BIBO 3304 and its effect on feeding in rodents. *Br. J. Pharmacol.* **125**, 549–555.

Wilding, J. P. H., Gilbey, S. G., Bailey, C. J., Batt, R. A., Williams, G., Ghatei, M. A., and Bloom, S. R. (1993). Increased neuropeptide-Y messenger ribonucleic acid (messenger RNA) and decreased neurotensin messenger RNA in the hypothalamus of the obese (ob/ob) mouse. *Endocrinology* **132**, 1939–1944.

Wyss, P., Levens, N., and Stricker-Krongrad, A. (1998a). Stimulation of feeding in lean but not in obese Zucker rats by a selective neuropeptide Y Y5 receptor agonist. *Neuroreport* **9**, 2675–2677.

Wyss, P., Stricker-Krongrad, A., Brunner, L., Miller, J., Crossthwaite, A., Whitebread, S., and Criscione, L. (1998b). The pharmacology of neuropeptide Y (NPY) receptor-mediated feeding in rats characterizes better Y5 than Y1, but not Y2 or Y4 subtypes. *Regul. Pept.* **75–6**, 363–371.

Xin, X. G., and Huang, X. F. (1998). Down-regulated NPY receptor subtype-5 mRNA expression in genetically obese mouse brain. *Neuroreport* **9**, 737–741.

Xu, B., Li, B. H., Rowland, N. E., and Kalra, S. P. (1995). Neuropeptide Y injection into the fourth cerebroventricle stimulates c-fos expression in the paraventricular nucleus and other nuclei in the forebrain: effect of food consumption. *Brain Res.* **698**, 227–231.

Yamamoto, Y., Ueta, Y., Date, Y., Nakazato, M., Hara, Y., Serino, R., Nomura, M., Sibuya, I., Matsukura, S., and Yamashita, H. (1999). Down regulation of the prepro-orexin gene expression in genetically obese mice. *Mol. Brain Res.* **65**, 14–22.

Yamamoto, Y., Ueta, Y., Serino, R., Nomura, M., Shibuya, I., and Yamashita, H. (2000a). Effects of food restriction on the hypothalamic prepro-orexin gene expression in genetically obese mice. *Brain Res. Bull.* **51**, 515–521.

Yamamoto, Y., Ueta, Y., Hara, Y., Serino, R., Nomura, M., Shibuya, I., Shirahata, A., and Yamashita, H. (2000b). Postnatal development of orexin/hypocretin in rats. *Mol. Brain Res.* **78**, 108–119.

Yamamoto, Y., Ueta, Y., Yamashita, H., Asayama, K., and Shirahata, A. (2002). Expressions of the prepro-orexin and orexin type 2 receptor genes in obese rat. *Peptides* **23**, 1689–1696.

Yamanaka, A., Kunii, K., Nambu, T., Tsujino, N., Sakai, A., Matsuzaki, I., Miwa, Y., Goto, K., and Sakurai, T.

(2000). Orexin-induced food intake involves neuropeptide Y pathway. *Brain Res.* **859**, 404–409.

Yang, B., Samson, W. K., and Ferguson, A. V. (2003). Excitatory effects of orexin-A on nucleus tractus solitarius neurons are mediated by phospholipase C and protein kinase C. *J. Neurosci.* **23**, 6215–6222.

Yokosuka, M., Dube, M. G., Kalra, P. S., and Kalra, S. P. (2001). The mPVN mediates blockade of NPY-induced feeding by a Y5 receptor antagonist: a c-FOS analysis. *Peptides* **22**, 507–514.

Zhang, J., Ren, P., Avsian-Kretchmer, O., Luo, C., Rauch, R., Klein, C., and Hsueh, A. (2005). Obestatin, a peptide encoded by the ghrelin gene, opposes ghrelin's effects on food intake. *Science* **310**, 996–999.

Zhang, Y., Proenca, R., Maffei, M., Barone, M., Leopold, L., and Friedman, J. M. (1994). Positional cloning of the mouse obese gene and its human homologue. *Nature* **372**, 425–432.

Zheng, H. Y., Corkern, M., Stoyanova, I., Patterson, L. M., Tian, R., and Berthoud, H. R. (2003). Peptides that regulate food intake—Appetite-inducing accumbens manipulation activates hypothalamic orexin neurons and inhibits POMC neurons. *Am. J. Physiol.* **284**, R1436–R1444.

Zheng, H. Y., Patterson, L. M., and Berthoud, H. R. (2005). Orexin-A projections to the caudal medulla and orexin-induced c-fos expression, food intake, and autonomic function. *J. Comp. Neurol.* **485**, 127–142.

Zukowska, Z., Pons, J., Lee, E., and Li, L. J. (2003). Neuropeptide Y: a new mediator linking sympathetic nerves, blood vessels and immune system? *Can J. Physiol. Pharmacol.* **81**, 89–94.

5

Brainstem-Hypothalamic Neuropeptides and the Regulation of Feeding

SIMON M. LUCKMAN
School of Biological Sciences, University of Manchester

I. The Control of Feeding
  A. The Basic Controls of Eating
  B. Brainstem Motor Systems Involved in the Control of Feeding
  C. The Visceromotor Control Column
  D. Positive Feed-Forward Stimulation in the Cephalic Phase of Eating
  E. Negative Sensory Feedback from the Gut
II. Neural Integration by the Dorsal Vagal Complex
III. Considerations in Assessing the Role of "Neuropeptides"
IV. "Humoral" Interactions with the Dorsal Vagal Complex
  A. Peptides from the Gut
  B. Peptides from the Pancreas
  C. Peptides from Adipose Tissue—Leptin
V. Intrinsic Peptidergic Neurons of the Dorsal Vagal Complex
  A. Neuropeptide Y
  B. Glucagon-like Peptides
  C. Prolactin-releasing Peptide
  D. Neuromedin U
  E. Melanocortins
  F. Galanin
VI. Descending Peptidergic Regulation of Brainstem Feeding Circuits
  A. CRH/Urocortin
  B. Oxytocin
  C. NPY and AGRP
  D. POMC and CART
  E. Orexin and MCH
  F. Thyrotrophin-releasing Hormone
VII. Summary and Potential for Drug Development
  References

The brainstem contains all the necessary circuitry to control the basic mechanisms of feeding and to regulate the size of individual meals. This is achieved through reciprocal connections with the whole length of the gastrointestinal tract. Feed-forward sensory information initiates a meal and feedback information terminates it. In order to modulate eating with respect to other facets of energy homeostasis, the brainstem detects neuronal and hormonal signals from the gut and other organs and integrates this information by communication with higher brain centers. Here, the role of neuropeptides is assessed with particular reference to the dorsal vagal complex. The term neuropeptide is treated in its widest context to include peptide transmitters and peptide hormones. Some neuropeptides that are intrinsic to the brainstem itself may have as important roles in energy homeostasis as their hypothalamic counterparts. Finally, the modulation of brainstem function by descending peptidergic pathways from higher brain centers suggests that the brainstem is a target for pharmaceutical intervention.

I. THE CONTROL OF FEEDING

A. The Basic Controls of Eating

Feeding behavior is regulated by a mixture of automated, subconscious motor behaviors, homeostatic control mechanisms, and higher cognitive functions. Traditionally the brainstem has been implicated in only the former, automated processes, whereby feed-forward and feedback information is sensed by cranial nerves that terminate within nuclei of the caudal brainstem (medulla oblongata and pons), leading to reflex regulation of efferents that control motor activity in the mouth, pharynx, and gastrointestinal tract. Other processes that include homeostatic integration to maintain a constant energy status, learning and decision making, necessarily must involve higher brain centers.

To a large extent this is true. The comprehensive and elegant experiments of Norgren, Kaplan, Grill, and others have demonstrated that decerebrate rats, in which all connections between the brainstem and higher brain centers are severed, are perfectly capable of eating and regulating the size of individual meals (Grill and Kaplan, 2002; Smith, 2000). The rats cannot procure food for themselves and they must be maintained by oral infusion. However, as long as food is introduced into the mouth, the decerebrate rat will lick/chew, swallow, and move the food along the gastrointestinal tract in a fashion that is indistinguishable from an intact animal. This complex ingestive behavior is dependent on positive, feed-forward orosensory information, and includes the sensation of food in the mouth. However, a point is reached at which the animal no longer swallows the food, indicating that satiation has occurred. Satiation is a negative feedback process that begins early in a meal and terminates when eating stops (Blundell and Latham, 1979; Smith, 2000). This is different from postprandial satiety, a state that begins at the end of one meal and lasts until the beginning of the next. Satiation is dependent on the detection of a postingestive load and of nutrients within the gut, influences that can be removed by "sham feeding" whereby a fistula is inserted in the animal's stomach, so that ingested food immediately passes out of the body (Smith, 2000).

Since the lower brainstem translates sensory information into the control of ingestive behavior, it must possess all the basic sensory, integratory, and motor elements to allow eating. However, decerebrate animals are unable to maintain metabolic stasis if energy consumption is altered, as can be demonstrated with a simple meal omission experiment. Both decerebrate and intact rats gain weight if fed three meals a day. However, if one of the meals is left out, the intact but not the decerebrate animals are able to compensate by eating more during the other two meals so that they do not lose weight (Le Magnen and Tallon, 1966; Seeley et al., 1994). Thus, connections with the forebrain are required to regulate ingestive behavior according to the metabolic state of the animal (Kaplan et al., 1993; Seeley et al., 1994). This fact led to the proposal that the direct controls of eating (i.e., the gut-brainstem-gut reflex arcs) are modified by a number of indirect controls (Smith, 2000). The indirect controls include metabolic effects (in order to maintain circulating glucose levels, the size of energy stores, or to induce a fever), diet preferences and aversions, food availability and foraging experience, rhythms (diurnal, seasonal, or ovarian), and social (both cultural or aesthetic). Importantly, each of these indirect controls requires the functioning of higher brain centers and, therefore, most have limited influence in a decerebrate animal.

B. Brainstem Motor Systems Involved in the Control of Feeding

The concept of direct and indirect controls of feeding forms the basis of our upcoming discussion on the role of neuropeptides in brainstem and hypothalamic circuits. The positive orosensory and negative postingestive feedback from the gut acts on motor behavioral systems that are located in neuronal columns similar to those that control other rhythmical motions, such as walking or breathing. In the case of eating, the final motor neurons that control rhythmical movements of the jaw, tongue, and facial muscles are in the trigeminal (V), facial (VII), and hypoglossal (XII) cranial nuclei. Motor neurons that project to the abdominal viscera to control gastric motility and secretion are in the dorsal motor nucleus of the vagus (DMX). Note that both the somatic motor neurons innervating striated muscles and preganglionic autonomic motor neurons use acetylcholine as a transmitter and can be considered as the "final common pathway." The motor neurons of the final pathway are themselves controlled in a hierarchical fashion by motor pattern generators, and in turn by motor pattern initiators and motor pattern controllers.

The vagus nerve carries motor fibers from the nucleus ambiguus to muscles of the rostral esophagus involved in the swallowing reflex. However, the DMX nerve is the primary source of efferent fibers to the abdominal viscera, including the gut, liver, and pancreas. Retrograde-tracer injections into individual organs of the gastrointestinal tract or branches of the subdiaphragmatic vagus of the rat have demonstrated a topographic organization of neurons in the DMX into longitudinal columns that control different parts of the gut (Fox and Powley, 1985). Coordination of digestion is affected by a number of vago-vagal reflexes, whereby sensory information from the gut activates vagal afferents that terminate in the nucleus of the tractus solitarius (NTS), which in turn project to efferent vagal preganglionic neurons in the DMX. These include the reflexes that transit food through the esophagus to the stomach, and from the stomach to the duodenum, as well as reflexes that control gastric and pancreatic secretions.

C. The Visceromotor Control Column

"Minimal circuit element" analysis has been used to attempt to place motor pattern generators, initiators, and controllers into basic hierarchies or behavioral control columns (Swanson, 2000). In this scheme, there is a visceromotor pattern generator centered in the medial region of the hypothalamus (Thompson and Swanson, 2003). Cell groups, including the dorsomedial nucleus, interconnect strongly with each other and particularly with the descending, nonneuroendocrine regions of the paraventricular and arcuate nuclei. Eating can be elicited pharmacologically from these nuclei (Leibowitz, 1978). The descending paraventricular nucleus, and other preautonomic regions of the hypothalamus, project directly to preganglionic parasympathetic and sympathetic cell groups in the brainstem and spinal cord. They also innervate visceral sensory nuclei in the brainstem (notably the NTS), cell groups associated with the relay of nociceptive information, and parts of the reticular core, including the central gray.

The hypothalamus is placed to receive information required for the homeostatic (indirect) regulation of feeding. Voluntary and cognitive inputs come from the prefrontal cortex, hippocampus, and amygdala via the bed nucleus of the stria terminalis and the ventral lateral septal nucleus (Petrovich et al., 2001). Behavioral state information comes from the suprachiasmatic nucleus, the tuberomammillary nucleus, the locus coeruleus, and the dorsal raphe (Saper et al., 2005; Thompson and Swanson, 1998). And finally, the loop is completed by viscerosensory and nociceptive information from the parabrachial nucleus (Bester et al., 1997), as well as the ventrolateral medulla and the NTS.

D. Positive Feed-Forward Stimulation in the Cephalic Phase of Eating

The presence of motor and sensory components, as well as central pattern generators for eating, all within the brainstem is analogous to the system required for basic stepping motion within the spinal cord. In both instances, the motor behavior is initiated following stimulation of the appropriate sensory surface: the mouth for eating or the soles of the feet for stepping (Travers et al., 1997). Consistent with this, a decerebrate animal will only initiate licking or chewing if food is placed directly in the mouth. Positive feed-forward information due to the presence of food in the mouth, and later the gut, will reach the brainstem reflex circuitry through the rostral and intermediate NTS, the sensory trigeminal complex, and parabrachial nucleus to acti-

vate or maintain ingestive behavior in a decerebrate animal. This is termed the cephalic phase of eating.

In intact animals, the process is modulated by feed-forward pathways from higher brain centers. The gustatory and visceral sensory information will be relayed to the cortex. These animals will also receive visual and olfactory information that will be processed by the orbitofrontal cortex, central nucleus of the amygdala, and lateral hypothalamus, all of which send projections directly to the brainstem to increase efferent activity in the cranial nerves.

E. Negative Sensory Feedback from the Gut

The major regions for physiological feedback from the gut are the intermediate, subpostrema, and caudal NTS that lie dorsal to the DMX in the medulla oblongata. Anterograde-transport studies following site-specific tracer injections into the gut or subdiaphragmatic vagal branches, as with retrograde tracing to the DMX, show topographically organized afferent termination fields within the NTS (Altschuler et al., 1989; Norgren and Smith, 1988; Rinaman and Schwartz, 2004). There are some vagal afferents that terminate directly on motor neurons of the DMX (Rinaman et al., 1989), although the relative importance of these is unknown. Topographic organization of the NTS and DMX allows for regional regulation of specific sites within the gastrointestinal tract.

Meal-related stimuli can be classified into mechanical or chemical and are detected by three distinct types of vagal afferent: mucosal terminals, intramuscular arrays, and interganglionic laminar endings. The latter two are mainly mechanosensitive, responding to tension and distortion. Mucosal terminals are sensitive to changes in pH and to chemical substances released by enteroendocrine cells, as well as mechanical distortion (Berthoud et al., 2001). Mechanosensitive vagal fibers in the stomach respond to changes in gastric load within the physiological range experienced following a meal, but independently of the nutrient quality (Davison and Clarke, 1988; Mathis et al., 1998; Schwartz et al., 1991). The distension of balloons in the rat stomach at rates and volumes designed to mimic the ingestion of food during a normal meal leads to the induction of the protein product of the c-*fos* gene (c-Fos) in the medial and commissural nuclei of the NTS (Fraser et al., 1995; Willing and Berthoud, 1997). Similar experiments utilizing closed pyloric cuffs demonstrate that a nonnutritive load confined to the stomach is capable of reducing food intake (Phillips and Powley, 1996, 1998; Powley and Phillips, 2004). Mice that are genetically deficient in either intramuscular arrays or interganglionic laminar endings in the small intestine have disrupted meal patterns, but normal body weight due to compensation by other, indirect controls (Chi and Powley, 2003; Fox et al., 2001).

Although distension of the duodenum can reduce feeding, rather than being dependent on stretch, negative feedback from the duodenum that causes satiation is mostly dependent on nutrient content. Amino acid–rich meals induce c-Fos activity mainly in the NTS rostral to or level with the area postrema, while glucose or lipids are more effective caudal to the area postrema, providing evidence for topographic NTS responses depending on the type of macronutrients ingested (Monnikes et al., 1997; Phifer and Berthoud, 1998; Zittel et al., 1994). Nutrients in the meal, and in particular lipids, are responsible for the release of cholecystokinin (CCK) from duodenal enteroendocrine cells (Liddle, 1994). CCK stimulates gall bladder contraction and pancreatic release of digestive enzymes, decreases gastric emptying, and reduces feeding behavior (Moran and McHugh, 1982). Capsaicin-sensitive vagal afferent fibers transmit the CCK signal to the brainstem (Smith et al., 1981, 1985), and this can be visualized by the induction of c-

Fos in medial and commissural subnuclei of the NTS following exogenous CCK (Fraser and Davison, 1992; Luckman, 1992; Luckman et al., 1993; Zittel et al., 1999). Capsaicin excitotoxically depolarizes primary sensory neurons through its interaction with the vanilloid receptor, VR-1 (Blackshaw et al., 2000). Afferents of the vagus nerve that form mucosal terminals contain VR-1 (Ward et al., 2003) and CCK-A receptors (Corp et al., 1993). In addition, they possess receptors for a number of peptides: PYY_{3-36} (Zhang et al., 1997), ghrelin (Date et al., 2002), and possibly other postprandially released gut peptides, including GLP-1 and gastrin-releasing peptide (Thorens and Larsen, 2004).

II. NEURAL INTEGRATION BY THE DORSAL VAGAL COMPLEX

Most information regarding neuropeptides in these feeding circuits relate to the dorsal vagal complex in the medulla oblongata and the reciprocal connections that exist between it and the hypothalamus. Thus, this section will concentrate on these areas. Rats that consume a normal meal to satiety express significant c-Fos in the area postrema and in the medial nucleus of the NTS, where gastrointestinal vagal afferents terminate (Fraser and Davison, 1993; Rinaman et al., 1998). Sham feeding significantly attenuates meal-induced c-Fos in medial and commissural NTS subnuclei, but not in more rostral NTS regions that receive input from oropharyngeal afferents (Fraser et al., 1995; Rinaman et al., 1998).

As noted already, there is some topographic organization of the NTS, although relatively little is known about the identity of second-order neurons and whether they are responsive to more than one sensory modality. Using electrophysiological techniques, Raybould et al. (1985) demonstrated that neurons in the medial subnucleus of the NTS that are sensitive to gastric disten-sion are also responsive to CCK injected locally into the gut vasculature. Zhang and colleagues (1995) suggested that the neurons in the medial subnucleus also can respond to duodenal distension and, thus, appear to be polymodal, secondary sensory neurons. Neurons sensitive to gastric dis-tension only are predominantly located in the gelatinous subnucleus, while those sen-sitive to duodenal, but not gastric, disten-sion are found in the subpostremal and commissural subnuclei (Zhang et al., 1995). Zittel et al. (1994) failed to show c-Fos activ-ity in the NTS following perfusion of the duodenum with HCl, whereas Zhang et al. (1998) show electrophsyiological responses. In the latter experiments, neurons were located in the subpostremal region, medial subnucleus, and gelatinous subnucleus. Finally, using a working viscera–brainstem preparation, Paton and co-workers (2000) recorded from NTS neurons in the medial subnucleus at the level of the area postrema that responded with excitation to stimula-tion of the subdiaphragmatic vagus nerve. These neurons failed to respond to stimula-tion of either vascular chemoreceptors or baroreceptors. Thus, cautiously, one can conclude that at least one class of neurons in the medial subnucleus of the NTS is responsive to a number of different sensory modalities deriving from gastrointestinal, vagally mediated signaling. They are, however, unresponsive to vagally mediated cardiovascular inputs. Other neurons in the subpostremal region may be polymodal, whereas a population of neurons in the gelatinous subnucleus are separate and are responsive to gastric, but not duodenal distension.

The first chemical phenotyping of cells related to gastric signaling was the demon-stration that approximately half of the NTS neurons expressing c-Fos after exogenous CCK are catecholaminergic, correspond-ing to neurons in the A2 noradrenergic group (Luckman, 1992). Catecholaminergic neurons are, by contrast, relatively unre-

sponsive to gastric distension (Vrang et al., 2003; Willing and Berthoud, 1997). One caveat from these immunohistochemical studies is that, because there was no retro-or anterograde identification, some of the NTS neurons will have been activated by long-loop descending inputs from the fore-brain. That is, some of the NTS neurons activated by CCK have been shown to project up to the hypothalamus that sends reciprocal connections back down to acti-vate more cells in the NTS (Onaka et al., 1995a; Rinaman et al., 1994, 1995). More recent evidence for a critical role for cate-cholaminergic neurons is that, if they are destroyed by targeting the toxin saporin selectively to neurons in the NTS that contain the synthetic enzyme dopamine β-hydroxylase, the animals no longer respond to exogenous CCK (Rinaman, 2003). Further, identification of NTS neurons according to their peptidergic phenotype will be discussed in Section V.

The NTS makes dense interconnections with the area postrema, the third structure in the dorsal vagal complex (Armstrong et al., 1981). Rats that eat a large meal to sati-ation express significantly more c-Fos in both the area postrema and the NTS when compared with unfed or nonsatiated rats (Fraser and Davison, 1993; Rinaman et al., 1998). Furthermore, rats with lesions of the area postrema have proportionately larger meals, suggesting that this is a site for feed-back to the brain (Edwards and Ritter, 1986). Importantly, both the area postrema and the NTS possess fenestrated capillaries (Broadwell and Sofroniew, 1993), which means that they are open to factors circu-lating in the bloodstream that will affect the regulation of feeding. These factors will include nutrients, such as glucose, but also hormones released from other organs including the gut. In fact, some of these hor-mones are released postprandially, such as insulin, amylin, peptide YY (PYY$_{3-36}$), and the glucagon-like peptides, and so might be considered as direct controls of feeding and are discussed in Section IV.

Many of the meal-related reflexes that involve the dorsal vagal complex are negative feedback mechanisms. Electrical recordings from NTS neurons made *in vivo* show that they are normally in an activated state, mainly due to glutamatergic gastrointestinal inputs (McCann and Rogers, 1992; Smith et al., 1998). However, in the same situation the vast majority of neurons in the DMX (and nucleus ambiguus), that is the efferent vagal output back to the digestive tract, are inhibited. Transmitters involved in these NTS inhibitory projections include GABA, catecholamines, and enkephalins.

Second-order NTS neurons send processed gastrointestinal afferent information to a variety of other structures, in the hindbrain and forebrain, to coordinate the direct and indirect controls of feeding (Norgren, 1978; Ricardo and Koh, 1978; Rinaman and Schwartz, 2004; Ter Horst et al., 1989). For example, the parabrachial nucleus integrates information from different cranial nerves, and from the area postrema, relating to palatability and aversions (Anini et al., 1999). The NTS projects to the amygdala (Ricardo and Koh, 1978), where the information is put into the context of situation and emotion (Swanson and Petrovich, 1998). There are important direct connections with preautonomic areas of the hypothalamus, the paraventricular and arcuate nuclei, where information on meals is integrated with that concerning overall energy status, or with the lateral hypothalamus that also receives information concerning arousal and reward (Saper et al., 2002).

III. CONSIDERATIONS IN ASSESSING THE ROLE OF "NEUROPEPTIDES"

The following sections will review the role of "neuropeptides" in the brainstem and hypothalamic axis involved in the regulation of feeding. As mentioned, the brain-stem contains many areas that are critical to feeding (e.g., motor nuclei of cranial nerves V, VII, and XII, the medullary reticular formation, and the parabrachial nucleus) and there are many higher brain centers in addition to the hypothalamus that send descending projections. Not all of these can be covered here. Instead the review will concentrate on peptides that relay sensory input to the dorsal vagal complex (here the "neuropeptides" may be hormonal), that project that information forward, or return commands to the brainstem from the hypothalamus.

Careful consideration should be made in interpreting the effects of any neuropeptide on feeding. Signaling peptides tend to be pleiotropic and can affect behavior through a number of interacting functions (e.g., appetitive drive, aversion, energy expenditure) that can change feeding either directly or indirectly. Different actions of a neuropeptide may predominate depending on dose and route of administration, and the dose chosen may not always reflect the likely active concentration at the target receptor. For a circulating peptide hormone, it would seem reasonable where possible to measure the levels that are in the bloodstream in normally replete animals or that are experienced around different feeding events and to try and mimic these in experiments. However, the same circulating peptides may be produced in the nervous system and have a very different range of effective doses. Furthermore, altering the level of a single factor by exogenous administration does not take into account the ambient levels of all other factors that can affect energy homeostasis. Would a small increase in ghrelin or PYY_{3-36} be expected to produce the equivalent effect in replete and fasted, or fat and thin, animals? Also, in order to get a measurable response in feeding behavior or brain activity, it may be necessary to first manipulate the status of the experimental animal. Here it would seem preferable, for example, to fast an animal overnight to reduce endogenous circulating leptin in order to examine the effects of the exogenous peptide, rather than to use an *ob/ob* mouse that has never experienced the peptide before.

The basic circuits that control feeding may be engaged by stimuli that are not deemed homeostatic: pain, illness, stress, or aversion. These factors still produce regulated responses, yet the neuropeptides mediating these processes (rather than those which can cause them as a side effect) are sometimes dismissed as not being physiologically relevant. In fact, one of the most important tasks in this field of research will be to determine whether the neuropeptides and their pathways are the same as those that control normal feeding behavior or whether they run parallel with them. An often-quoted example is that the pleasant feeling of being replete may be part of the same continuum of sensation as the nausea of being bloated. What is not clear is if the two stimulate exactly the same pathways at different intensities or whether the latter activates additional pathways.

These points are particularly relevant as many peptides have a physiological action directly on gut function and, thus, their effects on food intake may be secondary (Fujimiya and Inui, 2000). The slowing of gut motility following ingestion of a fatty meal serves to assist in normal digestion. This is a direct action of vago-vagal reflex circuitry to inhibit cholinergic excitation of the stomach, suppressing the transit and further delivery of ingested food to the intestine. However, over stimulation of this reflex by the application of exogenous peptides could lead to excessively delayed gastric emptying and abnormal stimulation of sensory receptors. It is then debatable if this is a physiological response normally associated with satiation, or instead a pain response triggered by a pathological blockage and mimicked by the experimental manipulation. If the latter, it is probably detected by mechanosensitive vagal fibers, but also pain sensory afferent fibers of the splanchnic nerve that synapse in the tho-

racic and lumbar spinal cord and of the pelvic nerve that synapse in the sacral cord (Cervero, 1994). Other adequate stimuli for the activation of visceral nociceptors include traction on the mesentery, ischemia, and inflammation, each of which will lead to nausea and/or pain (and a reduction in feeding). Many secondary sensory spinal neurons ascend and terminate in the NTS (Menetrey and Basbaum, 1987; Potts et al., 2002), where they may interact with the same circuits that reduce feeding during physiological satiation.

Finally, the use of transgenesis to selectively ablate brainstem neuropeptides, or their receptors, has not been used yet to the same extent as in the hypothalamus. Instead, surgical models have been utilized extensively in investigating the mode of action of neuropeptides on feeding. For example, many studies conclude an action at the level of the gut if a peptide is ineffective in animals following vagotomy. But this does not account for the severance of efferent, in addition to afferent vagal fibers. The problem of the relative contribution of different fiber types in the vagus is compounded by the fact that afferent fibers in long-term vagotomized animals possess regenerative potential, whereas efferent fibers tend not to (Phillips et al., 2003). Thus, it is difficult to interpret the true effect of vagotomy on the actions of a single neuropeptide without considering the time elapsed since surgery, as well as the specific technique, and the effects of surgery on other factors involved in energy homeostasis. In a similar way that vagotomy severs neural connections between the gut and the brainstem, midbrain transection severs connections between brainstem and forebrain. The use of the decerebrate model has proved powerfully that the brainstem possesses adequate circuitry to respond to most stimuli relevant to energy balance, but this does not mean that in the intact animal the brainstem responds to these stimuli in isolation (Grill and Kaplan, 2002; Harris et al., 2006). This and other chapters will highlight the role played by the whole brain in energy homeostasis.

IV. "HUMORAL" INTERACTIONS WITH THE DORSAL VAGAL COMPLEX

As alluded to earlier, the afferent, chemosensitive vagal fibers have receptors for a number of gut-released hormones, including CCK, ghrelin, and leptin (this hormone, normally associated with adipocytes, is produced in other tissues of the body, including the gastric mucosa (Bado et al., 1998). Many of these "hormones" also have receptors in the dorsal vagal complex and may even be produced within the brain itself. These complexities are often ignored in discussions on the regulation of appetite and body weight or, instead, they lead sometimes to apparently unresolvable disagreements as to the mode of action of individual factors. It is these very complexities that provide the necessary fine-tuning and redundancy that make a regulatory system functional. If we are to manipulate feeding systems pharmacologically we have to accept and understand that each factor may have a multitude of effects at multiple levels.

A. Peptides from the Gut

1. Cholecystokinin

The first factor to consider is an excellent example of a pleiotropic effector. One can imagine that peptides such as CCK have evolved to have a number of functions in complex body systems, but very often with similar end results. CCK is often quoted as being the archetypal satiety peptide. It is released from the gut by preabsorptive mechanisms in response to the nutrient content of the meal, and activates vago-vagal reflexes that help to coordinate the digestive process and ultimately to produce satiation and end the meal. As might be expected for a satiety factor, repeated peripheral administration of CCK reliably reduces meal size, although there is a resulting compensatory increase of meal frequency so that overall energy intake is not substantially altered (Crawley and Beinfeld, 1983; West et al., 1984). The explanation being that CCK released in the duodenum engages the direct controls of feeding, while the compensatory mechanism involves indirect controls. But what if CCK were also part of the indirect controls of feeding?

CCK_A receptors, normally considered as the peripheral type, are found in the gastrointestinal tract on vagal nerve endings, but also in the medial NTS and in the area postrema. CCK_B receptors are found in many areas of the brain, including in the lateral NTS and spinal tract of the trigeminal nerve, as well as in the paraventricular and ventromedial nuclei of the hypothalamus (Hill et al., 1987). CCK itself is found in various parts of the brain, including in the parabrachial nucleus, and colocalized with oxytocin and corticotrophin-releasing hormone (CRH) in the paraventricular hypothalamus (Mezey et al., 1986; Micevych et al., 1988; Vanderhaeghen et al., 1981). It is also present in vagal nerves (Broberger et al., 2001), so it could quite conceivably be active in a whole pathway involving vagal afferent, brainstem, and hypothalamic neurons, all which contain the same peptide. This could run parallel with other pathways that utilize CCK, for example, from the parabrachial nucleus in the pons to the ventromedial nucleus of the hypothalamus that is activated by painful stimuli to reduce feeding behavior (Malick et al., 2001). Grill and Smith (1988) found that chronic decerebrate rats still responded to CCK with a decrease in nutrient intake. However, it is interesting to note that although the activation of both brainstem and hypothalamus is completely blocked by a CCK_A-receptor antagonist, there is also a partial blockade of hypothalamic activity by a CCK_B-receptor antagonist, suggesting that there is some ascending information carried by CCK (Luckman et al., 1993).

Intravenous injection of CCK at supraphysiological levels induces c-Fos in the medial NTS and the area postrema and much of this activation could be via direct action on CCK_A receptors rather than through the vagal pathway (Luckman, 1992). It is true that high doses of CCK, like the other nausea-producing agents such as lithium chloride or apomorphine detected by the area postrema, can produce sickness-like behavior and, consequently, hypophagia (Stricker and Verbalis, 1991; Verbalis et al., 1986). This has led to the hypothesis that satiation and nausea are extremes of the same sensation. While the two may activate some pathways in common, there are subtle differences. High doses of CCK will produce a conditioned taste aversion (CTA; see review by Moran, 2000). However, there are differences between this and the CTA produced by lithium chloride (Ervin et al., 1995; Ervin and Teeter, 1986; Nachman and Ashe, 1973). Also, in the taste reactivity test, lithium chloride–treated rats are observed to decrease ingestive responses (mouth movements and tongue protrusions) and to increase aversive responses (gapes, chin/paw rubs, head shakes; Ossenkopp and Eckel, 1995; Parker, 1991, 1995). By contrast, intraperitoneal CCK decreases ingestive responses but does not increase aversive responses (Eckel and Ossenkopp, 1994).

2. Bombesin-like Peptides

Bombesin, an amphibian peptide, and its mammalian homologues, gastrin-releasing peptide (GRP) and neuromedin B (NMB), are potent anorexigens. The reduction of feeding following systemic injection of bombesin was first shown in rats by Gibbs and colleagues (1979) and this has been repeated in a wide range of species and by using the different bombesin-like peptides (Merali et al., 1999; Yamada et al., 2002). Systemic injections produce satiation without an indication of malaise or other

side effects. Whereas central injections are more potent, they can produce additional behavioral responses such as excessive grooming. However, these additional effects can be almost completely dissociated if the bombesin is injected directly into the NTS or the fourth ventricle (Johnston and Merali, 1988; Ladenheim and Ritter, 1988). Evidence for these peptides having a physiological role in appetite regulation is that receptor antagonists, injected into the fourth ventricle, can induce feeding in partially satiated rats (Flynn, 1993, 1997; Stratford et al., 1995). Furthermore, intracerebroventricular (i.c.v.) leptin enhances the suppression of food intake and the induction of c-Fos in the NTS following systemic injections of bombesin (Ladenheim et al., 2005). Finally, while neither GRP receptor- (Hampton et al., 1998) nor NMB receptor–knockout mice (Ohki-Hamazaki et al., 1999) display disrupted feeding behavior (but see Ladenheim et al., 2002), knockout of an orphan receptor in the family, bombesin receptor subtype-3, results in an obese phenotype (Ohki-Hamazaki et al., 1997).

The evidence for bombesin-like peptides having their effects through the dorsal vagal complex are strong. For example, bombesin is effective in chronically decerebrate rats (Flynn and Robillard, 1992), but not in rats in which the NTS/area postrema has been lesioned (Ladenheim and Ritter, 1993). Although bombesin does affect the firing of NTS neurons sensitive to gastric distension (Ewart et al., 1990), the evidence is less strong for these peptides being part of the afferent limb of the gut–brain axis. Receptors in the abdominal viscera appear to mediate the effects of peripherally administered bombesin (Kirkham et al., 1991). However, peripheral administration of receptor blockers attenuates the response to systemic, but not central injection of bombesin (Kirkham et al., 1994; Merali et al., 1988), and the effects of systemic bombesin are unaffected by subdiaphragmatic vagotomy (Schwartz et al., 1997;

Stuckey et al., 1985). Therefore, although bombesin-like peptides are produced in the gut (Panula, 1986), their actions in the brain are likely to result from an endogenous central source. There is dense bombesin-like immunoreactive fiber staining in the NTS of rat and human (Lynn et al., 1996), which may originate from different fore-brain areas: the prefrontal cortex, BNST, amygdala, and paraventricular, arcuate, and lateral regions of the hypothalamus (Costello et al., 1991; Merali and Kateb, 1993; Plamondon and Merali, 1997). These brain structures display c-Fos protein following injections of bombesin (Bonaz et al., 1993; Li and Rowland, 1996). Thus, although the brainstem is a major site of action for endogenous bombesin-like peptides it may be responding to descending pathways. Additionally, bombesin-like peptides may be acting at forebrain structures, such as the amygdala, to affect other aspects of appetitive behavior (Merali et al., 1998).

3. Glucagon-like Peptides

In the pancreas, the preproglucagon gene produces glucagon, whereas in the gut and, as we will see later, in the brainstem, the gene yields other products: glucagon-like peptide-1 (GLP-1), GLP-2, oxyntomodulin, and glicentin. These glucagon-like peptides are released from L-type enteroendocrine cells in response to fatty acids or carbohydrates. GLP-1 can inhibit gastric acid secretion and emptying, probably by acting locally, and also causes the postprandial release of insulin (Kreymann et al., 1987; Tourrel et al., 2001). Circulating GLP-1 has a very short half-life as it is rapidly inactivated by dipeptidyl peptidase IV (DPP-IV; Kieffer et al., 1995). Whereas oxyntomodulin may access the brain at the hypothalamic arcuate nucleus (Dakin et al., 2004), this does not appear to be the case for GLP-1. Systemic injection of GLP-1 activates neurons in the brainstem (Tang-Christensen et al., 2001) and other central autonomic control sites (Yamamoto et al., 2002), but it

is debatable whether this is a physiological site of action for circulating peptide and may instead reflect the presence of pre-proglucagon-derived peptides in neurons of the NTS, which will be discussed in Section V.A.

4. Peptide YY

Peptide YY (PYY) is produced by L-type enteroendocrine cells, mainly in the ileum and colon, in response to the caloric (especially lipid) content of the meal (Adrian et al., 1985; Grandt et al., 1994). The mature peptide PYY_{1-36} is cleaved by DPP-IV into PYY_{3-36}, but the two peptides are found in approximately equal concentrations in blood (Medeiros and Turner, 1995). There is much interest in PYY_{3-36}, an agonist at the neuropeptide Y (NPY)-family Y2 receptor (Y2-R), as it can reduce food intake in rodents, monkeys, and humans (Batterham et al., 2002, 2003; Challis et al., 2003; Moran et al., 2005). However, there has been also much discontented discussion as to whether PYY_{3-36} is a natural satiety factor and, if so, what is its mechanism of action (see Tschop et al., 2004). While some have recorded no obvious aversive responses or disruption of behavioral satiety following systemic PYY_{3-36} (Scott et al., 2005; Talsania et al., 2005), others have reported a conditioned taste aversion in mice, possibly due to a direct effect at the area postrema (Halatchev et al., 2004). Paradoxically, lesions of the area postrema can potentiate the effect of PYY_{3-36} to reduce food intake (Cox and Randich, 2004). In fact, there are many parallels with the literature on CCK and, again, one is left with the impression that PYY_{3-36} and possibly other Y2-receptor agonists can act at a number of different levels.

PYY release in the gut acts as an "ileal brake," a negative feedback response to inhibit gastric emptying and duodenal motility and to prevent further nutrient delivery (Moran et al., 2005; Pappas et al., 1985). This effect is dependent on receptors in the upper gastrointestinal tract (Laburthe et al., 1986) and Y2-receptors have been found in vagal endings (Koda et al., 2005). This would seem to be a sufficiently potent action of exogenous PYY_{3-36} to activate vagal afferents either directly, or indirectly by increasing distension-sensitive endings in the stomach, and to promote satiation. Gastric vagotomy is reported to block the anorectic effects of PYY_{3-36} by some (Abbott et al., 2005; Koda et al., 2005) but not other groups (Halatchev et al., 2004). While Chelikani and colleagues (2005) found PYY_{3-36} was effective in sham-feeding rats, which would argue its effect to reduce food intake is independent of its action to inhibit gastric emptying.

Equally, it has been proposed that PYY_{3-36} has actions on hypothalamic circuits. This could be due to projections from the brainstem, since the transection of ascending pathways can abolish reductions on feeding (Abbott et al., 2005; Koda et al., 2005). Systemic injections of PYY_{3-36} are poor inducers of c-Fos activity in the forebrain, with only a minor activation of neurons in the hypothalamic arcuate nucleus (Batterham et al., 2002; Neary et al., 2005), although it can significantly reduce the activation of putative NPYergic neurons by ghrelin (Scott and Luckman, unpublished). The original hypothesis that PYY_{3-36} has its main effect by causing the disinhibition of proopiomelanocortin (POMC) neurons in the arcuate (Batterham et al., 2002) appears to be incorrect. Notably, PYY_{3-36} is effective in mice with disrupted melanocortin signaling (Challis et al., 2004; Halatchev et al., 2004; Martin et al., 2004). Thus, the hypothalamic actions of PYY_{3-36} to reduce food intake are most probably due to a reduction of NPY neuron activity and a reduction of orexigenic drive (Acuna-Goycolea and van den Pol, 2005; Ghamari-Langroudi et al., 2005; Riediger et al., 2004).

5. Ghrelin

The one and, so far, only humoral factor to be produced in peripheral tissues that causes an increase in feeding is the stomach-produced, octanoylated peptide, ghrelin (Tschop et al., 2000). At first there was a remarkable consensus that ghrelin is secreted by the stomach and acts on NPY-ergic neurons of the hypothalamic arcuate nucleus, that express the ghrelin/growth hormone secretagogue (GHS) receptor, to induce feeding (Dickson and Luckman, 1997; Lawrence et al., 2002b; Nakazato et al., 2001; Willesen et al., 1999). The fact that circulating levels increase between meals or in fasted animals, and then rapidly reduce following a meal, suggested that ghrelin might even be a peripheral signal of hunger (Ariyasu et al., 2001; Cummings et al., 2001; Sugino et al., 2002; Tschop et al., 2001). However, soon additional information began to add complexity to this simple model.

Date and others (Asakawa et al., 2001; Date et al., 2002) demonstrated that subdiaphragmatic vagotomy or perivagal application of capsaicin, abolishes the feeding response to intravenous ghrelin. They also showed that ghrelin receptors are present in vagal neurone terminals, and provide some evidence that ghrelin suppresses firing in (CCK-excited) vagal afferents. Consistent with this demonstration in the rat, ghrelin does not stimulate food intake in human patients with surgical procedures involving vagotomy (le Roux et al., 2005). However, feeding in the rat induced by i.c.v. injection of ghrelin is not affected by vagotomy (Date et al., 2002), again suggesting that this peptide, like others, functions at different levels.

It is quite likely that circulating ghrelin or other GHS ligands can access the hypothalamus at the arcuate nucleus to activate NPY neurons that possess ghrelin receptor mRNA (Dickson and Luckman, 1997; Nakazato et al., 2001; Tamura et al., 2002; Willesen et al., 1999). Electrophysiological recordings suggest that although the majority of arcuate nucleus neurons are activated, a large proportion instead are inhibited (Dickson et al., 1995; Hewson et al., 1999) and these may include POMC neurons, some of which do contain ghrelin receptor mRNA (Willesen et al., 1999). Importantly, however, ghrelin receptors are expressed widely in the brain often in areas that would not have access to circulating ghrelin (Guan et al., 1997; Mitchell et al., 2001; Zigman et al., 2006). Indeed many more areas of the brain respond when ghrelin is administered centrally and more robust feeding effects are obtained (Lawrence et al., 2002b). This is intriguing, since there are reports on the existence of ghrelin-producing cells in the brain (Cowley et al., 2003; Lu et al., 2002; Sato et al., 2005). However, these reports remain controversial, not the least because different laboratories describe completely different distributions. As the available data are based on RT-PCR on hypothalamic tissue and/or immunohistochemistry, without confirmation by *in situ* hybridization histochemistry to confirm the location of cells, further studies are required.

There is good evidence for the action of ghrelin in the brainstem. For example, ghrelin receptor mRNA is present in all three regions of the dorsal vagal complex, the nucleus ambiguus and the facial motor nucleus, more rostrally in the parabrachial nucleus and raphe, as well as in dopaminergic neurons of the ventral tegmental area and substantia nigra (Zigman et al., 2006). Ghrelin is equally effective at increasing food intake if injected into the dorsal vagal complex or hindbrain fourth ventricle, as it is when injected into the forebrain third ventricle (Faulconbridge et al., 2003, 2005). As with the arcuate nucleus in the forebrain, both the area postrema and NTS presumably have access to circulating ghrelin, but this does not preclude the potential for a centrally produced ligand. Indeed there are gathering data that ghrelin is more than a "hunger" signal and affects additional brain pathways, perhaps including the mesocorticolimbic dopamine system, involved in motivational behaviors (Davidson et al., 2005; Naleid et al., 2005).

B. Peptides from the Pancreas

1. Pancreatic Polypeptide

Experiments in obese mice and humans suggest a role for factors secreted by the pancreas to act on satiety pathways (Berntson et al., 1993; Gates et al., 1974; Strautz, 1970). Pancreatic polypeptide (PP) is a 36-amino acid peptide, related to NPY and PYY, which is released from F cells in the islets of Langerhans during ingestion and perhaps in anticipation of a meal (Katsuura et al., 2002; Kimmel et al., 1975; Taylor et al., 1978). PP has a strong affinity for the NPY Y4 receptor, and lower affinity for the Y5 receptor, at least in nonrodent species (Bard et al., 1995; Kanatani et al., 2000; Yan et al., 1996). There is binding for PP, as well as functional PP receptors in the area postrema, NTS, and DMX that are the most likely target for PP within the circulation, resulting in the activation of efferent vagal nerves to control the gut (McTigue et al., 1997; McTigue and Rogers, 1995; Whitcomb et al., 1990). Indeed, intraperitoneal, intracisternal, or direct microinjection into the dorsal vagal complex increases gastric emptying, motility, and secretion (McTigue et al., 1993, 1995; Okumura et al., 1994)

The actions of PP on feeding and whether they are a normal physiological response are not clear. There are reports of reduced feeding following peripheral administration (Asakawa et al., 2003; Malaisse-Lagae et al., 1977; McLaughlin and Baile, 1981), which might seem counterintuitive for a stimulus that increases gastric emptying. However, these finding have been reiterated in a transgenic mouse that overexpresses circulating PP (Ueno et al., 1999). Other reports suggest peripherally administered PP is not a potent enough stimulus to overcome the strong orexigenic drives of fasted animals (Billington et al., 1983; Taylor and Garcia, 1985), while PP administered into the cerebral ventricles, by comparison, causes an increase in food intake (Asakawa et al., 1999a; Clark et al., 1984; Inui et al., 1991; Nakajima et al., 1994).

The latter effect may be due to the presence of Y4 receptors in the hypothalamus (Bard et al., 1995; Gerald et al., 1996).

2. Insulin

Circulating levels of insulin are proportional to body weight and, thus, the hormone may act as an adiposity signal to the brain, similar to leptin (Flier and Maratos-Flier, 1998; Niswender et al., 2004). However, unlike leptin, insulin is secreted from pancreatic β cells acutely in response to meals and so can code different information. Both hormones can enter the brain by receptor-mediated, saturable transport across brain capillary endothelial cells (Baura et al., 1993; Wu et al., 1997), although many of their direct actions are probably mediated by receptors located close to areas lacking a complete blood-brain barrier, such as in the mediobasal hypothalamus (Spanswick et al., 2000). Neuronal insulin receptors are certainly important in energy balance, as their selective knockout results in an obese animal (Bruning et al., 2000). Insulin reduces food intake and body weight in a dose-dependent manner when administered directly into the brain, although its main role may be to act as a permissive hormone for the action of other factors (Woods et al., 1979, 1996).

Insulin receptors and downstream effector molecules are present in the dorsal vagal complex (Folli et al., 1994; Unger et al., 1991; Werther et al., 1987). Although to date there is no conclusive evidence for a direct action here on eating behavior, insulin injected directly into the complex can modulate gastric motor function (Krowicki et al., 1998). In addition, insulin potentiates the ability of CCK to suppress feeding in both rats and baboons and this may be at multiple levels (Figlewicz et al., 1986, 1995; Riedy et al., 1995).

3. Amylin

Amylin (also known as islet amyloid polypeptide) is a 37-amino acid peptide cosecreted with insulin by pancreatic β cells in response to a meal (Cooper et al., 1987; Westermark et al., 1986). Amylin is also found in gut endocrine cells (Mulder et al., 1994, 1997), visceral sensory neurons (Mulder et al., 1995), the midbrain (D'Este et al., 2000), and the hypothalamus (Chance et al., 1991). Exogenous amylin potently reduces food intake (Arnelo et al., 1996, 1998), gastric emptying (Clementi et al., 1996; Kong et al., 1997), and gastric secretion (Guidobono et al., 1994; Rossowski et al., 1997). The main effect of amylin is to reduce the size of meals (Lutz et al., 1995b; Reidelberger et al., 2001), rather like other hormones thought to be involved in satiety signaling, such as CCK and bombesin-like peptides. Furthermore, since amylin is a much less potent inhibitor of sham compared with normal feeding (Asarian et al., 1998), it may inhibit food intake in part secondarily to its inhibition of gastric emptying. Amylin and CCK may have a synergistic effect on feeding (Bhavsar et al., 1998; Rushing et al., 2000b), although the amylin-receptor antagonist, AC253, has been reported to attenuate the anorectic effects of CCK and bombesin (Lutz et al., 2000).

High-affinity amylin binding is found throughout the brain at sites with and without a blood-brain barrier, including the nucleus accumbens, hypothalamus, area postrema, and lamina terminalis (Sexton et al., 1994; van Rossum et al., 1994). Central administration appears to be more potent than systemic administration for reducing food intake, plus the anorectic effects are not altered by subdiaphragmatic vagotomy or by destruction of capsaicin-sensitive nerve fibers (Lutz et al., 1995a, 1998a). Instead, the area postrema and NTS seem to play a primary role (Lutz et al., 1998b; Morley et al., 1994; Rowland and Richmond, 1999), possibly with amylin activating area postrema dopaminergic neurons that project to the NTS (Lutz et al., 2001).

Plasma levels of amylin, like those of leptin and insulin, reflect body adiposity

(Pieber et al., 1994) and, therefore, it may have longer-term effects on energy homeostasis. Repeat i.c.v. injections or infusions over several days of the amylin antagonist, AC187, increases food intake and body fat in rats (Arnelo et al., 1996; Rushing et al., 2000a, 2001), while mice with targeted deletion of the amylin gene develop obesity (Gebre-Medhin et al., 1998).

C. Peptides from Adipose Tissue—Leptin

A number of peptides are released from white adipose tissue, which together have been termed adipokines. It is likely that a number of these adipokines can affect appetite indirectly by their actions on body weight in general, or due to their inflammatory-like properties (Trayhurn and Wood, 2004). The adipokines will be a rich source for research in the near future although, to date, only leptin can be considered as a physiological regulator of appetite. When originally discovered, leptin was described as a satiety factor (Zhang et al., 1994) although, more rigorously, it is seen as a signal of negative energy balance by its absence (Friedman, 2002). In reality, leptin has a multitude of actions and imposes its influence on all parts of the body involved in appetite and body weight regulation. Thus, in its absence animals display hyperphagia, characterized by increased meal size rather than more frequent meals (Becker and Grinker, 1977; Ho and Chin, 1988; McLaughlin and Baile, 1981). Conversely, leptin reduces feeding specifically by reducing meal size, with no change in meal frequency (Eckel et al., 1998; Flynn et al., 1998; Hulsey et al., 1998). Thus, although it is not a satiety factor itself, it does interact with appetite-regulating systems that control meal size.

The most important effects of leptin on the brain that are related to energy balance appear to involve the long form of its receptor (Ob-Rb; Ahima and Flier, 2000). Mercer first reported the expression Ob-Rb mRNA in the brainstem dorsal vagal complex and parabrachial nucleus using in situ hybridization histology (Mercer et al., 1998b). This finding and the localization of receptor protein were confirmed by others (Ellacott et al., 2002; Elmquist et al., 1998; Grill and Kaplan, 2002; Shioda et al., 1998; Smedh et al., 1998). Although originally there was some dispute as to the selectivity of the antibodies used for the different receptor isoforms, the immunohistochemical results are generally agreed as accurate (Baskin et al., 1999; Breininger and Baskin, 2000; Grill and Kaplan, 2002). This is supported also by the phosphorylation of the transcription factor STAT3, one of the intracellular mediators of Ob-Rb, in the NTS following the intravenous administration of leptin (Hosoi et al., 2002). Neurons intrinsic to the NTS that contain Ob-Rb coexpress catecholamines (Hay-Schmidt et al., 2001), prolactin-releasing peptide (Ellacott et al., 2002) and glucagon-like peptides (Goldstone et al., 1997), but are unlikely to coexpress neuropeptide Y or POMC (Huo et al., 2006; Mercer et al., 1998a). Furthermore, leptin receptors are located in both afferent vagal neurons that have their cell bodies in the nodose ganglia (Burdyga et al., 2002; Buyse et al., 2001; Peiser et al., 2002) and in acetylcholinesterase-containing vagal motor efferents (Smedh et al., 1998).

A direct action of leptin on feeding behavior at the level of the brainstem has been suggested following injections of leptin into either the fourth ventricle or the dorsal vagal complex (Grill and Kaplan, 2002). The same routes of leptin administration suppress gastric emptying and this is blocked by subdiaphragmatic vagotomy (Smedh et al., 1998). Thus, the feeding effects may be secondary to the indirect actions of leptin on the gut. Equally, they may be secondary to leptin-sensitive projections descending from the forebrain that can modulate functioning in the brainstem and to increase sensitivity to gut–brain signals (see also Section VI).

The most widely studied interaction of leptin in satiety signaling is that with CCK.

Fasted animals, that have low levels of circulating leptin (McMinn et al., 2000) as well as rodents with genetically deficient leptin signaling (McLaughlin and Baile, 1981; Morton et al., 2005; Strohmayer and Smith, 1987) have attenuated responses to CCK. A series of experiments have shown synergistic effects of leptin and CCK on feeding and brain c-Fos activity (Barrachina et al., 1997; Emond et al., 1999; Matson et al., 2000; Matson and Ritter, 1999; Wang et al., 1998). These two peptide systems are able to interact at a number of different levels, ultimately dependent on the location of their receptors.

Thus, the first level of interaction between leptin and CCK is at sensory terminals of the vagus nerve (Burdyga et al., 2002; Buyse et al., 2001; Peiser et al., 2002) and could be dependent on leptin produced by the gastric mucosa and fundic glands rather than that derived from adipose tissue (Bado et al., 1998; Mix et al., 1999). Electrophysiology and imaging on *in vitro* preparations has demonstrated that vagal afferent neurons and secondary NTS targets are sensitive to CCK and/or leptin, and that coadministration of the two peptides can cause additive excitation (Peters et al., 2004; Wang et al., 1997; Yuan et al., 1999, 2000). An additional effect of leptin may be also to modulate the release of CCK itself (Guilmeau et al., 2003).

V. INTRINSIC PEPTIDERGIC NEURONS OF THE DORSAL VAGAL COMPLEX

Although lagging somewhat behind our knowledge of peptidergic neurons in the hypothalamus, progress is now being made in the identification of neuropeptides that are produced in the brainstem itself and that can affect eating behavior. As with most neuropeptides these messengers are pleiotropic and tend also to be expressed in more than one population in the brain. The usual criteria for determining if a factor has a truly physiological action still need to be proved for some of these neuropeptides. First, they should have a demonstrable effect of feeding that is not secondary to some other action. Second, they should be present in the areas of the brain known to be involved in appetite regulation and/or to act within these areas. Third, their expression should be modulated in the direction expected if the feeding/energy status of the animal is altered. Fourth, they would normally be expected to interact with other regulators of appetite and metabolism. Last, disruption of their action, either by receptor antagonism or gene knockout, should block the actions of the peptide and/or produce an energy imbalance phenotype.

A. Neuropeptide Y

Some of the earliest studies using c-Fos as a marker for neuronal activation demonstrated that exogenously applied CCK activated approximately half of the catecholaminergic cells in the NTS but that conversely, only half of the neurons activated in the NTS were catecholaminergic (Luckman, 1992). This suggested that subpopulations of catecholaminergic and noncatecholaminergic neurons were responding that could perhaps be identified by their peptide cotransmitters. Indeed, 39% of NPY-positive neurons in the NTS, but no neurotensin-positive neurons, are activated by CCK (Rinaman et al., 1993). At least some NPY neurons in the brainstem project to hypothalamic regions (including the preoptic, paraventricular, and dorsomedial nuclei) and could relay information regarding satiety signaling (Sahu et al., 1988; Sawchenko et al., 1985). However, this observation appears paradoxical, since NPY is a powerful orexigenic peptide (Clark et al., 1984; Levine and Morley, 1984) and descending NPYergic- or NPY-sensitive pathways can attenuate the response of NTS neurons to both CCK and gastric distension (McMinn et al., 2000; Schwartz and

Moran, 2002). Receptors are present in each region of the dorsal vagal complex (Kishi et al., 2005; Kopp et al., 2002; Wolak et al., 2003) and the orexigenic effect of NPY can be attained following fourth ventricular administration of the peptide (Corp et al., 1990, 2001; Faulconbridge et al., 2005). However, while NPY expression in the hypothalamic arcuate nucleus is increased by 48-h food deprivation, as expected for an orexigenic neuropeptide, it is not in the NTS (Ishizaki et al., 2003). So the paradox remains, but might be explained by NPY (possibly produced in the two brain areas) having opposing actions on different aspects of behavior. NPY increases appetitive behaviors, that is, the searching and acquisition of food, but can actually inhibit consummatory behavior (Ammar et al., 2005; Sederholm et al., 2002; Seeley et al., 1995). The fact that NPY and CCK can reduce consummatory ingestive behavior and induce c-Fos in the NTS to a similar extent and with no additivity, suggests that some of CCK's effects within the NTS may be mediated by NPY, a potential that will require careful further analysis.

B. Glucagon-like Peptides

As already discussed in Section IV.A.3., the glucagon-like peptides (GLP-1, GLP-2, oxyntomodulin, and glicentin) are produced in the gut (for review see Kieffer and Habener, 1999). However, they have a relatively short half-life in the circulation and many actions they have in the brain, remote from the gut, are likely to reflect their production within brainstem neurons. GLP neurons are located in the NTS, are noncatecholaminergic, and project to forebrain limbic structures, including the hypothalamus (Jin et al., 1988; Larsen et al., 1997; Merchenthaler et al., 1999). GLP receptor 1 (GLP-1R) mRNA is found in the same areas as GLP-1 fiber staining: nuclei of the hypothalamus and thalamus, bed nucleus of the stria terminalis, septum, zona incerta, amygdala, nucleus accumbens, parabra-

chial nucleus, raphe nuclei, NTS, DMX, and area postrema (Merchenthaler et al., 1999). Although presumably originating from the same NTS neurons, GLP-2 has a much more restricted distribution. Fibers are present in the hypothalamic paraventricular and arcuate nuclei, but particularly in the ventral dorsomedial nucleus (Tang-Christensen et al., 2000). Likewise, mRNA for GLP receptor 2 (GLP-2R) in the brain is found exclusively in the compact zone of the dorsomedial nucleus.

GLP-1 (Navarro et al., 1996; Tang-Christensen et al., 1996; Turton et al., 1996), GLP-2 (Tang-Christensen et al., 2000), and oxyntomodulin (Dakin et al., 2002) all cause a reduction in food intake when administered into the forebrain ventricles of rodents. These effects are blocked by previous administration of the receptor antagonist, exendin$_{9-39}$. This same antagonist has been used in otherwise naïve animals to cause an increase in food intake, which is clear evidence to suggest that GLPs are endogenous regulators of food intake (Meeran et al., 1999; Turton et al., 1996). However, as with all brainstem peptides, the difficulty is in determining whether GLPs act selectively under physiological conditions.

It is important to consider whether these neuropeptides are regulated accordingly by metabolic status and if they interact with other factors involved in energy homeostasis. Preproglucagon-containing neurons of the NTS possess mRNA for the long form of the leptin receptor (Goldstone et al., 1997), and leptin treatment can reverse the reduction of hypothalamic GLP-1 peptide content that follows a period of fasting (Goldstone et al., 2000). This is most likely due to an effect of leptin on NTS neuronal cell bodies, although a direct action of leptin or changes in the expression of preproglucagon mRNA according to energy status have not been measured. Repeat injections of GLP-1 receptor ligands have been found by some to reduce body weight (Larsen et al., 2001; Meeran et al., 1999), whereas others report no long-term effects (Donahey et al., 1998). Furthermore, GLP-1 receptor null mice display normal feeding behavior and body weight and respond to leptin normally, although they are glucose intolerant (Scrocchi et al., 1996, 2000; Scrocchi and Drucker, 1998).

Preproglucagon neurons in the NTS are activated, that is they are induced to express c-Fos protein, in response to a number of stimuli that can cause a reduction in feeding behavior: CCK, artificial distension of the stomach, lithium chloride, and the endotoxin, lipopolysaccharide (Rinaman, 1999a,b; Vrang et al., 2003). Perhaps surprisingly, the neurons are not activated by a satiating meal (Rinaman, 1999b), leading to the suggestion that GLP-1 may be mediating the effects of aversive stimuli, such as nausea, anxiety, and visceral pain, rather than normal satiety signals. Central GLP-1 administration can cause taste aversions in rodents (Seeley et al., 2000; Thiele et al., 1997; Van Dijk et al., 1996), and GLP receptor antagonists reduce the anorectic and aversive effects of lithium chloride (Kinzig et al., 2003; Rinaman, 1999a; Seeley et al., 2000; Thiele et al., 1997).

Thus, although the evidence that GLP-1 relays signals of interoceptive stress in rodents is strong, there remains the possibility that this and other preproglucagon-derived peptides still have a role in satiety signaling, and further investigations to understand the different signaling pathways are warranted. Central injection of GLP-1 induces c-Fos in various parts of the brain that may be targets for GLPergic signaling, including the hypothalamic paraventricular nucleus, central nucleus of the amygdala, parabrachial nucleus, NTS, and area postrema (Larsen et al., 1997; Rowland et al., 1997; Van Dijk et al., 1996). Injections of GLP-1 directly into the paraventricular nucleus reduces food intake and leads to the secretion of adrenocorticotrophic hormone without causing either anxiety or a taste aversion, while the opposite is true for injections into the central amygdala (Kinzig et al., 2003; McMahon and Wellman, 1998). One explanation is that the different responses are mediated by collaterals from the same neurons. GLP-2, is assumed to be produced by the same neurons, reduces food intake without taste aversion, and activates only the dorsomedial nucleus of the hypothalamus (Tang-Christensen et al., 2000).

C. Prolactin-releasing Peptide

Prolactin-releasing peptide (PrRP) is a member of the RFamide family of peptides, discovered by Hinuma and colleagues in 1998 in the hypothalamus and described as a hypophysiotropic factor positively regulating the secretion of prolactin from the anterior pituitary (Hinuma et al., 1998). However, it may have been misnamed since a number of subsequent studies have shown that PrRP is a poor prolactin secretagogue *in vivo*, or that it is only effective as such in very specific circumstances (Jarry et al., 2000; Matsumoto et al., 1999; Samson et al., 1998; Tokita et al., 1999). Further, although there is a small population of neurons in the hypothalamic dorsomedial nucleus, PrRP is produced mainly in two brainstem cell populations: one in the NTS and the other in the ventrolateral medulla (Maruyama et al., 1999; Roland et al., 1999). The NTS population is catecholaminergic and is distinct from that containing glucagon-like peptides (Chen et al., 1999; Morales et al., 2000). PrRP-immunoreactive fibers are present in the preoptic area, hypothalamic paraventricular, periventricular, and supraoptic nuclei, midline thalamus, bed nucleus of the stria terminalis, basolateral and central amygdala, septum, parabrachial nucleus, and dorsal vagal complex (Ibata et al., 2000; Iijima et al., 1999; Kalliomaki et al., 2004; Maruyama et al., 1999; Morales et al., 2000; Yamakawa et al., 1999). The PrRP receptor is expressed primarily in the hypothalamus, the thalamus, and the border of the NTS and area postrema (Ibata et al., 2000; Jarry et al., 2000; Lawrence et al., 2000; Roland et al., 1999).

PrRP does fit many of the criteria for being regarded as a homeostatic regulator of food intake. Central injections of PrRP cause a reduction in feeding without causing a taste aversion (Ellacott et al., 2002; Lawrence et al., 2000, 2002a; Seal et al., 2001). In a behavioral test it maintains the normal sequence of events following feeding but, like a true satiety factor, compresses the sequence (Lawrence et al., 2002a). As is usual for anorectic peptides in the brain, its expression is down regulated when an energy deficit is perceived in fasted, lactating, or Zucker rats (Ellacott et al., 2002; Lawrence et al., 2000). PrRP neurons possess leptin receptor immunoreactivity (in fact, PrRP neurons may be the major population of leptin-sensitive NTS neurons) and coadministration of PrRP and leptin has additive effects on appetite, body weight, and energy expenditure (Ellacott et al., 2002). Also, brainstem PrRP neurons are activated by CCK, but not by lithium chloride (Lawrence et al., 2002a). Results from repeated injections are more varied. Whereas Ellacott et al. (2003) report initial effects and tolerance developing within 4 days, Vergoni et al. (2002) note effects on feeding and body weight only after repeat dosing.

Rodents with mutated PrRP receptors have obese phenotypes. Transgenic mice with a mutated receptor gene develop obesity and decreased glucose tolerance in adulthood (Gu et al., 2004); however, the full phenotype of these mice has yet to be described. For example, it is not known if they have disrupted meal patterning. More recent results have suggested that PrRP-receptor knockout mice do not respond to anorectic doses of CCK given systemically (or PrRP given centrally), while they are still responsive to anorexia induced by systemic lipopolysaccharide (Bechtold and Luckman, 2006). Another strain of knockout mice does not respond to PrRP, while congenic wildtypes actually display an increase in feeding (Adams et al., 2005). The reasons for these differences are unknown, although further investigation is required into the possible interactions with other RFamide receptors (Engstrom et al., 2003). The RFamides are an interesting group of peptides with a role in feeding that appears to be conserved through evolution (Dockray, 2004). At least one other mammalian RFamide, neuropeptide FF (NPFF) is expressed in the NTS (Kalliomaki and Panula, 2004), and when injected at low doses, NPFF can reduce food intake in rats (Murase et al., 1996; Nicklous and Simansky, 2003; Sunter et al., 2001).

Perhaps one of the most intriguing rodent models is the Otsuka Long-Evans Tokushima Fatty (OLETF) rat strain which has long been known to have a mutation in the CCK_A receptor gene that was assumed to be responsible for its obese phenotype (Funakoshi et al., 1995; Moran et al., 1998). However, more recent studies have shown that the OLETF rat also has a mutation in the PrRP receptor gene (Watanabe et al., 2005). PrRP does not reduce food intake in OLETF rats unless they have been introduced with the wild-type allele by backcrossing with normal rats. Furthermore, such backcrossed strains are not obese, hyperphagic, or dyslipidemic like the congenic OLETF rats. However, again there may be strain differences, since other rats with the same start codon mutation (ATG to ATA) are not obese (Ellacott et al., 2005).

If PrRP is a physiological factor, then it may affect energy balance by different mechanisms, including a local action within the brainstem. First, injections of PrRP (or glutamate) into the dorsal vagal complex at the level of the area postrema leads to an increase in gastric tone, while the opposite is true if the injections are made caudal to the area postrema (Grabauskas et al., 2004). PrRP acts presynaptically to enhance excitatory glutamatergic currents in DMX neurons retrogradely labeled from the stomach. Thus, acute actions on feeding may be secondary to a reduction in gastric emptying. However, PrRP binding in the brainstem is more noticeably associated with the border of the NTS and area postrema, rather than the DMX (Ellacott et al., 2005). Second, PrRP neurons in the NTS project to the hypothalamic paraventricular nucleus (Morales et al., 2000) and the anorectic actions of PrRP can be partially blocked by administration of a CRH receptor antagonist (Lawrence et al., 2004). PrRP-immunoreactive fibers, but not receptor mRNA, are associated with the CRH neurons in the paraventricular nucleus (Lin et al., 2002), which may suggest another possible presynaptic effect. Finally, PrRP may have additional effects on body weight that are not dependent entirely on reduced feeding, and instead involve an increase in energy expenditure (Lawrence et al., 2000, 2004).

D. Neuromedin U

Another catecholaminergic cell type in the NTS that is activated by systemic administration of CCK is that which contains the peptide, neuromedin U [NMU; (Ivanov et al., 2004)]. However, there are three additional central populations of NMU-producing cells in the rat. In the brainstem, there are populations in the inferior olive and the sensory nucleus of the trigeminal nerve, and also there is a population in the *pars tuberalis* of the pituitary (Ivanov et al., 2002). The *pars tuberalis* ascends the pituitary stalk onto the mediobasal surface of the hypothalamus, and was previously misidentified as being part of the arcuate nucleus (Howard et al., 2000). NMU can overcome strong orexigenic drives to reduce both normal and fast-induced feeding, without supporting a taste aversion (Howard et al., 2000; Ivanov et al., 2002; Kojima et al., 2000; Nakazato et al., 2000). However, it is not clear which population might be responsible for these effects physiologically.

Knockout $NMU^{-/-}$ mice are obese, hyperphagic, and have decreased energy expen-

diture and locomotor activity (Hanada et al., 2004). Conversely, transgenic mice overexpressing NMU in the brain are hypophagic and lean (Kowalski et al., 2005). The assumption is that the populations of NMU-producing cells in the *pars tuberalis* or the NTS are responsible for effects on energy balance as NMU mRNA in these two areas, and not in the inferior olive or the sensory nucleus of the trigeminal nerve, is reduced during fasting and in animals with deficient leptin signaling (Howard et al., 2000; Ivanov et al., 2002). The cells that possess NMU receptor 2, that may respond to NMU produced from these two regions, are in the wall of the third ventricle (adjacent to the caudal hypothalamus), the hypothalamic paraventricular nucleus, and in the border between the NTS and area postrema (Graham et al., 2003; Guan et al., 2001; Howard et al., 2000; Ivanov et al., 2004). However, the phenotype of these cells has yet to be established.

E. Melanocortins

A large literature exists on melanocortins and the regulation of body weight, although most of this concentrates on the role of POMC-containing neurons in the arcuate nucleus of the hypothalamus (reviewed elsewhere; Cone, 1999). However, melanocortins are equally effective when administered in the hindbrain as they are when administered into the forebrain (Grill et al., 1998; Williams et al., 2000), and they can reduce the size of an individual meal, suggesting that they also may affect satiety signaling (Azzara et al., 2002; Williams et al., 2002). Although the potential exists for the regulation of brainstem circuits by descending melanocortin pathways from the arcuate nucleus (Joseph and Michael, 1988; Palkovits and Eskay, 1987), there are also POMC-positive neurons in the caudal NTS itself (Bronstein et al., 1992; Joseph et al., 1983). POMC-immunoreactive fibers, probably from both populations, as well as MC4 receptor mRNA (Kishi et al., 2003;

Mountjoy et al., 1994), are located in the NTS, ventrolateral medulla, and nucleus ambiguus within the brainstem. As with the other intrinsic NTS peptidergic populations discussed earlier, it is probable that melanocortin neurons do not just have local connections, but may also ascend to higher brain centers (Sim and Joseph, 1994).

POMC neurons in the NTS are neither catecholaminergic nor GLPergic, but approximately 30% of them in the mouse express c-Fos following systemic CCK (Fan et al., 2004). Furthermore, the reduction of food intake brought about in fasted and refed rats by CCK is attenuated by fourth ventricular injection of a melanocortin receptor antagonist and in MC4-receptor null mice (Fan et al., 2004). There is presently some debate as to whether these neurons are directly responsive to leptin (Ellacott et al., 2006; Huo et al., 2006). Other stimuli have yet to be fully characterized, although a small percentage of NTS POMC neurons express c-Fos after a meal in previously food-restricted rats (Fan et al., 2004). The NTS POMC neurons display an increase in excitatory postsynaptic potentials following stimulation of the solitary tract, which can be inhibited by enkephalins (Appleyard et al., 2005).

F. Galanin

Galanin is produced in many brain regions including the hypothalamus and brainstem (Melander et al., 1986a; Skofitsch and Jacobowitz, 1985). Within the caudal NTS it is localized within catecholaminergic cells, but more rostrally galanin cell bodies intermingle with but are distinct from the former (Melander et al., 1986b). It is not clear if galaninergic fibers in the NTS are all intrinsic, that is, originating in the NTS, or whether some belong to neurons descending from the hypothalamus or elsewhere. Receptor autoradiography has located galanin receptor binding in the NTS (Melander et al., 1988; Skofitsch et al., 1986), and galanin has a direct action there to

cause an increase in feeding (Koegler and Ritter, 1996, 1997, 1998; Koegler et al., 1999). The orexigenic action of a putatively intrinsic brainstem neuropeptide makes galanin a case that warrants further investigation.

VI. DESCENDING PEPTIDERGIC REGULATION OF BRAINSTEM FEEDING CIRCUITS

Brainstem circuits involved in the direct regulation of feeding are under the influence of higher brain centers that form the indirect regulators mentioned in Section I. These include descending pathways from the orbitofrontal cortex, amygdala, bed nucleus of the stria terminalis, and hypothalamus, thus forming reciprocal connections with the dorsal vagal complex (Rinaman et al., 2000; Rinaman and Schwartz, 2004; Strack and Loewy, 1990; van der Kooy et al., 1984; Wallace et al., 1992; Yang et al., 1999). Many of these pathways are peptidergic, although only those from the hypothalamus have been analyzed substantially to date. For example, the paraventricular nucleus is a major output of the hypothalamus and neurons project from here to the NTS (Sawchenko and Swanson, 1982; Toth et al., 1999), where they influence the firing of gut-sensitive neurons (Duan et al., 1999; Zhang et al., 1999). Systemic injection of CCK induces c-Fos in neurons of the paraventricular nucleus that include those containing CRH and oxytocin (Verbalis et al., 1991), and lesions of the paraventricular nucleus cause hyperphagia (Leibowitz et al., 1981), partly due to an attenuation of the satiating effects of CCK (Crawley and Kiss, 1985). These and some of the other descending peptidergic projections from the hypothalamus are considered here.

A. CRH/Urocortin

Corticotrophin-releasing hormone (CRH) and/or the urocortins not only activate the hypothalamo-pituitary-adrenal system, but also have central effects on feeding and on gastric emptying (Asakawa et al., 1999b; Gosnell et al., 1983; Spina et al., 1996; Turnbull and Rivier, 1997). CRH neurons are activated by systemic CCK (Verbalis et al., 1991) and may, thus, be involved in normal satiety signaling. However, these neurons also provide an obvious route whereby varied stressors mediate their effects on the direct regulators of feeding (Martinez et al., 1997; Tache et al., 1990). CRH neurons in the paraventricular nucleus project to the NTS (Sawchenko, 1987; Sawchenko and Swanson, 1982), while CRH, urocortin II, and CRH receptors are all present in the brainstem (Bittencourt and Sawchenko, 2000; Bittencourt et al., 1999; Reyes et al., 2001; Swanson et al., 1983; Van Pett et al., 2000). The i.c.v. injection of these peptides induce c-Fos in the NTS (Benoit et al., 2000). The hypophagic, gastric, and sympathoadrenal effects of CRH/urocortin can be induced by injecting the peptides into the fourth ventricle or into the dorsal vagal complex (Brown, 1986; Grill et al., 2000; Smedh et al., 1995), and urocortin is effective within the NTS of decrebrate rats (Daniels et al., 2004).

B. Oxytocin

As with CRH neurons, oxytocinergic neurons in the paraventricular nucleus (both the descending parvocellular neurons and the neurohypophyseal, magnocellular neurons) are activated by systemic administration of CCK (Luckman et al., 1993; Olson et al., 1992; Verbalis et al., 1991), and by other stimuli that cause hypophagia (McCann et al., 1989; Olson et al., 1991). This activation appears to be dependent on catecholaminergic projections from the brainstem, some of which are direct to the hypothalamus (Onaka et al., 1995a,b; Rinaman et al., 1995). It is well documented that parvocellular oxytocin neurons in the hypothalamic paraventricular nucleus form a reciprocal connection with the NTS

(Rinaman, 1998; Sawchenko and Swanson, 1982) and that oxytocin fibers and receptors are located in the latter (Barberis and Tribollet, 1996; Loup et al., 1989).

The i.c.v. injection of oxytocin can cause a reduction in feeding behavior and gastric motility (Arletti et al., 1989, 1990; Flanagan et al., 1992) and this correlates with an activation of neurons in the caudal brainstem (Olson et al., 1993). Oxytocinergic fibers are close to CCK-induced c-Fos in the NTS, and the reduction of feeding by exogenous CCK is attenuated by the fourth ventricular administration of an oxytocin receptor antagonist (Blevins et al., 2003). Furthermore, signaling by the descending oxytocinergic pathway from the hypothalamic paraventricular nucleus is potentiated by leptin (Blevins et al., 2004), providing a clear example of how signals are integrated by higher brain centers to modulate the direct regulators of feeding. This is not to infer that all the actions of CCK are mediated solely by oxytocin, and CCK is still effective in the oxytocin knockout mouse (Mantella et al., 2003).

C. NPY and AGRP

While leptin acts on descending neurons from the paraventricular nucleus, it has equally important actions on neurons of the arcuate nucleus, some of which may also send projections to the brainstem. Thus, the potentiating effect of third ventricular leptin on CCK-mediated neuronal activity in the NTS is partly dependent on leptin receptors in the hypothalamic arcuate nucleus (Morton et al., 2005). The obvious candidates for arcuate neurons are those that contain NPY or POMC. As already discussed, NPY-sensitive pathways attenuate the response of NTS neurons to both CCK and gastric distension (McMinn et al., 2000; Schwartz and Moran, 2002). The question is whether these actions relate to NPY produced by neurons intrinsic to the NTS (see details in Section V.A.) or by neurons in the arcuate nucleus. Although this cannot presently be answered directly, there is some indirect evidence against the latter. The majority of NPY neurons in the arcuate nucleus colocalize agouti gene-related peptide (AGRP), whereas other NPYergic populations do not. And, although AGRP-positive fibers are widespread in brain areas involved in energy homeostasis, very few are present within the brainstem (Broberger et al., 1998). It is also noteworthy that AGRP, although a ligand for melanocortin rather than NPY receptors, does not induce c-Fos acutely in the NTS, whereas NPY does (Hagan et al., 2000).

D. POMC and CART

As noted before, there are two populations of POMC neurons in the brain: in the hypothalamic arcuate nucleus and within the commissural NTS. POMC neurons in the arcuate project to a number of sites in the brainstem, including the periaqueductal gray, dorsal raphe nucleus, nucleus raphe magnus, nucleus raphe pallidus, locus coeruleus, parabrachial nucleus, nucleus reticularis gigantocellularis, NTS, and DMX (Chronwall, 1985; Palkovits and Eskay, 1987; Sim and Joseph, 1991; Zheng et al., 2005b). Many of these regions contain MC4 receptor mRNA (Kishi et al., 2003; Mountjoy et al., 1994) and are, therefore, potential sites for neuromodulation by POMC-derived peptides. Fourth ventricular or parenchymal injections of the melanocortin receptor agonist, melanotan II, produces hypophagia (Grill et al., 1998; Williams et al., 2000). The decrease in feeding is due to reductions in meal size rather than meal frequency and provides a mechanism by which a descending melanocortin pathway may mediate meal regulation by CCK (Sutton et al., 2005; Zheng et al., 2005b).

POMC neurons in the arcuate coexpress another neuropeptide, cocaine- and amphetamine-regulated transcript (CART; Kristensen et al., 1998; Vrang et al., 1999a). However, CART colocalizes with other peptides, like melanin-concentrating hormone in the lateral hypothalamic area (Cvetkovic et al., 2004). It is present also in the brainstem, intrinsic to the dorsal vagal complex, vagus nerve, and nodose ganglion (Broberger et al., 1999; Douglass et al., 1995; Dun et al., 2001; Koylu et al., 1998; Zheng et al., 2002). CART acts at the level of the brainstem to inhibit gastric emptying and reduce food intake (Okumura et al., 2000; Smedh and Moran, 2003; Zheng et al., 2001) and can activate neurons in the NTS (Vrang et al., 1999b; Zheng et al., 2002). However, it has been reported to affect motor and aversive behavior (Aja et al., 2001, 2002) and, thus, further information is required about the origin and role of CART in the brainstem. Interestingly, there are connections between CART-positive neurons in the NTS and neurons in the hypothalamus that contain CRH and thyrotrophin-releasing hormone (TRH; Fekete et al., 2005; Wittmann et al., 2004).

E. Orexin and MCH

Orexins (also termed hypocretins) are produced selectively in the lateral hypothalamus and project widely through the brain to affect sleep/arousal and energy homeostasis (De Lecea et al., 1998; Sakurai et al., 1998). Importantly, they form a link between limbic structures that have a role in homeostasis and higher centers in the brain (such as the nucleus accumbens and the prefrontal cortex) involved in cognitive functions, like reward and decision making (Kelley et al., 2005). Also, orexin-containing neurons project and terminate within the dorsal vagal complex, hypoglossal and trigeminal motor nuclei, ventrolateral medulla, nucleus ambiguus, and raphe nuclei in the brainstem (Berthoud et al., 2005; Ciriello et al., 2003; Date et al., 1999; Dergacheva et al., 2005; Fung et al., 2001; Guan et al., 2005; Zheng et al., 2005a), where receptors have been located (Krowicki et al., 2002; Marcus et al., 2001; Sunter et al., 2001). The direct application of orexin into the dorsal vagal complex (or onto slices) affects the activity of neurons with ingestive and autonomic functions (Berthoud et al., 2005; Dergacheva et al., 2005; Krowicki et al., 2002; Wu et al., 2004; Zheng et al., 2005a). Within the NTS, the effect of orexin tends to be excitatory (Smith et al., 2002; Yang et al., 2003), while in the DMX it excites some neurons projecting to the gastrointestinal tract, but enhances synaptic inhibition in others (Davis et al., 2003; Grabauskas and Moises, 2003; Hwang et al., 2001). Evidence for reciprocal connections with the brainstem exists since GLP-1 is capable of directly stimulating orexinergic neurons (Acuna-Goycolea and van den Pol, 2004; Harrison et al., 1999).

Melanin-concentrating hormone neurons are also located in the lateral hypothalamus, although they are distinct from those that express orexin. There is a good correlation between the expression of its receptors and MCH-immunolabeled fibers in different parts of the brain involved in arousal, motivated behavior, and feeding: including in the brainstem the locus coeruleus, facial, hypoglossal, trigeminal, and vagal motor nuclei (Saito et al., 2001). However, although there is a strong literature supporting an important role for MCH in feeding (Pissios and Maratos-Flier, 2003), there is not yet any direct evidence for its actions at the level of the brainstem.

F. Thyrotrophin-releasing Hormone

Receptors for TRH are found in the dorsal vagal complex (Manaker et al., 1985), but may not be targeted primarily by TRH from the paraventricular nucleus. There is a dense innervation of both the NTS and the DMX from the nucleus raphe obscurus that contains TRH (Palkovits et al., 1986; Rogers et al., 1980). This neuropeptide acts in concert with serotonin to potently inhibit NTS neurons and excite DMX neurons to counteract the basic vago-vagal negative feedback circuits (for reviews see Fujimiya and Inui, 2000; Rogers et al., 1996). The augmentation of gastrointestinal function by TRH may facilitate feeding (Kraly and Blass, 1976) and be part of this peptide's overall role in organizing metabolic and autonomic responses to the stress of cold exposure. TRH released from the hypothalamic paraventricular nucleus releases thyrotrophin from the pituitary to increase heat production, and TRH neurons of the medullo-spinal system terminate on preganglionic neurons to cause vasoconstriction and on skeletal motor neurons to cause shivering (Wei, 1981; White et al., 1989).

VII. SUMMARY AND POTENTIAL FOR DRUG DEVELOPMENT

Since the discovery of leptin (Zhang et al., 1994), there has been a surge in the study of neuropeptides related to body weight regulation and their potential as targets for drug development. Understandably, most interest has focused on the neuropeptides that function within the hypothalamus as this part of the brain is essential for homeostatic control. However, with the understanding that appetite and body weight are regulated by dispersed brain systems, attention to other parts of the nervous system has been regenerated, as this book attests. The brainstem plays an essential role in receiving inputs from the periphery and integrating these with information from higher brain centers. Thus, there are "neuropeptidergic" signals acting from many directions, as well as the neuropetides that are intrinsic to the brainstem itself. Part of the problem with studying neuropeptides is their multiple sources and sites of action. This belies how these systems evolved and should be of no surprise to the researcher. In fact, so long as the neuropetide always acts in the same direction (either anabolic or catabolic) in different parts of the brain they will remain suitable targets for drug development.

This chapter has reviewed the action of neuropeptides within the dorsal vagal complex and has only mentioned in passing other brainstem systems. This selectivity has at least enabled the discussion to consider the different modes of neuropetidergic signaling. Thus, there are peptidergic inputs from the periphery either carried neuronally or humorally. These signals are integrated by the NTS and projected back to the periphery or to other brain centers. The NTS is further modulated by descending peptidergic pathways. The importance of peptidergic neurons that are intrinsic to the NTS is beginning to be appreciated and they may yet prove to be as important in regulating appetite and body weight as some of their hypothalamic counterparts.

References

Abbott, C. R., Monteiro, M., Small, C. J., Sajedi, A., Smith, K. L., Parkinson, J. R., Ghatei, M. A., and Bloom, S. R. (2005). The inhibitory effects of peripheral administration of peptide YY(3–36) and glucagon-like peptide-1 on food intake are attenuated by ablation of the vagal-brainstem-hypothalamic pathway. *Brain Res.* **1044**, 127–131.

Acuna-Goycolea, C., and van den Pol, A. (2004). Glucagon-like peptide 1 excites hypocretin/orexin neurons by direct and indirect mechanisms: implications for viscera-mediated arousal. *J. Neurosci.* **24**, 8141–8152.

Acuna-Goycolea, C., and van den Pol, A. N. (2005). Peptide YY(3-36) inhibits both anorexigenic proopiomelanocortin and orexigenic neuropeptide Y neurons: implications for hypothalamic regulation of energy homeostasis. *J. Neurosci.* **25**, 10510–10519.

Adams, J. R., Cox-York, K., MacNeil, D. J., Reitman, M. L., and Marsh, D. J. (2005). Prolactin-releasing peptide exhibits orexigenic effects in mice that are not mediated by GPR10. *Soc. Neurosci. Abstr.* 764.7.

Adrian, T. E., Ferri, G. L., Bacarese-Hamilton, A. J., Fuessl, H. S., Polak, J. M., and Bloom, S. R. (1985). Human distribution and release of a putative new gut hormone, peptide YY. *Gastroenterology* **89**, 1070–1077.

Ahima, R. S., and Flier, J. S. (2000). Leptin. *Annu. Rev. Physiol.* **62**, 413–437.

Aja, S., Sahandy, S., Ladenheim, E. E., Schwartz, G. J., and Moran, T. H. (2001). Intracerebroventricular CART peptide reduces food intake and alters motor behavior at a hindbrain site. *Am. J. Physiol.* **281**, R1862–R1867.

Aja, S., Robinson, B. M., Mills, K. J., Ladenheim, E. E., and Moran, T. H. (2002). Fourth ventricular CART reduces food and water intake and produces a conditioned taste aversion in rats. *Behav. Neurosci.* **116**, 918–921.

Altschuler, S. M., Bao, X. M., Bieger, D., Hopkins, D. A., and Miselis, R. R. (1989). Viscerotopic representation of the upper alimentary tract in the rat: sensory ganglia and nuclei of the solitary and spinal trigeminal tracts. *J. Comp. Neurol.* **283**, 248–268.

Ammar, A. A., Nergardh, R., Fredholm, B. B., Brodin, U., and Sodersten, P. (2005). Intake inhibition by NPY and CCK-8: a challenge of the notion of NPY as an "Orexigen." *Behav. Brain Res.* **161**, 82–87.

Anini, Y., Fu-Cheng, X., Cuber, J. C., Kervran, A., Chariot, J., and Roz, C. (1999). Comparison of the postprandial release of peptide YY and proglucagon-derived peptides in the rat. *Pflugers Arch.* **438**, 299–306.

Appleyard, S. M., Bailey, T. W., Doyle, M. W., Jin, Y. H., Smart, J. L., Low, M. J., and Andresen, M. C. (2005). Proopiomelanocortin neurons in nucleus tractus solitarius are activated by visceral afferents: regulation by cholecystokinin and opioids. *J. Neurosci.* **25**, 3578–3585.

Ariyasu, H., Takaya, K., Tagami, T., Ogawa, Y., Hosoda, K., Akamizu, T., Suda, M., Koh, T., Natsui, K., Toyooka, S., Shirakami, G., Usui, T., Shimatsu, A., Doi, K., Hosoda, H., Kojima, M., Kangawa, K., and Nakao, K. (2001). Stomach is a major source of circulating ghrelin, and feeding state determines plasma ghrelin-like immunoreactivity levels in humans. *J. Clin. Endocrinol. Metab.* **86**, 4753–4758.

Arletti, R., Benelli, A., and Bertolini, A. (1989). Influence of oxytocin on feeding behavior in the rat. *Peptides* **10**, 89–93.

Arletti, R., Benelli, A., and Bertolini, A. (1990). Oxytocin inhibits food and fluid intake in rats. *Physiol. Behav.* **48**, 825–830.

Armstrong, D. M., Pickel, V. M., Joh, T. H., Reis, D. J., and Miller, R. J. (1981). Immunocytochemical localization of catecholamine synthesizing enzymes and neuropeptides in area postrema and medial nucleus tractus solitarius of rat brain. *J. Comp. Neurol.* **196**, 505–517.

Arnelo, U., Permert, J., Adrian, T. E., Larsson, J., Westermark, P., and Reidelberger, R. D. (1996). Chronic infusion of islet amyloid polypeptide causes anorexia in rats. *Am. J. Physiol.* **271**, R1654–R1659.

Arnelo, U., Reidelberger, R., Adrian, T. E., Larsson, J., and Permert, J. (1998). Sufficiency of postprandial plasma levels of islet amyloid polypeptide for suppression of feeding in rats. *Am. J. Physiol.* **275**, R1537–R1542.

Asakawa, A., Inui, A., Ueno, N., Fujimiya, M., Fujino, M. A., and Kasuga, M. (1999a). Mouse pancreatic polypeptide modulates food intake, while not influencing anxiety in mice. *Peptides* **20**, 1445–1448.

Asakawa, A., Inui, A., Ueno, N., Makino, S., Fujino, M. A., and Kasuga, M. (1999b). Urocortin reduces food intake and gastric emptying in lean and ob/ob obese mice. *Gastroenterology* **116**, 1287–1292.

Asakawa, A., Inui, A., Kaga, T., Yuzuriha, H., Nagata, T., Ueno, N., Makino, S., Fujimiya, M., Niijima, A., Fujino, M. A., and Kasuga, M. (2001). Ghrelin is an appetite-stimulatory signal from stomach with structural resemblance to motilin. *Gastroenterology* **120**, 337–345.

Asakawa, A., Inui, A., Yuzuriha, H., Ueno, N., Katsuura, G., Fujimiya, M., Fujino, M. A., Niijima, A., Meguid, M. M., and Kasuga, M. (2003). Characterization of the effects of pancreatic polypeptide in the regulation of energy balance. *Gastroenterology* **124**, 1325–1336.

Asarian, L., Eckel, L. A., and Geary, N. (1998). Behaviorally specific inhibition of sham feeding by amylin. *Peptides* **19**, 1711–1718.

Azzara, A. V., Sokolnicki, J. P., and Schwartz, G. J. (2002). Central melanocortin receptor agonist reduces spontaneous and scheduled meal size but does not augment duodenal preload-induced feeding inhibition. *Physiol. Behav.* **77**, 411–416.

Bado, A., Levasseur, S., Attoub, S., Kermorgant, S., Laigneau, J. P., Bortoluzzi, M. N., Moizo, L., Lehy, T., Guerre-Millo, M., Le Marchand-Brustel, Y., and Lewin, M. J. (1998). The stomach is a source of leptin. *Nature* **394**, 790–793.

Barberis, C., and Tribollet, E. (1996). Vasopressin and oxytocin receptors in the central nervous system. *Crit. Rev. Neurobiol.* **10**, 119–154.

Bard, J. A., Walker, M. W., Branchek, T. A., and Weinshank, R. L. (1995). Cloning and functional expression of a human Y4 subtype receptor for pancreatic polypeptide, neuropeptide Y, and peptide YY. *J. Biol. Chem.* **270**, 26762–26765.

Barrachina, M. D., Martinez, V., Wang, L., Wei, J. Y., and Tache, Y. (1997). Synergistic interaction between leptin and cholecystokinin to reduce short-term food intake in lean mice. *Proc. Natl. Acad. Sci. U.S.A.* **94**, 10455–10460.

Baskin, D. G., Schwartz, M. W., Seeley, R. J., Woods, S. C., Porte, D., Jr., Breininger, J. F., Jonak, Z., Schaefer, J., Krouse, M., Burghardt, C., Campfield, L. A., Burn, P., and Kochan, J. P. (1999). Leptin receptor long-form splice-variant protein expression in neuron cell bodies of the brain and co-localization with neuropeptide Y mRNA in the arcuate nucleus. *J. Histochem. Cytochem.* **47**, 353–362.

Batterham, R. L., Cowley, M. A., Small, C. J., Herzog, H., Cohen, M. A., Dakin, C. L., Wren, A. M., Brynes, A. E., Low, M. J., Ghatei, M. A., Cone, R. D., and Bloom, S. R. (2002). Gut hormone PYY(3-36) physiologically inhibits food intake. *Nature* **418**, 650–654.

Batterham, R. L., Cohen, M. A., Ellis, S. M., Le Roux, C. W., Withers, D. J., Frost, G. S., Ghatei, M. A., and Bloom, S. R. (2003). Inhibition of food intake in obese subjects by peptide YY3-36. *N. Engl. J. Med.* **349**, 941–948.

Baura, G. D., Foster, D. M., Porte, D., Jr., Kahn, S. E., Bergman, R. N., Cobelli, C., and Schwartz, M. W. (1993). Saturable transport of insulin from plasma into the central nervous system of dogs in vivo. A mechanism for regulated insulin delivery to the brain. *J. Clin. Invest* **92**, 1824–1830.

Bechtold, D. A., and Luckman, S. M. (2006). Prolactin-releasing peptide mediates CCK-induced satiety in mice. *Endocrinology*, ePub June 22.

Becker, E. E., and Grinker, J. A. (1977). Meal patterns in the genetically obese Zucker rat. *Physiol. Behav.* **18**, 685–692.

Benoit, S. C., Thiele, T. E., Heinrichs, S. C., Rushing, P. A., Blake, K. A., and Steeley, R. J. (2000). Comparison of central administration of corticotropin-releasing hormone and urocortin on food intake, conditioned taste aversion, and c-Fos expression. *Peptides* **21**, 345–351.

Berntson, G. G., Zipf, W. B., O'Dorisio, T. M., Hoffman, J. A., and Chance, R. E. (1993). Pancreatic polypeptide infusions reduce food intake in Prader-Willi syndrome. *Peptides* **14**, 497–503.

Berthoud, H. R., Lynn, P. A., and Blackshaw, L. A. (2001). Vagal and spinal mechanosensors in the rat stomach and colon have multiple receptive fields. *Am. J. Physiol.* **280**, R1371–R1381.

Berthoud, H. R., Patterson, L. M., Sutton, G. M., Morrison, C., and Zheng, H. (2005). Orexin inputs to caudal raphe neurons involved in thermal, cardiovascular, and gastrointestinal regulation. *Histochem. Cell Biol.* **123**, 147–156.

Bester, H., Besson, J. M., and Bernard, J. F. (1997). Organization of efferent projections from the parabrachial area to the hypothalamus: a *Phaseolus vulgaris*-leucoagglutinin study in the rat. *J. Comp. Neurol.* **383**, 245–281.

Bhavsar, S., Watkins, J., and Young, A. (1998). Synergy between amylin and cholecystokinin for inhibition of food intake in mice. *Physiol. Behav.* **64**, 557–561.

Billington, C. J., Levine, A. S., and Morley, J. E. (1983). Are peptides truly satiety agents? A method of testing for neurohumoral satiety effects. *Am. J. Physiol.* **245**, R920–R926.

Bittencourt, J. C., and Sawchenko, P. E. (2000). Do centrally administered neuropeptides access cognate receptors? An analysis in the central corticotropin-releasing factor system. *J. Neurosci.* **20**, 1142–1156.

Bittencourt, J. C., Vaughan, J., Arias, C., Rissman, R. A., Vale, W. W., and Sawchenko, P. E. (1999). Urocortin expression in rat brain: evidence against a pervasive relationship of urocortin-containing projections with targets bearing type 2 CRF receptors. *J. Comp. Neurol.* **415**, 285–312.

Blackshaw, L. A., Page, A. J., and Partosoedarso, E. R. (2000). Acute effects of capsaicin on gastrointestinal vagal afferents. *Neuroscience* **96**, 407–416.

Blevins, J. E., Eakin, T. J., Murphy, J. A., Schwartz, M. W., and Baskin, D. G. (2003). Oxytocin innervation of caudal brainstem nuclei activated by cholecystokinin. *Brain Res.* **993**, 30–41.

Blevins, J. E., Schwartz, M. W., and Baskin, D. G. (2004). Evidence that paraventricular nucleus oxytocin neurons link hypothalamic leptin action to caudal brain stem nuclei controlling meal size. *Am. J. Physiol.* **287**, R87–R96.

Blundell, J. E., and Latham, C. J. (1979). Serotonergic influences on food intake: effect of 5-hydroxytryptophan on parameters of feeding behaviour in deprived and free-feeding rats. *Pharmacol. Biochem. Behav.* **11**, 431–437.

Bonaz, B., De Giorgio, R., and Tache, Y. (1993). Peripheral bombesin induces c-fos protein in the rat brain. *Brain Res.* **600**, 353–357.

Breininger, J. F., and Baskin, D. G. (2000). Fluorescence in situ hybridization of scarce leptin receptor mRNA using the enzyme-labeled fluorescent substrate method and tyramide signal amplification. *J. Histochem. Cytochem.* **48**, 1593–1599.

Broadwell, R. D., and Sofroniew, M. V. (1993). Serum proteins bypass the blood-brain fluid barriers for extracellular entry to the central nervous system. *Exp. Neurol.* **120**, 245–263.

Broberger, C., Johansen, J., Johansson, C., Schalling, M., and Hokfelt, T. (1998). The neuropeptide Y/agouti gene-related protein (AGRP) brain circuitry in normal, anorectic, and monosodium glutamate-treated mice. *Proc. Natl. Acad. Sci. U.S.A.* **95**, 15043–15048.

Broberger, C., Holmberg, K., Kuhar, M. J., and Hokfelt, T. (1999). Cocaine- and amphetamine-regulated transcript in the rat vagus nerve: a putative mediator of cholecystokinin-induced satiety. *Proc. Natl. Acad. Sci. U.S.A.* **96**, 13506–13511.

Broberger, C., Holmberg, K., Shi, T. J., Dockray, G., and Hokfelt, T. (2001). Expression and regulation of cholecystokinin and cholecystokinin receptors in rat nodose and dorsal root ganglia. *Brain Res.* **903**, 128–140.

Bronstein, D. M., Schafer, M. K., Watson, S. J., and Akil, H. (1992). Evidence that beta-endorphin is synthesized in cells in the nucleus tractus solitarius: detection of POMC mRNA. *Brain Res.* **587**, 269–275.

Brown, M. (1986). Corticotropin releasing factor: central nervous system sites of action. *Brain Res.* **399**, 10–14.

Bruning, J. C., Gautam, D., Burks, D. J., Gillette, J., Schubert, M., Orban, P. C., Klein, R., Krone, W., Muller-Wieland, D., and Kahn, C. R. (2000). Role of brain insulin receptor in control of body weight and reproduction. *Science* **289**, 2122–2125.

Burdyga, G., Spiller, D., Morris, R., Lal, S., Thompson, D. G., Saeed, S., Dimaline, R., Varro, A., and Dockray, G. J. (2002). Expression of the leptin receptor in rat and human nodose ganglion neurones. *Neuroscience* **109**, 339–347.

Buyse, M., Ovesjo, M. L., Goiot, H., Guilmeau, S., Peranzi, G., Moizo, L., Walker, F., Lewin, M. J., Meister, B., and Bado, A. (2001). Expression and regulation of leptin receptor proteins in afferent and efferent neurons of the vagus nerve. *Eur. J. Neurosci.* **14**, 64–72.

Cervero, F. (1994). Sensory innervation of the viscera: peripheral basis of visceral pain. *Physiol. Rev.* **74**, 95–138.

Challis, B. G., Pinnock, S. B., Coll, A. P., Carter, R. N., Dickson, S. L., and O'Rahilly, S. (2003). Acute effects of PYY3-36 on food intake and hypothalamic neuropeptide expression in the mouse. *Biochem. Biophys. Res. Commun.* **311**, 915–919.

Challis, B. G., Coll, A. P., Yeo, G. S., Pinnock, S. B., Dickson, S. L., Thresher, R. R., Dixon, J., Zahn, D., Rochford, J. J., White, A., Oliver, R. L., Millington, G., Aparicio, S. A., Colledge, W. H., Russ, A. P., Carlton, M. B., and O'Rahilly, S. (2004). Mice lacking pro-opiomelanocortin are sensitive to high-fat feeding but respond normally to the acute anorectic effects of peptide-YY(3-36). *Proc. Natl. Acad. Sci. U.S.A.* **101**, 4695–4700.

Chance, W. T., Balasubramaniam, A., Zhang, F. S., Wimalawansa, S. J., and Fischer, J. E. (1991). Anorexia following the intrahypothalamic administration of amylin. *Brain Res.* **539**, 352–354.

Chelikani, P. K., Haver, A. C., and Reidelberger, R. D. (2005). Intravenous infusion of peptide YY(3-36) potently inhibits food intake in rats. *Endocrinology* **146**, 879–888.

Chen, C., Dun, S. L., Dun, N. J., and Chang, J. K. (1999). Prolactin-releasing peptide-immunoreactivity in A1 and A2 noradrenergic neurons of the rat medulla. *Brain Res.* **822**, 276–279.

Chi, M. M., and Powley, T. L. (2003). c-Kit mutant mouse behavioral phenotype: altered meal patterns and CCK sensitivity but normal daily food intake and body weight. *Am. J. Physiol.* **285**, R1170–R1183.

Chronwall, B. M. (1985). Anatomy and physiology of the neuroendocrine arcuate nucleus. *Peptides* **6** (Suppl. 2), 1–11.

Ciriello, J., McMurray, J. C., Babic, T., and de Oliveira, C. V. (2003). Collateral axonal projections from hypothalamic hypocretin neurons to cardiovascular sites in nucleus ambiguus and nucleus tractus solitarius. *Brain Res.* **991**, 133–141.

Clark, J. T., Kalra, P. S., Crowley, W. R., and Kalra, S. P. (1984). Neuropeptide Y and human pancreatic polypeptide stimulate feeding behavior in rats. *Endocrinology* **115**, 427–429.

Clementi, G., Caruso, A., Cutuli, V. M., de Bernardis, E., Prato, A., and Amico-Roxas, M. (1996). Amylin given by central or peripheral routes decreases gastric emptying and intestinal transit in the rat. *Experientia* **52**, 677–679.

Cone, R. D. (1999). The central melanocortin system and energy homeostasis. *Trends Endocrinol. Metab.* **10**, 211–216.

Cooper, G. J., Willis, A. C., Clark, A., Turner, R. C., Sim, R. B., and Reid, K. B. (1987). Purification and characterization of a peptide from amyloid-rich pancreases of type 2 diabetic patients. *Proc. Natl. Acad. Sci. U.S.A.* **84**, 8628–8632.

Corp, E. S., Melville, L. D., Greenberg, D., Gibbs, J., and Smith, G. P. (1990). Effect of fourth ventricular neuropeptide Y and peptide YY on ingestive and other behaviors. *Am. J. Physiol.* **259**, R317–R323.

Corp, E. S., McQuade, J., Moran, T. H., and Smith, G. P. (1993). Characterization of type A and type B CCK receptor binding sites in rat vagus nerve. *Brain Res.* **623**, 161–166.

Corp, E. S., McQuade, J., Krasnicki, S., and Conze, D. B. (2001). Feeding after fourth ventricular administration of neuropeptide Y receptor agonists in rats. *Peptides* **22**, 493–499.

Costello, J. F., Brown, M. R., and Gray, T. S. (1991). Bombesin immunoreactive neurons in the hypothalamic paraventricular nucleus innervate the dorsal vagal complex in the rat. *Brain Res.* **542**, 77–82.

Cowley, M. A., Smith, R. G., Diano, S., Tschop, M., Pronchuk, N., Grove, K. L., Strasburger, C. J., Bidlingmaier, M., Esterman, M., Heiman, M. L., Garcia-Segura, L. M., Nillni, E. A., Mendez, P., Low, M. J., Sotonyi, P., Friedman, J. M., Liu, H., Pinto, S., Colmers, W. F., Cone, R. D., and Horvath, T. L. (2003). The distribution and mechanism of action of ghrelin in the CNS demonstrates a novel hypothalamic circuit regulating energy homeostasis. *Neuron* **37**, 649–661.

Cox, J. E., and Randich, A. (2004). Enhancement of feeding suppression by PYY(3-36) in rats with area postrema ablations. *Peptides* **25**, 985–989.

Crawley, J. N., and Beinfeld, M. C. (1983). Rapid development of tolerance to the behavioural actions of cholecystokinin. *Nature* **302**, 703–706.

Crawley, J. N., and Kiss, J. Z. (1985). Paraventricular nucleus lesions abolish the inhibition of feeding induced by systemic cholecystokinin. *Peptides* **6**, 927–935.

Cummings, D. E., Purnell, J. Q., Frayo, R. S., Schmidova, K., Wisse, B. E., and Weigle, D. S. (2001). A preprandial rise in plasma ghrelin levels suggests a role in meal initiation in humans. *Diabetes* **50**, 1714–1719.

Cvetkovic, V., Brischoux, F., Jacquemard, C., Fellmann, D., Griffond, B., and Risold, P. Y. (2004). Characterization of subpopulations of neurons producing melanin-concentrating hormone in the rat ventral diencephalon. *J. Neurochem.* **91**, 911–919.

D'Este, L., Casini, A., Wimalawansa, S. J., and Renda, T. G. (2000). Immunohistochemical localization of amylin in rat brainstem. *Peptides* **21**, 1743–1749.

Dakin, C. L., Small, C. J., Park, A. J., Seth, A., Ghatei, M. A., and Bloom, S. R. (2002). Repeated ICV administration of oxyntomodulin causes a greater reduction in body weight gain than in pair-fed rats. *Am. J. Physiol.* **283**, E1173–E1177.

Dakin, C. L., Small, C. J., Batterham, R. L., Neary, N. M., Cohen, M. A., Patterson, M., Ghatei, M. A., and Bloom, S. R. (2004). Peripheral oxyntomodulin reduces food intake and body weight gain in rats. *Endocrinology* **145**, 2687–2695.

Daniels, D., Markison, S., Grill, H. J., and Kaplan, J. M. (2004). Central structures necessary and sufficient for ingestive and glycemic responses to urocortin I administration. *J. Neurosci.* **24**, 11457–11462.

Date, Y., Ueta, Y., Yamashita, H., Yamaguchi, H., Matsukura, S., Kangawa, K., Sakurai, T., Yanagisawa, M., and Nakazato, M. (1999). Orexins, orexigenic hypothalamic peptides, interact with autonomic, neuroendocrine and neuroregulatory systems. *Proc. Natl. Acad. Sci. U.S.A.* **96**, 748–753.

Date, Y., Murakami, N., Toshinai, K., Matsukura, S., Niijima, A., Matsuo, H., Kangawa, K., and Nakazato, M. (2002). The role of the gastric afferent vagal nerve in ghrelin-induced feeding and growth hormone secretion in rats. *Gastroenterology* **123**, 1120–1128.

Davidson, T. L., Kanoski, S. E., Tracy, A. L., Walls, E. K., Clegg, D., and Benoit, S. C. (2005). The interoceptive cue properties of ghrelin generalize to cues produced by food deprivation. *Peptides* **26**, 1602–1610.

Davis, S. F., Williams, K. W., Xu, W., Glatzer, N. R., and Smith, B. N. (2003). Selective enhancement of synaptic inhibition by hypocretin (orexin) in rat vagal motor neurons: implications for autonomic regulation. *J. Neurosci.* **23**, 3844–3854.

Davison, J. S., and Clarke, G. D. (1988). Mechanical properties and sensitivity to CCK of vagal gastric slowly adapting mechanoreceptors. *Am. J. Physiol.* **255**, G55–G61.

De Lecea, L., Kilduff, T. S., Peyron, C., Gao, X., Foye, P. E., Danielson, P. E., Fukuhara, C., Battenberg, E. L., Gautvik, V. T., Bartlett, F. S., Frankel, W. N., van den Pol, A. N., Bloom, F. E., Gautvik, K. M., and Sutcliffe, J. G. (1998). The hypocretins: hypothalamus-specific peptides with neuroexcitatory activity. *Proc. Natl. Acad. Sci. U.S.A.* **95**, 322–327.

Dergacheva, O., Wang, X., Huang, Z. G., Bouairi, E., Stephens, C., Gorini, C., and Mendelowitz, D. (2005). Hypocretin-1 (orexin-A) facilitates inhibitory and diminishes excitatory synaptic pathways to cardiac vagal neurons in the nucleus ambiguus. *J. Pharmacol. Exp. Ther.* **314**, 1322–1327.

Dickson, S. L., and Luckman, S. M. (1997). Induction of c-fos messenger ribonucleic acid in neuropeptide Y and growth hormone (GH)-releasing factor neurons in the rat arcuate nucleus following systemic injection of the GH secretagogue, GH-releasing peptide-6. *Endocrinology* **138**, 771–777.

Dickson, S. L., Leng, G., Dyball, R. E., and Smith, R. G. (1995). Central actions of peptide and non-peptide growth hormone secretagogues in the rat 4. *Neuroendocrinology* **61**, 36–43.

Dockray, G. J. (2004). The expanding family of -RFamide peptides and their effects on feeding behaviour. *Exp. Physiol.* **89**, 229–235.

Donahey, J. C., van Dijk, G., Woods, S. C., and Seeley, R. J. (1998). Intraventricular GLP-1 reduces short- but not long-term food intake or body weight in lean and obese rats. *Brain Res.* 779, 75–83.

Douglass, J., McKinzie, A. A., and Couceyro, P. (1995). PCR differential display identifies a rat brain mRNA that is transcriptionally regulated by cocaine and amphetamine. *J. Neurosci.* 15, 2471–2481.

Duan, Y. F., Kopin, I. J., and Goldstein, D. S. (1999). Stimulation of the paraventricular nucleus modulates firing of neurons in the nucleus of the solitary tract. *Am. J. Physiol.* 277, R403–R411.

Dun, S. L., Castellino, S. J., Yang, J., Chang, J. K., and Dun, N. J. (2001). Cocaine- and amphetamine-regulated transcript peptide-immunoreactivity in dorsal motor nucleus of the vagus neurons of immature rats. *Dev. Brain Res.* 131, 93–102.

Eckel, L. A., and Ossenkopp, K. P. (1994). Cholecystokinin reduces sucrose palatability in rats: evidence in support of a satiety effect. *Am. J. Physiol.* 267, R1496–R1502.

Eckel, L. A., Langhans, W., Kahler, A., Campfield, L. A., Smith, F. J., and Geary, N. (1998). Chronic administration of OB protein decreases food intake by selectively reducing meal size in female rats. *Am. J. Physiol.* 275, R186–R193.

Edwards, G. L., and Ritter, R. C. (1986). Area postrema lesions: cause of overingestion has not altered visceral nerve function. *Am. J. Physiol.* 251, R575–R581.

Ellacott, K. L., Lawrence, C. B., Rothwell, N. J., and Luckman, S. M. (2002). PRL-releasing peptide interacts with leptin to reduce food intake and body weight. *Endocrinology* 143, 368–374.

Ellacott, K. L., Lawrence, C. B., Pritchard, L. E., and Luckman, S. M. (2003). Repeated administration of the anorectic factor prolactin-releasing peptide leads to tolerance to its effects on energy homeostasis. *Am. J. Physiol.* 285, R1005–R1010.

Ellacott, K. L., Donald, E. L., Clarkson, P., Morten, J., Masters, D., Brennand, J., and Luckman, S. M. (2005). Characterization of a naturally-occurring polymorphism in the UHR-1 gene encoding the putative rat prolactin-releasing peptide receptor. *Peptides* 26, 675–681.

Ellacott, K. L., Halatchev, I. G., and Cone, R. D. (2006). Characterization of leptin responsive neurons in the caudal brainstem. *Endocrinology* 147, 3190–3195.

Elmquist, J. K., Bjorbaek, C., Ahima, R. S., Flier, J. S., and Saper, C. B. (1998). Distributions of leptin receptor mRNA isoforms in the rat brain. *J. Comp. Neurol.* 395, 535–547.

Emond, M., Schwartz, G. J., Ladenheim, E. E., and Moran, T. H. (1999). Central leptin modulates behavioral and neural responsivity to CCK. *Am. J. Physiol.* 276, R1545–R1549.

Engstrom, M., Brandt, A., Wurster, S., Savola, J. M., and Panula, P. (2003). Prolactin releasing peptide has high affinity and efficacy at neuropeptide FF2 receptors. *J. Pharmacol. Exp. Ther.* 305, 825–832.

Ervin, G. N., and Teeter, M. N. (1986). Cholecystokinin octapeptide and lithium produce different effects on feeding and taste aversion learning. *Physiol. Behav.* 36, 507–512.

Ervin, G. N., Birkemo, L. S., Johnson, M. F., Conger, L. K., Mosher, J. T., and Menius, J. A., Jr. (1995). The effects of anorectic and aversive agents on deprivation-induced feeding and taste aversion conditioning in rats. *J. Pharmacol. Exp. Ther.* 273, 1203–1210.

Ewart, W. R., Jones, M. V., and Primi, M. P. (1990). Bombesin changes excitability of rat brain stem neurons sensitive to gastric distension. *Am. J. Physiol.* 258, G841–G847.

Fan, W., Ellacott, K. L., Halatchev, I. G., Takahashi, K., Yu, P., and Cone, R. D. (2004). Cholecystokinin-mediated suppression of feeding involves the brainstem melanocortin system. *Nat. Neurosci.* 7, 335–336.

Faulconbridge, L. F., Cummings, D. E., Kaplan, J. M., and Grill, H. J. (2003). Hyperphagic effects of brainstem ghrelin administration. *Diabetes* 52, 2260–2265.

Faulconbridge, L. F., Grill, H. J., and Kaplan, J. M. (2005). Distinct forebrain and caudal brainstem contributions to the neuropeptide Y mediation of ghrelin hyperphagia. *Diabetes* 54, 1985–1993.

Fekete, C., Sarkar, S., and Lechan, R. M. (2005). Relative contribution of brainstem afferents to the cocaine- and amphetamine-regulated transcript (CART) innervation of thyrotropin-releasing hormone synthesizing neurons in the hypothalamic paraventricular nucleus (PVN). *Brain Res.* 1032, 171–175.

Figlewicz, D. P., Stein, L. J., West, D., Porte, D., Jr., and Woods, S. C. (1986). Intracisternal insulin alters sensitivity to CCK-induced meal suppression in baboons. *Am. J. Physiol.* 250, R856–R860.

Figlewicz, D. P., Sipols, A. J., Seeley, R. J., Chavez, M., Woods, S. C., and Porte, D., Jr. (1995). Intraventricular insulin enhances the meal-suppressive efficacy of intraventricular cholecystokinin octapeptide in the baboon. *Behav. Neurosci.* 109, 567–569.

Flanagan, L. M., Olson, B. R., Sved, A. F., Verbalis, J. G., and Stricker, E. M. (1992). Gastric motility in conscious rats given oxytocin and an oxytocin antagonist centrally. *Brain Res.* 578, 256–260.

Flier, J. S., and Maratos-Flier, E. (1998). Obesity and the hypothalamus: novel peptides for new pathways. *Cell* 92, 437–440.

Flynn, F. W. (1993). Fourth ventricular injection of selective bombesin receptor antagonists facilitates feeding in rats. *Am. J. Physiol.* 264, R218–R221.

Flynn, F. W. (1997). Bombesin receptor antagonists block the effects of exogenous bombesin but not of nutrients on food intake. *Physiol. Behav.* 62, 791–798.

Flynn, F. W., and Robillard, L. (1992). Inhibition of ingestive behavior following fourth ventricle bombesin injection in chronic decerebrate rats. *Behav. Neurosci.* 106, 1011–1014.

Flynn, M. C., Scott, T. R., Pritchard, T. C., and Plata-Salaman, C. R. (1998). Mode of action of OB protein (leptin) on feeding. *Am. J. Physiol.* **275**, R174–R179.

Folli, F., Bonfanti, L., Renard, E., Kahn, C. R., and Merighi, A. (1994). Insulin receptor substrate-1 (IRS-1) distribution in the rat central nervous system. *J. Neurosci.* **14**, 6412–6422.

Fox, E. A., and Powley, T. L. (1985). Longitudinal columnar organization within the dorsal motor nucleus represents separate branches of the abdominal vagus. *Brain Res.* **341**, 269–282.

Fox, E. A., Phillips, R. J., Baronowsky, E. A., Byerly, M. S., Jones, S., and Powley, T. L. (2001). Neurotrophin-4 deficient mice have a loss of vagal intraganglionic mechanoreceptors from the small intestine and a disruption of short-term satiety. *J. Neurosci.* **21**, 8602–8615.

Fraser, K. A., and Davison, J. S. (1992). Cholecystokinin-induced c-fos expression in the rat brain stem is influenced by vagal nerve integrity. *Exp. Physiol.* **77**, 225–228.

Fraser, K. A., and Davison, J. S. (1993). Meal-induced c-fos expression in brain stem is not dependent on cholecystokinin release. *Am. J. Physiol.* **265**, R235–R239.

Fraser, K. A., Raizada, E., and Davison, J. S. (1995). Oral-pharyngeal-esophageal and gastric cues contribute to meal-induced c-fos expression. *Am. J. Physiol.* **268**, R223–R230.

Friedman, J. M. (2002). The function of leptin in nutrition, weight, and physiology. *Nutr. Rev.* **60**, S1–S14.

Fujimiya, M., and Inui, A. (2000). Peptidergic regulation of gastrointestinal motility in rodents. *Peptides* **21**, 1565–1582.

Funakoshi, A., Miyasaka, K., Shinozaki, H., Masuda, M., Kawanami, T., Takata, Y., and Kono, A. (1995). An animal model of congenital defect of gene expression of cholecystokinin (CCK)-A receptor. *Biochem. Biophys. Res. Commun.* **210**, 787–796.

Fung, S. J., Yamuy, J., Sampogna, S., Morales, F. R., and Chase, M. H. (2001). Hypocretin (orexin) input to trigeminal and hypoglossal motoneurons in the cat: a double-labeling immunohistochemical study. *Brain Res.* **903**, 257–262.

Gates, R. J., Hunt, M. I., and Lazarus, N. R. (1974). Further studies on the amelioration of the characteristics of New Zealand Obese (NZO) mice following implantation of islets of Langerhans. *Diabetologia* **10**, 401–406.

Gebre-Medhin, S., Mulder, H., Pekny, M., Westermark, G., Tornell, J., Westermark, P., Sundler, F., Ahren, B., and Betsholtz, C. (1998). Increased insulin secretion and glucose tolerance in mice lacking islet amyloid polypeptide (amylin). *Biochem. Biophys. Res. Commun.* **250**, 271–277.

Gerald, C., Walker, M. W., Criscione, L., Gustafson, E. L., Batzl-Hartmann, C., Smith, K. E., Vaysse, P., Durkin, M. M., Laz, T. M., Linemeyer, D. L., Schaffhauser, A. O., Whitebread, S., Hofbauer, K. G.,

Taber, R. I., Branchek, T. A., and Weinshank, R. L. (1996). A receptor subtype involved in neuropeptide-Y-induced food intake. *Nature* **382**, 168–171.

Ghamari-Langroudi, M., Colmers, W. F., and Cone, R. D. (2005). PYY3-36 inhibits the action potential firing activity of POMC neurons of arcuate nucleus through postsynaptic Y2 receptors. *Cell. Metab.* **2**, 191–199.

Gibbs, J., Fauser, D. J., Rowe, E. A., Rolls, B. J., Rolls, E. T., and Maddison, S. P. (1979). Bombesin suppresses feeding in rats. *Nature* **282**, 208–210.

Goldstone, A. P., Mercer, J. G., Gunn, I., Moar, K. M., Edwards, C. M., Rossi, M., Howard, J. K., Rasheed, S., Turton, M. D., Small, C., Heath, M. M., O'Shea, D., Steere, J., Meeran, K., Ghatei, M. A., Hoggard, N., and Bloom, S. R. (1997). Leptin interacts with glucagon-like peptide-1 neurons to reduce food intake and body weight in rodents. *FEBS Lett.* **415**, 134–138.

Goldstone, A. P., Morgan, I., Mercer, J. G., Morgan, D. G., Moar, K. M., Ghatei, M. A., and Bloom, S. R. (2000). Effect of leptin on hypothalamic GLP-1 peptide and brain-stem pre-proglucagon mRNA. *Biochem. Biophys. Res. Commun.* **269**, 331–335.

Gosnell, B. A., Morley, J. E., and Levine, A. S. (1983). A comparison of the effects of corticotropin releasing factor and sauvagine on food intake. *Pharmacol. Biochem. Behav.* **19**, 771–775.

Grabauskas, G., and Moises, H. C. (2003). Gastrointestinal-projecting neurones in the dorsal motor nucleus of the vagus exhibit direct and viscerotopically organized sensitivity to orexin. *J. Physiol.* **549**, 37–56.

Grabauskas, G., Zhou, S. Y., Das, S., Lu, Y., Owyang, C., and Moises, H. C. (2004). Prolactin-releasing peptide affects gastric motor function in rat by modulating synaptic transmission in the dorsal vagal complex. *J. Physiol.* **561**, 821–839.

Graham, E. S., Turnbull, Y., Fotheringham, P., Nilaweera, K., Mercer, J. G., Morgan, P. J., and Barrett, P. (2003). Neuromedin U and neuromedin U receptor-2 expression in the mouse and rat hypothalamus: effects of nutritional status. *J. Neurochem.* **87**, 1165–1173.

Grandt, D., Schimiczek, M., Beglinger, C., Layer, P., Goebell, H., Eysselein, V. E., and Reeve, J. R., Jr. (1994). Two molecular forms of peptide YY (PYY) are abundant in human blood: characterization of a radioimmunoassay recognizing PYY 1-36 and PYY 3-36. *Regul. Pept.* **51**, 151–159.

Grill, H. J., and Smith, G. P. (1988). Cholecystokinin decreases sucrose intake in chronic decerebrate rats. *Am. J. Physiol.* **254**, R836–R853.

Grill, H. J., and Kaplan, J. M. (2002). The neuroanatomical axis for control of energy balance. *Front. Neuroendocrinol.* **23**, 2–40.

Grill, H. J., Ginsberg, A. B., Seeley, R. J., and Kaplan, J. M. (1998). Brainstem application of melanocortin receptor ligands produces long-lasting effects on feeding and body weight. *J. Neurosci.* **18**, 10128–10135.

Grill, H. J., Markison, S., Ginsberg, A., and Kaplan, J. M. (2000). Long-term effects on feeding and body weight after stimulation of forebrain or hindbrain CRH receptors with urocortin. *Brain Res.* **867**, 19–28.

Gu, W., Geddes, B. J., Zhang, C., Foley, K. P., and Stricker-Krongrad, A. (2004). The prolactin-releasing peptide receptor (GPR10) regulates body weight homeostasis in mice. *J. Mol. Neurosci.* **22**, 93–103.

Guan, J. L., Wang, Q. P., Kageyama, H., Kita, T., Takenoya, F., Hori, T., and Shioda, S. (2005). Characterization of orexin A immunoreactivity in the rat area postrema. *Regul. Pept.* **129**, 17–23.

Guan, X. M., Yu, H., Palyha, O. C., McKee, K. K., Feighner, S. D., Sirinathsinghji, D. J., Smith, R. G., Van der Ploeg, L. H., and Howard, A. D. (1997). Distribution of mRNA encoding the growth hormone secretagogue receptor in brain and peripheral tissues. *Mol. Brain Res.* **48**, 23–29.

Guan, X. M., Yu, H., Jiang, Q., Van Der Ploeg, L. H., and Liu, Q. (2001). Distribution of neuromedin U receptor subtype 2 mRNA in the rat brain. *Brain Res. Gene Expr. Patterns* **1**, 1–4.

Guidobono, F., Coluzzi, M., Pagani, F., Pecile, A., and Netti, C. (1994). Amylin given by central and peripheral routes inhibits acid gastric secretion. *Peptides* **15**, 699–702.

Guilmeau, S., Buyse, M., Tsocas, A., Laigneau, J. P., and Bado, A. (2003). Duodenal leptin stimulates cholecystokinin secretion: evidence of a positive leptin-cholecystokinin feedback loop. *Diabetes* **52**, 1664–1672.

Hagan, M. M., Rushing, P. A., Pritchard, L. M., Schwartz, M. W., Strack, A. M., Van Der Ploeg, L. H., Woods, S. C., and Seeley, R. J. (2000). Long-term orexigenic effects of AgRP-(83-132) involve mechanisms other than melanocortin receptor blockade. *Am. J. Physiol.* **279**, R47–R52.

Halatchev, I. G., Ellacott, K. L., Fan, W., and Cone, R. D. (2004). Peptide YY3-36 inhibits food intake in mice through a melanocortin-4 receptor-independent mechanism. *Endocrinology* **145**, 2585–2590.

Hampton, L. L., Ladenheim, E. E., Akeson, M., Way, J. M., Weber, H. C., Sutliff, V. E., Jensen, R. T., Wine, L. J., Arnheiter, H., and Battey, J. F. (1998). Loss of bombesin-induced feeding suppression in gastrin-releasing peptide receptor-deficient mice. *Proc. Natl. Acad. Sci. U.S.A.* **95**, 3188–3192.

Hanada, R., Teranishi, H., Pearson, J. T., Kurokawa, M., Hosoda, H., Fukushima, N., Fukue, Y., Serino, R., Fujihara, H., Ueta, Y., Ikawa, M., Okabe, M., Murakami, N., Shirai, M., Yoshimatsu, H., Kangawa, K., and Kojima, M. (2004). Neuromedin U has a novel anorexigenic effect independent of the leptin signaling pathway. *Nat. Med.* **10**, 1067–1073.

Harris, R. B., Kelso, E. W., Flatt, W. P., Bartness, T. J., and Grill, H. J. (2006). Energy expenditure and body composition of chronically maintained decerebrate rats in the fed and fasted condition. *Endocrinology* **147**, 1365–1376.

Harrison, T. A., Chen, C. T., Dun, N. J., and Chang, J. K. (1999). Hypothalamic orexin A-immunoreactive neurons project to the rat dorsal medulla. *Neurosci. Lett.* **273**, 17–20.

Hay-Schmidt, A., Helboe, L., and Larsen, P. J. (2001). Leptin receptor immunoreactivity is present in ascending serotonergic and catecholaminergic neurons of the rat. *Neuroendocrinology* **73**, 215–226.

Hewson, A. K., Viltart, O., McKenzie, D. N., Dyball, R. E., and Dickson, S. L. (1999). GHRP-6-induced changes in electrical activity of single cells in the arcuate, ventromedial and periventricular nucleus neurones of a hypothalamic slice preparation in vitro. *J. Neuroendocrinol.* **11**, 919–923.

Hill, D. R., Campbell, N. J., Shaw, T. M., and Woodruff, G. N. (1987). Autoradiographic localization and biochemical characterization of peripheral type CCK receptors in rat CNS using highly selective nonpeptide CCK antagonists. *J. Neurosci.* **7**, 2967–2976.

Hinuma, S., Habata, Y., Fujii, R., Kawamata, Y., Hosoya, M., Fukusumi, S., Kitada, C., Masuo, Y., Asano, T., Matsumoto, H., Sekiguchi, M., Kurokawa, T., Nishimura, O., Onda, H., and Fujino, M. (1998). A prolactin-releasing peptide in the brain. *Nature* **393**, 272–276.

Ho, A., and Chin, A. (1988). Circadian feeding and drinking patterns of genetically obese mice fed solid chow diet. *Physiol. Behav.* **43**, 651–656.

Hosoi, T., Kawagishi, T., Okuma, Y., Tanaka, J., and Nomura, Y. (2002). Brain stem is a direct target for leptin's action in the central nervous system. *Endocrinology* **143**, 3498–3504.

Howard, A. D., Wang, R., Pong, S. S., Mellin, T. N., Strack, A., Guan, X. M., Zeng, Z., Williams, D. L., Jr., Feighner, S. D., Nunes, C. N., Murphy, B., Stair, J. N., Yu, H., Jiang, Q., Clements, M. K., Tan, C. P., McKee, K. K., Hreniuk, D. L., McDonald, T. P., Lynch, K. R., Evans, J. F., Austin, C. P., Caskey, C. T., Van der Ploeg, L. H., and Liu, Q. (2000). Identification of receptors for neuromedin U and its role in feeding. *Nature* **406**, 70–74.

Hulsey, M. G., Lu, H., Wang, T., Martin, R. J., and Baile, C. A. (1998). Intracerebroventricular (i.c.v.) administration of mouse leptin in rats: behavioral specificity and effects on meal patterns. *Physiol. Behav.* **65**, 445–455.

Huo, L., Grill, H. J., and Bjorbaek, C. (2006). Divergent regulation of proopiomelanocortin neurons by leptin in the nucleus of the solitary tract and in the arcuate hypothalamic nucleus. *Diabetes* **55**, 567–573.

Hwang, L. L., Chen, C. T., and Dun, N. J. (2001). Mechanisms of orexin-induced depolarizations in rat dorsal motor nucleus of vagus neurones in vitro. *J. Physiol.* **537**, 511–520.

Ibata, Y., Iijima, N., Kataoka, Y., Kakihara, K., Tanaka, M., Hosoya, M., and Hinuma, S. (2000). Morphological survey of prolactin-releasing peptide and its receptor with special reference to their functional roles in the brain. *Neurosci. Res.* **38**, 223–230.

Iijima, N., Kataoka, Y., Kakihara, K., Bamba, H., Tamada, Y., Hayashi, S., Matsuda, T., Tanaka, M., Honjyo, H., Hosoya, M., Hinuma, S., and Ibata, Y. (1999). Cytochemical study of prolactin-releasing peptide (PrRP) in the rat brain. *Neuroreport* **10**, 1713–1716.

Inui, A., Okita, M., Nakajima, M., Inoue, T., Sakatani, N., Oya, M., Morioka, H., Okimura, Y., Chihara, K., and Baba, S. (1991). Neuropeptide regulation of feeding in dogs. *Am. J. Physiol.* **261**, R588–R594.

Ishizaki, K., Honma, S., Katsuno, Y., Abe, H., Masubuchi, S., Namihira, M., and Honma, K. (2003). Gene expression of neuropeptide Y in the nucleus of the solitary tract is activated in rats under restricted daily feeding but not under 48-h food deprivation. *Eur. J. Neurosci.* **17**, 2097–2105.

Ivanov, T. R., Lawrence, C. B., Stanley, P. J., and Luckman, S. M. (2002). Evaluation of neuromedin U actions in energy homeostasis and pituitary function. *Endocrinology* **143**, 3813–3821.

Ivanov, T. R., Le Rouzic, P., Stanley, P. J., Ling, W. Y., Parello, R., and Luckman, S. M. (2004). Neuromedin U neurones in the rat nucleus of the tractus solitarius are catecholaminergic and respond to peripheral cholecystokinin. *J. Neuroendocrinol.* **16**, 612–619.

Jarry, H., Heuer, H., Schomburg, L., and Bauer, K. (2000). Prolactin-releasing peptides do not stimulate prolactin release in vivo. *Neuroendocrinology* **71**, 262–267.

Jin, S. L., Han, V. K., Simmons, J. G., Towle, A. C., Lauder, J. M., and Lund, P. K. (1988). Distribution of glucagonlike peptide I (GLP-I), glucagon, and glicentin in the rat brain: an immunocytochemical study. *J. Comp. Neurol.* **271**, 519–532.

Johnston, S. A., and Merali, Z. (1988). Specific neuroanatomical and neurochemical correlates of grooming and satiety effects of bombesin. *Peptides* **9** (Suppl. 1), 233–244.

Joseph, S. A., and Michael, G. J. (1988). Efferent ACTH-IR opiocortin projections from nucleus tractus solitarius: a hypothalamic deafferentation study. *Peptides* **9**, 193–201.

Joseph, S. A., Pilcher, W. H., and Bennett-Clarke, C. (1983). Immunocytochemical localization of ACTH perikarya in nucleus tractus solitarius: evidence for a second opiocortin neuronal system. *Neurosci. Lett.* **38**, 221–225.

Kalliomaki, M. L., and Panula, P. (2004). Neuropeptide FF, but not prolactin-releasing peptide, mRNA is differentially regulated in the hypothalamic and medullary neurons after salt loading. *Neuroscience* **124**, 81–87.

Kalliomaki, M. L., Pertovaara, A., Brandt, A., Wei, H., Pietila, P., Kalmari, J., Xu, M., Kalso, E., and Panula, P. (2004). Prolactin-releasing peptide affects pain, allodynia and autonomic reflexes through medullary mechanisms. *Neuropharmacology* **46**, 412–424.

Kanatani, A., Ishihara, A., Iwaasa, H., Nakamura, K., Okamoto, O., Hidaka, M., Ito, J., Fukuroda, T., MacNeil, D. J., Van der Ploeg, L. H., Ishii, Y., Okabe, T., Fukami, T., and Ihara, M. (2000). L-152,804: orally active and selective neuropeptide Y Y5 receptor antagonist. *Biochem. Biophys. Res. Commun.* **272**, 169.

Kaplan, J. M., Seeley, R. J., and Grill, H. J. (1993). Daily caloric intake in intact and chronic decerebrate rats. *Behav. Neurosci.* **107**, 876–881.

Katsuura, G., Asakawa, A., and Inui, A. (2002). Roles of pancreatic polypeptide in regulation of food intake. *Peptides* **23**, 323–329.

Kelley, A. E., Baldo, B. A., Pratt, W. E., and Will, M. J. (2005). Corticostriatal-hypothalamic circuitry and food motivation: integration of energy, action and reward. *Physiol. Behav.* **86**, 773–795.

Kieffer, T. J., and Habener, J. F. (1999). The glucagon-like peptides. *Endocr. Rev.* **20**, 876–913.

Kieffer, T. J., McIntosh, C. H., and Pederson, R. A. (1995). Degradation of glucose-dependent insulinotropic polypeptide and truncated glucagon-like peptide 1 in vitro and in vivo by dipeptidyl peptidase IV. *Endocrinology* **136**, 3585–3596.

Kimmel, J. R., Hayden, L. J., and Pollock, H. G. (1975). Isolation and characterization of a new pancreatic polypeptide hormone. *J. Biol. Chem.* **250**, 9369–9376.

Kinzig, K. P., D'Alessio, D. A., Herman, J. P., Sakai, R. R., Vahl, T. P., Figueiredo, H. F., Murphy, E. K., and Seeley, R. J. (2003). CNS glucagon-like peptide-1 receptors mediate endocrine and anxiety responses to interoceptive and psychogenic stressors. *J. Neurosci.* **23**, 6163–6170.

Kirkham, T. C., Gibbs, J., and Smith, G. P. (1991). Satiating effect of bombesin is mediated by receptors perfused by celiac artery. *Am. J. Physiol.* **261**, R614–R618.

Kirkham, T. C., Walsh, C. A., Gibbs, J., Smith, G. P., Leban, J., and McDermed, J. (1994). A novel bombesin receptor antagonist selectively blocks the satiety action of peripherally administered bombesin. *Pharmacol. Biochem. Behav.* **48**, 809–811.

Kishi, T., Aschkenasi, C. J., Lee, C. E., Mountjoy, K. G., Saper, C. B., and Elmquist, J. K. (2003). Expression of melanocortin 4 receptor mRNA in the central nervous system of the rat. *J. Comp. Neurol.* **457**, 213–235.

Kishi, T., Aschkenasi, C. J., Choi, B. J., Lopez, M. E., Lee, C. E., Liu, H., Hollenberg, A. N., Friedman, J. M., and Elmquist, J. K. (2005). Neuropeptide Y Y1 receptor mRNA in rodent brain: distribution and colocalization with melanocortin-4 receptor. *J. Comp. Neurol.* **482**, 217–243.

Koda, S., Date, Y., Murakami, N., Shimbara, T., Hanada, T., Toshinai, K., Niijima, A., Furuya, M., Inomata, N., Osuye, K., and Nakazato, M. (2005). The role of the vagal nerve in peripheral PYY3–36-

induced feeding reduction in rats. *Endocrinology* **146**, 2369–2375.

Koegler, F. H., and Ritter, S. (1996). Feeding induced by pharmacological blockade of fatty acid metabolism is selectively attenuated by hindbrain injections of the galanin receptor antagonist, M40. *Obes. Res.* **4**, 329–336.

Koegler, F. H., and Ritter, S. (1997). Aqueduct occlusion does not impair feeding induced by either third or fourth ventricle galanin injection. *Obes. Res.* **5**, 262–267.

Koegler, F. H., and Ritter, S. (1998). Galanin injection into the nucleus of the solitary tract stimulates feeding in rats with lesions of the paraventricular nucleus of the hypothalamus. *Physiol. Behav.* **63**, 521–527.

Koegler, F. H., York, D. A., and Bray, G. A. (1999). The effects on feeding of galanin and M40 when injected into the nucleus of the solitary tract, the lateral parabrachial nucleus, and the third ventricle. *Physiol. Behav.* **67**, 259–267.

Kojima, M., Haruno, R., Nakazato, M., Date, Y., Murakami, N., Hanada, R., Matsuo, H., and Kangawa, K. (2000). Purification and identification of neuromedin U as an endogenous ligand for an orphan receptor GPR66 (FM3). *Biochem. Biophys. Res. Commun.* **276**, 435–438.

Kong, M. F., King, P., Macdonald, I. A., Stubbs, T. A., Perkins, A. C., Blackshaw, P. E., Moyses, C., and Tattersall, R. B. (1997). Infusion of pramlintide, a human amylin analogue, delays gastric emptying in men with IDDM. *Diabetologia* **40**, 82–88.

Kopp, J., Xu, Z. Q., Zhang, X., Pedrazzini, T., Herzog, H., Kresse, A., Wong, H., Walsh, J. H., and Hokfelt, T. (2002). Expression of the neuropeptide Y Y1 receptor in the CNS of rat and of wild-type and Y1 receptor knock-out mice. Focus on immunohistochemical localization. *Neuroscience* **111**, 443–532.

Kowalski, T. J., Spar, B. D., Markowitz, L., Maguire, M., Golovko, A., Yang, S., Farley, C., Cook, J. A., Tetzloff, G., Hoos, L., Del Vecchio, R. A., Kazdoba, T. M., McCool, M. F., Hwa, J. J., Hyde, L. A., Davis, H., Vassileva, G., Hedrick, J. A., and Gustafson, E. L. (2005). Transgenic overexpression of neuromedin U promotes leanness and hypophagia in mice. *J. Endocrinol.* **185**, 151–164.

Koylu, E. O., Couceyro, P. R., Lambert, P. D., and Kuhar, M. J. (1998). Cocaine- and amphetamine-regulated transcript peptide immunohistochemical localization in the rat brain. *J. Comp. Neurol.* **391**, 115–132.

Kraly, F. S., and Blass, E. M. (1976). Mechanisms for enhanced feeding in the cold in rats. *J. Comp. Physiol. Psychol.* **90**, 714–726.

Kreymann, B., Williams, G., Ghatei, M. A., and Bloom, S. R. (1987). Glucagon-like peptide-1 7-36: a physiological incretin in man. *Lancet* **2**, 1300–1304.

Kristensen, P., Judge, M. E., Thim, L., Ribel, U., Christjansen, K. N., Wulff, B. S., Clausen, J. T.,

Jensen, P. B., Madsen, O. D., Vrang, N., Larsen, P. J., and Hastrup, S. (1998). Hypothalamic CART is a new anorectic peptide regulated by leptin. *Nature* **393**, 72–76.

Krowicki, Z. K., Nathan, N. A., and Hornby, P. J. (1998). Gastric motor and cardiovascular effects of insulin in dorsal vagal complex of the rat. *Am. J. Physiol.* **275**, G964–G972.

Krowicki, Z. K., Burmeister, M. A., Berthoud, H. R., Scullion, R. T., Fuchs, K., and Hornby, P. J. (2002). Orexins in rat dorsal motor nucleus of the vagus potently stimulate gastric motor function. *Am. J. Physiol.* **283**, G465–G472.

Laburthe, M., Chenut, B., Rouyer-Fessard, C., Tatemoto, K., Couvineau, A., Servin, A., and Amiranoff, B. (1986). Interaction of peptide YY with rat intestinal epithelial plasma membranes: binding of the radioiodinated peptide. *Endocrinology* **118**, 1910–1917.

Ladenheim, E. E., and Ritter, R. C. (1988). Low-dose fourth ventricular bombesin selectively suppresses food intake. *Am. J. Physiol.* **255**, R988–R993.

Ladenheim, E. E., and Ritter, R. C. (1993). Caudal hindbrain participation in the suppression of feeding by central and peripheral bombesin. *Am. J. Physiol.* **264**, R1229–R1234.

Ladenheim, E. E., Hampton, L. L., Whitney, A. C., White, W. O., Battey, J. F., and Moran, T. H. (2002). Disruptions in feeding and body weight control in gastrin-releasing peptide receptor deficient mice. *J. Endocrinol.* **174**, 273–281.

Ladenheim, E. E., Emond, M., and Moran, T. H. (2005). Leptin enhances feeding suppression and neural activation produced by systemically administered bombesin. *Am. J. Physiol.* **289**, R473–R477.

Larsen, P. J., Tang-Christensen, M., and Jessop, D. S. (1997). Central administration of glucagon-like peptide-1 activates hypothalamic neuroendocrine neurons in the rat. *Endocrinology* **138**, 4445–4455.

Larsen, P. J., Fledelius, C., Knudsen, L. B., and Tang-Christensen, M. (2001). Systemic administration of the long-acting GLP-1 derivative NN2211 induces lasting and reversible weight loss in both normal and obese rats. *Diabetes* **50**, 2530–2539.

Lawrence, C. B., Celsi, F., Brennand, J., and Luckman, S. M. (2000). Alternative role for prolactin-releasing peptide in the regulation of food intake. *Nat. Neurosci.* **3**, 645–646.

Lawrence, C. B., Ellacott, K. L., and Luckman, S. M. (2002a). PRL-releasing peptide reduces food intake and may mediate satiety signaling. *Endocrinology* **143**, 360–367.

Lawrence, C. B., Snape, A. C., Baudoin, F. M., and Luckman, S. M. (2002b). Acute central ghrelin and GH secretagogues induce feeding and activate brain appetite centers. *Endocrinology* **143**, 155–162.

Lawrence, C. B., Liu, Y. L., Stock, M. J., and Luckman, S. M. (2004). Anorectic actions of prolactin-releasing peptide are mediated by corticotropin-releasing hormone receptors. *Am. J. Physiol.* **286**, R101–R107.

Le Magnen, J., and Tallon, S. (1966). The spontaneous periodicity of ad libitum food intake in white rats. *J. Physiol. (Paris)* **58**, 323–349.

Le Roux, C. W., Neary, N. M., Halsey, T. J., Small, C. J., Martinez-Isla, A. M., Ghatei, M. A., Theodorou, N. A., and Bloom, S. R. (2005). Ghrelin does not stimulate food intake in patients with surgical procedures involving vagotomy. *J. Clin. Endocrinol. Metab.* **90**, 4521–4524.

Leibowitz, S. F. (1978). Paraventricular nucleus: a primary site mediating adrenergic stimulation of feeding and drinking. *Pharmacol. Biochem. Behav.* **8**, 163–175.

Leibowitz, S. F., Hammer, N. J., and Chang, K. (1981). Hypothalamic paraventricular nucleus lesions produce overeating and obesity in the rat. *Physiol. Behav.* **27**, 1031–1040.

Levine, A. S., and Morley, J. E. (1984). Neuropeptide Y: a potent inducer of consummatory behavior in rats. *Peptides* **5**, 1025–1029.

Li, B. H., and Rowland, N. E. (1996). Peripherally and centrally administered bombesin induce Fos-like immunoreactivity in different brain regions in rats. *Regul. Pept.* **62**, 167–172.

Liddle, R. A. (1994). Regulation of cholecystokinin synthesis and secretion in rat intestine. *J. Nutr.* **124**, 1308S–1314S.

Lin, S. H., Leslie, F. M., and Civelli, O. (2002). Neurochemical properties of the prolactin releasing peptide (PrRP) receptor expressing neurons: evidence for a role of PrRP as a regulator of stress and nociception. *Brain Res.* **952**, 15–30.

Loup, F., Tribollet, E., Dubois-Dauphin, M., Pizzolato, G., and Dreifuss, J. J. (1989). Localization of oxytocin binding sites in the human brainstem and upper spinal cord: an autoradiographic study. *Brain Res.* **500**, 223–230.

Lu, S., Guan, J. L., Wang, Q. P., Uehara, K., Yamada, S., Goto, N., Date, Y., Nakazato, M., Kojima, M., Kangawa, K., and Shioda, S. (2002). Immunocytochemical observation of ghrelin-containing neurons in the rat arcuate nucleus. *Neurosci. Lett.* **321**, 157–160.

Luckman, S. M. (1992). Fos expression in the brainstem of the rat following peripheral administration of cholecystokinin. *J. Neuroendocrinol.* **4**, 149–152.

Luckman, S. M., Hamamura, M., Antonijevic, I., Dye, S., and Leng, G. (1993). Involvement of cholecystokinin receptor types in pathways controlling oxytocin secretion. *Br. J. Pharmacol.* **110**, 378–384.

Lutz, T. A., Del Prete, E., and Scharrer, E. (1995a). Subdiaphragmatic vagotomy does not influence the anorectic effect of amylin. *Peptides* **16**, 457–462.

Lutz, T. A., Geary, N., Szabady, M. M., Del Prete, E., and Scharrer, E. (1995b). Amylin decreases meal size in rats. *Physiol. Behav.* **58**, 1197–1202.

Lutz, T. A., Althaus, J., Rossi, R., and Scharrer, E. (1998a). Anorectic effect of amylin is not transmitted by capsaicin-sensitive nerve fibers. *Am. J. Physiol.* **274**, R1777–R1782.

Lutz, T. A., Senn, M., Althaus, J., Del Prete, E., Ehrensperger, F., and Scharrer, E. (1998b). Lesion of the area postrema/nucleus of the solitary tract (AP/NTS) attenuates the anorectic effects of amylin and calcitonin gene-related peptide (CGRP) in rats. *Peptides* **19**, 309–317.

Lutz, T. A., Tschudy, S., Rushing, P. A., and Scharrer, E. (2000). Attenuation of the anorectic effects of cholecystokinin and bombesin by the specific amylin antagonist AC 253. *Physiol. Behav.* **70**, 533–536.

Lutz, T. A., Tschudy, S., Mollet, A., Geary, N., and Scharrer, E. (2001). Dopamine D(2) receptors mediate amylin's acute satiety effect. *Am. J. Physiol.* **280**, R1697–R1703.

Lynn, R. B., Hyde, T. M., Cooperman, R. R., and Miselis, R. R. (1996). Distribution of bombesin-like immunoreactivity in the nucleus of the solitary tract and dorsal motor nucleus of the rat and human: colocalization with tyrosine hydroxylase. *J. Comp. Neurol.* **369**, 552–570.

Malaisse-Lagae, F., Carpentier, J. L., Patel, Y. C., Malaisse, W. J., and Orci, L. (1977). Pancreatic polypeptide: a possible role in the regulation of food intake in the mouse. Hypothesis. *Experientia* **33**, 915–917.

Malick, A., Jakubowski, M., Elmquist, J. K., Saper, C. B., and Burstein, R. (2001). A neurohistochemical blueprint for pain-induced loss of appetite. *Proc. Natl. Acad. Sci. U.S.A.* **98**, 9930–9935.

Manaker, S., Winokur, A., Rostene, W. H., and Rainbow, T. C. (1985). Autoradiographic localization of thyrotropin-releasing hormone receptors in the rat central nervous system. *J. Neurosci.* **5**, 167–174.

Mantella, R. C., Rinaman, L., Vollmer, R. R., and Amico, J. A. (2003). Cholecystokinin and D-fenfluramine inhibit food intake in oxytocin-deficient mice. *Am. J. Physiol* **285**, R1037–R1045.

Marcus, J. N., Aschkenasi, C. J., Lee, C. E., Chemelli, R. M., Saper, C. B., Yanagisawa, M., and Elmquist, J. K. (2001). Differential expression of orexin receptors 1 and 2 in the rat brain. *J. Comp. Neurol.* **435**, 6–25.

Martin, N. M., Small, C. J., Sajedi, A., Patterson, M., Ghatei, M. A., and Bloom, S. R. (2004). Pre-obese and obese agouti mice are sensitive to the anorectic effects of peptide YY(3-36) but resistant to ghrelin. *Int. J. Obes. Relat. Metab. Disord.* **28**, 886–893.

Martinez, V., Rivier, J., Wang, L., and Tache, Y. (1997). Central injection of a new corticotropin-releasing factor (CRF) antagonist, astressin, blocks CRF- and stress-related alterations of gastric and colonic motor function. *J. Pharmacol. Exp. Ther.* **280**, 754–760.

Maruyama, M., Matsumoto, H., Fujiwara, K., Kitada, C., Hinuma, S., Onda, H., Fujino, M., and Inoue, K. (1999). Immunocytochemical localization of prolactin-releasing peptide in the rat brain. *Endocrinology* **140**, 2326–2333.

Mathis, C., Moran, T. H., and Schwartz, G. J. (1998). Load-sensitive rat gastric vagal afferents encode volume but not gastric nutrients. *Am. J. Physiol.* **274**, R280–R286.

Matson, C. A., and Ritter, R. C. (1999).

Matson, C. A., Reid, D. F., Cannon, T. A., and Ritter, R. C. (2000). Cholecystokinin and leptin act synergistically to reduce body weight. *Am. J. Physiol.* **278**.

Matsumoto, H., Noguchi, J., Horikoshi, Y., Kawamata, Y., Kitada, C., Hinuma, S., Onda, H., Nishimura, O., and Fujino, M. (1999). Stimulation of prolactin release by prolactin-releasing peptide in rats. *Biochem. Biophys. Res. Commun.* **259**, 321–324.

McCann, M. J., and Rogers, R. C. (1992). Impact of antral mechanoreceptor activation on the vago-vagal reflex in the rat: functional zonation of responses. *J. Physiol.* **453**, 401–411.

McCann, M. J., Verbalis, J. G., and Stricker, E. M. (1989). LiCl and CCK inhibit gastric emptying and feeding and stimulate OT secretion in rats. *Am. J. Physiol.* **256**, R463–R468.

McLaughlin, C. L., and Baile, C. A. (1981). Obese mice and the satiety effects of cholecystokinin, bombesin and pancreatic polypeptide. *Physiol. Behav.* **26**, 433–437.

McMahon, L. R., and Wellman, P. J. (1998). PVN infusion of GLP-1-(7-36) amide suppresses feeding but does not induce aversion or alter locomotion in rats. *Am. J. Physiol.* **274**, R23–R29.

McMinn, J. E., Sindelar, D. K., Havel, P. J., and Schwartz, M. W. (2000). Leptin deficiency induced by fasting impairs the satiety response to cholecystokinin. *Endocrinology* **141**, 4442–4448.

McTigue, D. M., and Rogers, R. C. (1995). Pancreatic polypeptide stimulates gastric acid secretion through a vagal mechanism in rats. *Am. J. Physiol.* **269**, R983–R987.

McTigue, D. M., Edwards, N. K., and Rogers, R. C. (1993). Pancreatic polypeptide in dorsal vagal complex stimulates gastric acid secretion and motility in rats. *Am. J. Physiol.* **265**, G1169–G1176.

McTigue, D. M., Chen, C. H., Rogers, R. C., and Stephens, R. L., Jr. (1995). Intracisternal rat pancreatic polypeptide stimulates gastric emptying in the rat. *Am. J. Physiol.* **269**, R167–R172.

McTigue, D. M., Hermann, G. E., and Rogers, R. C. (1997). Effect of pancreatic polypeptide on rat dorsal vagal complex neurons. *J. Physiol.* **499**, 475–483.

Medeiros, M. D., and Turner, A. J. (1995). Metabolic stability of some tachykinin analogues to cell-surface peptidases: roles for endopeptidase-24.11 and aminopeptidase N. *Peptides* **16**, 441–447.

Meeran, K., O'Shea, D., Edwards, C. M., Turton, M. D., Heath, M. M., Gunn, I., Abusnana, S., Rossi, M., Small, C. J., Goldstone, A. P., Taylor, G. M., Sunter, D., Steere, J., Choi, S. J., Ghatei, M. A., and Bloom, S. R. (1999). Repeated intracerebroventricular administration of glucagon-like peptide-1-(7-36) amide or exendin-(9-39) alters body weight in the rat. *Endocrinology* **140**, 244–250.

Melander, T., Hokfelt, T., and Rokaeus, A. (1986a). Distribution of galanin-like immunoreactivity in the rat central nervous system. *J. Comp. Neurol.* **248**, 475–517.

Melander, T., Hokfelt, T., Rokaeus, A., Cuello, A. C., Oertel, W. H., Verhofstad, A., and Goldstein, M. (1986b). Coexistence of galanin-like immunoreactivity with catecholamines, 5-hydroxytryptamine, GABA and neuropeptides in the rat CNS. *J. Neurosci.* **6**, 3640–3654.

Melander, T., Kohler, C., Nilsson, S., Hokfelt, T., Brodin, E., Theodorsson, E., and Bartfai, T. (1988). Autoradiographic quantitation and anatomical mapping of 125I-galanin binding sites in the rat central nervous system. *J. Chem. Neuroanat.* **1**, 213–233.

Menetrey, D., and Basbaum, A. I. (1987). Spinal and trigeminal projections to the nucleus of the solitary tract: a possible substrate for somatovisceral and viscerovisceral reflex activation. *J. Comp. Neurol.* **255**, 439–450.

Merali, Z., and Kateb, C. C. (1993). Rapid alterations of hypothalamic and hippocampal bombesin-like peptide levels with feeding status. *Am. J. Physiol.* **265**, R420–R425.

Merali, Z., Merchant, C. A., Crawley, J. N., Coy, D. H., Heinz-Erian, P., Jensen, R. T., and Moody, T. W. (1988). (D-Phe12) bombesin and substance P analogues function as central bombesin receptor antagonists. *Synapse* **2**, 282–287.

Merali, Z., McIntosh, J., Kent, P., Michaud, D., and Anisman, H. (1998). Aversive and appetitive events evoke the release of corticotropin-releasing hormone and bombesin-like peptides at the central nucleus of the amygdala. *J. Neurosci.* **18**, 4758–4766.

Merali, Z., McIntosh, J., and Anisman, H. (1999). Role of bombesin-related peptides in the control of food intake. *Neuropeptides* **33**, 376–386.

Mercer, J. G., Moar, K. M., Findlay, P. A., Hoggard, N., and Adam, C. L. (1998a). Association of leptin receptor (OB-Rb), NPY and GLP-1 gene expression in the ovine and murine brainstem. *Regul. Pept.* **75-76**, 271–278.

Mercer, J. G., Moar, K. M., and Hoggard, N. (1998b). Localization of leptin receptor (Ob-R) messenger ribonucleic acid in the rodent hindbrain. *Endocrinology* **139**, 29–34.

Merchenthaler, I., Lane, M., and Shughrue, P. (1999). Distribution of pre-pro-glucagon and glucagon-like peptide-1 receptor messenger RNAs in the rat central nervous system. *J. Comp. Neurol.* **403**, 261–280.

Mezey, E., Reisine, T. D., Skirboll, L., Beinfeld, M., and Kiss, J. Z. (1986). Role of cholecystokinin in corticotropin release: coexistence with vasopressin and corticotropin-releasing factor in cells of the rat hypothalamic paraventricular nucleus. *Proc. Natl. Acad. Sci. U.S.A.* **83**, 3510–3512.

Micevych, P., Akesson, T., and Elde, R. (1988). Distribution of cholecystokinin-immunoreactive cell bodies in the male and female rat: II. Bed nucleus of the stria terminalis and amygdala. *J. Comp. Neurol.* **269**, 381–391.

Mitchell, V., Bouret, S., Beauvillain, J. C., Schilling, A., Perret, M., Kordon, C., and Epelbaum, J. (2001). Comparative distribution of mRNA encoding the growth hormone secretagogue-receptor (GHS-R) in *Microcebus murinus* (Primate, lemurian) and rat forebrain and pituitary. *J. Comp. Neurol.* **429**, 469–489.

Mix, H., Manns, M. P., Wagner, S., Widjaja, A., and Brabant, G. (1999). Expression of leptin and its receptor in the human stomach. *Gastroenterology* **117**, 509.

Monnikes, H., Lauer, G., Bauer, C., Tebbe, J., Zittel, T. T., and Arnold, R. (1997). Pathways of Fos expression in locus ceruleus, dorsal vagal complex, and PVN in response to intestinal lipid. *Am. J. Physiol.* **273**, R2059–R2071.

Morales, T., Hinuma, S., and Sawchenko, P. E. (2000). Prolactin-releasing peptide is expressed in afferents to the endocrine hypothalamus, but not in neurosecretory neurones. *J. Neuroendocrinol.* **12**, 131–140.

Moran, T. H. (2000). Cholecystokinin and satiety: current perspectives. *Nutrition* **16**, 858–865.

Moran, T. H., and McHugh, P. R. (1982). Cholecystokinin suppresses food intake by inhibiting gastric emptying. *Am. J. Physiol.* **242**, R491–R497.

Moran, T. H., Katz, L. F., Plata-Salaman, C. R., and Schwartz, G. J. (1998). Disordered food intake and obesity in rats lacking cholecystokinin A receptors. *Am. J. Physiol.* **274**, R618–R625.

Moran, T. H., Smedh, U., Kinzig, K. P., Scott, K. A., Knipp, S., and Ladenheim, E. E. (2005). Peptide YY(3-36) inhibits gastric emptying and produces acute reductions in food intake in rhesus monkeys. *Am. J. Physiol.* **288**, R384–R388.

Morley, J. E., Flood, J. F., Horowitz, M., Morley, P. M., and Walter, M. J. (1994). Modulation of food intake by peripherally administered amylin. *Am. J. Physiol.* **267**, R178–R184.

Morton, G. J., Blevins, J. E., Williams, D. L., Niswender, K. D., Gelling, R. W., Rhodes, C. J., Baskin, D. G., and Schwartz, M. W. (2005). Leptin action in the forebrain regulates the hindbrain response to satiety signals. *J. Clin. Invest.* **115**, 703–710.

Mountjoy, K. G., Mortrud, M. T., Low, M. J., Simerly, R. B., and Cone, R. D. (1994). Localization of the melanocortin-4 receptor (MC4-R) in neuroendocrine and autonomic control circuits in the brain. *Mol. Endocrinol.* **8**, 1298–1308.

Mulder, H., Lindh, A. C., Ekblad, E., Westermark, P., and Sundler, F. (1994). Islet amyloid polypeptide is expressed in endocrine cells of the gastric mucosa in the rat and mouse. *Gastroenterology* **107**, 712–719.

Mulder, H., Leckstrom, A., Uddman, R., Ekblad, E., Westermark, P., and Sundler, F. (1995). Islet amyloid

polypeptide (amylin) is expressed in sensory neurons. *J. Neurosci.* **15**, 7625–7632.

Mulder, H., Ekelund, M., Ekblad, E., and Sundler, F. (1997). Islet amyloid polypeptide in the gut and pancreas: localization, ontogeny and gut motility effects. *Peptides* **18**, 771–783.

Murase, T., Arima, H., Kondo, K., and Oiso, Y. (1996). Neuropeptide FF reduces food intake in rats. *Peptides* **17**, 353–354.

Nachman, M., and Ashe, J. H. (1973). Learned taste aversions in rats as a function of dosage, concentration, and route of administration of LiCl. *Physiol. Behav.* **10**, 73–78.

Nakajima, M., Inui, A., Teranishi, A., Miura, M., Hirosue, Y., Okita, M., Himori, N., Baba, S., and Kasuga, M. (1994). Effects of pancreatic polypeptide family peptides on feeding and learning behavior in mice. *J. Pharmacol. Exp. Ther.* **268**, 1010–1014.

Nakazato, M., Hanada, R., Murakami, N., Date, Y., Mondal, M. S., Kojima, M., Yoshimatsu, H., Kangawa, K., and Matsukura, S. (2000). Central effects of neuromedin U in the regulation of energy homeostasis. *Biochem. Biophys. Res. Commun.* **277**, 191–194.

Nakazato, M., Murakami, N., Date, Y., Kojima, M., Matsuo, H., Kangawa, K., and Matsukura, S. (2001). A role for ghrelin in the central regulation of feeding. *Nature* **409**, 194–198.

Naleid, A. M., Grace, M. K., Cummings, D. E., and Levine, A. S. (2005). Ghrelin induces feeding in the mesolimbic reward pathway between the ventral tegmental area and the nucleus accumbens. *Peptides* **26**, 2274–2279.

Navarro, M., Rodriquez, de Fonseca, F., Alvarez, E., Chowen, J. A., Zueco, J. A., Gomez, R., Eng, J., and Blazquez, E. (1996). Colocalization of glucagon-like peptide-1 (GLP-1) receptors, glucose transporter GLUT-2, and glucokinase mRNAs in rat hypothalamic cells: evidence for a role of GLP-1 receptor agonists as an inhibitory signal for food and water intake. *J. Neurochem.* **67**, 1982–1991.

Neary, N. M., Small, C. J., Druce, M. R., Park, A. J., Ellis, S. M., Semjonous, N. M., Dakin, C. L., Filipsson, K., Wang, F., Kent, A. S., Frost, G. S., Ghatei, M. A., and Bloom, S. R. (2005). Peptide YY3-36 and glucagon-like peptide-17-36 inhibit food intake additively. *Endocrinology* **146**, 5120–5127.

Nicklous, D. M., and Simansky, K. J. (2003). Neuropeptide FF exerts pro- and anti-opioid actions in the parabrachial nucleus to modulate food intake. *Am. J. Physiol.* **285**, R1046–R1054.

Niswender, K. D., Baskin, D. G., and Schwartz, M. W. (2004). Insulin and its evolving partnership with leptin in the hypothalamic control of energy homeostasis. *Trends Endocrinol. Metab.* **15**, 362–369.

Norgren, R. (1978). Projections from the nucleus of the solitary tract in the rat. *Neuroscience* **3**, 207–218.

Norgren, R., and Smith, G. P. (1988). Central distribution of subdiaphragmatic vagal branches in the rat. *J. Comp. Neurol.* **273**, 207–223.

Ohki-Hamazaki, H., Watase, K., Yamamoto, K., Ogura, H., Yamano, M., Yamada, K., Maeno, H., Imaki, J., Kikuyama, S., Wada, E., and Wada, K. (1997). Mice lacking bombesin receptor subtype-3 develop metabolic defects and obesity. *Nature* **390**, 165–169.

Ohki-Hamazaki, H., Sakai, Y., Kamata, K., Ogura, H., Okuyama, S., Watase, K., Yamada, K., and Wada, K. (1999). Functional properties of two bombesin-like peptide receptors revealed by the analysis of mice lacking neuromedin B receptor. *J. Neurosci.* **19**, 948–954.

Okumura, T., Pappas, T. N., and Taylor, I. L. (1994). Intracisternal injection of pancreatic polypeptide stimulates gastric emptying in rats. *Neurosci. Lett.* **178**, 167–170.

Okumura, T., Yamada, H., Motomura, W., and Kohgo, Y. (2000). Cocaine- and amphetamine-regulated transcript (CART) acts in the central nervous system to inhibit gastric acid secretion via brain corticotropin-releasing factor system. *Endocrinology* **141**, 2854–2860.

Olson, B. R., Drutarosky, M. D., Stricker, E. M., and Verbalis, J. G. (1991). Brain oxytocin receptor antagonism blunts the effects of anorexigenic treatments in rats: evidence for central oxytocin inhibition of food intake. *Endocrinology* **129**, 785–791.

Olson, B. R., Hoffman, G. E., Sved, A. F., Stricker, E. M., and Verbalis, J. G. (1992). Cholecystokinin induces c-fos expression in hypothalamic oxytocinergic neurons projecting to the dorsal vagal complex. *Brain Res.* **569**, 238–248.

Olson, B. R., Freilino, M., Hoffman, G. E., Stricker, E. M., Sved, A. F., and Verbalis, J. G. (1993). c-Fos expression in the rat brain and brainstem nuclei in response to treatments that alter food intake and gastric motility. *Mol. Cell. Neurosci.* **4**, 93–106.

Onaka, T., Luckman, S. M., Antonijevic, I., Palmer, J. R., and Leng, G. (1995a). Involvement of the noradrenergic afferents from the nucleus tractus solitarii to the supraoptic nucleus in oxytocin release after peripheral cholecystokinin octapeptide in the rat. *Neuroscience* **66**, 403–412.

Onaka, T., Luckman, S. M., Guevera-Guzman, R., Ueta, Y., Kendrick, K., and Leng, G. (1995b). Presynaptic actions of morphine; blockade of cholecystokinin-induced noradrenaline release and Fos expression in the rat supraoptic nucleus. *J. Physiol.* **482**, 69–79.

Ossenkopp, K. P., and Eckel, L. A. (1995). Toxin-induced conditioned changes in taste reactivity and the role of the chemosensitive area postrema. *Neurosci. Biobehav. Rev.* **19**, 99–108.

Palkovits, M., and Eskay, R. L. (1987). Distribution and possible origin of beta-endorphin and ACTH in discrete brainstem nuclei of rats. *Neuropeptides* **9**, 123–137.

Palkovits, M., Mezey, E., Eskay, R. L., and Brownstein, M. J. (1986). Innervation of the nucleus of the solitary tract and the dorsal vagal nucleus by thyrotropin-releasing hormone-containing raphe neurons. *Brain Res.* **373**, 246–251.

Panula, P. (1986). Histochemistry and function of bombesin-like peptides. *Med. Biol.* **64**, 177–192.

Pappas, T. N., Debas, H. T., Goto, Y., and Taylor, I. L. (1985). Peptide YY inhibits meal-stimulated pancreatic and gastric secretion. *Am. J. Physiol.* **248**, G118–G123.

Parker, L. A. (1991). Taste reactivity responses elicited by reinforcing drugs: a dose–response analysis. *Behav. Neurosci.* **105**, 955–964.

Parker, L. A. (1995). Rewarding drugs produce taste avoidance, but not taste aversion. *Neurosci. Biobehav. Rev.* **19**, 143–157.

Paton, J. F., Li, Y. W., Deuchars, J., and Kasparov, S. (2000). Properties of solitary tract neurons receiving inputs from the sub-diaphragmatic vagus nerve. *Neuroscience* **95**, 141–153.

Peiser, C., Springer, J., Groneberg, D. A., McGregor, G. P., Fischer, A., and Lang, R. E. (2002). Leptin receptor expression in nodose ganglion cells projecting to the rat gastric fundus. *Neurosci. Lett.* **320**, 41–44.

Peters, J. H., Karpiel, A. B., Ritter, R. C., and Simasko, S. M. (2004). Cooperative activation of cultured vagal afferent neurons by leptin and cholecystokinin. *Endocrinology* **145**, 3652–3657.

Petrovich, G. D., Canteras, N. S., and Swanson, L. W. (2001). Combinatorial amygdalar inputs to hippocampal domains and hypothalamic behavior systems. *Brain Res. Rev.* **38**, 247–289.

Phifer, C. B., and Berthoud, H. R. (1998). Duodenal nutrient infusions differentially affect sham feeding and Fos expression in rat brain stem. *Am. J. Physiol.* **274**, R1725–R1733.

Phillips, R. J., and Powley, T. L. (1996). Gastric volume rather than nutrient content inhibits food intake. *Am. J. Physiol.* **271**, R766–R769.

Phillips, R. J., and Powley, T. L. (1998). Gastric volume detection after selective vagotomies in rats. *Am. J. Physiol.* **274**, R1626–R1638.

Phillips, R. J., Baronowsky, E. A., and Powley, T. L. (2003). Long-term regeneration of abdominal vagus: efferents fail while afferents succeed. *J. Comp. Neurol.* **455**, 222–237.

Pieber, T. R., Roitelman, J., Lee, Y., Luskey, K. L., and Stein, D. T. (1994). Direct plasma radioimmunoassay for rat amylin-(1-37): concentrations with acquired and genetic obesity. *Am. J. Physiol.* **267**, E156–E164.

Pissios, P., and Maratos-Flier, E. (2003). Melanin-concentrating hormone: from fish skin to skinny mammals. *Trends Endocrinol. Metab.* **14**, 243–248.

Plamondon, H., and Merali, Z. (1997). Anorectic action of bombesin requires receptor for corticotropin-releasing factor but not for oxytocin. *Eur. J. Pharmacol.* **340**, 99–109.

Potts, J. T., Lee, S. M., and Anguelov, P. I. (2002). Tracing of projection neurons from the cervical dorsal horn to the medulla with the anterograde tracer biotinylated dextran amine. *Auton. Neurosci.* **98**, 64–69.

Powley, T. L., and Phillips, R. J. (2004). Gastric satiation is volumetric, intestinal satiation is nutritive. *Physiol. Behav.* **82**, 69–74.

Raybould, H. E., Gayton, R. J., and Dockray, G. J. (1985). CNS effects of circulating CCK8: involvement of brainstem neurones responding to gastric distension. *Brain Res.* **342**, 187–190.

Reidelberger, R. D., Arnelo, U., Granqvist, L., and Permert, J. (2001). Comparative effects of amylin and cholecystokinin on food intake and gastric emptying in rats. *Am. J. Physiol.* **280**, R605–R611.

Reyes, T. M., Lewis, K., Perrin, M. H., Kunitake, K. S., Vaughan, J., Arias, C. A., Hogenesch, J. B., Gulyas, J., Rivier, J., Vale, W. W., and Sawchenko, P. E. (2001). Urocortin II: a member of the corticotropin-releasing factor (CRF) neuropeptide family that is selectively bound by type 2 CRF receptors. *Proc. Natl. Acad. Sci. U.S.A.* **98**, 2843–2848.

Ricardo, J. A., and Koh, E. T. (1978). Anatomical evidence of direct projections from the nucleus of the solitary tract to the hypothalamus, amygdala, and other forebrain structures in the rat. *Brain Res.* **153**, 1–26.

Riediger, T., Bothe, C., Becskei, C., and Lutz, T. A. (2004). Peptide YY directly inhibits ghrelin-activated neurons of the arcuate nucleus and reverses fasting-induced c-Fos expression. *Neuroendocrinology* **79**, 317–326.

Riedy, C. A., Chavez, M., Figlewicz, D. P., and Woods, S. C. (1995). Central insulin enhances sensitivity to cholecystokinin. *Physiol. Behav.* **58**, 755–760.

Rinaman, L. (1998). Oxytocinergic inputs to the nucleus of the solitary tract and dorsal motor nucleus of the vagus in neonatal rats. *J. Comp. Neurol.* **399**, 101–109.

Rinaman, L. (1999a). A functional role for central glucagon-like peptide-1 receptors in lithium chloride-induced anorexia. *Am. J. Physiol.* **277**, R1537–R1540.

Rinaman, L. (1999b). Interoceptive stress activates glucagon-like peptide-1 neurons that project to the hypothalamus. *Am. J. Physiol.* **277**, R582–R590.

Rinaman, L. (2003). Hindbrain noradrenergic lesions attenuate anorexia and alter central cFos expression in rats after gastric viscerosensory stimulation. *J. Neurosci.* **23**, 10084–10092.

Rinaman, L., and Schwartz, G. (2004). Anterograde transneuronal viral tracing of central viscerosensory pathways in rats. *J. Neurosci.* **24**, 2782–2786.

Rinaman, L., Card, J. P., Schwaber, J. S., and Miselis, R. R. (1989). Ultrastructural demonstration of a gastric monosynaptic vagal circuit in the nucleus of the solitary tract in rat. *J. Neurosci.* **9**, 1985–1996.

Rinaman, L., Verbalis, J. G., Stricker, E. M., and Hoffman, G. E. (1993). Distribution and neurochemical phenotypes of caudal medullary neurons activated to express cFos following peripheral

administration of cholecystokinin. *J. Comp. Neurol.* **338**, 475–490.

Rinaman, L., Hoffman, G. E., Stricker, E. M., and Verbalis, J. G. (1994). Exogenous cholecystokinin activates cFos expression in medullary but not hypothalamic neurons in neonatal rats. *Dev. Brain Res.* **77**, 140–145.

Rinaman, L., Hoffman, G. E., Dohanics, J., Le, W. W., Stricker, E. M., and Verbalis, J. G. (1995). Cholecystokinin activates catecholaminergic neurons in the caudal medulla that innervate the paraventricular nucleus of the hypothalamus in rats. *J. Comp. Neurol.* **360**, 246–256.

Rinaman, L., Baker, E. A., Hoffman, G. E., Stricker, E. M., and Verbalis, J. G. (1998). Medullary c-Fos activation in rats after ingestion of a satiating meal. *Am. J. Physiol.* **275**, R262–R268.

Rinaman, L., Levitt, P., and Card, J. P. (2000). Progressive postnatal assembly of limbic-autonomic circuits revealed by central transneuronal transport of pseudorabies virus. *J. Neurosci.* **20**, 2731–2741.

Rogers, R. C., Kita, H., Butcher, L. L., and Novin, D. (1980). Afferent projections to the dorsal motor nucleus of the vagus. *Brain Res. Bull.* **5**, 365–373.

Rogers, R. C., McTigue, D. M., and Hermann, G. E. (1996). Vagal control of digestion: modulation by central neural and peripheral endocrine factors. *Neurosci. Biobehav. Rev.* **20**, 57–66.

Roland, B. L., Sutton, S. W., Wilson, S. J., Luo, L., Pyati, J., Huvar, R., Erlander, M. G., and Lovenberg, T. W. (1999). Anatomical distribution of prolactin-releasing peptide and its receptor suggests additional functions in the central nervous system and periphery. *Endocrinology* **140**, 5736–5745.

Rossowski, W. J., Jiang, N. Y., and Coy, D. H. (1997). Adrenomedullin, amylin, calcitonin gene-related peptide and their fragments are potent inhibitors of gastric acid secretion in rats. *Eur. J. Pharmacol.* **336**, 51–63.

Rowland, N. E., and Richmond, R. M. (1999). Area postrema and the anorectic actions of dexfenfluramine and amylin. *Brain Res.* **820**, 86–91.

Rowland, N. E., Crews, E. C., and Gentry, R. M. (1997). Comparison of Fos induced in rat brain by GLP-1 and amylin. *Regul. Pept.* **71**, 171–174.

Rushing, P. A., Hagan, M. M., Seeley, R. J., Lutz, T. A., and Woods, S. C. (2000a). Amylin: a novel action in the brain to reduce body weight. *Endocrinology* **141**, 850–853.

Rushing, P. A., Lutz, T. A., Seeley, R. J., and Woods, S. C. (2000b). Amylin and insulin interact to reduce food intake in rats. *Horm. Metab. Res.* **32**, 62–65.

Rushing, P. A., Hagan, M. M., Seeley, R. J., Lutz, T. A., D'Alessio, D. A., Air, E. L., and Woods, S. C. (2001). Inhibition of central amylin signaling increases food intake and body adiposity in rats. *Endocrinology* **142**, 5035.

Sahu, A., Kalra, S. P., Crowley, W. R., and Kalra, P. S. (1988). Evidence that NPY-containing neurons in the brainstem project into selected hypothalamic nuclei: implication in feeding behavior. *Brain Res.* **457**, 376–378.

Saito, Y., Cheng, M., Leslie, F. M., and Civelli, O. (2001). Expression of the melanin-concentrating hormone (MCH) receptor mRNA in the rat brain. *J. Comp. Neurol.* **435**, 26–40.

Sakurai, T., Amemiya, A., Ishii, M., Matsuzaki, I., Chemelli, R. M., Tanaka, H., Williams, S. C., Richarson, J. A., Kozlowski, G. P., Wilson, S., Arch, J. R., Buckingham, R. E., Haynes, A. C., Carr, S. A., Annan, R. S., McNulty, D. E., Liu, W. S., Terrett, J. A., Elshourbagy, N. A., Bergsma, D. J., and Yanagisawa, M. (1998). Orexins and orexin receptors: a family of hypothalamic neuropeptides and G protein-coupled receptors that regulate feeding behavior. *Cell* **92**, 1.

Samson, W. K., Resch, Z. T., Murphy, T. C., and Chang, J. K. (1998). Gender-biased activity of the novel prolactin releasing peptides: comparison with thyrotropin releasing hormone reveals only pharmacologic effects. *Endocrine* **9**, 289–291.

Saper, C. B., Chou, T. C., and Elmquist, J. K. (2002). The need to feed: homeostatic and hedonic control of eating. *Neuron* **36**, 199–211.

Saper, C. B., Lu, J., Chou, T. C., and Gooley, J. (2005). The hypothalamic integrator for circadian rhythms. *Trends Neurosci.* **28**, 152–157.

Sato, T., Fukue, Y., Teranishi, H., Yoshida, Y., and Kojima, M. (2005). Molecular forms of hypothalamic ghrelin and its regulation by fasting and 2-deoxy-D-glucose administration. *Endocrinology* **146**, 2510–2516.

Sawchenko, P. E. (1987). Evidence for differential regulation of corticotropin-releasing factor and vasopressin immunoreactivities in parvocellular neurosecretory and autonomic-related projections of the paraventricular nucleus. *Brain Res.* **437**, 253–263.

Sawchenko, P. E., and Swanson, L. W. (1982). Immunohistochemical identification of neurons in the paraventricular nucleus of the hypothalamus that project to the medulla or to the spinal cord in the rat. *J. Comp. Neurol.* **205**, 260–272.

Sawchenko, P. E., Swanson, L. W., Grzanna, R., Howe, P. R., Bloom, S. R., and Polak, J. M. (1985). Colocalization of neuropeptide Y immunoreactivity in brainstem catecholaminergic neurons that project to the paraventricular nucleus of the hypothalamus. *J. Comp. Neurol.* **241**, 138–153.

Schwartz, G. J., and Moran, T. H. (2002). Leptin and neuropeptide Y have opposing modulatory effects on nucleus of the solitary tract neurophysiological responses to gastric loads: implications for the control of food intake. *Endocrinology* **143**, 3779–3784.

Schwartz, G. J., McHugh, P. R., and Moran, T. H. (1991). Integration of vagal afferent responses to gastric loads and cholecystokinin in rats. *Am. J. Physiol.* **261**, R64–R69.

Schwartz, G. J., Moran, T. H., White, W. O., and Ladenheim, E. E. (1997). Relationships between gastric motility and gastric vagal afferent responses to CCK and GRP in rats differ. *Am. J. Physiol.* **272**, R1726–R1733.

Scott, V., Kimura, N., Stark, J. A., and Luckman, S. M. (2005). Intravenous peptide YY3-36 and Y2 receptor antagonism in the rat: effects on feeding behaviour. *J. Neuroendocrinol.* **17**, 452–457.

Scrocchi, L. A., and Drucker, D. J. (1998). Effects of aging and a high fat diet on body weight and glucose tolerance in glucagon-like peptide-1 receptor −/− mice. *Endocrinology* **139**, 3127–3132.

Scrocchi, L. A., Brown, T. J., MaClusky, N., Brubaker, P. L., Auerbach, A. B., Joyner, A. L., and Drucker, D. J. (1996). Glucose intolerance but normal satiety in mice with a null mutation in the glucagon-like peptide 1 receptor gene. *Nat. Med.* **2**, 1254–1258.

Scrocchi, L. A., Hill, M. E., Saleh, J., Perkins, B., and Drucker, D. J. (2000). Elimination of glucagon-like peptide 1R signaling does not modify weight gain and islet adaptation in mice with combined disruption of leptin and GLP-1 action. *Diabetes* **49**, 1552–1560.

Seal, L. J., Small, C. J., Dhillo, W. S., Stanley, S. A., Abbott, C. R., Ghatei, M. A., and Bloom, S. R. (2001). PRL-releasing peptide inhibits food intake in male rats via the dorsomedial hypothalamic nucleus and not the paraventricular hypothalamic nucleus. *Endocrinology* **142**, 4236–4243.

Sederholm, F., Ammar, A. A., and Sodersten, P. (2002). Intake inhibition by NPY: role of appetitive ingestive behavior and aversion. *Physiol. Behav.* **75**, 567–575.

Seeley, R. J., Grill, H. J., and Kaplan, J. M. (1994). Neurological dissociation of gastrointestinal and metabolic contributions to meal size control. *Behav. Neurosci.* **108**, 347–352.

Seeley, R. J., Benoit, S. C., and Davidson, T. L. (1995). Discriminative cues produced by NPY do not generalize to the interoceptive cues produced by food deprivation. *Physiol. Behav.* **58**, 1237–1241.

Seeley, R. J., Blake, K., Rushing, P. A., Benoit, S., Eng, J., Woods, S. C., and D'Alessio, D. (2000). The role of CNS glucagon-like peptide-1 (7-36) amide receptors in mediating the visceral illness effects of lithium chloride. *J. Neurosci.* **20**, 1616–1621.

Sexton, P. M., Paxinos, G., Kenney, M. A., Wookey, P. J., and Beaumont, K. (1994). In vitro autoradiographic localization of amylin binding sites in rat brain. *Neuroscience* **62**, 553–567.

Shioda, S., Funahashi, H., Nakajo, S., Yada, T., Maruta, O., and Nakai, Y. (1998). Immunohistochemical localization of leptin receptor in the rat brain. *Neurosci. Lett.* **243**, 41–44.

Sim, L. J., and Joseph, S. A. (1991). Arcuate nucleus projections to brainstem regions which modulate nociception. *J. Chem. Neuroanat.* **4**, 97–109.

Sim, L. J., and Joseph, S. A. (1994). Efferents of the opiocortin-containing region of the commissural nucleus tractus solitarius. *Peptides* **15**, 169–174.

Skofitsch, G., and Jacobowitz, D. M. (1985). Immuno-histochemical mapping of galanin-like neurons in the rat central nervous system. *Peptides* **6**, 509–546.

Skofitsch, G., Sills, M. A., and Jacobowitz, D. M. (1986). Autoradiographic distribution of 125I-galanin binding sites in the rat central nervous system. *Peptides* **7**, 1029–1042.

Smedh, U., and Moran, T. H. (2003). Peptides that regulate food intake: separable mechanisms for dorsal hindbrain CART peptide to inhibit gastric emptying and food intake. *Am. J. Physiol.* **284**, R1418–R1426.

Smedh, U., Uvnas-Moberg, K., Grill, H. J., and Kaplan, J. M. (1995). Fourth ventricle injection of corticotropin-releasing factor and gastric emptying of glucose during gastric fill. *Am. J. Physiol.* **269**, G1000–G1003.

Smedh, U., Hakansson, M. L., Meister, B., and Uvnas-Moberg, K. (1998). Leptin injected into the fourth ventricle inhibits gastric emptying. *Neuroreport* **9**, 297–301.

Smith, B. N., Dou, P., Barber, W. D., and Dudek, F. E. (1998). Vagally evoked synaptic currents in the immature rat nucleus tractus solitarii in an intact in vitro preparation. *J. Physiol.* **512**, 149–162.

Smith, B. N., Davis, S. F., Van Den Pol, A. N., and Xu, W. (2002). Selective enhancement of excitatory synaptic activity in the rat nucleus tractus solitarius by hypocretin 2. *Neuroscience* **115**, 707–714.

Smith, G. P. (2000). The controls of eating: a shift from nutritional homeostasis to behavioral neuroscience. *Nutrition* **16**, 814–820.

Smith, G. P., Jerome, C., Cushin, B. J., Eterno, R., and Simansky, K. J. (1981). Abdominal vagotomy blocks the satiety effect of cholecystokinin in the rat. *Science* **213**, 1036–1037.

Smith, G. P., Jerome, C., and Norgren, R. (1985). Afferent axons in abdominal vagus mediate satiety effect of cholecystokinin in rats. *Am. J. Physiol.* **249**, R638–R641.

Spanswick, D., Smith, M. A., Mirshamsi, S., Routh, V. H., and Ashford, M. L. (2000). Insulin activates ATP-sensitive K+ channels in hypothalamic neurons of lean, but not obese rats. *Nat. Neurosci.* **3**, 757–758.

Spina, M., Merlo-Pich, E., Chan, R. K., Basso, A. M., Rivier, J., Vale, W., and Koob, G. F. (1996). Appetite-suppressing effects of urocortin, a CRF-related neuropeptide. *Science* **273**, 1561–1564.

Strack, A. M., and Loewy, A. D. (1990). Pseudorabies virus: a highly specific transneuronal cell body marker in the sympathetic nervous system. *J. Neurosci.* **10**, 2139–2147.

Stratford, T. R., Gibbs, J., Coy, D. H., and Smith, G. P. (1995). Fourth ventricular injection of the bombesin receptor antagonist [D-Phe6]bombesin(6-13)methyl ester, but not BW2258U89, increases food intake in rats. *Pharmacol. Biochem. Behav.* **50**, 463–471.

Strautz, R. L. (1970). Studies of hereditary-obese mice (obob) after implantation of pancreatic islets in Millipore filter capsules. *Diabetologia* **6**, 306–312.

Stricker, E. M., and Verbalis, J. G. (1991). Caloric and noncaloric controls of food intake. *Brain Res. Bull.* **27**, 299–303.

Strohmayer, A. J., and Smith, G. P. (1987). A sex difference in the effect of CCK-8 on food and water intake in obese (ob/ob) and lean (+/+) mice. *Peptides* **8**, 845–848.

Stuckey, J. A., Gibbs, J., and Smith, G. P. (1985). Neural disconnection of gut from brain blocks bombesin-induced satiety. *Peptides* **6**, 1249–1252.

Sugino, T., Hasegawa, Y., Kikkawa, Y., Yamaura, J., Yamagishi, M., Kurose, Y., Kojima, M., Kangawa, K., and Terashima, Y. (2002). A transient ghrelin surge occurs just before feeding in a scheduled meal-fed sheep. *Biochem. Biophys. Res. Commun.* **295**, 255–260.

Sunter, D., Hewson, A. K., Lynam, S., and Dickson, S. L. (2001). Intracerebroventricular injection of neuropeptide FF, an opioid modulating neuropeptide, acutely reduces food intake and stimulates water intake in the rat. *Neurosci. Lett.* **313**, 145–148.

Sutton, G. M., Duos, B., Patterson, L. M., and Berthoud, H. R. (2005). Melanocortinergic modulation of cholecystokinin-induced suppression of feeding through extracellular signal-regulated kinase signaling in rat solitary nucleus. *Endocrinology* **146**, 3739–3747.

Swanson, L. W. (2000). Cerebral hemisphere regulation of motivated behavior. *Brain Res.* **886**, 113–164.

Swanson, L. W., and Petrovich, G. D. (1998). What is the amygdala? *Trends Neurosci.* **21**, 323–331.

Swanson, L. W., Sawchenko, P. E., Rivier, J., and Vale, W. W. (1983). Organization of ovine corticotropin-releasing factor immunoreactive cells and fibers in the rat brain: an immunohistochemical study. *Neuroendocrinology* **36**, 165–186.

Tache, Y., Garrick, T., and Raybould, H. (1990). Central nervous system action of peptides to influence gastrointestinal motor function. *Gastroenterology* **98**, 517–528.

Talsania, T., Anini, Y., Siu, S., Drucker, D. J., and Brubaker, P. L. (2005). Peripheral exendin-4 and peptide YY(3-36) synergistically reduce food intake through different mechanisms in mice. *Endocrinology* **146**, 3748–3756.

Tamura, H., Kamegai, J., Shimizu, T., Ishii, S., Sugihara, H., and Oikawa, S. (2002). Ghrelin stimulates GH but not food intake in arcuate nucleus ablated rats. *Endocrinology* **143**, 3268–3275.

Tang-Christensen, M., Larsen, P. J., Goke, R., Fink-Jensen, A., Jessop, D. S., Moller, M., and Sheikh, S. P. (1996). Central administration of GLP-1-(7-36) amide inhibits food and water intake in rats. *Am. J. Physiol.* **271**, R848–R856.

Tang-Christensen, M., Larsen, P. J., Thulesen, J., Romer, J., and Vrang, N. (2000). The proglucagon-derived peptide, glucagon-like peptide-2, is a neurotransmitter involved in the regulation of food intake. *Nat. Med.* **6**, 802–807.

Tang-Christensen, M., Vrang, N., and Larsen, P. J. (2001). Glucagon-like peptide containing pathways in the regulation of feeding behaviour. *Int. J. Obes. Relat. Metab. Disord.* **25** (Suppl. 5), S42–S47.

Taylor, I. L., and Garcia, R. (1985). Effects of pancreatic polypeptide, caerulein, and bombesin on satiety in obese mice. *Am. J. Physiol.* **248**, G277–G280.

Taylor, I. L., Feldman, M., Richardson, C. T., and Walsh, J. H. (1978). Gastric and cephalic stimulation of human pancreatic polypeptide release. *Gastroenterology* **75**, 432–437.

Ter Horst, G. J., de Boer, P., Luiten, P. G., and van Willigen, J. D. (1989). Ascending projections from the solitary tract nucleus to the hypothalamus. A *Phaseolus vulgaris* lectin tracing study in the rat. *Neuroscience* **31**, 785–797.

Thiele, T. E., Van Dijk, G., Campfield, L. A., Smith, F. J., Burn, P., Woods, S. C., Bernstein, I. L., and Seeley, R. J. (1997). Central infusion of GLP-1, but not leptin, produces conditioned taste aversions in rats. *Am. J. Physiol.* **272**, R726–R730.

Thompson, R. H., and Swanson, L. W. (1998). Organization of inputs to the dorsomedial nucleus of the hypothalamus: a reexamination with Fluorogold and PHAL in the rat. *Brain Res. Rev.* **27**, 89–118.

Thompson, R. H., and Swanson, L. W. (2003). Structural characterization of a hypothalamic visceromotor pattern generator network. *Brain Res. Rev.* **41**, 153–202.

Thorens, B., and Larsen, P. J. (2004). Gut-derived signaling molecules and vagal afferents in the control of glucose and energy homeostasis. *Curr. Opin. Clin. Nutr. Metab. Care* **7**, 471–478.

Tokita, R., Nakata, T., Katsumata, H., Konishi, S., Onodera, H., Imaki, J., and Minami, S. (1999). Prolactin secretion in response to prolactin-releasing peptide and the expression of the prolactin-releasing peptide gene in the medulla oblongata are estrogen dependent in rats. *Neurosci. Lett.* **276**, 103–106.

Toth, Z. E., Gallatz, K., Fodor, M., and Palkovits, M. (1999). Decussations of the descending paraventricular pathways to the brainstem and spinal cord autonomic centers. *J. Comp. Neurol.* **414**, 255–266.

Tourrel, C., Bailbe, D., Meile, M. J., Kergoat, M., and Portha, B. (2001). Glucagon-like peptide-1 and exendin-4 stimulate beta-cell neogenesis in streptozotocin-treated newborn rats resulting in persistently improved glucose homeostasis at adult age. *Diabetes* **50**, 1562–1570.

Travers, J. B., Dinardo, L. A., and Karimnamazi, H. (1997). Motor and premotor mechanisms of licking. *Neurosci. Biobehav. Rev.* **21**, 631–647.

Trayhurn, P., and Wood, I. S. (2004). Adipokines: inflammation and the pleiotropic role of white adipose tissue. *Br. J. Nutr.* **92**, 347–355.

Tschop, M., Smiley, D. L., and Heiman, M. L. (2000). Ghrelin induces adiposity in rodents. *Nature* **407**, 908–913.

Tschop, M., Wawarta, R., Riepl, R. L., Friedrich, S., Bidlingmaier, M., Landgraf, R., and Folwaczny, C. (2001). Post-prandial decrease of circulating human ghrelin levels. *J. Endocrinol. Invest.* **24**, RC19.

Tschop, M., Castaneda, T. R., Joost, H. G., Thone-Reineke, C., Ortmann, S., Klaus, S., Hagan, M. M., Chandler, P. C., Oswald, K. D., Benoit, S. C., Seeley, R. J., Kinzig, K. P., Moran, T. H., Beck-Sickinger, A. G., Koglin, N., Rodgers, R. J., Blundell, J. E., Ishii, Y., Beattie, A. H., Holch, P., Allison, D. B., Raun, K., Madsen, K., Wulff, B. S., Stidsen, C. E., Birringer, M., Kreuzer, O. J., Schindler, M., Arndt, K., Rudolf, K., Mark, M., Deng, X. Y., Whitcomb, D. C., Halem, H., Taylor, J., Dong, J., Datta, R., Culler, M., Craney, S., Flora, D., Smiley, D., and Heiman, M. L. (2004). Physiology: does gut hormone PYY3-36 decrease food intake in rodents? *Nature* **430**, 1.

Turnbull, A. V., and Rivier, C. (1997). Corticotropin-releasing factor (CRF) and endocrine responses to stress: CRF receptors, binding protein, and related peptides. *Proc. Soc. Exp. Biol. Med.* **215**, 1–10.

Turton, M. D., O'Shea, D., Gunn, I., Beak, S. A., Edwards, C. M., Meeran, K., Choi, S. J., Taylor, G. M., Heath, M. M., Lambert, P. D., Wilding, J. P., Smith, D. M., Ghatei, M. A., Herbert, J., and Bloom, S. R. (1996). A role for glucagon-like peptide-1 in the central regulation of feeding. *Nature* **379**, 69–72.

Ueno, N., Inui, A., Iwamoto, M., Kaga, T., Asakawa, A., Okita, M., Fujimiya, M., Nakajima, Y., Ohmoto, Y., Ohnaka, M., Nakaya, Y., Miyazaki, J. I., and Kasuga, M. (1999). Decreased food intake and body weight in pancreatic polypeptide-overexpressing mice. *Gastroenterology* **117**, 1427–1432.

Unger, J. W., Moss, A. M., and Livingston, J. N. (1991). Immunohistochemical localization of insulin receptors and phosphotyrosine in the brainstem of the adult rat. *Neuroscience* **42**, 853–861.

Van der Haeghen, J. J., Lotstra, F., Vierendeels, G., Gilles, C., Deschepper, C., and Verbanck, P. (1981). Cholecystokinins in the central nervous system and neurohypophysis. *Peptides* **2** (Suppl. 2), 81–88.

van der Kooy, D., Koda, L. Y., McGinty, J. F., Gerfen, C. R., and Bloom, F. E. (1984). The organization of projections from the cortex, amygdala, and hypothalamus to the nucleus of the solitary tract in rat. *J. Comp. Neurol.* **224**, 1–24.

Van Dijk, G., Thiele, T. E., Donahey, J. C., Campfield, L. A., Smith, F. J., Burn, P., Bernstein, I. L., Woods, S. C., and Seeley, R. J. (1996). Central infusions of leptin and GLP-1-(7-36) amide differentially stimulate c-FLI in the rat brain. *Am. J. Physiol.* **271**, R1096–R1100.

Van Pett, K., Viau, V., Bittencourt, J. C., Chan, R. K., Li, H. Y., Arias, C., Prins, G. S., Perrin, M., Vale, W., and Sawchenko, P. E. (2000). Distribution of mRNAs encoding CRF receptors in brain and pituitary of rat and mouse. *J. Comp. Neurol.* **428**, 191–212.

van Rossum, D., Menard, D. P., Fournier, A., St-Pierre, S., and Quirion, R. (1994). Autoradiographic distribution and receptor binding profile of [125I]Bolton Hunter-rat amylin binding sites in the rat brain. *J. Pharmacol. Exp. Ther.* **270**, 779–787.

Verbalis, J. G., McCann, M. J., McHale, C. M., and Stricker, E. M. (1986). Oxytocin secretion in response to cholecystokinin and food: differentiation of nausea from satiety. *Science* **232**, 1417–1419.

Verbalis, J. G., Stricker, E. M., Robinson, A. G., and Hoffman, G. E. (1991). Cholecystokinin activates C-Fos expression in hypothalamic oxytocin and corticotrophin-releasing hormone neurons. *J. Neuroendocrinol.* **3**, 205–213.

Vergoni, A. V., Watanobe, H., Guidetti, G., Savino, G., Bertolini, A., and Schioth, H. B. (2002). Effect of repeated administration of prolactin releasing peptide on feeding behavior in rats. *Brain Res.* **955**, 207–213.

Vrang, N., Larsen, P. J., Clausen, J. T., and Kristensen, P. (1999a). Neurochemical characterization of hypothalamic cocaine-amphetamine-regulated transcript neurons. *J. Neurosci.* **19**, RC5.

Vrang, N., Tang-Christensen, M., Larsen, P. J., and Kristensen, P. (1999b). Recombinant CART peptide induces c-Fos expression in central areas involved in control of feeding behaviour. *Brain Res.* **818**, 499–509.

Vrang, N., Phifer, C. B., Corkern, M. M., and Berthoud, H. R. (2003). Gastric distension induces c-Fos in medullary GLP-1/2-containing neurons. *Am. J. Physiol.* **285**, R470–R478.

Wallace, D. M., Magnuson, D. J., and Gray, T. S. (1992). Organization of amygdaloid projections to brainstem dopaminergic, noradrenergic, and adrenergic cell groups in the rat. *Brain Res. Bull.* **28**, 447–454.

Wang, L., Martinez, V., Barrachina, M. D., and Tache, Y. (1998). Fos expression in the brain induced by peripheral injection of CCK or leptin plus CCK in fasted lean mice. *Brain Res.* **791**, 157–166.

Wang, Y. H., Tache, Y., Sheibel, A. B., Go, V. L., and Wei, J. Y. (1997). Two types of leptin-responsive gastric vagal afferent terminals: an in vitro single-unit study in rats. *Am. J. Physiol.* **273**, R833–R837.

Ward, S. M., Bayguinov, J., Won, K. J., Grundy, D., and Berthoud, H. R. (2003). Distribution of the vanilloid receptor (VR1) in the gastrointestinal tract. *J. Comp. Neurol.* **465**, 121–135.

Watanabe, T. K., Suzuki, M., Yamasaki, Y., Okuno, S., Hishigaki, H., Ono, T., Oga, K., Mizoguchi-Miyakita, A., Tsuji, A., Kanemoto, N., Wakitani, S., Takagi, T., Nakamura, Y., and Tanigami, A. (2005). Mutated G-protein-coupled receptor GPR10 is responsible for the hyperphagia/dyslipidaemia/obesity locus of Dmo1 in the OLETF rat. *Clin. Exp. Pharmacol. Physiol.* **32**, 355–366.

Wei, E. T. (1981). Pharmacological aspects of shaking behavior produced by TRH, AG-3-5, and morphine withdrawal. *Fed. Proc.* **40**, 1491–1496.

Werther, G. A., Hogg, A., Oldfield, B. J., McKinley, M. J., Figdor, R., Allen, A. M., and Mendelsohn, F. A. (1987). Localization and characterization of insulin receptors in rat brain and pituitary gland using in vitro autoradiography and computerized densitometry. *Endocrinology* **121**, 1562–1570.

West, D. B., Fey, D., and Woods, S. C. (1984). Cholecystokinin persistently suppresses meal size but not food intake in free-feeding rats. *Am. J. Physiol.* **246**, R776–R787.

Westermark, P., Wernstedt, C., Wilander, E., and Sletten, K. (1986). A novel peptide in the calcitonin gene related peptide family as an amyloid fibril protein in the endocrine pancreas. *Biochem. Biophys. Res. Commun.* **140**, 827–831.

Whitcomb, D. C., Taylor, I. L., and Vigna, S. R. (1990). Characterization of saturable binding sites for circulating pancreatic polypeptide in rat brain. *Am. J. Physiol.* **259**, G687–G691.

White, S. R., Crane, G. K., and Jackson, D. A. (1989). Thyrotropin-releasing hormone (TRH) effects on spinal cord neuronal excitability. *Ann. N.Y. Acad. Sci.* **553**, 337–350.

Willesen, M. G., Kristensen, P., and Romer, J. (1999). Co-localization of growth hormone secretagogue receptor and NPY mRNA in the arcuate nucleus of the rat. *Neuroendocrinology* **70**, 306–316.

Williams, D. L., Kaplan, J. M., and Grill, H. J. (2000). The role of the dorsal vagal complex and the vagus nerve in feeding effects of melanocortin-3/4 receptor stimulation. *Endocrinology* **141**, 1332–1337.

Williams, D. L., Grill, H. J., Weiss, S. M., Baird, J. P., and Kaplan, J. M. (2002). Behavioral processes underlying the intake suppressive effects of melanocortin 3/4 receptor activation in the rat. *Psychopharmacology (Berlin)* **161**, 47–53.

Willing, A. E., and Berthoud, H. R. (1997). Gastric distension-induced c-fos expression in catecholaminergic neurons of rat dorsal vagal complex. *Am. J. Physiol.* **272**, R59–R67.

Wittmann, G., Liposits, Z., Lechan, R. M., and Fekete, C. (2004). Medullary adrenergic neurons contribute to the cocaine- and amphetamine-regulated transcript-immunoreactive innervation of thyrotropin-releasing hormone synthesizing neurons in the hypothalamic paraventricular nucleus. *Brain Res.* **1006**, 1–7.

Wolak, M. L., DeJoseph, M. R., Cator, A. D., Mokashi, A. S., Brownfield, M. S., and Urban, J. H. (2003). Comparative distribution of neuropeptide Y Y1 and Y5 receptors in the rat brain by using immunohistochemistry. *J. Comp. Neurol.* **464**, 285–311.

Woods, S. C., Lotter, E. C., McKay, L. D., and Porte, D., Jr. (1979). Chronic intracerebroventricular infusion of insulin reduces food intake and body weight of baboons. *Nature* **282**, 503–505.

Woods, S. C., Chavez, M., Park, C. R., Riedy, C., Kaiyala, K., Richardson, R. D., Figlewicz, D. P., Schwartz, M. W., Porte, D., Jr. and Seeley, R. J. (1996).

The evaluation of insulin as a metabolic signal influencing behavior via the brain. *Neurosci. Biobehav. Rev.* **20**, 139–144.

Wu, D., Yang, J., and Pardridge, W. M. (1997). Drug targeting of a peptide radiopharmaceutical through the primate blood-brain barrier in vivo with a monoclonal antibody to the human insulin receptor. *J. Clin. Invest.* **100**, 1804–1812.

Wu, X., Gao, J., Yan, J., Owyang, C., and Li, Y. (2004). Hypothalamus–brain stem circuitry responsible for vagal efferent signaling to the pancreas evoked by hypoglycemia in rat. *J. Neurophysiol.* **91**, 1734–1747.

Yamada, K., Wada, E., Santo-Yamada, Y., and Wada, K. (2002). Bombesin and its family of peptides: prospects for the treatment of obesity. *Eur. J. Pharmacol.* **440**, 281–290.

Yamakawa, K., Kudo, K., Kanba, S., and Arita, J. (1999). Distribution of prolactin-releasing peptide-immunoreactive neurons in the rat hypothalamus. *Neurosci. Lett.* **267**, 113–116.

Yamamoto, H., Lee, C. E., Marcus, J. N., Williams, T. D., Overton, J. M., Lopez, M. E., Hollenberg, A. N., Baggio, L., Saper, C. B., Drucker, D. J., and Elmquist, J. K. (2002). Glucagon-like peptide-1 receptor stimulation increases blood pressure and heart rate and activates autonomic regulatory neurons. *J. Clin. Invest.* **110**, 43–52.

Yan, H., Yang, J., Marasco, J., Yamaguchi, K., Brenner, S., Collins, F., and Karbon, W. (1996). Cloning and functional expression of cDNAs encoding human and rat pancreatic polypeptide receptors. *Proc. Natl. Acad. Sci. U.S.A.* **93**, 4661–4665.

Yang, B., Samson, W. K., and Ferguson, A. V. (2003). Excitatory effects of orexin-A on nucleus tractus solitarius neurons are mediated by phospholipase C and protein kinase C. *J. Neurosci.* **23**, 6215–6222.

Yang, M., Zhao, X., and Miselis, R. R. (1999). The origin of catecholaminergic nerve fibers in the subdiaphragmatic vagus nerve of rat. *J. Auton. Nerv. Syst.* **76**, 108–117.

Yuan, C. S., Attele, A. S., Wu, J. A., Zhang, L., and Shi, Z. Q. (1999). Peripheral gastric leptin modulates brain stem neuronal activity in neonates. *Am. J. Physiol.* **277**, G626–G630.

Yuan, C. S., Attele, A. S., Dey, L., and Xie, J. T. (2000). Gastric effects of cholecystokinin and its interaction with leptin on brainstem neuronal activity in neonatal rats. *J. Pharmacol. Exp. Ther.* **295**, 177–182.

Zhang, X., Fogel, R., and Renehan, W. E. (1995). Relationships between the morphology and function of gastric- and intestine-sensitive neurons in the nucleus of the solitary tract. *J. Comp. Neurol.* **363**, 37–52.

Zhang, X., Shi, T., Holmberg, K., Landry, M., Huang, W., Xiao, H., Ju, G., and Hokfelt, T. (1997). Expression and regulation of the neuropeptide Y Y2 receptor in sensory and autonomic ganglia. *Proc. Natl. Acad. Sci. U.S.A.* **94**, 729–734.

Zhang, X., Renehan, W. E., and Fogel, R. (1998). Neurons in the vagal complex of the rat respond to mechanical and chemical stimulation of the GI tract. *Am. J. Physiol.* **274**, G331–G341.

Zhang, X., Fogel, R., and Renehan, W. E. (1999). Stimulation of the paraventricular nucleus modulates the activity of gut-sensitive neurons in the vagal complex. *Am. J. Physiol.* **277**, G79–G90.

Zhang, Y., Proenca, R., Maffei, M., Barone, M., Leopold, L., and Friedman, J. M. (1994). Positional cloning of the mouse obese gene and its human homologue. *Nature* **372**, 425–432.

Zheng, H., Patterson, C., and Berthoud, H. R. (2001). Fourth ventricular injection of CART peptide inhibits short-term sucrose intake in rats. *Brain Res.* **896**, 153–156.

Zheng, H., Patterson, L. M., and Berthoud, H. R. (2002). CART in the dorsal vagal complex: sources of immunoreactivity and effects on Fos expression and food intake. *Brain Res.* **957**, 298–310.

Zheng, H., Patterson, L. M., and Berthoud, H. R. (2005a). Orexin-A projections to the caudal medulla and orexin-induced c-Fos expression, food intake, and autonomic function. *J. Comp. Neurol.* **485**, 127–142.

Zheng, H., Patterson, L. M., Phifer, C. B., and Berthoud, H. R. (2005b). Brain stem melanocortinergic modulation of meal size and identification of hypothalamic POMC projections. *Am. J. Physiol.* **289**, R247–R258.

Zigman, J. M., Jones, J. E., Lee, C. E., Saper, C. B., and Elmquist, J. K. (2006). Expression of ghrelin receptor mRNA in the rat and the mouse brain. *J. Comp. Neurol.* **494**, 528–548.

Zittel, T. T., De Giorgio, R., Sternini, C., and Raybould, H. E. (1994). Fos protein expression in the nucleus of the solitary tract in response to intestinal nutrients in awake rats. *Brain Res.* **663**, 266–270.

Zittel, T. T., Glatzle, J., Kreis, M. E., Starlinger, M., Eichner, M., Raybould, H. E., Becker, H. D., and Jehle, E. C. (1999). C-fos protein expression in the nucleus of the solitary tract correlates with cholecystokinin dose injected and food intake in rats. *Brain Res.* **846**, 1–11.

6

The Gut–Brain Axis in the Control of Eating

THOMAS A. LUTZ
Institute of Veterinary Physiology, University of Zurich
NORI GEARY
Institute of Animal Sciences, Institute of Technology,
Zurich (ETHZ)

I. Introduction
II. Gastric Mechanoreception
    A. Physiological Role in the Control of Eating
    B. Gastric Mechanoreception and Obesity Treatment
III. Intestinal Cholecystokinin (CCK)
    A. Gut Peptides and the Science of Eating
    B. Physiological Relevance of Intestinal CCK in the Control of Eating
    C. Peripheral and Central Mechanisms
    D. Interactions of CCK and Other Signals Controlling Eating
    E. CCK Satiation and Obesity Treatment
IV. Amylin
    A. Physiological Relevance of Amylin in the Control of Eating
    B. Amylin As an Adiposity Signal
    C. Peripheral and Central Mechanisms of Amylin Action
    D. Interactions of Amylin and Other Signals Controlling Eating
    E. Amylin and Obesity Treatment
V. Ghrelin
    A. Physiological Relevance of Ghrelin in the Control of Eating
    B. Ghrelin As an Adiposity Signal
    C. Peripheral and Central Mechanisms of Ghrelin Action
    D. Ghrelin and Obesity Treatment
VI. Potentials and Problems of Gut–Brain Axis Signals in the Treatment of Obesity
    References

This chapter reviews the physiological controls of eating that arise in the gut, that is, the abdominal viscera. Almost the entire gut is involved in the processing of ingested food, from preabsorptive actions of ingesta and food-related exocrine secretions in the gastrointestinal tract to metabolic effects of the absorbed nutrients, gut endocrine secretions, etc., in the liver and beyond. Information related to many of these

TABLE 1   Gut Peptides Potentially Involved in the Physiological Control of Eating

Peptide	Site(s) of production	Site(s) of critical receptors	References
Amylin	Pancreatic B cell	Area postrema (AP)	Lutz (2005)
Apolipoprotein A IV	Villus epithelia	?	Tso and Liu (2004)
Bombesin-like peptides (GRP; neuromedin B)	Stomach	Vagal and spinal visceral afferents	Yamada et al. (2000)
CCK	Proximal small intestine	Vagal afferents in pyloric area; pyloric smooth muscle; liver	Moran (2004)
Enterostatin	Exocrine pancreas	?	Berger et al. (2004)
Ghrelin	Stomach	Arcuate nucleus (Arc); AP; vagal afferents	Cummings et al. (2005)
GLP-1	Ileum	Brainstem; hypothalamus	Gutzwiller et al. (2004b)
Glucagon	Pancreatic A cell	Liver	Geary (1999)
Insulin	Pancreatic B cell	Arc	Woods et al. (2006)
(gastric) Leptin	Stomach	Arc; vagal afferents; brainstem	Berthoud (2005); Woods (2005)
PYY_{3-36}	Ileum	Hypothalamus	leRoux and Bloom (2005)

processes is relayed to the brain and affects subsequent eating. The principal means of communication between the gut and brain are hormonal, that is, gastrointestinal and pancreatic peptides whose release is affected by eating, and neural, that is, vagal and spinal visceral afferent neural signals originating in the gut.

I. INTRODUCTION

The list of gut signals that have been implicated in the control of eating is long. Table 1 lists gut peptides that presently appear to have signal functions controlling eating, and Table 2 lists potential neural controls arising from the stimulation of gut mechanoreceptors, chemoreceptors, and thermoreceptors. Because it is clearly beyond the scope of this chapter to review each of these signals in depth, we focus on four examples: one neural signal; gastric fill, and three peptide hormones; cholecystokinin (CCK), amylin, and ghrelin. These are chosen in order to illustrate (1) the basic science issues facing any hypothesized gut signal in the control of food intake, including (i) its physiological relevance in the initiation, maintenance, or termination of eating during meals, especially in humans, and (ii) the site of the receptors initiating the eating effect as well as its peripheral and central signaling mechanisms, and (2) issues related to the clinical potential of manipulation of these gut signals for the treatment of obesity, where possible, including (i) potential functional interactions with other signals, which we emphasize because of their potential pharmacological significance, (ii) effects of the signal in obesity models, (iii) effects of chronic stimulation of the signal, and (iv) side effects and risk factors that may limit the therapeutic use of these signals. Tables 1 and 2 include references to key papers or reviews for each signal. In addition, reviews of the burgeoning list of candidate gut controls of eating appear regularly (e.g., Stanley et al., 2005; Strader and Woods, 2005); several useful conceptual approaches to the problems and progress in this have been described as well (e.g., Geary, 2004a; Kaplan and Moran, 2004; Schwartz, 2000; Smith, 1999; Woods, 2004, 2005).

TABLE 2   Gut Neural Receptors Potentially Involved in the
Physiological Control of Eating

Receptor site and characteristics	References
Gastric mechanoreceptors	Kaplan and Moran (2004); Kral and Rolls (2004)
Hepatic chemoreceptors[a]	Langhans (2003); Scharrer (1999)
Hepatic thermoreceptors	Di Bella et al. (1981a,b); De Vries et al. (1993)
Intestinal chemoreceptors[a]	Schwartz (2000); Savastano et al. (2005)
Intestinal mechanoreceptors	Schwartz (2000)
Intestinal osmoreceptors	Houpt et al. (1983)

[a]Note: We define chemoreceptors broadly to include specific nutrients (e.g., glucose) or their digestive and metabolic products or consequences (e.g., lactate, changes in hepatocyte membrane potential, or changes in pH).

II. GASTRIC MECHANORECEPTION

A. Physiological Role in the Control of Eating

Electrophysiological and anatomical studies reveal that the upper gastrointestinal tract is richly innervated with mechanoreceptors that are stimulated by various aspects of gut loading during and after meals and that signal the brain via both vagal and splanchnic visceral afferents. The effects of gastric mechanoreceptor signaling on eating in rats have been elegantly analyzed using gastric cannulae in combination with chronic pyloric cuffs that can be inflated during meals to limit food stimuli to the oropharynx and stomach (Eisen et al., 2001; Kaplan and Moran, 2004; Phillips and Powley, 1996). The key findings are (1) when gastric cannulae are used to prevent ingested liquid food from accumulating in the stomach, meal size is dramatically increased; (2) when ingested food is prevented from entering the intestines by inflating pyloric cuffs, meal size is about normal; (3) when fluid loads are infused into the stomach of rats with closed pyloric cuffs, eating is inhibited in proportion to the volume infused, and (4) the effect of gastric

fill on eating is identical whether nutrient or nonnutrient loads are used. These data indicate that gastric volume is an adequate stimulus for mechanoreceptors that can contribute to the control of eating. The normal contribution of this signal to eating, however, appears to be small. First, intragastric infusions inhibit eating in rats with closed pyloric cuffs only when the total gastric fill (ingesta plus infusion) is markedly larger than the control meal size. Second, the dose–response relation between infusion volume and amount less eaten is relatively flat, such that the behavior does not nearly compensate for the infusion. And third, because normal gastric emptying is prevented in the cuff-closed condition, the gastric volume at meal end is markedly larger than the gastric volume at the end of a similarly sized meal in the cuff-open (normal) condition. Indeed, gastric emptying usually proceeds at a surprisingly brisk pace during the meal, at least when liquids are ingested. In both rats and rhesus monkeys, the intrameal rate of gastric emptying of liquid diet is about five times the postmeal rate (Kaplan and Moran, 2004).

Because of its key role in the distribution of ingesta in the gastrointestinal tract, gastric emptying is a highly regulated and adaptable function with numerous hormonal and neural controls. The fact that some of these controls affect eating independently of their influence on gastric emptying (e.g., CCK), complicates analysis of the effects of gastric emptying per se on eating. The influence of gastric emptying on eating is further complicated by the obvious fact that gastric emptying both decreases the intensity of gastric signals and increases the intensity of intestinal and postabsorptive signals. Importantly, the fact that at meal end there are significant amounts of ingesta both in the stomach and at postgastric sites suggests the possibility that gastric and postgastric signals interact in the normal control of eating. Evidence for such interactions is considered in the following sections.

Because the pyloric cuff technique cannot be applied in humans, direct evidence of the contribution of gastric volume to the control of human eating is lacking. There is, however, indirect evidence. In an elegant series of experiments using oral and intragastrically infused preloads, Rolls and her colleagues (Kral and Rolls, 2004) demonstrated that the energy density and the volume of ingested food make independent contributions to the size of meals in human volunteers. The relative potencies of gastric and postgastric receptors to these effects remain uncertain, of course. However, a role for gastric volume is suggested by the Sturm et al. (2004) report (1) that the cross-sectional area of the gastric antrum before meals correlates well with perceived hunger and subsequent meal size and, (2) that antral cross-sectional area at meal end correlates with preceding meal size.

B. Gastric Mechanoreception and Obesity Treatment

There are no data directly relating gastric mechanoreception to the control of eating in obesity. Nevertheless, the smaller meals, reduced total food intake, and weight loss in obese patients following gastric reduc-tion surgery strongly suggest that gastric mechanoreception can initiate powerful and therapeutically useful controls of eating (Korenkov et al., 2005), although the long-term outcome may be less than other types of bariatric surgery. Rolls and Barnett (2000) have published an interesting diet plan based on exploiting gastric mechanoreception and gastric emptying, but the causal mechanism underlying any efficacy this program may have in obesity therapy has not been clarified. As reviewed later in this chapter, there is good reason to believe that gastric mechanoreceptors may also interact with gut hormones to control normal eating, so that the potential of combination therapies should not be neglected.

III. INTESTINAL CHOLECYSTOKININ (CCK)

A. Gut Peptides and the Science of Eating

CCK is synthesized by I cells that are situated in the mucosa of the proximal small intestine such that their apical surfaces are exposed to intraluminal stimuli and that CCK is secreted through their basal membrane on the serosal side. CCK secreted during and after meals has long been considered an essential physiological control of gastric emptying and exocrine pancreatic secretion. The classic report of Gibbs et al. (1973) that intraperitoneal injections of CCK selectively inhibit eating established the control of food intake as another potential physiological function of CCK. CCK has remained the paradigmatic example of a gut peptide which acts as a signal to control eating because of the intense focus on its actions by a number of research groups (e.g., Beglinger and colleagues, Moran and McHugh, Ritter and colleagues, Reidelberger and colleagues; and Smith and Gibbs). This sustained effort has had an extremely important influence on appetite science. First, it has shifted focus from the problem of the initiation of eating by

TABLE 3   Criteria for the Physiological Action of Hormonal Controls of Eating[a]

1. Secretion	Eating leads to a change in the secretion of the hormone.
2. Receptor	Receptors are expressed at the primary site of action.
3. Physiological dose	Mimicking the secretion pattern of the endogenous hormone by administration of the hormone reproduces the eating effect, perhaps only in synergy with other endogenous stimuli.
4. Removal, replacement	Removing the hormone or the critical receptors prevents the eating effect, and appropriate hormone replacement or receptor rescue reproduces the effect.
5. Antagonism	Selectively antagonizing hormone signaling prevents the eating effect of the endogenous hormone and prevents the eating effect of physiological doses of the exogenous hormone.

[a]Geary (2004a).

postabsorptive, metabolic signals to the problem of the termination of eating ("satiation") by preabsorptive signals, especially gut peptide signals (see also, Geary, 2004b; Smith, 1999; Smith and Gibbs, 1984). Second, it has facilitated the development of explicit empirical criteria, modeled on classic endocrinological concepts, for the evaluation of the physiological roles of endocrine signals in the control of eating. The first set of criteria proposed by Gibbs et al. (1973) has been revised several times, as new methodologies have emerged. Table 3 gives a more recent version, which we will use in considering the effects on eating of each of the gut peptides reviewed here. Even if a peptide does not appear to play a likely physiological control of eating according to the criteria referred to in Table 3, this should not preclude it from being a potential pharmaceutical target, as long as eating and body weight can be influenced without triggering major side effects.

B. Physiological Relevance of Intestinal CCK in the Control of Eating

In both experimental animals and humans, administration of CCK at meal onset decreases meal size with little effect on the following intermeal interval (Geary, 2004a; Little, et al., 2005; Moran and Kinzig, 2004). As demonstrated initially by Gibbs et al. (1973) and by countless subsequent reports, except at high doses, CCK's action

on eating is highly selective and not aversive or toxic. For example, CCK administration decreases the intake of liquid food in hungry rats but not of water in thirsty rats. Similarly, in humans, CCK's eating-inhibitory action occurs without subjective or physical side effects. Thus, CCK seems to be a highly selective satiation signal.

The physiological status of this action has been investigated intensively. Some of the key evidence that, in both rats and humans, CCK meets the criteria summarized in Table 3 for a physiological satiation signal are outlined here.

1. *Secretion.* Plasma CCK concentration increases within minutes of meal onset in humans, but has rarely been shown to increase within the time frame of meals in rats (Geary, 2004a). The failure of CCK to appear in the blood during rats' meals coupled with the relatively greater potency of intraperitoneal versus intravenous CCK administration to inhibit eating and the ability of prandial injection of CCK antagonists to increase meal size (see later) suggest that CCK may act as a paracrine rather than an endocrine signal in rats. This paracrine action suggests that CCK secreted from neurons (Larsson and Rehfeld, 1978) may also be important in CCK-induced satiation.

2. *Receptors.* Selective antagonist studies indicate that the satiating effect of CCK

is mediated by low affinity CCK$_A$ receptors in rats and humans (Beglinger and Degen, 2004; Geary, 2004a; Moran and Kinzig, 2004). In animals, tests of close-arterial infusions, pylorectomies, and selective vagotomies (e.g., Moran and Kinzig, 2004; Moran et al., 1992) indicate that the satiating action of CCK is mediated by CCK$_A$ receptors in vagal afferents and in the pyloric circular muscle. CCK$_A$ receptors in the liver may also be involved, especially in mediating the satiating effects of larger molecular forms of CCK (Eisen et al., 2005). CCK$_A$ receptors also occur in several brain areas, including the nucleus of the solitary tract (NTS), but whether CCK released from the intestines during meals reaches these receptors remains uncertain.

3. *Physiological dose.* Verification of the paracrine hypothesis of CCK satiation in rats would require measurements of prandial changes in CCK concentration at the site of the apparently critical receptors, which has not yet been attempted. In contrast, systemic venous levels may provide a crucial test in humans, and there are two reports that physiological doses of CCK are sufficient to inhibit eating (Ballinger et al., 1995; Lieverse et al., 1993). The many negative reports, however, suggest that the positive findings may not be entirely general. One explanation for this, as considered later, is that CCK may synergize with other eating control signals, so that test conditions may be crucial.

4. *Removal and replacement.* This criterion is difficult to test because CCK-producing cells are distributed widely, so surgical removal would require removal of much of the small intestine. The effects of spontaneous mutations in the CCK$_A$ receptor, however, have been described in three species and lend mixed support to the criterion (reviewed in Geary, 2004a, and Moran and Kinzig, 2004). In rats, meal size, total daily food intake, and body weight are all increased, exogenous CCK has no effect on eating, and the satiating effect of gastrointestinal nutrient delivery is decreased. The phenotype of CCK$_A$ receptor knockout mice, however, is markedly different— in these animals, adult body weight and daily food intake are unchanged (Noble and Roques, 2002). Unfortunately, meal patterns have not yet been determined. The reason for this species difference is not yet clear but may be due to different receptor distributions, in particular the presence of CCK$_A$ receptors in the dorsomedial hypothalamus of the rat (Bi and Moran, 2002). Finally, in humans, spontaneous mutations of the CCK$_A$ receptor gene are associated with overeating and obesity.

5. *Antagonism.* This criterion has been repeatedly and successfully investigated in rats and humans. CCK receptor antagonists have been shown to increase meal size and to block the satiating effect induced by intraduodenal infusions of fat, in which CCK plays a significant role, in both species and to increase the perception of hunger in humans (Beglinger and Degen, 2004).

To summarize, CCK released from the small intestine during meals most likely plays a significant and physiological role in meal termination. It acts rapidly and selectively. According to Geary's (2004a) scheme, CCK exemplifies a fully coupled endocrine satiating signal, that is, the adequate stimulus (food in the small intestine) almost immediately leads to hormone secretion, which in turn affects eating within minutes. This tight linkage would seem to be an advantage both for the analysis of physiological mechanisms and for the development of pharmacotherapy.

C. Peripheral and Central Mechanisms

Each of the peripheral CCK$_A$ receptor populations described previously appears to signal the brain via the vagus nerve, with all the subdiaphragmatic vagal branches apparently involved (note that all four subdiaphragmatic vagal branches innervate the pyloric area; Berthoud et al., 2004). Some pharmacological reports also suggest the involvement of CCK$_A$ and CCK$_B$ receptors in the dorsal vagal complex (DVC) or area postrema (AP), but this is likely to reflect local actions of neural CCK, not hormonal actions of intestinal CCK. The same is true of the CCK$_A$ receptors in the dorsomedial hypothalamus. Indeed, this situation reflects the fact that the same molecules are released from peripheral nerves or glands as well as from neurons in the brain, with apparently variable degrees of segregation between the effects of the two pools. Thus, in the case of CCK, central receptors, perhaps in the AP, may mediate the decreases in food intake and conditioned taste aversions that can be elicited by larger doses of CCK (i.e., intraperitoneal injection of >8 µg/kg CCK-8), even in vagotomized rats.

As yet, relatively little is known about the processing of individual eating control signals in the brain or the extensive convergence of the many signals that apparently are simultaneously involved. For example, it is not certain whether or how the identities of individual peripheral signals are maintained during CNS processing, even during the initial stages of central neural processing. Neural signals from the gut project directly (in the case of vagally mediated signals such as CCK) or indirectly (in the case of spinal visceral afferents) to the NTS. Exogenous CCK has been shown to stimulate catecholaminergic and glucagon-like peptide-1 (GLP-1) releasing neurons in the NTS. NTS neurons that are activated by CCK have also been shown to express the melanocortin-4 receptor (MC4R). These neurons apparently originate in the hypothalamus and modulate hindbrain neuronal activity (Zheng et al., 2005). At least in females, CCK also activates NTS neurons expressing estrogen receptor-α (ERα) (Asarian and Geary, 2006). CCK satiation also seems to depend on brain serotonin 5-HT$_1$ receptors (Poeschla et al., 1993; Stallone et al., 1989), although the site remains uncertain, and on forebrain dopamine receptors (Bednar et al., 1995). Norepinephrine seems not to be necessary (Cannon and Palmiter, 2003). And, as mentioned before, CCK receptors both in the caudal brainstem and hypothalamus may be involved. One important issue regarding these initial clues to the central network functions mediating CCK satiation is that most of this work has involved tests of exogenous CCK, and it is not clear whether the neural processing of exogenous CCK is identical to that of endogenous CCK. For example, it seems likely that smaller doses of CCK or endogenous CCK, which signal via the vagus, and larger doses of CCK, which signal at least in part independent of the vagus, do not activate identical neural networks.

D. Interactions of CCK and Other Signals Controlling Eating

If many control signals operate simultaneously to control a single behavior, they must interact. As it seems likely that combination therapies for obesity may have the same advantages of increased potency and decreased side effects that have been demonstrated in many other areas of pharmacological therapeutics, the study of interactions is an especially important area in the context of the therapeutic potential. Basic research in this direction is relatively well advanced in the case of CCK, which has been reported to interact functionally with several other peripheral signals that control eating. The most important of these are listed in Table 4. Note that in Table 4 we define synergy as a further decrease in eating when CCK and another signal are

TABLE 4   Functional Interactions Between CCK and Other Peripheral
Signals Controlling Eating

Signal	Nature of interaction[a]	References
Amylin	Synergy	Bhavsar et al. (1998) Mollet et al. (2003)
Estradiol	Synergy	Clegg and Woods (2004)
Gastric load	Synergy	Kissileff et al. (2003) Moran and Kinzig (2004)
GLP-1	None	Gutzwiller et al. (2004a,b) Brennan et al. (2005)
Pancreatic glucagon	Antagonism	Geary et al. (1992)
Insulin (central)	Synergy	Woods (2005)
Leptin	Synergy	Woods (2005)

[a]Synergy is defined as a further decrease in eating when CCK and the other signal are applied simultaneously in comparison to CCK alone; and antagonism is a smaller decrease in eating when CCK and the other signal are applied simultaneously. As many of the cited reports used alternative definitions, in particular definitions based on infra- or supra-additivity of effects, their conclusions might differ.

applied simultaneously as compared to CCK alone; no interaction is no change in the effect of CCK in the presence of the additional signal; and antagonism is a smaller decrease in eating when CCK and the other signal are applied simultaneously. As many of the original reports used alternative definitions, in particular definitions based on infra- or supra-additivity of effects, their conclusions might differ. It has been argued persuasively that the use of effect–additivity is mathematically and physiologically incoherent when agents have different dose–response curves (Berenbaum, 1989). Finally, as most investigations of synergy involve only a few dose combinations and test conditions, neither positive nor negative reports should be considered conclusive.

Evidence of functional synergy between CCK and gastric loads has been reported in rats, primates, and, significantly, humans (Kissileff et al., 2003; Moran and Kinzig, 2004). Some progress has been made in identifying the mechanisms underlying these synergies. For example, in rats the electrophysiological activity of many vagal afferent neurons is increased both by CCK and by gastrointestinal mechanostimulation, and the response to combined stimuli often appears quantitatively or temporally synergistic (Schwartz, 2000; Schwartz and Moran, 1996). Additionally, pharmacological studies indicate that this interaction may depend on activation of 5-HT$_3$ receptors in the periphery (Hayes et al., 2004) and N-methyl-D-aspartate (NMDA) receptors in the DVC (Covasa et al., 2004).

The Kissileff et al. (2003) report of synergy between CCK and gastric load is the only report we know of a synergistic interaction between two gut signals controlling eating in humans. This together with progress in identifying its mechanistic basis makes it an excellent candidate for further preclinical research. Few other signal combinations have been tried in humans. CCK plus GLP-1 failed to produce a synergistic decrease in eating in humans (Brennan et al., 2005; Gutzwiller et al., 2004b), although in one test the perception of hunger was synergistically reduced (Gutzwiller et al., 2004b). CCK plus pancreatic glucagon also failed to synergige in humans (Geary et al., 1992), despite that in sham-feeding rats, CCK and pancreatic glucagon clearly synergized (Le Sauter and Geary, 1987).

E. CCK Satiation and Obesity Treatment

1. Animal Models

The efficacy of acute administration of CCK to reduce test meal size has been verified in several animal models of obesity, including in rats with large lesions of the ventromedial hypothalamus (which include the arcuate nucleus), *ob/ob* (leptin deficient) mice, Zucker *fa/fa* (partially leptin receptor deficient) rats, Koletsky fa^k/fa^k (wholly leptin receptor deficient) rats, and rats with dietary-induced obesity (Chandler et al., 2004; Smith and Gibbs, 1985; Strohmayer and Smith, 1986; Wildman et al., 2000). These data suggest that CCK can be acutely effective in a wide variety of models of obesity. The data also suggest that defects in CCK sensitivity do not account for the hyperphagia displayed in these models, with the possible exception of the Koletsky fa^k/fa^k rat, in which Morton et al. (2005) reported that CCK decreased meal size less than in lean control rats. Interestingly, in that study the satiating effect of one dose of exogenous CCK, but not the increased basal test meal size, was normalized by restoration of leptin receptor function by adenoviral gene therapy into the arcuate nucleus, suggesting that leptin may control meal size through an obligatory interaction with CCK, consistent with Smith's (1996) theory of the direct and indirect controls of meal size.

The effects of chronic administration of CCK on eating and body weight have been less extensively studied. In a widely cited study, West et al. (1984) infused CCK intraperitoneally prior to each spontaneous meal for six days in rats. Meal size was reduced throughout the study with no sign of tolerance, but meal frequency increased

by day four sufficiently to return total daily food intake to control level; body weight gain was reduced for only one day. Similar results as well as evidence of tolerance to CCK satiation have been reported in free-feeding rats when CCK was continuously infused for several days and when CCK_A receptor agonists were tested chronically (Simmons et al., 1999). In contrast, when CCK was intraperitoneally injected before each of three scheduled daily 30-min meals for 3 weeks, CCK consistently reduced meal size and led to significant weight loss (reviewed in Smith and Gibbs, 1985). Thus, the rat data suggest that, at least when meal frequency is controlled (as in people trying to lose body weight who often restrict the number of meals), CCK can effectively reduce meal size, total food intake, and body weight.

Finally, the issue of whether chronically increasing endogenous CCK by exposing rats to a high fat diet leads to a decrease in the satiating potency of exogenous CCK remains controversial (reviewed in Torregrossa and Smith, 2003). This issue is of potential relevance to humans consuming high fat diets.

2. CCK and Human Obesity

Tests of acute adminstration of CCK or CCK_A receptor agonists in obese humans have produced promising results. Intravenous administration of CCK-8 significantly reduced test meal size by an average of 13% in eight obese men (Pi-Sunyer et al., 1982), intravenous administration of CCK-33 nonsignificantly reduced meal size by an average of 20% in eight obese women (Lieverse et al., 1995), and a CCK_A receptor agonist produced a dose-dependent reduction in 24-h food intake in obese women (Little et al., 2005). Another strategy, indirectly increasing prandial CCK secretion by oral administration of a proteinase inhibitor, which slows degradation of intraluminal CCK releasing factor, increased plasma CCK levels and decreased food intake in lean men (Hill et al., 1990). Despite these encouraging results, the current status of CCK in obesity therapy remains as Smith and Gibbs (1985) stated over 20 years ago—existing experiments "are not definitive studies and their primary value is heuristic" (p. 419).

IV. AMYLIN

Amylin (or islet amyloid polypeptide) is synthesized by the pancreatic B cells and cosecreted with insulin. It is considered a necessary complementary factor to insulin in the control of nutrient flux and postprandial glucose concentration (Weyer et al., 2001). Among the functions that contribute to this role are inhibitions of gastric acid secretion, gastric emptying, pancreatic glucagon secretion, digestive enzyme secretion, and eating. The best investigated of these functions is amylin's role as a hormonal control of meal-ending satiation (Lutz, 2005; Lutz et al., 1994, 1995). Acute intraportal or intraperitoneal injection of amylin rapidly and dose-dependently inhibits eating and reduces meal size with no effect on the following intermeal interval. The effect is behaviorally specific and of brief duration of action. In addition, unlike CCK, amylin also may act as an adiposity signal, such as leptin and insulin are hypothesized to do. That is, basal blood concentration of these hormones are each correlated with the total amount of body fat, and there is evidence that the brain senses these concentrations via specific receptors in distinct brain regions and that this information is used as a feedback signal controlling energy homeostasis, that is, food intake and energy expenditure (Woods, 2005; Woods et al., 2006).

A. Physiological Relevance of Amylin in the Control of Eating

Amylin has been shown to meet most of the criteria summarized in Table 3 for a physiological satiation signal in rats.

1. *Secretion.* In humans, rats, and other species, food intake results in a marked increase in plasma amylin concentration that begins within 5 min. The magnitude of this increase is correlated with the amount eaten (Butler et al., 1990; Lutz et al., 1997).

2. *Receptors.* Functional amylin receptors consist of the calcitonin receptor (CT-R) as a core receptor, together with several receptor-activity modifying proteins (RAMPs) that confer amylin affinity and selectivity to the receptor complex. As described later, amylin appears to act in the AP to inhibit eating, and all the components of the amylin receptor complex appear to be expressed in the AP.

3. *Physiological dose.* The lowest dose of exogenous amylin that significantly inhibited eating yielded plasma amylin levels that were about two times higher than those measured postprandially (Arnelo et al., 1998). Strictly speaking, therefore, this criterion has not been fulfilled. The difference, however, is not large and may be artifactual. For example, the different kinetics of the plasma amylin concentrations after exogenous amylin delivery and after endogenous secretion may have affected the results because relative changes in the plasma amylin concentration rather than absolute concentrations may affect eating.

4. *Removal and replacement.* This criterion has been difficult to test because neither surgical (pancreatectomy) nor chemical (the B-cell toxin streptozotocin) removal of pancreatic B cells is specific. Indirect evidence is provided by the amylin knockout mouse, which shows the expected phenotype of overeating and increased body weight (Devine and Young, 1998; Lutz, 2005). Replacement of physiological doses of amylin in these mice has not yet been performed.

5. *Antagonism.* Peripherally or centrally delivered amylin antagonists produce

an effect opposite to that of amylin, that is, an increase in meal size. The most conclusive of these studies used the amylin receptor antagonist AC187 (Mollet et al., 2004; Reidelberger et al., 2004). When AC187 was administered into the AP, it blocked the anorectic effect of intraperitoneally injected amylin and, when administered alone stimulated eating and intake by an increase in meal size (Mollet et al., 2004). In summary, amylin appears to act physiologically as a satiating hormone in rats and, like CCK, has a fully coupled endocrine action. Amylin's role in satiation in humans has not yet been investigated in detail.

B. Amylin As an Adiposity Signal

Amylin appears to meet all the main criteria for an adiposity signal in rats (Lutz, 2005; Woods et al., 2006): (1) The basal plasma level of amylin is positively correlated with body weight or, more accurately, with body adiposity (Pieber et al., 1994). (2) Chronic continuous amylin infusion reduces food intake and body weight (Arnelo et al., 1996; Isaksson et al., 2005; Lutz et al., 2001a, Rushing et al., 2001), which occurred mainly because spontaneous meal size was decreased (Lutz et al., 2001a; but also see Arnelo et al., 1996). (3) Chronic peripheral or central infusion of amylin antagonists increased body weight and body fat mass, with little effect on lean body mass (Lutz et al., 2001a; Reidelberger et al., 2004; Rushing et al., 2001). In addition, body weight is significantly higher in the amylin knockout mice than in wild-type controls (Devine and Young, 1998; Lutz, 2005). Therefore, amylin may be considered a putative adiposity signal, like leptin or insulin.

C. Peripheral and Central Mechanisms of Amylin Action

As mentioned previously, the functional amylin receptor involves the CT-R as a core receptor and several RAMPs, which participate in ligand binding and seem to regulate both the transport of the core receptors to the cell surface and their glycosylation state, thereby determining ligand specificity (Fischer et al., 2002; McLatchie et al., 1998; Muff et al., 2004). The typical amylin receptor arises from the interaction of RAMP1 or RAMP3 with the CT-R. Amylin binding sites are widely distributed in the central nervous system and occur in high densities in the AP (Sexton et al., 1994). Both RAMP1 and RAMP3 mRNA have been discovered in the mouse AP (Ueda et al., 2001), and amylin-induced Fos mRNA and RAMP3 mRNA expression are colocalized in rat AP cells (Barth et al., 2004), as are amylin-sensitive AP neurons and neurons carrying the CT-R (Becskei et al., 2004).

A wealth of experimental evidence indicates that the effects of peripheral amylin administration on eating are mediated by amylin receptors in the AP (Lutz et al., 1998a, 2001a; Mollet et al., 2004; Riediger et al., 2002). Blood-borne amylin has easy access to the AP due to the lack of a functional blood-brain barrier. The effects of endogenous amylin on eating also appear to arise from AP amylin receptors because direct infusion of the amylin receptor antagonist AC187 into the AP increased food intake (Mollet et al., 2004). The anorectic actions of both acute and chronic peripheral amylin in rats were eliminated in rats with lesions in the AP/NTS region (Lutz et al., 1998a, 2001a). Vagal and nonvagal visceral afferents, in contrast, do not seem to be involved (Lutz et al., 1998b; Morley et al., 1994). Further, both peripheral amylin and postdeprivation refeeding increase neuronal activation in the AP, as gauged by expression of Fos protein, and both responses were effectively blocked by AC187 (Riediger et al., 2001, 2002, 2004). Finally, direct application of amylin onto AP neurons in slice preparations led to dose-dependent increases in their activity (Riediger et al., 2001, 2002).

Amylin receptors located more rostrally may also contribute to the anorectic action of amylin. Amylin infusion into the third cerebral ventricle produces a potent and long-lasting reduction in feeding, and third ventricular infusion of AC187 increases food intake and body weight in rats (Rushing et al., 2001). Whether humoral amylin accesses these receptors, however, remains unclear.

Peripheral amylin elicits a positive Fos response in a number of sites rostral to the AP, including the NTS, the lateral parabrachial nucleus (lPBN), the central nucleus of the amygdala (CeNA), and the bed nucleus of the stria terminalis (BNST) (Riediger et al., 2004; Rowland and Richmond, 1999). The AP, NTS, and the lPBN are necessary to convey amylin's anorectic signal to higher brain structures. The activation of these areas is secondary to an action of amylin in the AP because no Fos response was observed in AP-lesioned animals (Riediger et al., 2004; Rowland and Richmond, 1999).

The same areas have been implicated in the processing of other anorectic signals, such as CCK. Different neurotransmitters, and therefore at least partially different neuronal networks, however, may mediate the effects of the different peptides. For example, whereas CCK satiation involves the central serotoninergic system, the dopaminergic and histaminergic systems have been implicated in amylin signaling (Lutz et al., 1996b, 2001b; Mollet et al., 2001). Finally, the anorectic action of amylin may be due at least in part to reduced expression of orexin and melanin-concentrating hormone in the lateral hypothalamic area (Barth et al., 2003).

D. Interactions of Amylin and Other Signals Controlling Eating

The synergistic interaction between CCK and amylin noted in Table 4 may reflect a necessary part of CCK signaling because amylin antagonists attenuated CCK's anorectic action in rats (Lutz et al., 1996a, 2000a). Indeed, in the complete absence of endogenous amylin, that is, in amylin knockout mice, the anorectic effect of CCK was almost completely abolished, but could be restored by small doses of amylin (Mollet et al., 2003). In contrast, CCK signaling seems not to be required for amylin satiation as CCK antagonists had no effect on amylin-induced satiation (Morley et al., 1994). With the exception of a report that amylin synergizes with insulin in the inhibition of eating (Rushing et al., 2000), few other functional interactions with amylin have been tested.

E. Amylin and Obesity Treatment

1. Animal Models

Amylin's effect to reduce food intake has been tested in a few animal models of obesity, including the *ob/ob* mouse, the obese Zucker *fa/fa* rat, and the melanocortin-4 receptor knockout mouse (Eiden et al., 2002; Grabler and Lutz, 2004; Morley et al., 1994). In each model, obese animals still ate less after peripheral administration of amylin or its agonist, salmon calcitonin (sCT). In one (Eiden et al., 2002) but not another (Morley et al., 1994) study of *ob/ob* mice, amylin was less sensitive in the knockout than in the wild-type controls. In addition, obese Zucker *fa/fa* rats ate more after peripheral AC187 (Grabler and Lutz, 2004). The latter effect is especially interesting given that these rats display basal hyperamylinemia, probably in parallel to hyperinsulinemia induced by insulin resistance. These promising data indicate that further basic research on amylin in obesity is warranted. Indeed, amylin seems to be an especially interesting pharmacological target for obesity because it shares characteristics of typical meal-related, satiating hormones and also of adiposity signals.

2. Amylin and Human Obesity

Studies suggest that amylin can reduce food intake and body weight in overweight patients. Much of this work arises from the recognition that type 1 and severe type 2 diabetes mellitus is a disease not only of insulin but also of amylin deficiency, so that supplementing amylin as well as insulin might provide better control of glucose metabolism than insulin treatment alone. Amylin has three effects that suggest its usefulness in the treatment of diabetes mellitus: first, amylin inhibits eating; second, amylin inhibits glucagon secretion, which corrects for postprandial hyperglucagonemia in diabetics; and third, amylin reduces the rate of gastric emptying, which is increased in diabetes (Weyer et al., 2001; Young and Denaro, 1998). Chronic amylin treatment per se, however, is contraindicated as it leads to formation of amyloid plaques and type 2 diabetes. Pramlintide is a nonamyloidogenic analog that has been used in several clinical studies (Weyer et al., 2001) and is now approved for treatment of human diabetes.

In many patients with type 2 diabetes mellitus, insulin therapy to produce near-normoglycemia leads to an increase in body weight. When such patients were treated with insulin plus pramlintide, less insulin was required, better glycemic control was achieved, and patients lost weight. Body weight was decreased most in the most seriously obese patients (body mass index (BMI) >40; Hollander et al., 2004). Another study indicates that acute pramlintide treatment decreases the size of test meals by about 20% in both type 2 diabetics and non-diabetic obese individuals without disturbing normal perceptions of fullness or causing nausea or other side effects (Chapman et al., 2005). Pramlintide did produce some reports of nausea in another study (Hollander et al., 2004), but this was transient (i.e., <4 weeks) and mild. Chronic pramlintide treatment did not produce signs of cardiovascular, hepatic, or renal toxicity, nor did it have adverse effects on the counterregulatory hormonal, metabolic, and symptomatic responses to hypoglycemia in type 1 diabetics (Amiel et al., 2005). This is consistent with studies in rats showing that amylin did not decrease

gastric emptying or inhibit glucagon secretion during hypoglycemia (Gedulin and Young, 1998; Young, 1997). These data suggest that pramlintide might be appropriate and effective in the treatment of obesity in both nondiabetic and diabetic patients.

Salmon calcitonin (sCT) is another potentially useful amylin agonist. sCT, unlike mammalian CT, binds irreversibly to amylin receptors and has a markedly prolonged anorectic action in rats compared with amylin (Fischer et al., 2002; Lutz et al., 2000b; Riediger et al., 2001). sCT is also a potent agonist of human calcitonin for its effects on bone metabolism, and it has been used therapeutically for many years in the treatment of osteoporosis. We know of no reports that sCT causes anorexia as a side effect, however, perhaps because patients receiving sCT treatment for osteoporosis are typically very lean elderly women. At least in rats, chronic amylin or sCT mainly decreased body fat mass with no effect on lean body mass (e.g., Rushing et al., 2001).

RAMPs, described earlier, may be another potential avenue for development of amylin-related obesity therapy. Even though these proteins may also play a role in other G-protein coupled receptors (Hay et al., 2006), specific manipulation of RAMP signaling may be a means to enhance the action of amylin with relatively little side effects.

V. GHRELIN

The 28-amino acid polypeptide ghrelin is expressed mainly in stomach mucosal cells. Ghrelin was discovered in 1999 during a search for an endogenous ligand for the growth hormone secretagogue receptor (GHSR). Since then, the biological functions of ghrelin have been intensively investigated (Kangawa and Kojima, 2005; Kojima et al., 1999; Ueno et al., 2005; van der Lely et al., 2004). Perhaps the peptide's most interesting aspect regards the control of eating and energy balance (Geary, 2004a; Kojima et al., 1999; Kojima and Kangawa, 2005; Ueno et al., 2005; van der Lely et al., 2004; Williams and Cummings, 2005). Ghrelin is unique in that, first, it is the only gut peptide whose secretion is stimulated by a reduction in gastrointestinal contents and inhibited by eating (van der Lely et al., 2004; Williams and Cummings, 2005); and second, it is the only gut peptide whose administration stimulates eating, which it does in both rats and humans (Inui, 2001; Nakazato et al., 2001; Wren et al., 2001).

A. Physiological Relevance of Ghrelin in the Control of Eating

Ghrelin potently stimulates food intake and induces weight gain following peripheral or central administration (Inui, 2001; Tschöp et al., 2000; Wren et al., 2000). The increase in food intake seems to be mainly due to a decrease in the latency to eat and an increase in meal number. Ghrelin's effect on food intake appears more pronounced in younger than older rats (Gilg and Lutz, 2006), suggesting a role in coordinating the extensive need for energy and nutrients during phases of rapid body growth.

The physiological status of ghrelin action has not yet been fully investigated.

1. *Secretion*. In humans, rats, and other species, the synthesis and secretion of ghrelin is significantly affected by meals, such that plasma ghrelin levels rise shortly before meals are initiated and fall rapidly when food is consumed (although there is also a nocturnal rise and fall in plasma ghrelin that is not associated with meals). A close correspondence between the postprandial profiles of plasma ghrelin levels and hunger ratings has been reported (Cummings et al., 2004). The stimuli that lead to the premeal rise in plasma ghrelin are unknown but it is well-established that introduction of nutrients into the gastrointestinal tract causes plasma ghrelin to decrease (Cummings et al., 2001; Shiiya et al., 2002; Toshinai et al., 2001). This seems to involve postgastric stimuli, and possibly glucose, insulin, and perhaps amylin contribute to this down regulation.

2. *Receptors*. The orexigenic effect of ghrelin is presumed to be mediated by ghrelin interaction with the GHSRs. These receptors occur in several brain areas which have been implicated in ghrelin's effect on food intake such as the hypothalamic arcuate nucleus and the brainstem (Tschöp et al., 2000; Yokote et al., 1998), but they may also be present in peripheral vagal afferents which were originally presumed to mediate ghrelin's effects on eating (Asakawa et al., 2001).

3. *Physiological dose*. The physiological role of ghrelin in controlling food intake has not yet been defined because it is unknown whether mimicking physiological ghrelin levels and especially the physiological preprandial rise in circulating ghrelin will trigger eating. Ghrelin knockout mice, however, do not show the expected phenotype (De Smet et al., 2006; Sun et al., 2003). The matter of mimicking the physiological situation is complicated in that gastric mucosa cells are not the only source of ghrelin. Hypothalamic ghrelinergic neurons have been described (Cowley et al., 2003) but their contribution to the control of eating by ghrelin is unknown. Mimicking their release pattern at the site of action would be technically very difficult.

4. *Removal and replacement*. The only available method to remove peripheral ghrelin secreting cells is gastrectomy, which clearly has numerous other effects. The ghrelin knockout mouse did not exhibit the expected lean phenotype (Sun et al., 2003). Thus, this criterion remains unmet for ghrelin.

5. *Antagonism*. Studies regarding the physiological relevance of ghrelin's effect have also been hampered by the lack of potent and specific ghrelin antagonists (Zhao, 2005). An interesting more recent approach is the use of specific ghrelin RNA Spiegelmers (SPM)—modified RNA oligonucleotides. These SPM, presumably by binding circulating ghrelin and hence lowering free circulating ghrelin, blocked the orexigenic action of ghrelin (Helmling et al., 2004; Kobelt et al., 2006). The SPM strategy may therefore constitute a powerful tool for future studies on the physiological role of ghrelin.

To summarize, although ghrelin has been shown to trigger eating in part by reducing the latency to eat, the physiological importance of these findings remains unclear. Recently, new tools have emerged which may help to address the open questions regarding the role of endogenous ghrelin in the control of eating.

B. Ghrelin As an Adiposity Signal

Ghrelin is also implicated in the control of body weight: (1) Endogenous ghrelin levels are inversely correlated with body weight in humans and rats, although diet composition also appears to play an important role in determining ghrelin levels independent of body weight (Cummings et al., 2002; Williams and Cummings, 2005); (2) Chronic ghrelin administration increases adiposity and body weight in mice and rats, although the relative contributions of eating and energy expenditure remain unclear. (Nakazato et al., 2001; Tschöp et al., 2000). Ghrelin may interact with both leptin and insulin as an adiposity signal (Riediger et al., 2003; Träbert et al., 2002; Willesen et al., 1999; Williams and Cummings, 2005). Interestingly, ghrelin levels are often dissociated from body weight after bariatric surgery, depending on the technique used (Williams and Cummings, 2005). For example, patients who lost about 35% body weight after Roux-en-Y bypass surgery had plasma ghrelin levels about one-fourth or less of those in normal weight controls and showed no prandial rhythms (Cummings et al., 2002). In summary, the hypothesis that ghrelin is an adiposity signal controlling eating is interesting but as yet unproved.

C. Peripheral and Central Mechanisms of Ghrelin Action

Ghrelin's primary site of action remains controversial. Whether and how communication between the three putative sites (vagal afferents, Arc, brainstem) is important for the action of ghrelin remains completely unclear at present. In the original studies describing the orexigenic effect of ghrelin, subdiaphragmatic vagotomy or chemical vagal deafferentation with capsaicin has been reported to block the ability of peripherally administered ghrelin to stimulate food intake (Asakawa et al., 2001; Date et al., 2002). These data suggest a peripheral site of action and vagal mediation of the peptide's action. Consistent with this view are demonstrations of the presence of GHS-R within the vagus and of ghrelin-induced suppression of overall activity within the gastric vagal branches (Asakawa et al., 2001). Subsequently, support for a central site of action has come from experiments demonstrating that central ghrelin administration stimulates food intake at markedly lower doses than does peripheral ghrelin (Kobelt et al., 2006; Kojima et al., 2001; Tschöp et al., 2000). Further, more recent studies employing a selective surgical lesioning method to ablate abdominal vagal afferents did not demonstrate a difference between lesioned and nonlesioned animals in ghrelin's stimulatory effect on food intake (Arnold et al., in press). Hence, these studies lend support to the idea that ghrelin has a direct central mode of action, most likely via the arcuate nucleus (Arc).

The Arc is not the only brain site containing GHSRs is involved in the control of food intake. The receptors have also been found in the hindbrain, which is consistent with studies showing that hindbrain administration of ghrelin strongly stimulates food intake (Faulconbridge et al., 2003, 2005; own unpublished observation). Ghrelin activates neurons in the AP that contain GHSRs (Lawrence et al., 2002; Yokote et al., 1998), and a specific lesion of the AP may prevent peripheral ghrelin from increasing food intake at least under certain experimental conditions (Gilg and Lutz, 2006). An interesting observation in the latter study was that though an AP lesion completely blocked the orexigenic effect of ghrelin, the ghrelin-induced increase in body weight was still present in AP-lesioned rats. This seems contradictory at first sight, but it may indicate that ghrelin controls food intake and energy expenditure differentially. While both effects may contribute to the ghrelin-induced increase in body weight, it could be hypothesized that the AP at least in part mediates the orexigenic effect of ghrelin whereas a direct effect of ghrelin, perhaps on the Arc, underlies its effect on energy expenditure. To summarize, at least three potential sites of action are under discussion regarding ghrelin's stimulatory effect on food intake. Because ghrelin is also expressed in the brain, the effects of ghrelin that have been observed after central administration may involve different mechanisms than those following peripheral administration.

D. Ghrelin and Obesity Treatment

1. Animal Models

Whether chronic ghrelin antagonism reduces food intake and body weight in animals remains unknown because of the lack of specific and potent ghrelin antagonists (Zhao, 2005). However, the specific ghrelin SPM (Kobelt et al., 2006) discussed previously may prove useful for future chronic studies. Another strategy that has produced promising results is active anti-ghrelin immunization using ghrelin

coupled to virus-like particles (Lutz et al., 2005). Vaccinated rats and mice showed slightly decreased rate of body weight gain (Lutz et al., 2005). Interestingly, in vaccinated rats, the orexigenic effect of peripheral, but not central, ghrelin was blocked. Hence, it is possible that blockade of the endogenous ghrelin system can be used therapeutically to reverse obesity.

2. Ghrelin and Human Obesity

Although human obesity is usually associated with low blood levels of ghrelin, hyperphagic, obese patients with Prader-Willi syndrome have highly elevated ghrelin levels (Del Parigi et al., 2002; Haqq et al., 2003; van der Lely et al., 2004). Whether elevated ghrelin causes the increased appetite and obesity in these patients or whether anti-ghrelin treatment can effectively treat human obesity, however, remains unclear.

VI. POTENTIALS AND PROBLEMS OF GUT–BRAIN AXIS SIGNALS IN THE TREATMENT OF OBESITY

The first sections of this chapter described progress in elucidating the physiological roles of four gut–brain signals in the control of eating and briefly reviewed research aimed at establishing their utility in the treatment of obesity. The four signals were chosen to illustrate some general principles of the role of the gut–brain axis in the control of eating as well as its therapeutic potential in the treatment of obesity. We conclude here with a discussion of several more general issues related to the gut–brain axis as well as other potential physiological bases for obesity therapy.

First, given the physiological perspective emphasized here, it should be reemphasized that progress in physiology and in pharmacotherapy need not go hand in hand. Purely *physiological* relevance of a particular signal either in normal eating or in the pathophysiology of overeating and obesity is not necessarily a deciding factor in determining its *pharmacotherapeutic* utility. Related to that point, numerous examples make clear that results from one species cannot be applied to others with certainty. For example, species-specific differences in the distribution of CCK receptors greatly limit the generalizability of animal models of CCK's role in the control of eating and adiposity.

Another general point, relevant to all the gut–brain signals reviewed and many other controls as well, is that clinical research is not far advanced. In particular, no gut–brain axis control applied to obesity has yet fulfilled the classic clinical issues of efficacy, safety, and potency of long-term manipulations in various patient groups. In regard to the control of eating, it is easy to imagine that innate or learned (especially social or cultural practices) controls of eating may antagonize particular physiological or pharmacological treatments in at least some patient groups, especially over the long term.

Nevertheless, gut–brain axis signals, as well as other peripheral signals, offer several advantages and their potentials should not be overlooked. First, many details of the physiology of several gut–brain controls have been established, at least in animals (the physiology of CCK satiation is perhaps most advanced in humans; Geary, 2004a). This means that at least slightly more educated choices can be made regarding how to manipulate these controls therapeutically and that the types and severity of side effects might be more predictable. Peripheral side effects are often more tractable to treatment as well. In addition, peripheral signals can be manipulated peripherally, which is in general more accessible, selective, predictable, and perhaps more safe than manipulation of central neurotransmitter systems.

Gut–brain axis manipulation often causes gastrointestinal dysfunction and nausea as side effects. CCK, amylin, and other gut–brain signals certainly can,

however, at least acutely reduce eating in the absence of such side effects, and the amylinergic drug pramlintide has been shown to reduce eating and improve metabolism chronically without lasting gastrointestinal side effects. Whether the same is true of other gut–brain axis hormones such as GLP-1 or its analogs, for example, exendin-4, or peptide YY (PYY) is less clear. Although GLP-1, exendin-4, and PYY potently reduce food intake, this effect is often accompanied by presumably gastrointestinal side effects leading to conditioned taste aversions in animals or nausea in humans (e.g., Boggiano et al., 2005; Degen et al., 2005).

Another advantage of the hormonal gut–brain signals reviewed here (as well as many others) is that they are what have been called fully coupled hormonal controls (Geary, 2004a). This means that the hormones are released in tight temporal linkage with a particular eating-related stimulus and that their behavioral action occurs in similarly tight linkage with their release. This indicates that ideal timing of agonistic or antagonistic therapies can be predicted—for example, perhaps a ghrelin antagonist should be applied several hours before meals, whereas a CCK or amylin agonist should be applied immediately before eating.

A potential disadvantage of brain–gut axis signals mediated by peptide hormones is that the peptide or peptidergic agonists or antagonists cannot be administered orally due to the efficient digestion of peptides. This problem can be countered by the increasing number of nonpeptide drugs that affect peptide receptors or by the development of other routes of administration, for example, transdermal or intranasal routes (e.g., Simmons et al., 1994). Increasing recognition of the medical severity of obesity may soon be sufficient to justify injections as a route of administration.

Not all gut–brain signals, however, act peripherally. Amylin's action in the AP does not pose a special challenge, as this site has a very permeable blood-brain barrier. The action of ghrelin in the Arc, in contrast, means that manipulations aimed at ghrelin receptors must penetrate the blood-brain barrier (the blood-brain barrier of the Arc is not as leaky as once supposed, and insulin, leptin, and many other peripheral signals acting in the Arc reach it by special transport systems (Banks, 2003, Banks et al., 2004)). This is at once a challenge and an opportunity—the latter because manipulation of blood-brain barrier transport mechanisms can be an effective way to increase or decrease the penetration of hormones into the brain.

The controls of eating in animals and humans are famously multifactorial and redundant. The implications of this for pharmacotherapy are ambiguous. On the one hand, it may be that manipulations of individual signals will be ineffective because of functionally antagonistic, adaptive responses of others, as perhaps in the case of chronic CCK administration in free-feeding rats. On the other hand, the multiplicity of signals creates the potential for therapies based on simultaneous manipulation of several signals. We have emphasized numerous such synergistic interactions here. "Cocktail" therapies based on such interactions, besides potentially increasing the influence on a single aspect of eating, such as meal size, may provide a means to counteract side effects such as the sorts of antagonistic adaptive responses mentioned earlier. For example, perhaps a ghrelin-based therapy could be used to prevent increases in meal frequency from neutralizing decreases in meal size produced by a CCK- or amylin-based therapy.

Meal size in particular seems to be under the control of many signals (Geary 2004a; Smith, 1999; Stanley et al., 2005; Strader and Woods 2005; Woods, 2004, 2005; Woods et al., 2006). One potential reason for this is that avoiding an excessive meal size is very important. Meals, especially large meals, are clearly major physiological stresses

(Woods, 1991) and at some point may be unadaptive. This reasoning suggests that the controls of meal size, including many gut–brain controls, may indeed be very potent and are likely to provide very efficacious therapeutic levers. It is also interesting to note that this consideration is quite the reverse of the more frequent argument that adiposity signals rather than satiation signals or other meal controls provide the better obesity therapy. A related reason not to underemphasize the potential of meal controls in obesity therapy is the increasing evidence that the (neo)classic weight-regulatory system based on leptin, insulin, and other adiposity signals seems in fact more designed to stimulate eating and conserve energy when adiposity levels are too low than to inhibit eating and expend energy when adiposity increases (reviewed in Schwartz et al., 2003). The former situation (low level of adiposity) is likely to have prevailed throughout most periods of our evolution, while the latter situation (high level of adiposity) is considered to be a relatively new phenomenon for which our phylogeny has prepared us poorly.

References

Amiel, S. A., Heller, S. R., Macdonald, I. A., Schwartz, S. L., Klaff, L. J., Ruggles, J. A., Weyer, C., Kolterman, O. G., and Maggs, D. G. (2005). The effect of pramlintide on hormonal, metabolic or symptomatic responses to insulin-induced hypoglycaemia in patients with type 1 diabetes. *Diabetes Obes. Metab.* 7, 504–516.

Arnelo, U., Permert, J., Adrian, T. E., Larsson, J., Westermark, P., and Reidelberger, R. D. (1996). Chronic infusion of IAPP causes anorexia in rats. *Am. J. Physiol.* 271, R1654–R1659.

Arnelo, U., Reidelberger, R., Adrian, T. E., Larsson, J., and Permert, J. (1998). Sufficiency of postprandial plasma levels of islet amyloid polypeptide for suppression of feeding in rats. *Am. J. Physiol.* 275, R1537–R1542.

Arnold, M., Mura, A., Langhans, W., and Geary, N. Subdiaphragmatic vagal afferents are not necessary for the feeding-stimulatory effect of intraperitoneally administered ghrelin. *J. Neurosci.* (in press). Gut vagal offerents are not necessary for the eating-stimulatory effect of intraperitoneally injected ghrelin in the rat.

Asakawa, A., Inui, A. A., Kaga, T., Yuzuriha, H., Nagata, T., Ueno, N., Makino, S., Fujimiya, M., Niijima, A., Fujino, M. A., and Kasuga, M. (2001). Ghrelin is an appetite-stimulatory signal from stomach with structural resemblance to motilin. *Gastroenterology* **120**, 337–345.

Asarian, L., and Geary, N. (2006). Modulation of appetite by gonadal steroid hormones. *Philos. Trans. R. Soc. B: Biol. Sci.* **361**, 1251–1263.

Ballinger, A., McLoughlin, L., Medbank, S., and Clark, M. (1995). Cholecystokinin is a satiety hormone in humans at physiological post-prandial concentrations. *Clin. Sci.* **89**, 375–381.

Banks, W. A. (2003). Is obesity a disease of the blood-brain barrier? Physiological, pathological, and evolutionary considerations. *Curr. Pharmaceut. Design* **9**, 801–809.

Banks, W. A., Coon, A. B., Robinson, S. M., Moinuddin, A., Shultz, J. M., Nakaoke, R., and Morley, J. E. (2004). Triglycerides induce leptin resistance at the blood-brain barrier. *Diabetes* **53**(5), 1253–1260.

Barth, S. W., Riediger, T., Lutz, T. A., and Rechkemmer, G. (2003). Differential effects of amylin and salmon calcitonin on neuropeptide gene expression in the lateral hypothalamic area and the arcuate nucleus of the rat. *Neurosci. Lett.* **341**, 131–134.

Barth, S. W., Riediger, T., Lutz, T. A., and Rechkemmer, G. (2004). Peripheral amylin activates circumventricular organs expressing calcitonin receptor a/b subtypes and receptor activity modifying proteins in the rat. *Brain Res.* **997**, 97–102.

Becskei, C., Riediger, T., Zünd, D., Wookey, P., and Lutz, T. A. (2004). Immunohistochemical mapping of calcitonin receptors in the adult rat brain. *Brain Res.* **1030**, 221–233.

Bednar, I., Carrer, H., Qureshi, G. A., and Södersten, P. (1995). Dopamine D1 or D2 antagonists enhance inhibition of consummatory ingestive behavior by CCK8. *Am. J. Physiol.* **269**, R896–R903.

Beglinger, C., and Degen, L. (2004). Fat in the intestine as a regulator of appetite—role of CCK. *Physiol. Behav.* **83**, 617–621.

Berenbaum, M. C. (1989). What is synergy? *Pharmacol. Rev.* **41**, 93–141.

Berger, K., Winzell, M. S., Mei, J., and Erlanson-Albertsson, C. (2004). Enterostatin and its target mechanisms during regulation of fat intake. *Physiol. Behav.* **83**, 623–630.

Berthoud, H. R. (2005). A new role for leptin as a direct satiety signal from the stomach. *Am. J. Physiol.* **288**, R796–R797.

Berthoud, H. R., Blackshaw, L. A., Brookes, S. J., and Grundy, D. (2004). Neuroanatomy of extrinsic afferents supplying the gastrointestinal tract. *Neurogastroenterol. Motil.* **16**(Suppl. 1), 28–33.

Bhavsar, S., Watkins, J., and Young, A. (1998). Synergy between amylin and cholecystokinin for inhibition of food intake in mice. *Physiol. Behav.* **64**, 557–561.

Bi, S., and Moran, T. H. (2002). Actions of CCK in the controls of food intake and body weight: lessons from the CCK-A receptor deficient OLETF rat. *Neuropeptides* **36**, 171–181.

Boggiano, M. M., Chandler, P. C., Oswald, K. D., Rodgers, R. J., Blundell, J. E., Ishii, Y., Beattie, A. H., Holch, P., Allison, D. B., Schindler, M., Arndt, K., Rudolf, K., Mark, M., Schoelch, C., Joost, H. G., Klaus, S., Thone-Reineke, C., Benoit, S. C., Seeley, R. J., Beck-Sickinger, A. G., Koglin, N., Raun, K., Madsen, K., Wulff, B. S., Stidsen, C. E., Birringer, M., Kreuzer, O. J., Deng, X. Y., Whitcomb, D. C., Halem, H., Taylor, J., Dong, J., Datta, R., Culler, M., Ortmann, S., Castaneda, T. R., and Tschop, M. (2005). PYY3-36 as an anti-obesity drug target. *Obes. Rev.* **6**, 307–322.

Brennan, I. M., Feltrin, K. L., Horowitz, M., Smout, A. J., Meyer, J. H., Wishart, J., and Feinle-Bisset, C. (2005). Evaluation of interactions between CCK and GLP-1 in their effects on appetite, energy intake, and antropyloroduodenal motility in healthy men. *Am. J. Physiol. Regul. Integr. Comp. Physiol.* **288**, R1477–R1485.

Butler, P. C., Chou, J., Carter, W. B., Wang, Y. N., Bu, B. H., Chang, P., Chang, J. K., and Rizza, R. A. (1990). Effects of meal ingestion on plasma amylin concentration in NIDDM and nondiabetic humans. *Diabetes* **39**, 752–765.

Cannon, C. M., and Palmiter, R. D. (2003). Norepinephrine is not required for reduction of feeding induced by cholecystokinin. *Am. J. Physiol.* **284**, R1384–R1388.

Chandler, P. C., Wauford, P. K., Oswald, K. D., Maldonado C. R., and Hagan, M. M. (2004). Change in CCK-8 response after diet-induced obesity and MC3/4-receptor blockade. *Peptides* **25**, 299–306.

Chapman, I., Parker, B., Doran, S., Feinle-Bisset, C., Wishart, J., Strobel, S., Wang, Y., Burns, C., Lush, C., Weyer, C., and Horowitz, M. (2005). Effect of pramlintide on satiety and food intake in obese subjects. *Diabetologia* **48**, 838–848.

Clegg, D. J., and Woods, S. C. (2004). The physiology of obesity. *Clin. Obstetr. Gynecol.* **47**, 967–979.

Covasa, M., Ritter, R. C., and Burns, G. A. (2004). NMDA receptor blockade attenuates CCK-induced reduction of real feeding but not sham feeding. *Am. J. Physiol.* **286**, R826–R831.

Cowley, M. A., Smith, R. G., Diano, S., Tschop, M., Pronchuk, N., Grove, K. L., Strasburger, C. J., Bidlingmaier, M., Esterman, M., Heiman, M. L., Garcia-Segura, L. M., Nillni, E. A., Mendez, P., Low M. J., Sotonyi, P., Friedman, J. M., Liu, H., Pinto, S., Colmers, W. F., Cone, R. D., and Horvath, T. L. (2003). *Neuron.* Feb 20, **37**(4), 649–661.

Cummings, D. E., Purnell, J. Q., Frayo, R. S., Schmidova, K., Wisse, B. E., and Weigle, D. S. A. (2001). Preprandial rise in plasma ghrelin levels suggests a role in meal initiation in humans. *Diabetes* **50**, 1714–1719.

Cummings, D. E., Weigle, D. S., Frayo, R. S., Breen, P. A., Ma, M. K., Dellinger, E. P., and Purnell, J. Q. (2002). Plasma ghrelin levels after diet-induced weight loss or gastric bypass surgery. *N. Engl. J. Med.* **346**, 1623–1630.

Cummings, D. E., Frayo, R. S., Marmonier, C., Aubert, R., and Chapelot, D. (2004). Plasma ghrelin levels and hunger scores in humans initiating meals voluntarily without time- and food-related cues. *Am. J. Physiol. Endocrinol. Metab.* **287**, E297–E304.

Cummings, D. E., Foster-Schubert, K. E., and Overduin, J. (2005). Ghrelin and energy balance: focus on current controversies. *Curr. Drug Targets* **6**, 153–169.

Date, Y., Murakami, N., Toshinai, K., Matsukura, S., Niijima, A., Matsuo, H., Kangawa, K., and Nakazato, M. (2002). The role of the gastric afferent vagal nerve in ghrelin-induced feeding and growth hormone secretion in rats. *Gastroenterology* **123**, 1120–1128.

Degen, L., Oesch, S., Casanova, M., Graf, S., Ketterer, S., Drewe, J., and Beglinger, C. (2005). Effect of peptide YY(3-36) on food intake in humans. *Gastroenterology* **129**, 1430–1436.

Del Parigi, A., Tschop, M., Heiman, M. L., Salbe, A. D., Vozarova, B., Sell, S. M., Bunt, J. C., and Tataranni, P. A. (2002). High circulating ghrelin: a potential cause for hyperphagia and obesity in Prader-Willi syndrome. *J. Clin. Endocrinol. Metab.* **87**, 5461–5464.

De Smet, B., Depoortere, I., Moechars, D., Swennen, Q., Moreaux, B., Cryns, K., Tack, J., Buyse, J., Coulie, B., and Peeters, T. (2006). Energy homeostasis and gastric emptying in ghrelin knockout mice. *J. Pharmacol. Exp. Ther.* Oct. 3, **316**, 431–439.

Devine, E., and Young, A. A. (1998). Weight gain in male and female mice with amylin gene knockout. *Diabetes* **47**, A317.

De Vries, J., Strubbe, J. H., Wildering, W. C., Gorter, J. A., and Prins, A. J. A. (1993). Patterns of body temperature during feeding in rats under varying ambient temperatures. *Physiol. Behav.* **53**, 229–235.

Di Bella, L., Tarozzi, G., Rossi, M. T., and Scalera, G. (1981a). Effect of liver temperature increase on food intake. *Physiol. Behav.* **26**, 45–51.

Di Bella, L., Tarozzi, G., Rossi, M. T., and Scalera, G. (1981b). Behavioral patterns proceeding from liver thermoreceptors. *Physiol. Behav.* **26**, 53–59.

Eiden, S., Daniel, C., Steinbrueck, A., Schmidt, I., and Simon, E. (2002). Salmon calcitonin—a potent inhibitor of food intake in states of impaired leptin signaling in laboratory rodents. *J. Physiol.* **541**, 1041–1048.

Eisen, S., Davis, J. D., Rauhofer, E., and Smith, G. P. (2001). Gastric negative feedback produced by volume and nutrient during a meal in rats. *Am. J. Physiol.* **281**, R1201–R1214.

Eisen, S., Phillips, R. J., Geary, N., Baronowsky, E. A., Powley, T. L., and Smith, G. P. (2005). Inhibitory effects on intake of cholecystokinin-8 and cholecystokinin-33 in rats with hepatic proper or common hepatic branch vagal innervation. *Am. J. Physiol.* **289**, R456–R462.

Faulconbridge, L. F., Cummings, D. E., Kaplan, J. M., and Grill, H. J. (2003). Hyperphagic effects of brainstem ghrelin administration. *Diabetes* **52**, 2260–2265.

Faulconbridge, L. F., Grill, H. J., and Kaplan, J. M. (2005). Distinct forebrain and caudal brainstem contributions to the neuropeptide Y mediation of ghrelin hyperphagia. *Diabetes* **54**, 1985–1993.

Fischer, J. A., Muff, R., and Born, W. (2002). Functional relevance of G-protein-coupled-receptor-associated proteins, exemplified by receptor-activity-modifying proteins (RAMPs). *Biochem. Soc. Trans.* **30**, 455–460.

Geary, N. (1999). Effects of glucagon, insulin, amylin and CGRP on feeding. *Neuropeptides* **33**, 400–405.

Geary, N. (2004a). Endocrine controls of eating: CCK, leptin, and ghrelin. *Physiol. Behav.* **81**, 719–733.

Geary, N. (2004b). On Gerard P. Smith's scientific character and thought. *Physiol. Behav.* **82**, 159–166.

Geary, N., Kissileff, H. R., Pi-Sunyer, F. X., and Hinton, V. (1992). Individual, but not simultaneous, glucagon and cholecystokinin infusions inhibit feeding in men. *Am. J. Physiol.* **262**, R975–R980.

Gedulin, B. R., and Young, A. A. (1998). Hypoglycemia overrides amylin-mediated regulation of gastric emptying in rats. *Diabetes* **47**, 93–97.

Gibbs, J., Young, R. C., and Smith, G. P. (1973). Cholecystokinin decreases food intake in rats. *J. Comp. Physiol. Psychol.* **84**, 488–495.

Gilg, S., and Lutz, T. A. (2006). The orexigenic effect of peripheral ghrelin differs between rats of different age and with different baseline food intake, and it may in part be mediated by the area postrema. *Physiol. Behav.* **87**, 353–359.

Grabler, V., and Lutz, T. A. (2004). Chronic infusion of the amylin antagonist AC 187 increases feeding in Zucker fa/fa rats but not in lean controls. *Physiol. Behav.* **81**, 481–488.

Gutzwiller, J. P., Degen, L., Matzinger, D., Prestin, S., and Beglinger, C. (2004a). Interaction between GLP-1 and CCK-33 in inhibiting food intake and appetite in men. *Am. J. Physiol.* **287**, 562–567.

Gutzwiller, J. P., Degen, L., Heuss, L., and Beglinger, C. (2004b). Glucagon-like peptide 1 (GLP-1) and eating. *Physiol. Behav.* **82**, 17–19.

Haqq, A. M., Farooqi, I. S., O'Rahilly, S., Stadler, D. D., Rosenfeld, R. G., Pratt, K. L., LaFranchi, S. H., and Purnell, J. Q. (2003). Serum ghrelin levels are inversely correlated with body mass index, age, and insulin concentrations in normal children and are markedly increased in Prader-Willi syndrome. *J. Clin. Endocrinol. Metab.* **88**, 174–178.

Hay, D. L., Poyner, D. R., and Sexton, P. M. (2006). GPCR modulation by RAMPs. Pharmacol. *Therapeut.* **109**, 173–197.

Hayes, M. R., Moore, R. L., Shah, S. M., and Covasa, M. (2004). 5-HT3 receptors participate in CCK-induced suppression of food intake by delaying gastric emptying. *Am. J. Physiol.* **287**, R817–R823.

Helmling, S., Maasch, C., Eulberg, D., Buchner, K., Schroder, W., Lange, C., Vonhoff, S., Wlotzka, B., Tschop, M. H., Rosewicz, S., and Klussmann, S. (2004). Inhibition of ghrelin action in vitro and in vivo by an RNA-Spiegelmer. *Proc. Natl. Acad. Sci. U.S.A.* **7**(101), 13174–13179.

Hill, A. J., Peikin, S. R., Ryan, C. A., and Blundell, J. E. (1990). Oral administration of proteinase inhibitor II from potatoes reduces energy intake in man. *Physiol. Behav.* **48**, 241–246.

Hollander, P., Maggs, D. G., Ruggles, J. A., Fineman, M., Shen, L., Kolterman, O. G., and Weyer, C. (2004). *Obes. Res.* **12**, 661–668.

Houpt, T. R., Houpt, K. A., and Swan, A. A. (1983). Duodenal osmoconcentration and food intake in pigs after ingestion of hypertonic nutrients. *Am. J. Physiol.* **245**, R181–R189.

Inui, A. (2001). Ghrelin: an orexigenic and somatotrophic signal from the stomach. *Nat. Rev.* **2**, 551–561.

Isaksson, B., Wang, F., Permert, J., Olsson, M., Fruin, B., Herrington, M. K., Enochsson, L., Erlanson-Albertsson, C., and Arnelo, U. (2005). Chronically administered islet amyloid polypeptide in rats serves as an adiposity inhibitor and regulates energy homeostasis. *Pancretology* **5**, 29–36.

Kaplan, J. M., and Moran, T. H. (2004). Gastrointestinal signaling in the control of food intake. *In* "Handbook of Behavioral Neurobiology, Volume 14, Neurobiology of Food and Fluid Intake" (E. Stricker and S. C. Woods, eds.), 2nd ed., pp. 275–306. Kluwer Academic, Plenum Publishers, New York.

Kissileff, H. R., Carretta, J. C., Geliebter, A., and Pi-Sunyer, F. X. (2003). Cholecystokinin and stomach distension combine to reduce food intake in humans. *Am. J. Physiol.* **285**, R992–R998.

Kobelt, P., Helmling, S., Stengel, A., Wlotzka, B., Andresen, V., Klapp, B. F., Wiedenmann, B., Klussmann, S., and Monnikes, H. (2006). Anti-ghrelin Spiegelmer NOX-B11 inhibits neurostimulatory and orexigenic effects of peripheral ghrelin in rats. *Gut.* **55**, 788–792.

Kojima, M., and Kangawa, K. (2005). Ghrelin: structure and function. *Physiol. Rev.* **85**, 495–522.

Kojima, M., Hosoda, H., Date, Y., Nakazato, M., Matsuo, H., and Kangawa K. (1999). Ghrelin is a GH-releasing acylated peptide from stomach. *Nature* **402**, 656–660.

Kojima, M., Hosoda, H., Matsuo, H., and Kangawa, K. (2001). Ghrelin: discovery of the natural endogenous ligand for the growth hormone secretagogue receptor. *Trends Endocrinol. Metab.* **12**, 118–122.

Korenkov, M., Sauerland, S., and Junginger, T. (2005). Surgery for obesity. *Curr. Opin. Gastroenterol.* **21**, 679–683.

Kral, T. V., and Rolls, B. J. (2004). Energy density and portion size: their independent and combined effects on energy intake. *Physiol. Behav.* **82**, 131–138.

Langhans, W. (2003). Role of the liver in the control of glucose-lipid utilization and body weight. *Curr. Opin. Clin. Nutr. Metab. Care* **6**, 449–455.

Larsson, L. I., and Rehfeld, J. F. (1978). Distribution of gastrin and CCK cells in the rat gastrointestinal tract. Evidence for the occurrence of three distinct cell types storing COOH-terminal gastrin immunoreactivity. *Histochemistry* **58**, 23–31.

Lawrence, C. B., Snape, A. C., Baudoin, F. M., and Luckman, S. M. (2002). Acute central ghrelin and GH secretagogues induce feeding and activate brain appetite centers. *Endocrinology* **143**, 155–162.

Le Roux, C. W., and Bloom, S. R. (2005). Peptide YY, appetite and food intake. *Proc. Nutr. Soc.* **64**, 213–216.

Le Sauter, J., and Geary, N. (1987). Pancreatic glucagon and cholecystokinin synergistically inhibit sham feeding in rats. *Am. J. Physiol.* **253**, R719–R725.

Lieverse, R. J., Jansen, J. M. B., van de Zwan, A., Samson, L., Masclee, A. M., and Lamers, C. B. H. W. (1993). Effects of a physiological dose of cholecystokinin on food intake and postprandial satiation in humans. *Regul. Pept.* **43**, 83–89.

Lieverse, R. J., Jansen, J. B., Masclee, A. A., and Lamers, C. B. (1995). Satiety effects of a physiological dose of cholecystokinin in humans. *Gut* **36**, 176–179.

Little, T. J., Horowitz, M., and Feinle-Bisset, C. (2005). Role of cholecystokinin in appetite control and body weight regulation. *Obes. Rev.* **6**, 297–306.

Lutz, T. A. (2005). Pancreatic amylin as a centrally acting satiating hormone. *Curr. Drug Targets* **6**, 181–189.

Lutz, T. A., Del Prete, E., and Scharrer, E. (1994). Reduction of food intake in rats by intraperitoneal injection of low doses of amylin. *Physiol. Behav.* **55**, 891–895.

Lutz, T. A., Geary, N., Szabady, M. M., Del Prete, E., and Scharrer, E. (1995). Amylin decreases meal size in rats. *Physiol. Behav.* **58**, 1197–1202.

Lutz, T. A., Del Prete, E., Szabady, M. M., and Scharrer, E. (1996a). Attenuation of the anorectic effects of glucagon, cholecystokinin, and bombesin by the amylin receptor antagonist CGRP(8-37). *Peptides* **17**, 119–124.

Lutz, T. A., Del Prete, E., Walzer, B., and Scharrer, E. (1996b). The histaminergic, but not the serotoninergic, system mediates amylin's anorectic effect. *Peptides* **17**, 1317–1322.

Lutz, T. A., Pieber, T. R., Walzer, B., Del Prete, E., and Scharrer, E. (1997). Different influence of CGRP (8-37), an amylin and CGRP antagonist, on the anorectic effects of cholecystokinin and bombesin in diabetic and normal rats. *Peptides* **18**, 643–649.

Lutz, T. A., Senn, M., Althaus, J., Del Prete, E., Ehrensperger, F., and Scharrer, E. (1998a). Lesion of the area postrema/nucleus of the solitary tract (AP/NTS) attenuates the anorectic effects of amylin and calcitonin gene-related peptide (CGRP) in rats. *Peptides* **19**, 309–317.

Lutz, T. A., Althaus, J., Rossi, R., and Scharrer, E. (1998b). Anorectic effect of amylin is not transmitted by capsaicin-sensitive nerve fibres. *Am. J. Physiol.* **274**, R1777–R1782.

Lutz, T. A., Tschudy, S., Rushing, P. A., and Scharrer, E. (2000a). Attenuation of the anorectic effects of cholecystokinin and bombesin by the specific amylin antagonist AC 253. *Physiol. Behav.* **70**, 533–536.

Lutz, T. A., Tschudy, S., Rushing, P. A., and Scharrer E. (2000b). Amylin receptors mediate the anorectic action of salmon calcitonin (sCT). *Peptides* **21**, 233–238.

Lutz, T. A., Mollet, A., Rushing, P. A., Riediger, T., and Scharrer, E. (2001a). The anorectic effect of a chronic peripheral infusion of amylin is abolished in area postrema/nucleus of the solitary tract (AP/NTS)-lesioned rats. *Int. J. Obes.* **25**, 1005–1011.

Lutz, T. A., Tschudy, S., Mollet, A., Geary, N., and Scharrer, E. (2001b). Dopamine D_2-receptors mediate amylin's acute satiety effect. *Am. J. Physiol.* **280**, R1697–R1703.

Lutz, T. A., Osto, M., Cettuzzi, B., Walser, N., Fulurija, A., Sladko, K., Saudan, P., and Bachmann, M. (2005). Ghrelin Immunodrug$^{(TM)}$ reduces body weight gain in male mice. *Appetite* **44**, 365.

McLatchie, L. M., Fraser, N. J., Main, M. J., Wise, A., Brown, J., Thompson, N., Solari, R., Lee, M. G., and Foord, S. M. (1998). RAMPs regulate the transport and ligand specificity of the calcitonin-receptor-like receptor. *Nature* **393**, 333–339.

Mollet, A., Lutz, T. A., Meier, S., Riediger, T., Rushing, P. A., and Scharrer E. (2001). Histamine H_1 receptors mediate the anorectic action of the pancreatic hormone amylin. *Am. J. Physiol.* **281**, R1442–R1448.

Mollet, A., Meier, S., Grabler, V., Gilg, S., Scharrer, E., and Lutz, T. A. (2003). Endogenous amylin contributes to the anorectic effects of cholecystokinin and bombesin. *Peptides* **24**, 91–98.

Mollet, A., Gilg, S., Riediger, T., and Lutz, T. A. (2004). Infusion of the amylin antagonist AC 187 into the area postrema increases food intake in rats. *Physiol. Behav.* **81**, 149–155.

Moran, T. H. (2004). Gut peptides in the control of food intake: 30 years of ideas. *Physiol. Behav.* **82**, 175–180.

Moran, T. H., and Kinzig, K. P. (2004). Gastrointestinal satiety signals. II. Cholecystokinin. *Am. J. Physiol.* **286**, 183–188.

Moran, T. H., Ameglio, P. J., Schwartz, G. J., and McHugh, P. R. (1992). Blockade of type A, but not type B, CCK receptors attenuates satiety actions of exogenous and endogenous CCK. *Am. J. Physiol.* **262**, R46–R50.

Morley, J. E., Flood, J. F., Horowitz, Morley, M. P. M. K., and Walter, M. J. (1994). Modulation of food intake by peripherally administered amylin. *Am. J. Physiol.* **267**, R178–R184.

Morton, G. J., Blevins, J. E., Williams, D. L., Niswender, K. D., Gelling, R. W., Rhodes, C. J., Baskin, D. G., and Schwartz, M. W. (2005). Leptin action in the forebrain regulates the hindbrain response to satiety signals. *J. Clin. Invest.* **115**, 703–710.

Muff, R., Born, W., Lutz, T. A., and Fischer J. A. (2004). Biological importance of the peptides of the calcitonin family as revealed by disruption and transfer of corresponding genes. *Peptides* **25**, 2027–2038.

Nakazato, M., Murakami, N., Date, Y., Kojima, M., Matsuo, H., Kangawa, K., and Matsukura, S. (2001). A role for ghrelin in the central regulation of feeding. *Nature* **409**, 194–198.

Noble, F., and Roques, B. P. (2002). Phenotypes of mice with invalidation of cholecystokinin (CCK1 or CCK2) receptors. *Neuropeptides* **36**, 157–170.

Phillips, R. J., and Powley, T. L. (1996). Gastric volume rather than nutrient content inhibits food intake. *Am. J. Physiol.* **271**, R766–R769.

Pieber, T. R., Roitelman, J., Lee, Y., Luskey, K. L., and Stein, D. T. (1994). Direct plasma radioimmunoassay for rat amylin-(1–37): concentrations with acquired and genetic obesity. *Am. J. Physiol.* **267**, E156–E164.

Pi-Sunyer, X., Kissileff, H. R., Thornton, J., and Smith, G. P. (1982). C-terminal octapeptide of cholecystokinin decreases food intake in obese men. *Physiol. Behav.* **29**, 627–630.

Poeschla, B. J., Gibbs, K. J., Simansky, K., Greenberg, D., and Smith, G. P. (1993). Cholecystokinin-induced satiety depends on activation of 5-HT1C receptors. *Am. J. Physiol.* **264**, R62–R64.

Reidelberger, R. D., Haver, A. C., Arnelo, U., Smith, D. D., Schaffert, C. S., and Permert, J. (2004). Amylin receptor blockade stimulates food intake in rats. *Am J. Physiol.* **287**, R568–R574.

Riediger, T., Schmid, H. A., Lutz, T. A., and Simon, E. (2001). Amylin potently activates AP neurons possibly via formation of the excitatory second-messenger cGMP. *Am. J. Physiol.* **281**, R1833–R1843.

Riediger, T., Schmid, H. A., Lutz, T. A., and Simon, E. (2002). Amylin and glucose coactivate area postrema neurons of the rat. *Neurosci. Lett.* **328**, 121–124.

Riediger, T., Traebert, M., Schmid, H. A., Scheel, C., Lutz, T. A., and Scharrer, E. (2003). Site-specific effects of ghrelin on the neuronal activity in the hypothalamic arcuate nucleus. *Neurosci. Lett.* **341**, 151–155.

Riediger, T., Zünd, D., Becskei, C., and Lutz, T. A. (2004). The anorectic hormone amylin contributes to feeding-related changes of neuronal activity in key structures of the gut-brain axis. *Am. J. Physiol.* **286**, R114–R122.

Rolls, B., and Barnett, R. A. (2000). "The Volumetrics Weight-Control Plan: Feel Full on Fewer Calories." Quill Publishers, New York.

Rowland, N. E., and Richmond, R. M. (1999). Area postrema and the anorectic actions of dexfenfluramine and amylin. *Brain Res.* **820**, 86–91.

Rushing, P. A., Lutz, T. A., Seeley, R. J., and Woods, S. C. (2000). Amylin and insulin interact to reduce food intake in rats. *Horm. Metabol. Res.* **32**, 62–65.

Rushing, P. A., Hagan, M. M., Seeley, R. J., Lutz, T. A., D'Alession, D. A., Air, E. L., and Woods, S. C. (2001). Inhibition of central amylin signaling increases food intake and body adiposity in rats. *Endocrinology* **142**, 5035–5038.

Savastano, D. M., Carelle, M., and Covasa, M. (2005). Serotonin-type 3 receptors mediate intestinal polycose- and glucose-induced suppression of intake. *Am. J. Physiol.* **288**, R1499–R1508.

Scharrer, E. (1999). Control of food intake by fatty acid oxidation and ketogenesis. *Nutrition* **15**, 704–714.

Schwartz, G. J. (2000). The role of gastrointestinal vagal afferents in the control of food intake: current prospects. *Nutrition* **16**, 866–873.

Schwartz, G. J., and Moran, T. H. (1996). Subdiaphragmatic vagal afferent integration of meal-related gastrointestinal signals. *Neurosci. Biobehav. Rev.* **20**, 47–56.

Schwartz, M. W., Woods, S. C., Seeley, R. J., Barsh, G. S., Baskin, D. G., and Leibel, R. L. (2003). Is the energy homeostasis system inherently biased toward weight gain? *Diabetes* **52**, 232–238.

Sexton, P. M., Paxinos, G., Kenney, M. A., Wookey, P. J., and Beaumont, K. (1994). In vitro autoradiographic localization of amylin binding sites in rat brain. *Neuroscience* **62**, 553–567.

Shiiya, T., Nakazato, M., Mizuta, M., Date, Y., Mondal, M. S., Tanaka, M., Nozoe, S., Hosoda, H., Kangawa, K., and Matsukura, S. (2002). Plasma ghrelin levels in lean and obese humans and the effect of glucose on ghrelin secretion. *J. Clin. Endocrinol. Metab.* **87**, 240–244.

Simmons, R. D., Blosser, J. C., and Rosamond, J. R. (1994). FPL 14294: a novel CCK-8 agonist with potent intranasal anorectic activity in the rat. *Pharmacol. Biochem. Behav.* **47**, 701–708.

Simmons, R. D., Kaiser, F. C., and Hudzik, T. J. (1999). Behavioral effects of AR-R 15849, a highly selective CCK-A agonist. *Pharmacol. Biochem. Behav.* **62**, 549–557.

Smith, G. P. (1996). The direct and indirect controls of meal size. *Neurosci. Biobehav. Rev.* **20**, 41–46.

Smith, G. P. (1999). Introduction to the reviews on peptides and the control of food intake and body weight. *Neuropeptide* **33**, 323–328.

Smith, G. P., and Gibbs, J. (1984). Gut peptides and postprandial satiety. *Fed. Proc.* **43**, 2889–2892.

Smith, G. P., and Gibbs, J. (1985). The satiety effect of cholecystokinin. Recent progress and current problems. *Ann. N.Y. Acad. Sci.* **448**, 417–423.

Stallone, D. S., Nicolaidis, S., and Gibbs, J. (1989). Cholecystokinin-induced anorexia depends on serotoninergic function. *Am. J. Physiol.* **256**, R1138–R1141.

Stanley, S., Wynne, K., McGowan, B., and Bloom, S. (2005). Hormonal regulation of food intake. *Physiol. Rev.* **85**, 1131–1158.

Strader, A. D., and Woods, S. C. (2005). Gastrointestinal hormones and food intake. *Gastroenterology* **128**, 175–191.

Strohmayer, A. J., and Smith, G. P. (1986). Obese male mice (ob/ob) are normally sensitive to the satiating effect of CCK-8. *Brain Res. Bull.* **17**, 571–573.

Sturm, K., Parker, B., Wishart, J., Feinle-Bisset, C., Jones, K. L., Chapman, I., and Horowitz, M. (2004). Energy intake and appetite are related to antral area in healthy young and older subjects. *Am. J. Clin. Nutr.* **80**,656–667.

Sun, Y., Ahmed, S., and Smith, R. G. (2003). Deletion of ghrelin impairs neither growth nor appetite. *Mol. Cell. Biol.* **22**, 7973–7981.

Torregrossa, A. M., and Smith, G. P. (2003). Two effects of high-fat diets on the satiating potency of cholecystokinin-8. *Physiol. Behav.* **78**, 19–25.

Toshinai, K., Mondal, M. S., Nakazato, M., Date, Y., Murakami, N., Kojima, M., Kangawa, K., and Matsukura, S. (2001). Upregulation of ghrelin expression in the stomach upon fasting, insulin-induced hypoglycemia, and leptin administration. *Biochem. Biophys. Res. Commun.* **281**, 1220–1225.

Träbert, M., Riediger, T., Whitebread, S., Scharrer, E., and Schmid, H. A. (2002). Ghrelin acts on leptin-responsive neurones in the rat arcuate nucleus. *J. Neuroendocrinol.* **14**, 580–586.

Tschöp, M., Smiley, D. L., and Heimann, M. L. (2000). Ghrelin induces adiposity in rodents. *Nature* **407**, 908 913.

Tso, P., and Liu, M. (2004). Apolipoprotein A-IV, food intake, and obesity. *Physiol. Behav.* **83**, 631–643.

Ueda, T., Ugawa, S., Saishin, Y., and Shimada, S. (2001). Expression of receptor-activity modifying protein (RAMP) mRNAs in the mouse brain. *Mol. Brain Res.* **93**, 36–45.

Ueno, H., Yamaguchi, H., Kangawa, K., and Nakazato, M. (2005). Ghrelin: a gastric peptide that regulates food intake and energy homeostasis. *Regul. Pept.* **126**, 11–19.

Van der Lely, A. J., Tschöp, M., Heiman, M. L., and Ghigo, E. (2004). Biological, physiological, pathophysiological, and pharmacological aspects of ghrelin. *Endocr. Rev.* **25**, 426–457.

West, D. B., Fey, D., and Woods, S. C. (1984). Cholecystokinin persistently suppresses meal size but not food intake in free-feeding rats. *Am. J. Physiol.* **246**, R776–R787.

Weyer, C., Maggs, D. G., Young, A. A., and Kolterman, O. G. (2001). Amylin replacement with pramlintide as an adjunct to insulin therapy in type 1 and type 2 diabetes mellitus: a physiologic approach toward improved metabolic control. *Curr. Pharmaceut. Design* **7**, 1353–1373.

Wildman, H. F., Chua, S., Jr., Leibel, R. L., and Smith, G. P. (2000). Effects of leptin and cholecystokinin in rats with a null mutation of the leptin receptor Lepr(fak). *Am. J. Physiol. Regul. Integr. Comp. Physiol.* **278**, R1518–R1523.

Willesen, M. G., Kristensen, P., and Romer, J. (1999). Co-localization of growth hormone secretagogue receptor and NPY mRNA in the arcuate nucleus of the rat. *Neuroendocrinology* **70**, 306–316.

Williams, D. L., and Cummings, D. E. (2005). Regulation of ghrelin in physiologic and pathophysiologic states. *J. Nutr.* **135**, 1320–1325.

Woods, S. C. (1991). The eating paradox: how we tolerate food. *Psychol. Rev.* **98**, 488–505.

Woods, S. C. (2004). Gastrointestinal satiety signals. I. An overview of gastrointestinal signals that influence food intake. *Am. J. Physiol.* **286**, G7–G13.

Woods, S. C. (2005). Signals that influence food intake and body weight. *Physiol. Behav.* **86**, 709–716.

Woods, S. C., Lutz, T. A., Geary, N., and Langhans, W. (2006). Pancreatic signals controlling food intake—insulin, glucagon and amylin. *Philos. Trans. R. Soc. B: Biol. Sci.* **361**, 1219–1235.

Wren, A. M., Small, C. J., Ward, H. L., Murphy, K. G., Dakin, C. L., Taheri, S., Kennedy, A. R., Roberts, G. H., Morgan, D. G. A., Ghatei, M. A., and Bloom, S. R. (2000). The novel hypothalamic peptide ghrelin stimulates food intake and growth hormone secretion. *Endocrinology* **141**, 4325–4328.

Wren, A. M., Small, C. J., Abbott, C. R., Dhillo, W. S., Seal, L. J., Cohen, M. A., Batterham, R. L., Taheri, S., Stanley, S. A., Ghatei, M. A., and Bloom, S. R. (2001).

Ghrelin causes hyperphagia and obesity in rats. *Diabetes* **50**, 2540–2547.

Yamada, K., Wada, E., and Wada, K. (2000). Bombesin-like peptides: studies on food intake and social behaviour with receptor knock-out mice. *Ann. Med.* **32**, 519–529.

Yokote, R., Sato, M., Matsubara, S., Ohye, H., Niimi, M., Murao, K., and Takahara, J. (1998). Molecular cloning and gene expression of growth hormone-releasing peptide receptor in rat tissues. *Peptides* **19**, 15–50.

Young, A. A. (1997). Amylin's physiology and its role in diabetes. *Curr. Opin. Endocrinol. Diabetes* **4**, 282–290.

Young, A., and Denaro, M. (1998). Roles of amylin in diabetes and in regulation of nutrient load. *Nutrition* **14**, 524–527.

Zhao, H. (2005). Growth hormone secretagogue receptor antagonists as potential therapeutic agents for obesity. *Drug Dev. Res.* **65**, 50–54.

Zheng, H., Patterson, L. M., Phifer, C. B., and Berthoud, H. R. (2005). Brain stem melanocortinergic modulation of meal size and identification of hypothalamic POMC projections. *Am. J. Physiol.* **289**, R247–R258.

Integration of Peripheral Adiposity Signals and Psychological Controls of Appetite

DIANNE FIGLEWICZ LATTEMANN
Metabolism Endocrinology, VA Puget Sound Health Care System
NICOLE M. SANDERS
Metabolism Endocrinology, VA Puget Sound Health Care System
AMY MacDONALD NALEID
Department of Psychiatry and Behavioral Sciences,
University of Washington
ALFRED J. SIPOLS
Institute for Experimental and Clinical Medicine,
Department of Medicine, University of Latvia

   I. Introduction and Overview
  II. Mesolimbic Dopamine Circuitry and Energy Regulatory Signals
 III. Brain Opioid Systems and Energy Regulatory Signals
 IV. Endocannabinoids and Energy Regulatory Signals
  V. LHA Circuitry and Energy Regulatory Signals
 VI. Other CNS Sites: Target for Future Studies?
VII. Human and Clinical Studies: At the Forefront of Our Knowledge
VIII. Concluding Remarks
     Acknowledgments
     References

Psychological modulation of feeding involves taste hedonics and preferences, and the motivating or rewarding aspects of food. The brain circuitries implicated in stimulus reward, and in the regulation of energy balance, have traditionally been considered as separate. However more recently accumulated evidence suggests that there is both anatomical and functional crosstalk between these sets of CNS circuitry. Adding to the potential crosstalk is evidence for modulation by the peripheral adiposity signals insulin and leptin. These findings open the possibility for more extensive interaction between the two circuitries mediated by endogenous CNS neurotransmitters: an expanding frontier for research into the basic physiology of regulation of food intake, and the potential for new therapeutic approaches. This chapter provides an overview of the major anatomical and neurochemical participants in brain reward circuitry and the (limited) evidence available to date which supports the hypothesis that energy regulatory signals can modulate food reward.

I. INTRODUCTION AND OVERVIEW

Obesity is recognized as a significant risk factor for diabetes, cardiovascular disease, and several cancers, as well as shortened life span (Hill et al., 2003; Mokdad et al., 2001; Pi-Sunyer, 2003). The long-term regulation of body weight and body fat stores (adiposity) has thus been a major focus for basic and clinical research over the past 50 years. Because of the rapid intragenerational increase in obesity incidence in Westernized societies over the past 15 years, there is a new focus on environmental and psychological factors which contribute to overeating and perturbed energy homeostasis. A major psychological factor which modulates food intake is the palatability and hedonic evaluation of food (i.e., food reward).

In 1979, Woods, Porte, and colleagues (Woods et al., 1979) proposed a model of a negative feedback loop between the brain and peripheral energy regulatory signals (circulating factors such as insulin and leptin whose concentrations reflect the size of adipose stores, and that signal this information to the CNS), which has received substantial experimental support (Baskin et al., 1999). While there also is support for the concept that feeding behavior can be modified by the rewarding aspect(s) of food, the concept that the perceived rewarding value of food in turn may be regulated is still somewhat novel. The purpose of this chapter is to provide an overview of current knowledge regarding CNS mediation of food reward and the evidence in support of its potential regulation or modulation,

including discussion of neurochemical and neuroanatomical substrates for crosstalk between CNS energy-regulatory and CNS reward circuitry. The meaning of "reward" has been controversial among psychologists (Berridge and Robinson, 1998; Robbins and Everitt, 1996; Wise, 2002; Wise and Hoffman, 1992). For the purposes of this chapter, we will define "reward" functionally: The "rewarding" aspect of food is gauged by the function of its being sought out and consumed, be it in an animal experiment or a free-choice setting for humans.

Studies evaluating the physiological defense of caloric intake by the CNS have focused on the medial hypothalamus as a major anatomical site of integration (Saper et al., 2002; Williams et al., 2001) and have evaluated the actions of hormones and neurotransmitters experimentally using highly controlled and stimulus-deprived environments for these tests. While this experimental approach has been necessary and is correct for these sorts of studies, the applicability of the findings to human eating behavior has been challenged. Hill and colleagues (2003) have pointed out that the current obesity epidemic may be ascribed to an environment of convenient and economically affordable food that is both highly palatable and high in caloric density and fat content. If, in fact, the medial hypothalamic circuitry acts as the final common arbiter of caloric intake, then why cannot caloric intake (and in adults, body adiposity) remain constant and appropriate in the face of whatever foodstuffs are available? One response to this query is that the data we have gleaned regarding the calorie-regulatory circuitry of the hypothalamus have been obtained in circumstances where there are no environmental challenges or choices, and perhaps limited activation of the CNS reward/motivational circuitry. That is, the majority of data on food intake regulation by energy regulatory signals have been collected from animals feeding in their home cages on commercial rodent "chow": a bland, monotonous, and relatively low fat diet which is presented in abundance. Hence, the animal needs minimal engagement of motor systems in order to eat and can generally expect that there will be as much to eat as it wants. This situation offers an almost-perfect model for studying the regulation of caloric need by neural and endocrine factors. However, in 1988 a key study from the laboratory of Bray made it clear that the function of the CNS–energy-regulatory signal feedback loop can be altered by an environmental intervention, that is, changing the fat content of the diet. They demonstrated that rats fed a high fat diet lost less body weight in response to a direct CNS infusion of insulin, than rats maintained on standard lab chow (Arase et al., 1988). This finding was replicated subsequently by Chavez, Woods, and colleagues, who demonstrated that the effect was "dose-dependent" on the concentration of fat in the diet (Chavez et al., 1996). While the precise mechanism(s) of this effect remain unclear, these studies made the point that diet composition has a major impact on the function of the calorie-regulatory CNS circuitry.

Historically, study of the CNS and reward has focused anatomically on the lateral hypothalamic area (LHA) and midbrain dopaminergic (DA) cell bodies and their projection sites, and functionally on paradigms such as brain self-stimulation or self-administration of various neurally active substances (Olds, 1962; Wise, 1988). Not surprisingly, it has become appreciated that additional CNS sites have a role in mediating the rewarding aspects of stimuli. As an anatomical basis for potential crosstalk between energy regulatory circuitry and reward circuitry, it must be appreciated that the medial hypothalamic nuclei are extensively connected with the CNS regions that mediate reward and motivation. For example, the LHA is a major relay area for projections from the mediobasal hypothalamus and thus could serve as a critical integrator for signals from both reward circuitry and calorie regulatory circuitry. The limbic reward system can be functionally defined as those CNS struc-

tures that mediate the rewarding, reinforcing, and emotional aspects of stimuli. From an anatomical perspective, there is a general consensus that the LHA, amygdala, select regions of the cerebral cortex, the ventral tegmental area (VTA), and ventral striatum or nucleus accumbens (NAc) are components of this circuitry (De Olmos and Heimer, 1999; Everitt et al., 1999). Reciprocal synaptic connections exist between the amygdala and cortex, and between the nucleus accumbens and cortex, and there are substantial efferent projections from the amygdala to the hypothalamus and the VTA/substantia nigra pars compacta (SNc). There appear to be limited forebrain inputs directly to the paraventricular nucleus of the hypothalamus (PVN) although PVN efferent projections to the LHA are abundant. Rather, the arcuate nucleus of the mediobasal hypothalamus appears to receive critical limbic inputs: projections from the LHA containing the feeding-stimulatory peptide orexin, as well as input from the central nucleus of the amygdala. In turn, there are direct projections from the arcuate nucleus to the LHA. The central nucleus of the amygdala also projects to the LHA, and LHA and amygdala receive direct taste inputs from the nucleus of the solitary tract (NTS, the critical primary-to-secondary relay site for the taste pathway). Other relevant synaptic connections include reciprocal projections from the NAc to the VTA, and projections from the NAc to the LHA. For more detailed anatomical discussion, the reader is referred to reviews by Berthoud or Kelley (Berthoud, 2002; Kelley et al., 2002).

In addition to anatomical links between the reward circuitry and energy regulatory circuitry, there are functional links as well. This is reflected in the observation that fasting or food restriction have marked behavioral and neurochemical effects on both. Fasting or food restriction activate, or enhance the activation of, reward circuitry as evaluated in several different behavioral paradigms (Carr, 2002; Shizgal et al., 2001). In one of the most striking illustrations of this effect, Carroll and colleagues studied rats allowed to self-administer a threshold dose of cocaine (Carroll and Meisch, 1984). In Carroll's study, rats were fasted or fed on an alternating day schedule prior to having access to the cocaine. When tested on days after they were fed overnight, they self-administered almost no cocaine. When tested on days after they had been fasted overnight, rats robustly self-administered cocaine. This result demonstrates that reward circuitry in the CNS is rapidly responsive to changes in metabolic status. The finding has been replicated with food restriction rather than fasting and has been observed in a somewhat different self-administration paradigm: food restriction will enhance the propensity to drug-taking relapse in rats that have extinguished drug self-administration (Shalev et al., 2002).

II. MESOLIMBIC DOPAMINE CIRCUITRY AND ENERGY REGULATORY SIGNALS

Within the reward circuitry, the collection of mesocorticolimbic (VTA) dopamine neurons, which project to the ventral striatum or NAc, and to the prefrontal cortex, has been viewed as a central neuroanatomical substrate for reward and motivation (Ikemoto and Panksepp, 1996; McBride et al., 1999). Activation of VTA dopamine neurons, and release of dopamine within the NAc, have long been viewed as indicative of reward enhancement. What does activation of the VTA–NAc pathway reflect in terms of food reward? Although this remains a topic of lively debate (Hoebel et al., 1989; Salamone et al., 2003; Schultz, 2002, 2004), Berridge and colleagues have demonstrated that this activity is not correlated with enhanced hedonic value of food, as evaluated in the "taste reactivity" paradigm, and thus dopaminergic activation does not reflect an increase in the animal's "liking" of the food (Berridge, 1996). Rather

they have proposed that mesocorticolimbic dopamine activity reflects an increase in the "incentive salience" of a stimulus, including food. This property can be modulated by the nutritional status of an animal. In the schema of Berridge, with food deprivation, the food stimulus would be more relevant and more motivating. It has been documented that with repeated training to gain access to a diet in a defined physical environment, initial exposure leads to increased release of dopamine in the NAc shell whereas subsequent exposure leads to either no increase of dopamine (Richardson and Gratton, 1996); an increase of dopamine in anticipation of the presentation of food (Kiyatkin, 1995); increased release of dopamine within a different part of the NAc (Bassareo and DiChiara, 1997); and sustained release of dopamine in the prefrontal cortex (Bassareo and DiChiara, 1997, 1999). Although interpretation of these findings remains controversial, the concept that the contextual stimuli themselves (odor or visual cues) become salient, and can elicit dopamine release with repeated exposure to food in the same context, seems experimentally validated. However, if dopamine release specifically in the NAc shell reflects "reward," the habitual presentation of the same food could be predicted to result in a loss of its primary reward value.

The behavioral task known as conditioned place preference (CPP) assesses the strength of a learned association between the perceived reward value of a stimulus (such as a drug treatment or food) and the location in which the animal received the stimulus (Bardo and Bevins, 2000). The strength of the conditioning can be sensitive to the nutritional status of the animal. Conditioning of a place preference by cocaine is enhanced by food restriction (Bell et al., 1997). Place preference also can be conditioned by food and, perhaps not surprisingly, the strength of the CPP is enhanced by food restriction (Agmo et al., 1995; Figlewicz et al., 2001; Lepore et al., 1995; Papp, 1988; Swerdlow et al., 1983). Place preference conditioning by food is dependent on intact dopaminergic activation, as development of the CPP is blocked by administration of dopamine receptor antagonists during training sessions (Agmo et al., 1995; Figlewicz et al., 2001).

Given the central role of the VTA dopamine neurons in reward circuitry, it has been hypothesized that the effect of food restriction is due to enhanced activation of these neurons. Indeed, Wilson and colleagues demonstrated that food-restricted rats trained to drink a palatable liquid food had greater dopamine release in the NAc than free-feeding rats (Wilson et al., 1995). One question, then, is whether these dopamine neurons are a target for neural or endocrine factors that change in association with fasting and food restriction. The Carroll study suggests that neural or humoral factors modulating these phenomena must be able to change with a time course of several hours.

Collectively, studies of the effects of glucocorticoids, insulin, and leptin support the conclusion that a neuroendocrine milieu exists in fasted animals that would bias them toward enhanced dopaminergic function. Adrenal glucocorticoid levels are elevated with fasting. Piazza and colleagues have provided evidence that glucocorticoids can facilitate dopamine release and dopamine-mediated behaviors (Marinelli and Piazza, 2002). Additionally, both insulin and leptin levels rapidly decrease in association with food restriction or fasting (Havel, 2000). Both insulin and leptin inhibit performance in food reward behavioral tasks that are dopamine-dependent (e.g., Figlewicz et al., 2004; Shalev et al., 2001). As summarized in a more recent review (Figlewicz, 2003), there is evidence for effects of insulin at the level of dopamine reuptake: *in vivo*, intracerebroventricular (i.c.v.) insulin increases steady-state mRNA levels of the dopamine reuptake transporter (DAT). *In vitro*, insulin administration (at a physiological concentration) facilitates striatal dopamine

reuptake, which should in turn curtail dopaminergic signaling. The functional consequence of decreased dopamine signaling should be that insulin decreases the rewarding aspect of stimuli. Consistent with the possibility that modulation of DAT function has behavioral consequences, Pecina et al. (2003) demonstrated that DAT-knockdown mice eat more, and have decreased latency to onset for eating food treats. Krugel et al. (2003) have demonstrated that i.c.v. leptin administration decreases both baseline and food-stimulated NAc dopamine release. Receptors for both insulin and leptin have been identified on the VTA dopamine cell bodies (Figlewicz et al., 2003), supporting the possibility that insulin and leptin may act directly at the VTA and/or on NAc nerve terminals to blunt dopaminergic activity and its contribution to food reward. It is of course possible that insulin and leptin effects at the MH may contribute to modulation of CNS neurotransmitters or projections between the hypothalamus and the VTA/NAc. Few studies have examined the role of other putative neurotransmitters or neuroendocrines in the modulation of VTA/NAc-mediated food reward. Acute i.c.v. administration of the α-MSH agonist MT II enhances the reward potentiating effect of *d*-amphetamine (Cabeza de Vaca et al., 2002), which was also observed following more chronic MT II administration (Cabeza de Vaca et al., 2005). Additionally, the orexigenic peptide ghrelin has more recently been shown to potently stimulate food intake when administered either into the VTA or the NAc (Naleid et al., 2005).

III. BRAIN OPIOID SYSTEMS AND ENERGY REGULATORY SIGNALS

Brain opioid systems have served as a focus for investigation as endogenous opioid neural networks appear to play a role in the regulation of food intake, food hedonics, and food choice (Glass et al.,

1999; Levine et al., 2003). Although experimental evidence demonstrates that DA and the opioids play somewhat different roles in the mediation of food reward, the neuroanatomical circuitry that is implicated in opioid effects overlaps significantly with the VTA/NAc reward circuitry. Opioidergic activation may mediate the hedonic valuation of foods, whereas activation of the VTA/NAc mediates the rewarding (motivating, reinforcing, incentive salient) properties of food. The potential interaction of opioidergic and dopaminergic systems seems obvious, as one would predict that a "more pleasing" food would be more rewarding. A compelling reason for targeting the opioids for continued investigation by basic scientists is the observation that endogenous opioids may play a role in hedonic valuation of foods in human subjects. In one report, the opiate antagonist nalmefene decreased fat and protein intake from a standardized buffet meal in nonobese subjects (Yeomans et al., 1990). Drewnowksi and colleagues reported that both taste preferences for, and intake of, sweet high fat foods (such as cookies or chocolate) were decreased by treatment with the opioid antagonist naloxone in binge eaters but not in nonbinge eaters (Drewnowski et al., 1992, 1995). This finding suggests that some endogenous opioid systems may be (more) active in association with food binges.

It is clear that opioids, endogenous and synthetic, can enhance food intake. From the first report of increased feeding on daily injection of morphine in addicted rats (Martin et al., 1963), to countless studies showing that opioid antagonists inhibit various aspects of feeding behavior (Frisina and Sclafani, 2002; Jarosz and Metzger, 2002; Kirkham and Blundell, 1986; Lang et al., 1981; Marks-Kaufman et al., 1984; Yu et al., 1997, 1999) mediated by all of the opioid receptor subtypes (Arjune et al., 1990, 1991; Carr et al., 1991; Islam and Bodnar, 1990; Levine et al., 1990, 1991), it has become generally accepted that opioids play some role in food intake. The nature of this role is less

clear, as are the mechanisms by which opioids respond to and modulate food intake and body weight. Gene expression studies demonstrate that opioid expression in some opioidergic neuronal populations decreases when animals are deprived of food, suggesting that these opioids may not be involved in hunger-induced feeding (Brady et al., 1990; Kim et al., 1996). Furthermore, when animals gained weight due to overconsumption of palatable food, opioid mRNA levels increased in the hypothalamus (Welch et al., 1996), which suggests that opioids are generated in response to consumption of palatable food, rather than prior to consumption. Another set of studies demonstrated that opioid antagonists inhibit ingestion of sucrose solutions in spite of the presence of a gastric fistula, indicating that opioid control of food intake does not require absorption of nutrients, but may actually be a response to the orosensory aspects of consumption. In fact, these studies showed that opioid antagonism had the same effect on ingestion as dilution of the sucrose solution, essentially decreasing the palatability of the solution (Kirkham and Cooper, 1988a,b).

One ongoing question is whether opioids are responsible for intake of food in general or only of pleasurable food intake, and this issue is confounded by the apparent plasticity of palatability. That is, palatability is subjective—based on individual preference, nutritional status, previous experience, and presence of multiple choices, among other factors. Cleary and colleagues (1996) showed that motivation to work for sucrose solutions increases with increasing sucrose concentration (thereby palatability) and that naloxone was most effective at inhibiting intake of the highest (most palatable) concentrations. Giraudo et al. (1993) found that naloxone inhibited intake of chocolate chip cookies to a greater extent than that of regular chow or less-palatable high-fiber chow. A very recent study has addressed the role of opioids and dopamine in feeding behavior, taking into account deprivation state and palatability

(Barbano and Cador, 2005). The authors presented both sated and deprived animals with chow or a palatable chocolate cereal and tested the consummatory, motivational, and anticipatory aspects of feeding with injections of naloxone or the dopamine antagonist flupenthixol. A relatively low dose of naloxone decreased motivation for, and consumption of, only the palatable food in sated rats and had no effect on deprived rats' intake of either diet, while flupenthixol only affected anticipatory behavior in hungry rats expecting the palatable cereal. Thus, opioids appear to signal hedonic value of a food when all nutritional needs have been met.

Site-specific microinjection techniques have allowed investigators to determine the role of individual sites and networks on food intake, and this method has revealed much about the opioid feeding network. Opioids, especially the μ agonist, DAMGO, injected into either the VTA or the NAc can stimulate 4-h food intake (Badiani et al., 1995; Noel and Wise, 1995), and this is blocked by prior injection of opioid or GABA antagonists (Lamonte et al., 2002a,b). One noteworthy study, using microinjection and microdialysis techniques showed that, as with morphine injection, when rats ate a palatable snack, DA release in the NAc increased. Most interestingly, this DA release was blocked by opioid antagonism in the VTA (Tanda and DiChiara, 1998). This study revealed that in the mesolimbic reward pathway, palatable food exerted the same effects as drugs of abuse; and that there is functional crosstalk in this pathway between opioids and dopamine.

One persistent hypothesis has been that opioids in one site may have a different function in feeding behavior than opioids in a different site. For example, when animals were deprived of food for 24h, then given a choice between a preferred diet and a nonpreferred diet, the general opioid antagonist, naltrexone (NTX), injected into the hypothalamic paraventricular nucleus (PVN) decreased intake of both diets, but NTX injected into the amygdala decreased intake only of each animal's preferred diet (Glass et al., 2000). Since the PVN plays a larger role in energy homeostasis, and the amygdala mediates portions of the emotional response to feeding, the study concluded that opioids affect intake for different reasons in different sites. In many early studies, animals received standard rodent chow. However, in studies in which animals could choose between a high fat and a high carbohydrate food, results have varied extensively and long-standing disagreements have developed about the role of opioids in diet choice.

Some have argued that opioids enhance intake of fat, while neuropeptide Y (NPY) triggers intake of carbohydrate, deemed the "one peptide, one nutrient" hypothesis (Levine et al., 2003). Indeed, many studies support a role for opioids in fat appetite (Glass et al., 1996; Islam and Bodnar, 1990; Kelley et al., 2002; Weldon et al., 1996; Yanovski and Yanovski, 2002; Zhang et al., 1998). Zhang and colleagues (1998) investigated the role of opioids in the shell region of the nucleus accumbens (NAcSh) on diet choice in rats presented with both high fat and high carbohydrate diets. This study showed that when given a two-diet choice, DAMGO, a μ opioid agonist, stimulated intake of the high fat diet, regardless of the animals' initial preference for fat or carbohydrate. Furthermore, the authors found that while 24-h deprivation caused animals to eat both fat and carbohydrate to similar extents, a peripherally injected general opioid antagonist, NTX, selectively blocked fat intake without affecting carbohydrate intake. These results strongly suggest that opioids in the AcbSh mediate fat intake to a greater extent than carbohydrate intake.

Other studies, however, suggest that opioids modulate intake of an animal's preferred food, regardless of nutrient content (Glass et al., 1996; Gosnell and Krahn, 1992; Levine et al., 2003; Marks-Kaufman et al., 1985). Moreover, when given a choice of two types of fat or two types of carbohydrate, opioids appear to enhance intake of the more palatable source (i.e., sucrose versus cornstarch) (Glass et al., 1997, 1999; Pomonis et al., 1997). When these findings are taken into account, they call into question the universality of the findings of Zhang and colleagues (1998), who used a mixture of cornstarch and dextrin, with little sucrose, as the carbohydrate choice. In two studies mentioned previously, opioids affected intake of sucrose but did not affect intake of cornstarch, when animals had a choice (Glass et al., 1997; Pomonis et al., 1997), indicating that sucrose intake may be more subject to control by opioids than less-palatable, isocaloric counterparts. Perhaps providing a more palatable carbohydrate source would alter the finding that opioids selectively enhance intake of fat, as described by Zhang et al. (1998). The conclusion so far can only be that there are more factors at play than originally thought. Macronutrient content, individual preference, source of macronutrient, and perhaps differing brain chemistry may all contribute to the choice an animal makes during these experiments.

An animal's energy state can have an impact on activity of the opioid system and the behaviors mediated by opioids. In a progressive ratio test of deprived and satiated rats, deprivation increased responding for sucrose pellets, and this responding was decreased by injection of the opioid antagonist, naloxone (Rudski et al., 1994). Naloxone also has been shown to decrease intake of normal chow in deprived rats to a greater extent than sweet chow, and in sated rats, naloxone only decreased intake of sweet chow (Levine et al., 1995), implying opioid-based enhancement of palatability of normal chow in conditions of deprivation.

To more specifically address the potential mechanisms by which metabolic status can alter the perception of palatability, insulin and leptin have been shown to decrease reward responding, which is essentially the opposite of the effects of food deprivation. For example, insulin injection into the cerebral ventricles diminishes sucrose pellet intake stimulated by a ventricularly injected κ-opioid agonist, and acts cooperatively with a subthreshold dose of a κ-opioid antagonist to decrease baseline intake of sucrose pellets (Sipols et al., 2002). Similarly, sucrose pellet intake stimulated by intra-VTA injection of the μ-opioid agonist DAMGO, can be inhibited by concurrent injection of insulin into the same site (Figlewicz, 2003). Conditioned place preference (CPP which develops when animals learn to associate a reward with the place it was encountered) can develop with food as the reward, is enhanced by food deprivation (Lepore et al., 1995), and is blocked with intraventricular injection of insulin or leptin (Figlewicz et al., 2004). Furthermore, CPP enhanced by food deprivation can be blocked with physiological doses of peripheral leptin (Figlewicz et al., 2001). Considering that CPP development to palatable food likely has an endogenous opioid component (Lu et al., 2003; McBride et al., 1999; Tszchentke, 1998), this is further evidence of the interaction between metabolic factors and opioid-modulated reward pathways. With regard to CNS-intrinsic energy regulatory signals, as reviewed by Olszewski and Levine (2004), the opioid nociceptin may enhance or sustain feeding by interacting with feeding-termination neuropeptide pathways such as α-MSH, oxytocin, or CRH (corticotrophin-releasing hormone).

Anatomical evidence supports this interaction as well. Food restricted rats show significant reductions in μ- and increases in κ-opioid receptor binding in several forebrain areas related to food reward and in the hindbrain parabrachial nucleus, in quantitative autoradiography studies (Wolinsky et al., 1994, 1996a,b), indicating that deprivation (and its accompanying decreases in circulating insulin and leptin) can alter opioid receptor distribution in key reward sites. Conversely, μ-opioid receptor binding is increased in reward-related sites in animals made obese (and therefore resistant to the actions of insulin and leptin) on a high fat diet (Smith et al., 2002). At the very least, these studies indicate an interaction between nutritional state and the opioid system in reward-related sites. Pooled with future studies involving leptin and insulin directly, these data may reveal a pattern to the interaction that explains aberrant food intake behavior such as binging.

IV. ENDOCANNABINOIDS AND ENERGY REGULATORY SIGNALS

Evidence supports the role of the endogenous cannabinoids (endocannabinoids, or ECs) anandamide and 2-arachidonoyl glycerol (2-AG) in food intake. Please see more recent reviews for detailed discussion of this system; evidence supports both forebrain and hindbrain sites of efficacy to enhance palatable food intake (Berry and Mechoulam, 2002; Fride, 2002; Harrold and Williams, 2003; Kirkham, 2005; Kirkham and Williams, 2004; Miller et al., 2004). One conclusion that may be made with the current status of knowledge is that the ECs may be at the interface of CNS energy-regulatory systems and the reward system (see, e.g., discussion in Ravinet Trillou et al., 2003). There is some evidence demonstrating interaction with CNS leptin effects, and also interaction with the mesocorticolimbic dopamine system. Genetic models of leptin (*ob/ob* mouse) or leptin receptor (Zucker *fa/fa* rat; *db/db* mouse) deficiency have higher hypothalamic levels of the EC, 2-AG. Exogenous (i.p.) administration of leptin decreases ECs in the hypothalamus but not cerebellum of rats. However, ECs were not measured in the limbic circuitry in that study (Di Marzo et al., 2001), although EC content is higher in limbic circuitry (brainstem, striatum, and hippocampus) than in the diencephalon (Fride, 2002). Measurements in normal rats demonstrate increases in limbic content of both anandamide and 2-AG with food deprivation, with a modest increase of 2-AG in the hypothalamus (and no change of hypo-

thalamic anandamide content). Further, direct administration of 2-AG into the NAc shell significantly stimulates feeding (Kirkham et al., 2002). Endocannabinoid type 1 (CB1) receptor content, assessed by quantitative autoradiography, is decreased in the hippocampus and NAc of rats fed a palatable diet for 10 weeks, and the energy intake from the diet was inversely correlated with CB1 receptor density, interpreted as enhanced receptor–ligand interaction, that is, increased EC activity (Harrold et al., 2002).

Protocols evaluating motivation or reward (as opposed to free-feeding measurements) certainly implicate ECs in feeding. Thus the CB1 antagonist SR141716 (rimonabant) decreases place preference conditioned by food (Chaperon et al., 1998), as well as self-administration of food (Arnone et al., 1997). Duarte et al. (2004) demonstrated that relapse to food intake is enhanced by a D3 agonist and this effect is blocked by the CB1 receptor antagonist, suggesting some synergy between EC and dopamine pathways. Further, "progressive ratios" performance for sucrose self-administration is decreased in CB1 knockout mice (Sanchis-Segura et al., 2004). Finally, LHA-stimulation-induced feeding is increased by (exogenous) tetrahydrocannabinol (Trojniar and Wise, 1991), and decreased by CB1 antagonism (Deroche-Gamonet et al., 2001). Together these studies argue for a role of the ECs in the motivational aspect(s) of feeding.

Other than the limited studies examining possible leptin–EC interaction, little is known regarding the impact of energy regulatory signals on EC function, or vice versa. Clearly this is an area that warrants additional study. A more recent study has examined EC interaction with the novel orexigen, ghrelin (Cummings and Shannon, 2003). Tucci et al. (2004) show that a dose of CB1 antagonist which does not decrease food intake on its own will reverse feeding induced by ghrelin administration into the PVN. This finding supports a role of the

ECs in ghrelin feeding effects within the medial hypothalamus.

V. LHA CIRCUITRY AND ENERGY REGULATORY SIGNALS

As discussed before, a central neuroanatomical substrate for coordinating both reward inputs and energy circuitry inputs may be the LHA. The anatomical basis for this concept also is well-established, as the LHA receives direct and indirect limbic inputs and direct projections from the arcuate nucleus of the hypothalamus (which is a major target for candidate adiposity signals) as well as numerous intrahypothalamic and neuroendocrine inputs (Elias et al., 1998). Lesion experiments conducted over 50 years ago by Anand and Brobeck revealed a critical role for the LHA in the regulation of ingestive behavior. LHA lesions reduce food intake and body weight and reduce motivation for pleasurable stimuli (Anand and Brobeck, 1951). On the other hand, electrical stimulation (Delgado and Anand, 1953) or glutamate receptor stimulation within the LHA (Stanley et al., 1993) potently stimulate food intake. Beginning with the lesion studies in the 1950s, additional research confirms that the LHA is an important neural substrate in the control of food intake today (for review see, Bernardis and Bellinger, 1993). Not long after the lesioning studies of Anand and Brobeck, Olds and co-workers (1958) brought attention to the LHA as an important hypothalamic site mediating reward. They reported that rats robustly self-stimulate (SS) electrical current delivered within the LHA. The LHA sites that elicited SS overlapped with sites that also stimulated food intake in response to electrical stimulation (Margules and Olds, 1962). Collectively, these early lesioning and electrical SS studies have defined a role for the LHA in both feeding behavior and reward. The early LHA lesioning and electrical SS studies have converged. The initial studies conducted by Margules and Olds (1962) reported that the majority of animals SS to a lower level of current after 24 h of food deprivation demonstrating, for the first time, that nutritional status influences the reinforcing and rewarding effect of lateral hypothalamic self-stimulation (LHSS). Several investigators since have also reported enhanced LHSS as a function of food deprivation (Blundell and Herberg, 1973) or experimental diabetes (Carr, 1994). Likewise, the reinforcing effect of drug self-administration (SA) is also augmented in a food deprived state (Carroll and Meisch, 1984).

While food deprivation, food restriction, and experimental diabetes (metabolic circumstances of low insulin or low leptin) all induce more global metabolic changes, metabolic challenges that induce specific alterations in nutrient availability also influence LHSS. Low doses of insulin that induce mild hypoglycemia robustly increase LHSS while systemic administration of glucagon, which increases blood glucose levels, significantly reduces LHSS (Balagura and Hoebel, 1967). Subsequent studies, however, using pharmacological systemic blockade of glucose utilization, reported no change in LHSS (Cabeza de Vaca et al., 1998). These disparate findings could be due to the acute lethargy induced by severe glucoprivation or a result of the pronounced hyperglycemic response elicited by pharmacological blockade of glycolysis. Nonetheless, deficits in blood glucose availability may stimulate food intake, in part, by targeting the LHA feeding and reward circuitry, a mechanism that would ensure adequate food intake to correct hypoglycemia, a life-threatening condition. This hypothesis is supported by the findings that glucose sensing neurons are located in the LHA (Oomura et al., 1969, 1974) and LHA lesions block hypoglycemia-induced food intake (Epstein and Teitelbaum, 1967). Thus, there is a significant literature demonstrating that metabolic need, both nonspecific and specific, potentiates reward-related behaviors and the motivation to feed.

The question of which neurochemical or neuroendocrine substrate(s) mediate(s) restriction-enhanced LHSS has been pursued with some interesting results. Studies have focused on physiological commonalities that occur in response to food deprivation, food restriction, and uncontrolled diabetes as possible mediators of enhanced reward efficacy. One such physiological adaptation that occurs is elevated plasma corticosterone. However, blockade of corticosterone synthesis in food-restricted rats failed to reverse the sensitization of LHSS (Abrahamsen and Carr, 1996). Enhanced opioidergic activity may be the intrinsic neurochemical change that mediates the shift in current threshold since administration of naltrexone into the lateral ventricles can reverse the effect of fasting on threshold current shift within individual rats (Carr and Wolinsky, 1993). More recently, it has been shown that i.c.v. leptin administration reverses chronic food restriction-induced enhancement of LHSS (Fulton et al., 2000) as does i.c.v. insulin (Carr et al., 2000); and receptors for both insulin and leptin have been identified by immunocytochemistry in the LHA (Figlewicz, 2003). Shizgal and colleagues have also systematically evaluated whether the orexigenic neuropeptide Y (NPY), or the anorexic peptide corticotropin-releasing factor (CRF), can modulate food restriction-sensitive LHA sites of self-stimulation (Shizgal et al., 2001). Their studies demonstrate that NPY does not modulate LHA stimulation reward, and CRH only affects LHA sites that are insensitive to chronic food restriction (Fulton et al., 2002a,b, 2004).

The neural circuitry underlying the interaction between nutritional status and reward is beginning to be elucidated. One neuropeptide in particular, melanin-concentrating hormone (MCH), uniquely expressed in the LHA (Bittencourt et al., 1992), has emerged as a potentially critical neuropeptide at the interface of both feeding behavior and reward mechanisms. The majority of MCH neurons are localized to the perifornical region of the LHA, a site from which SS can be robustly elicited (Olds, 1958; Shizgal et al., 2001). Administered centrally, MCH potently stimulates food intake in satiated animals (Qu et al., 1996) and overexpression of MCH increases fat intake and body weight (Ludwig et al., 2001), whereas MCH knockout mice are hypophagic and lean (Shimada et al., 1998), reminiscent of LHA lesioned animals. LHA MCH neurons are strategically positioned to directly or indirectly receive input from signals, such as leptin, that relate peripheral energy balance to the CNS. Indeed, LHA MCH mRNA expression is sensitive to changes in nutritional status. Acute food deprivation, acute pharmacological glucoprivation (Presse et al., 1996; Sergeyev et al., 2000), and acute insulin-induced hypoglycemia (Presse et al., 1996) all increase MCH mRNA expression. MCH neurons receive dense projections from chemically defined cell groups within the hypothalamic arcuate nucleus that receive both leptin and insulin signals. For example, LHA MCH neurons receive innervation from arcuate nucleus NPY, agouti-gene related protein (AGRP) (Broberger et al., 1998; Elias et al., 1998), and from α-MSH immunoreactive fibers (Elias et al., 1998). Thus, MCH neurons receive input from hypothalamic sites critical to the overall regulation of food intake and body weight. In addition, MCH neurons express the leptin receptor (Hakanson et al., 1998), thus LHA MCH neurons may serve as one anatomical substrate that mediates the ability of leptin to impair LHSS and conditioned place preference to palatable foods. In turn, MCH neurons project to widespread brain regions, including nuclei that participate in feeding behavior and reward (Bittencourt et al., 1992).

The shell of the nucleus accumbens (NAcSh), long appreciated for its involvement in motivated behaviors and reward (Koob and Bloom, 1988; Robbins and Koob, 1980; Robbins et al., 1989), is also critical to feeding behavior via its connectivity with the LHA. MCH receptors are heavily expressed within the NAcSh (Saito et al., 2001) and the LHA exhibits a reciprocal and functional projection to the NAcSh (Heimer et al., 1991; Kirouac and Ganguly, 1995; Phillipson and Griffiths, 1985), in which both MCH and orexin/hypocretin (Zheng et al., 2003) are implicated. Stimulation of GABA$_A$ or GABA$_B$ receptors within the NAcSh (Basso and Kelley, 1999; Stratford and Kelley, 1997) or excitatory amino acid receptor antagonism (Maldonado-Irizarry et al., 1995) potently stimulates food intake (with an increased preference for palatable foods) in satiated rats, a response that is blocked by LHA inactivation (Maldonado-Irizarry et al., 1995). Further supporting a functional relationship between the NAcSh and LHA, GABA agonists or opioid stimulation (Bakshi and Kelley, 1993; Zhang and Kelley, 2000), both of which stimulate food intake, significantly increase fos-immunoreactivity, a cellular marker of activation, in the LHA, a response that is also blocked by the temporary inactivation of the LHA (Stratford and Kelley, 1999). Thus, it has been hypothesized that MCH signaling between the LHA and NAcSh may play an important role in enhancing the rewarding aspects of food intake (Saper et al., 2002). In support of this hypothesis, it was more recently demonstrated that MCH, delivered directly into the NAcSh, potently stimulates food intake in satiated rats, an effect that is blocked by a selective MCH receptor antagonist (Georgescu et al., 2005).

VI. OTHER CNS SITES: TARGET FOR FUTURE STUDIES?

Other forebrain structures are implicated in both limbic function and food reward/motivation. Studies from Baunez and colleagues have demonstrated that the subthalamic nucleus plays a critical role in modulating food-related motivation (Baunez et al., 2002). This nucleus is a component of the basal ganglia circuitry, within a functional loop that includes the NAc and ventral pallidum. The subthalamic nucleus is connected to the prefrontal cortex through a two-synapse pathway (subthalamic nucleus—substantia nigra—cortex) as well as through the pallidal loop (Maurice et al., 1999). Baunez and colleagues observed that bilateral lesion of the subthalamic nucleus results in an increased rate of eating food pellets, an increase in performance in the "progressive ratios" paradigm, and increased reinforcing properties of food-associated stimuli. These effects were situation-dependent and therefore not due to a nonspecific enhancement of motor responding (Baunez et al., 2005). Additionally, specific subcomponents of the cerebral cortex are integrally involved in taste recognition, taste memory and valuation, and executive function in initiating ingestive decisions based on visual and olfactory cues (Rolls, 2004). Primary taste cortex (i.e., agranular insular cortex) has efferent connections to the orbitofrontal cortex (OFC) and the other major limbic areas, including NAc, LHA, and amygdala. Additionally, there are direct projections to the NTS and autonomic motor CNS structures. The OFC receives multimodal inputs including gustatory, olfactory, visual, and somatosensory information. For example, some OFC neurons respond to the oral texture of fat (Rolls et al., 1999). Outputs from this region of the cortex project to the striatum, the ventral midbrain, and the sympathetic nervous system (Berthoud, 2002). In rats, electrical stimulation of the OFC initiates feeding (Bielajew and Trzcinska, 1994), and infusion of various neuropeptides or neurotransmitters into the OFC can alter respiratory quotient and energy expenditure as thermogenesis (McGregor et al., 1990a,b; Westerhaus and Lowy, 2001). Unfortunately, to date no studies have examined potential direct effects of insulin or leptin (or their medial hypothalamic target peptides) on subthalamic nucleus or gustatory cortex function.

VII. HUMAN AND CLINICAL STUDIES: AT THE FOREFRONT OF OUR KNOWLEDGE

Understanding the integration of peripheral adiposity signals and psychological controls of feeding in humans requires a body of knowledge that is still in its infancy, though eagerly anticipated. The term interaction "might" be more appropriate at this stage, with a number of more recent human studies addressing predominantly psychologically driven disorders of appetite and feeding from a more physiological perspective, either in etiology or treatment. However, an interaction of peripheral adiposity signals with psychological factors in the regulation of food intake in humans can at most be only suggested at this point, although insulin and leptin measurements in populations with eating disorders are being reported with increasing frequency. As a result, two different approaches have been used in an attempt to assess the relationship between adiposity signals and psychological factors.

The first approach involves assessment of patients with eating disorders regarding onset, development, and treatment of the particular disease with respect to physiological and psychological factors. In a review of two forms of disordered eating associated with obesity, Stunkard and Allison (2003) compared binge eating disorder (BED) with night eating syndrome (NES), concluding that BED is primarily a psychological disorder that responds well to weight reduction programs, whereas NES is a stress-related eating, sleeping, and mood disorder that responds to selective serotonin reuptake inhibitors (SSRIs). More importantly, NES appears to have a unique neuroendocrine profile. While plasma leptin levels in overweight subjects are higher than those in normal weight subjects, both for NES and controls, Birketvedt et al. (1999) found that only NES patients, both normal weight and obese, did not display the rise in plasma leptin levels at night seen in obese and normal weight controls. This finding suggests that the normal nocturnal rise in leptin may play a role in night time anorexia, and that when this rise is blunted (as in NES), the anorexia likewise diminishes. In addition, the elevated circadian plasma cortisol levels in NES patients supports the notion that NES is stress-induced, and the blunted CRH-induced ACTH and cortisol responses (Birketvedt et al., 2002) likely represent an exhaustion of the hypothalamic–pituitary–adrenal axis.

A similarly distinctive role for leptin in BED has not yet been identified, although Monteleone et al. (2000) found that serum leptin concentrations were elevated in women with BED in comparison to healthy age-matched controls, suggestive of leptin resistance. Other studies, however, have not replicated these findings. Karhunen et al. (1997) and Adami et al. (2002) reported that leptin levels were similar in obese binging and nonbinging women, and that serum leptin concentrations were positively correlated with BMI. These equivocal findings indicate that BED is likely a more complicated disease than previously thought, with multiple forms of the disorder that might differentially affect leptin levels. Branson et al. (2003) reported that mutations of the melanocortin-4 receptor, which is downstream of leptin, are found in 5% of severely obese patients with BED, which could explain the elevated leptin levels in the context of leptin resistance.

Given the physiological evidence of leptin involvement in NES and perhaps some forms of BED, the question of interaction between leptin levels and psychological controls of appetite remains unanswered. BED has a pronounced psychological component, hence this disorder can be treated with some success using psychotherapy, resulting in a decrease in eating disorder symptoms, though without appreciable weight loss (de Zwaan, 2001). Dawe and Loxton (2004) have characterized BED as a disorder of impulsivity, and it could be argued that elevated leptin levels in nonbinging obese individuals (which comprise approximately 75% of the severely obese population) might well serve as an inhibitor of impulsive eating, whereas leptin resistance in the remaining 25% affords those individuals no such protection. The facility with which NES lends itself to pharmacological, rather than psychological, intervention is illustrated by the number of studies reporting that SSRIs decrease NES behaviors without decreasing the often accompanying depression (Stunkard and Allison, 2003). The role of leptin in this case might not be as a modulator of psychological factors, but simply as a promoter of anorexia. It would certainly be informative to artificially elevate leptin levels in NES patients to determine the effectiveness of leptin in moderating night eating.

Although Hebebrand et al. (1997) and other investigators have found that serum leptin concentrations are reduced in acute anorexia nervosa (AN) patients in comparison to age- and weight-matched healthy controls, the unequivocal relationship between serum leptin concentrations and body weight or percentage of body fat in AN patients before, during, and after recovery (Monteleone et al., 2000) suggests that lowered body fat mass is the major determinant of decreased leptin levels in AN. This disease is also striking in that its major psychological characteristics are distorted self-image and pathological fear of weight gain, with depression or other affective disorders quite minimal. Thus it would be expected that the interaction of body adiposity signals with psychological controls of appetite would similarly be minimal. It should be noted that notwithstanding the fact most AN patients continue to report hunger and cravings throughout their weight reduction endeavors, these highly motivated individuals are nevertheless able to resist and ignore such signals in their relentless quest for thinness.

Bulimia nervosa (BN) patients also exhibit binge eating behaviors but, unlike AN patients, maintain relatively stable body weights over time due in large part to purging (voluntary vomiting or excessive laxative use) and absence of restrictive diets. Monteleone et al. (2000) reported that serum leptin concentrations were positively correlated with BMI in BN. Interestingly, BN patients mirror AN patients in that they also have reduced serum leptin concentrations in comparison to age- and weight-matched healthy controls (Jimerson et al., 2000), though generally not as low as in patients with AN. Leptin levels decrease, however, with chronicity and severity of BN (Monteleone et al., 2002), indicating that leptin secretion mechanisms may also have a body weight–independent component. One hallmark of BN is the nearly universally observed depression that is always pharmacologically treated during or even before behavioral modification of the BN repertoire. Similarly to NES, decreased leptin levels may lead to decreased anorexia (food avoidance) and, consequently, result in even more severe and more frequent binging episodes. Hence, leptin's role as a modifier of psychological controls of appetite might be limited in BN.

The second approach involves functional magnetic resonance imaging (fMRI) studies investigating the localization of food reward in the CNS under normal and energy-challenged conditions, including subjects with eating disorders, and the overlap of these sites with loci involved in drug reward behavior. The medial prefrontal cortex, thalamus, and hypothalamus appear to be central to the rewarding and motivating aspects of food stimuli, whereas the amygdala and OFC respond to food cues without regard to caloric (i.e., reward) value, indicating a more general role in energy regulation (Frank et al., 2003; Gautier et al., 2000; Gottfried et al., 2003). Whether insulin or leptin play a direct role in these mechanisms in humans is at present unknown, although animal studies (see earlier) confirm insulin and leptin action in many CNS areas to regulate or modulate feeding.

A number of more recent human imaging studies involving taste and ingestion have similarly extended our knowledge of neural substrates involved with food affinity. In a study of brain activity in response to glucose taste, Frank et al. (2003) used fMRI in a double-blind protocol in which they observed increased right medial OFC activation in healthy normal-weight adult women, relative to the response to artificial saliva. The marked OFC activation suggests that tastes with greater hedonic or emotional value are represented preferentially in this brain region. Gottfried et al. (2003) used fMRI to study hungry volunteers who were first presented with picture–odor pairings and then fed a meal specific to the presented odor until sated, simultaneously lessening the subject's reported value of the picture. This devaluation was associated with strikingly decreased neural responses in the left dorsomedial amygdala and OFC to the picture after the meal in comparison to those just before the meal, and modest decreases in the ventral striatum, insula, and cingulate. A speculative interpretation of this study is that satiety signals generated during feeding contribute to the decrease in the value of a food-related sensory cue, whether the cue is primary or secondary to a learned association. OFC activation has been observed to be low in obese (relative to lean) men in response to a satiating meal (Gautier et al., 2000). This latter study strongly suggests a role of the OFC in human energy balance.

Studies investigating localization of neurotransmitter signaling during meal consumption have also yielded noteworthy insights concerning food reward and reinforcement. Small et al. (2003) used labeled raclopride positron emission tomography (PET) scanning following a 16-h fast and a favorite meal in normal volunteers to measure regional dopamine (DA) binding. Reduced DA binding was observed in the full versus hungry state in the dorsal striatum, indicating DA release on food consumption. The reduction in raclopride binding was correlated with meal pleasantness, but not with hunger before eating or satiety following the meal, indicating that the amount of dorsal striatal DA released correlates with pleasure. Using similar imaging methodology, Volkow et al. (2003) correlated DA release with eating behavior survey results and food stimulation (smell and taste) in normal volunteers. Eating restraint scores were positively correlated with DA release to food stimulation, and emotionality scores were negatively correlated with baseline D_2 receptors—all in the dorsal, not ventral, striatum. Hence, dorsal striatal DA is likely implicated in at least two different neurobiological aspects of eating behavior.

One obvious clinical application of these new imaging techniques is the assessment of neural substrates of body weight dysregulation and eating disorders. In a study using PET scanning to investigate brain DA involvement in pathologically obese individuals, Wang et al. (2001) found that D_2 receptor availability in these subjects was inversely correlated with body weight (in contrast to normal-weight controls), suggesting that decreased brain DA activity in the obese may well predispose them to excessive food intake (i.e., the greater the BMI, the fewer the DA receptors). The same investigators (Volkow et al., 2002) found an increase in dorsal striatal extracellular DA following nonconsumed food display in normal-weight fasting subjects, further implicating the dorsal striatum as the neural substrate in the incentive properties of ingestion.

It is becoming abundantly clear that the reward circuitry involved in feeding appetitive behaviors is also intimately associated with, if not identical to, the neural substrate involved in reinforcement of many addictive behaviors. For example, assessing 90 female eating disorder patients and 115 healthy female controls, Shinohara et al. (2004) found that binge eating was associated with a defect (short allele) in the dopamine transporter gene similar to that observed in nicotine, cocaine, and alcohol abuse, suggesting that dysregulation of DA reuptake which results in overstimulation of DA receptors may be a mechanism common to binging and substance abuse. Furthermore, the same reductions in brain DA receptors observed by Wang et al. (2001) in the obese have also been reported in abusers of cocaine, methamphetamine, alcohol, and heroin (Wang et al., 2002). Davis et al. (2003) tested sensitivity to reward (STR) using food reward in normal, overweight, and obese women. STR was positively correlated with measures of emotional overeating, and overweight women (>25 BMI) scored higher on STR than normal-weight women; however, obese women (>30 BMI) were more anhedonic (i.e., lower on the STR scale) than overweight women. These findings suggest that activity of reward circuits in the brain may be implicated in the initial stages of intake-driven obesity, but on achieving frank obesity, some neuroadaptation to brain reward circuit overactivity may occur.

The valence of various rewards of different modalities (e.g., food and drugs) and how these rewards interact using at least part of the same neural substrate has also been the subject of numerous studies. In a comparison of carbohydrate snack (CHO) preference to money rewards during nicotine deprivation in female smokers, Spring et al. (2004) found that abstinent smokers worked harder for CHO rewards relative to money in comparison to nonsmokers. They also worked harder for CHO during nicotine deprivation than when smoking, indicating that deprivation of one reward may increase the reinforcing value of the other. Whether there is an overlap in the neural circuitry of these rewards of differing modality remains to be addressed. However, the genetic basis of such interaction was the subject of a study by Lerman et al. (2004), who observed the effect of bupropion (a DA reuptake blocker) and the DA D_2 receptor gene (DRD2) on food

reward in smokers. Subjects underwent a test of food reward before bupropion or abstinence, and again after 3 weeks of bupropion or 1 week of abstinence. It was found that DRD2 A1 allele carriers exhibited greater food reward value after abstinence, which was attenuated by bupropion. Higher food reward levels predicted a 6-month weight increase in the placebo, but not bupropion group. Since the A1 allele renders D_2 receptors less able to bind DA, lowered neuronal DA activity in the face of reinforcement (nicotine) deprivation is thought to increase the rewarding value of food, leading to increased body weight. By promoting DA activity, bupropion could well decrease the value of food reward, hence lessen hedonically driven overeating and subsequent weight gain. In a more general study, Hodgkins et al. (2004) investigated the relationship between drug abstinence and body weight change in adolescents in a treatment facility, finding a significant body weight and BMI increase with abstinence. These results suggest that patients seeking drug treatment may be substituting food for their drug of choice, leading to obesity. It then should follow that individuals who rely on food's rewarding aspects to maintain signaling in neural reward circuits should not need to use drugs for such signaling, hence exceptional drug use ought not to be a problem in an obese population. This was borne out in a study by Kleiner et al. (2004), who found a significant inverse relationship between BMI and alcohol use in obese females, concluding that overeating may compete with alcohol for brain reward sites, making alcohol less reinforcing.

The collective point of these studies is that the multisensory experience of feeding and food choice in humans strongly activates the limbic forebrain, and activation patterns differ depending on the nutritional status (physiology) and degree of obesity (pathophysiology) of human subjects. These findings not only support conclusions obtained from animal studies, but also shed light on which neural substrates should be investigated further in animal studies.

VIII. CONCLUDING REMARKS

In closing, we return to the point made early in this chapter: that energy regulatory circuitry is intricately linked with reward circuitry. As such, it would seem to be a logical, rather than a radical, proposition that energy regulatory signals communicate directly with reward circuitry and vice versa. The concept has been put forth that energy regulatory circuitry is part of a negative feedback loop which includes the generation of peripheral signals that reflect body adipose stores, and these signals act primarily at the medial hypothalamus to regulate the efferent components of this feedback loop, specifically food intake and energy balance. However, the CNS anatomy suggests that reward circuitry ultimately should not be viewed as functionally separate from energy regulatory circuitry but as part of the loop. Inputs from reward circuitry may not just be "modulatory input" but are undoubtedly one critical component of the total CNS network that regulates food intake. Although currently, data that suggest that food reward is regulated are limited, the potential clinical and public health significance of understanding how this circuitry functions *in toto* is sufficient to justify future investigation.

Acknowledgments

Dianne Figlewicz Lattemann is supported by the Merit Review Program of the Dept. of Veterans Affairs and NIH Grant RO1-DK40963. Alfred J. Sipols is supported by the University of Latvia, Riga, Latvia. Nicole M. Sanders is supported by a Transition Faculty Grant from the American Diabetes Association. Amy MacDonald Naleid is supported by NIH Training Grant T32-AA007455, "Psychology Training in Alcohol Research."

References

Abrahamsen, G. C., and Carr, K. D. (1996). Effects of corticosteroid synthesis inhibitors on the sensitization of reward by food restriction. *Brain Res.* **726**, 39–48.

Adami, G., Campostano, A., Cella, F., and Fernandes, G. (2002). Serum leptin level and restrained eating—Study with Eating Disorder Examination. *Physiol. Behav.* **75**, 189–192.

Agmo, A., Galvan, A., and Talamantes, B. (1995). Reward and reinforcement produced by drinking sucrose: two processes that may depend on different neurotransmitters. *Pharmacol. Biochem. Behav.* **52**, 403–414.

Anand, B. K., and Brobeck, J. R. (1951). Hypothalamic control of food intake in rats and cats. *Yale J. Biol. Med.* **24**, 123–146.

Arase, K., Fisler, J. S., Shargill, N. S., York, D. A., and Bray, G. A. (1988). Intracerebroventricular infusions of 3-OHB and insulin in a rat model of dietary obesity. *Am. J. Physiol.* **255**, R974–R981.

Arjune, D., Standifer, K. M., Pasternak, G. W., and Bodnar, R. J. (1990). Reduction by central beta-funaltrexamine of food intake in rats under freely-feeding, deprivation and glucoprivic conditions. *Brain Res.* **535**, 101–109.

Arjune, D., Bowen, W. D., and Bodnar, R. J. (1991). Ingestive behavior following central [D-Ala2, Leu5,Cys6]-enkephalin (DALCE), a short-acting agonist and long-acting antagonist at the delta opioid receptor. *Pharmacol. Biochem. Behav.* **39**, 429–436.

Arnone, M., Maruani, J., Chaperon, F., Thiebot, M. H., Poncelet, M., Soubrie, P., and LeFur, G. (1997). Selective inhibition of sucrose and ethanol intake by SR141716, an antagonist of central cannabinoid (CB1) receptors. *Psychopharmacology* **132**, 104–106.

Badiani, A., Leone, P., Noel, M. B., and Stewart, J. (1995). Ventral tegmental area opioid mechanisms and modulation of ingestive behavior. *Brain Res.* **670**, 264–276.

Bakshi, V. P., and Kelley, A. E. (1993). Feeding induced by opioid stimulation of the ventral striatum: role of opiate receptor subtypes. *J. Pharmacol. Exp. Ther.* **265**, 1253–1260.

Balagura, S., and Hoebel, B. G. (1967). Self-stimulation of the lateral hypothalamus modified by insulin and glucagons. *Physiol. Behav.* **2**, 337–340.

Barbano, M. F., and Cador, M. (2005). Differential regulation of the consummatory, motivational, and anticipatory aspects of feeding behavior by dopaminergic and opioidergic drugs. *Neuropsychopharmacology*, e-Pub Oct. 2005.

Bardo, M. T., and Bevins, R. A. (2000). Conditioned place preference: what does it add to our preclinical understanding of drug reward? *Psychopharmacology* **153**, 31–43.

Baskin, D. G., Figlewicz Lattemann, D., Seeley, R. J., Woods, S. C., Porte, D., Jr., and Schwartz, M. W. (1999). Insulin and leptin: dual adiposity signals to the brain for the regulation of food intake and body weight. *Brain Res.* **848**, 114–123.

Bassareo, V., and DiChiara, G. (1997). Differential influence of associative and nonassociative learning mechanisms on the responsiveness of prefrontal and accumbal dopamine transmission to food stimulus in rats fed ad libitum. *J. Neurosci.* **17**, 851–861.

Bassareo, V., and DiChiara, G. (1999). Modulation of feeding-induced activation of mesolimbic dopamine transmission by appetitive stimuli and its relation to motivational state. *Eur. J. Neurosci.* **11**, 4389–4397.

Basso, A. M., and Kelley, A. E. (1999). Feeding induced by GABA (A) receptor stimulation within the nucleus accumbens shell: regional mapping and characterization of macronutrient and taste preference. *Behav. Neurosci.* **113**, 324–336.

Baunez, C., Amalric, M., and Robbins, T. W. (2002). Enhanced food-related motivation after bilateral lesions of the subthalamic nucleus. *J. Neurosci.* **22**, 562–568.

Baunez, C., Dias, C., Cador, M., and Amalric, M. (2005). The subthalamic nucleus exerts opposite control on cocaine and 'natural' rewards. *Nat. Neurosci.* **8**, 484–489.

Bell, S. M., Stewart, R. B., Thompson, S. C., and Meisch, R. A. (1997). Food deprivation increases cocaine-induced conditioned place preference and locomotor activity in rats. *Psychopharmacology (Berlin)* **131**, 1–8.

Bernardis, L. L., and Bellinger, L. L. (1993). The lateral hypothalamic area revisited: neuroanatomy, body weight regulation, neuroendocrinology and metabolism. *Neurosci. Biobehav. Rev.* **17**, 141–193.

Berridge, K. C. (1996). Food reward: brain substrates of wanting and liking. *Neurosci. Biobehav. Rev.* **28**, 309–369.

Berridge, K. C., and Robinson, T. E. (1998). What is the role of dopamine in reward: hedonic impact, reward learning, or incentive salience? *Brain Res. Rev.* **28**, 309–369.

Berry, E. M., and Mechoulam, R. (2002). Tetrahydro-cannabinol and endocannabinoids in feeding and appetite. *Pharmacol. Ther.* **95**, 185–190.

Berthoud, H. R. (2002). Multiple neural systems controlling food intake and body weight. *Neurosci. Biobehav. Rev.* **26**, 393–428.

Bielajew, C., and Trzcinska, M. (1994). Characteristics of stimulation-induced feeding sites in the sulcal prefrontal cortex. *Behav. Brain Res.* **61**, 29–35.

Birketvedt, G., Florholmen, J., Sundsfjord, J., Osterud, B., Dinges, D., Bilker, W., and Stunkard, A. J. (1999).

Behavioral and neuroendocrine characteristics of the night-eating syndrome. *JAMA* **282**, 657–663.

Birketvedt, G., Sundsfjord, J., and Florholmen, J. R. (2002). Hypothalamic-pituitary-adrenal axis in the night-eating syndrome. *Am. J. Physiol. Endocrinol. Metab.* **282**, E366–E369.

Bittencourt, J. C., Presse, F., Arias, C., Peto, C., Vaughan, J., Nahon, J. L., Vale, W., and Sawchenko, P. E. (1992). The melanin-concentrating hormone system of the rat brain: an immuno- and hybridization histochemical characterization. *J. Comp. Neurol.* **319**, 218–245.

Blundell, J. E., and Herberg, L. J. (1973). Effectiveness of lateral hypothalamic stimulation, arousal, and food deprivation in the intiation of hoarding behavior in naïve rats. *Physiol. Behav.* **4**, 763–764.

Brady, L. C., Smith, M. A., Gold, P. W., and Herkenham, M. (1990). Altered expression of hypothalamic neuropeptide mRNAs in food-restricted and food-deprived rats. *Neuroendocrinology* **52**, 441–447.

Branson, R., Potoczna, N., Kral, J. G., Lenter, K. U., Hoehe, M. R., and Horber, F. F. (2003). Binge eating as a major phenotype of melanocortin 4 receptor gene mutations. *N. Engl. J. Med.* **348**, 1096–1103.

Broberger, C., DeLecea, L., Sutcliffe, J. G., and Hokfelt, T. (1998). Hypocretin/orexin- and melanin-concentrating hormone-expressing cells form distinct populations in the rodent lateral hypothalamus: relationship to the neuropeptide Y and agouti gene-related protein systems. *J. Comp. Neurol.* **402**, 460–474.

Cabeza de Vaca, S., Holiman, S., and Carr, K. D. (1998). A search for the metabolic signal that sensitizes lateral hypothalamic self-stimulation in food-restricted rats. *Physiol. Behav.* **64**, 251–260.

Cabeza de Vaca, S., Kim G. Y., and Carr, K. D. (2002). The melanocortin receptor agonist MT II augments the rewarding effect of amphetamine in ad-libitum and food-restricted rats. *Psychopharmacology (Berlin)* **161**, 77–85.

Cabeza de Vaca, S., Hao, J., Afroz, T., Krahne, L. L., and Carr, K. D. (2005). Feeding, body weight, and sensitivity to non-ingestive reward stimuli during and after 12-day continuous central infusions of melanocortin receptor ligands. *Peptides* **26**, 2314–2321.

Carr, K. D. (1994). Streptozotocin-induced diabetes produces a naltrexone-reversible lowering of threshold for lateral hypothalamic self-stimulation. *Brain Res.* **664**, 211–214.

Carr, K. D. (2002). Augmentation of drug reward by chronic food restriction: behavioral evidence and underlying mechanisms. *Physiol. Behav.* **76**, 353–364.

Carr, K. D., and Wolinsky, T. D. (1993). Chronic food restriction and weight loss produce opioid facilitation of perifornical hypothalamic self-stimulation. *Brain Res.* **607**, 141–148.

Carr, K. D., Aleman, D. O., Bak, T. H., and Simon, E. J. (1991). Effects of parabrachial opioid antagonism on stimulation-induced feeding. *Brain Res.* **545**, 283–286.

Carr, K. D., Kim, G. Y., and Cabeza de Vaca, S. (2000). Hypoinsulinemia may mediate the lowering of self-stimulation thresholds by food restriction and streptozotocin-induced diabetes. *Brain Res.* **863**, 160–168.

Carroll, M. E., and Meisch, R. A. (1984). Increased drug-reinforced behavior due to food deprivation. *Adv. Behav. Pharmacol.* **4**, 47–88.

Chaperon, F., Soubrie, P., Puech, A. J., and Thiebot, M. H. (1998). Involvement of central cannabinoid (CB1) receptors in the establishment of place conditioning in rats. *Psychopharmacology (Berlin)* **135**, 324–332.

Chavez, M., Riedy, C. A., Van Dijk, G. V., and Woods, S. C. (1996). Central insulin and macronutrient intake in the rat. *Am. J. Physiol.* **271**, R727–R731.

Cleary, J., Weldon, D. T., O'Hare, E., Billington, C., and Levine, A. S. (1996). Naloxone effects on sucrose-motivated behavior. *Psychopharmacology (Berlin)* **126**, 110–114.

Cummings, D. E., and Shannon, M. H. (2003). Roles for ghrelin in the regulation of appetite and body weight. *Arch. Surg.* **138**, 389–396.

Davis, C., Strachan, S., and Berkson, M. (2003). Sensitivity to reward: implications for overeating and overweight. *Appetite* **42**, 131–138.

Dawe, S., and Loxton, N. J. (2004). The role of impulsivity in the development of substance use and eating disorders. *Eat. Behav.* **4**, 343–351.

Delgado, J., and Anand, B. K. (1953). Increase of food intake induced by electrical stimulation of the lateral hypothalamus. *J. Comp. Physiol. Psychol.* **172**, 162–168.

De Olmos, J. S., and Heimer, L. (1999). The concepts of the ventral striatopallidal system and extended amygdala. *Ann. N.Y. Acad. Sci.* **877**, 1–32.

Deroche-Gamonet, V., LeMoal, M. Piazza, P. V., and Soubrie, P. (2001). SR141716, a CB1 receptor antagonist, decreases the sensitivity to the reinforcing effects of electrical brain stimulation in rats. *Psychopharmacology (Berlin)* **157**, 254–259.

de Zwaan, M. (2001). Binge eating disorder and obesity. *Int. J. Obes. Rel. Metab. Disord.* **25**(Suppl. 1), S51–S55.

Di Marzo, V., Goparaju, S. K., Wang, L., Liu, J., Batkal, S., Jaral, Z., Fezza, F., Miura, G. I., Palmiter, R. D., Sugiura, T., and Kunos, G. (2001). Leptin-regulated endocannabinoids are involved in maintaining food intake. *Nature* **410**, 822–825.

Drewnowski, A., Krahn, D. D., Demitrack, M. A., Nairn, K., and Gosnell, B. A. (1992). Taste responses and preferences for sweet high fat foods: evidence of opioid involvement. *Physiol. Behav.* **51**, 371–379.

Drewnowski, A., Krahn, D. D., Demitrack, M. A., Nairn, K., and Gosnell, B. A. (1995). Naloxone, an opiate blocker, reduces the consumption of sweet high fat foods in obese and lean female binge eaters. *Am. J. Clin. Nutr.* **61**, 1206–1212.

Duarte, C., Alonso, R., Bichet, N., Cohen, C., Soubrie, P., and Thiebot, M.-H. (2004). Blockade by the cannabinoid CB1 receptor antagonist, Rimonabant (SR141716), of the potentiation by quinelorane of food-primed reinstatement of food-seeking behavior. *Neuropsychopharmacology* **29**, 911–920.

Elias, C. F., Saper, C. B., Maratos-Flier, E., Tritos, N. A., Lee, C., Kelly, J., Tatro, J. B., Hoffman, G. E., Ollman, M. M., Barsh, G. S., Sakurai, T., Yanagisawa, M., and Elmquist, J. K. (1998). Chemically defined projections linking the mediobasal hypothalamus and the lateral hypothalamic area. *J. Comp. Neurol.* **402**, 442–459.

Epstein, A. N., and Teitelbaum, P. (1967). Specific loss of the hypoglycemic control of feeding in recovered lateral rats. *Am. J. Physiol.* **213**, 1159–1167.

Everitt, B. J., Parkinson, J. A., Olmstead, M. C., Arroyo, M., Robledo, P., and Robbins, T. W. (1999). Associative processes in addiction and reward. The role of amygdala-ventral striatal subsystems. *Ann. N.Y. Acad. Sci.* **877**, 412–438.

Figlewicz, D. P. (2003). Adiposity signals and food reward: expanding the CNS roles of insulin and leptin. *Am. J. Physiol.* **284**, R882–R892.

Figlewicz, D. P., Higgins, M. S., Ng-Evans, S. B., and Havel, P. J. (2001). Leptin reverses sucrose-conditioned place preference in food-restricted rats. *Physiol. Behav.* **73**, 229–234.

Figlewicz, D. P., Evans, S. B., Murphy, J., Hoen, M., and Baskin, D. G. (2003). Expression of receptors for insulin and leptin in the ventral tegmental area/substantia nigra (VTA/SN) of the rat. *Brain Res.* **964**, 107–115.

Figlewicz, D. P., Bennett, J., Evans, S. B., Kaiyala, K., Sipols, A. J., and Benoit, S. B. (2004). Intraventricular insulin and leptin reverse place preference conditioned with high-fat diet in rats. *Behav. Neurosci.* **118**, 479–487.

Frank, G. K., Kaye, W. H., Carter, C. S., Brooks, S., May, C., Fissell, K., and Stenger, V. A. (2003). The evaluation of brain activity in response to taste stimuli—a pilot study and method for central taste activation as assessed by event-related fMRI. *J. Neurosci. Methods* **131**, 99–105.

Fride, E. (2002). Endocannabinoids in the central nervous system—an overview. *Prostaglandins, Leukot. Essent. Fatty Acids* **66**, 221–233.

Frisina, P. G., and Sclafani, A. (2002). Naltrexone suppresses the late but not early licking response to a palatable sweet solution: opioid hedonic hypothesis reconsidered. *Pharmacol. Biochem. Behav.* **74**, 163–172.

Fulton, S., Woodside, B., and Shizgal, P. (2000). Modulation of brain reward circuitry by leptin. *Science* **287**, 125–128.

Fulton, S., Richard, D., Woodside, B., and Shizgal, P. (2002a). Interaction of CRH and energy balance in the modulation of brain stimulation reward. *Behav. Neurosci.* **116**, 651–659.

Fulton, S., Woodside, B., and Shizgal, P. (2002b). Does neuropeptide Y contribute to the modulation of brain stimulation reward by chronic food restriction? *Behav. Brain Res.* **134**, 157–164.

Fulton, S., Richard, D., Woodside, B., and Shizgal, P. (2004). Food restriction and leptin impact brain reward circuitry in lean and obese Zucker rats. *Behav. Brain Res.* **155**, 319–329.

Gautier, J. F., Chen, K., Salbe, A. D., Bandy, D., Pratley, R. E., Heiman, M., Ravussin, E., Reiman, E. M., and Tataranni, P. A. (2000). Differential brain responses to satiation in obese and lean men. *Diabetes* **49**, 838–846.

Georgescu, D., Sears, R. M., Hommel, J. D., Barrot, M., Bolanos, C. A., Marsh, D. J., Bednarek, M. A., Bibb, J. A., Maratos-Flier, E., Nestler, E. J., and DiLeone, R. J. (2005). The hypothalamic neuropeptide melanin-concentrating hormone acts in the nucleus accumbens to modulate feeding behavior and forced-swim performance. *J. Neurosci.* **25**, 2933–2940.

Giraudo, S. Q., Grace, M. K., Welch, C. C., Billington, C. J., and Levine, A. S. (1993). Naloxone's anorectic effect is dependent upon the relative palatability of food. *Pharmacol. Biochem. Behav.* **46**, 917–921.

Glass, M. J., Grace, M., Cleary, J. P., Billington, C. J., and Levine, A. S. (1996). Potency of naloxone's anorectic effect in rats is dependent on diet preference. *Am. J. Physiol.* **271**, R217–R221.

Glass, M. J., Cleary, J. P., Billington, C. J., and Levine, A. S. (1997). Role of carbohydrate type on diet selection in neuropeptide Y-stimulated rats. *Am. J. Physiol.* **273**, R2040–R2045.

Glass, M. J., Billington, C. J., and Levine, A. S. (1999). Role of lipid type on morphine-stimulated diet selection in rats. *Am. J. Physiol.* **277**, R1345–R1350.

Glass, M. J., Billington, C. J., and Levine, A. S. (2000). Naltrexone administered to central nucleus of amygdale or PVN: neural dissociation of diet and energy. *Am. J. Physiol.* **279**, R86–R92.

Gosnell, B. A., and Krahn, D. D. (1992). The effects of continuous naltrexone infusions on diet preferences are modulated by adaption to the diets. *Physiol. Behav.* **51**, 239–244.

Gottfried, J. A., O'Doherty, J., and Dolan, R. J. (2003). Encoding predictive reward value in human amygdala and orbitofrontal cortex. *Science* **301**, 1104–1107.

Hakansson, M. L., Brown, H., Ghilardi, N., Skoda, R. C., and Meister, B. (1998). Leptin receptor immunoreactivity in chemically defined target neurons of the hypothalamus. *J. Neurosci.* **18**, 559–572.

Harrold, J. A., and Williams, G. (2003). The cannabinoid system: a role in both the homeostatic and hedonic control of eating? *Br. J. Nutr.* **90**, 729–734.

Harrold, J. A., Elliott, J. C., King, P. J., Widdowson, P. S., and Williams, G. (2002). Downregulation of cannabinoid-1 (CB-1) receptors in specific extrahy-

pothalamic regions of rats with dietary obesity: a role for endogenous cannabinoids in driving appetite for palatable food? *Brain Res.* **952**, 232–238.

Havel, P. J. (2000). Role of adipose tissue in body-weight regulation: mechanisms regulating leptin production and energy balance. *Proc. Nutr. Soc.* **59**, 359–371.

Hebebrand, J., Blum, W. F., Barth, N., Coners, H., Enlargo, P., Juul, A., Ziegler, A., Warnke, A., Rascher, W., and Remschmidt, H. (1997). Leptin levels in patients with anorexia nervosa are reduced in the acute stage and elevated upon short-term weight restoration. *Mol. Psychiatry* **2**, 330–334.

Heimer, L., Zahm, D. S., Churchill, L., Kalivas, P. W., and Wohltmann, C. (1991). Specificity in the projection patterns of accumbal core and shell in the rat. *Neuroscience* **41**, 89–125.

Hill, J. O., Wyatt, H. R., Reed, G. W., and Peters, J. C. (2003). Obesity and the environment: where do we go from here? *Science* **299**, 853–855.

Hodgkins, C., Cahill, K. S., Seraphine, A. E., Frost-Pineda, K., and Gold, M. S. (2004). Adolescent drug addiction treatment and weight gain. *J. Addict. Dis.* **23**, 55–65.

Hoebel, B., Hernandez, L., Schwartz, D. H., Mark, G. P., and Hunter, G. A. (1989). Microdialysis studies of brain norepinephine, serotonin, and dopamine release during ingestive behavior. Theoretical and clinical implications. *Ann. N.Y. Acad. Sci.* **575**, 171–191.

Ikemoto, S., and Panksepp, J. (1996). Dissociations between appetitive and consummatory responses by pharmacological manipulations of reward-relevant brain regions. *Behav. Neurosci.* **100**, 331–345.

Islam, A. K., and Bodnar, R. J. (1990). Selective opioid receptor antagonist effects upon intake of a high-fat diet in rats. *Brain Res.* **508**, 293–296.

Jarosz, P. A., and Metzger, B. L. (2002). The effect of opioid antagonism on food intake behavior and body weight in a biobehavioral model of obese binge eating. *Biol. Res. Nurs.* **3**, 198–209.

Jimerson, D. C., Mantzoros, C., Wolfe, B. E., and Metzger, E. D. (2000). Decreased serum leptin in bulimia nervosa. *J. Clin. Endocrinol. Metabol.* **85**, 4511–4514.

Karhunen, L., Haffner, S., Lappalainen, R., Turpeinen, A., Miettinen, H., and Uusitupa, M. (1997). Serum leptin and short-term regulation of eating in obese women. *Clin. Sci.* **92**, 573–578.

Kelley, A. E., Bakshi, V. P., Haber, S. N., Steininger, T. L., Will, M. J., and Zhang, M. (2002). Opioid modulation of taste hedonics within the ventral striatum. *Physiol. Behav.* **76**, 365–377.

Killgore, W. D. S., Young, A. D., Femia, L. A., Bogorodzki, P., Rogowska, J., and Yurgelun-Todd, D. A. (2003). Cortical and limbic activation during viewing of high- versus low-calorie foods. *Neuroimage* **19**, 1381–1394.

Kim, E. M., Welch, C. C., Grace, M. K., Billington, C. J., and Levine, A. S. (1996). Chronic food restriction and acute food deprivation decrease mRNA levels of opioid peptides in arcuate nucleus. *Am. J. Physiol.* **270**, R1019–R1024.

Kirkham, T. C. (2005). Endocannabinoids in the regulation of appetite and body weight. *Behav. Pharmacol.* **16**, 297–313.

Kirkham, T. C., and Blundell, J. E. (1986). Effect of naloxone and naltrexone on the development of satiation measured in the runway: comparisons with *d*-amphetamine and *d*-fenfluramine. *Pharmacol. Biochem. Behav.* **25**, 123–128.

Kirkham, T. C., and Cooper, S. J. (1988a). Attenuation of sham feeding by naloxone is stereospecific: evidence for opioid mediation of orosensory reward. *Physiol. Behav.* **43**, 845–857.

Kirkham, T. C., and Cooper, S. J. (1988b). Naloxone attenuation of sham feeding is modified by manipulation of sucrose concentration. *Physiol. Behav.* **44**, 491–494.

Kirkham, T. C., and Williams, C. M. (2004). Endocannabinoid receptor antagonists. *Treat. Endocrinol.* **3**, 1–16.

Kirkham, T. C., Williams, C. M., Fezza, F., and Di Marzo, V. (2002). Endocannabinoid levels in rat limbic forebrain and hypothalamus in relation to fasting, feeding and satiation: stimulation of eating by 2-arachidonoyl glycerol. *Br. J. Pharmacol.* **136**, 550–557.

Kirouac, G. J., and Ganguly, P. K. (1995). Topographical organization in the nucleus accumbens of afferents from the basolateral amygdala and efferents to the lateral hypothalamus. *Neuroscience* **67**, 625–630.

Kiyatkin, A. E. (1995). Functional significance of mesolimbic dopamine. *Neurosci. Biobehav. Rev.* **19**, 573–598.

Kleiner, K. D., Gold, M. S., Frost-Pineda, K., Lenz-Brunsman, B., Perri, M. G., and Jacobs, W. S. (2004). Body mass index and alcohol use. *J. Addict. Dis.* **23**, 105–118.

Koob, G. F., and Bloom, F. (1988). Cellular and molecular mechanisms of drug dependence. *Science* **242**, 715–723.

Kringelbach, M. L., O'Doherty, J., Rolls, E. T., and Andrews, C. (2003). Activation of the human orbitofrontal cortex to a liquid food stimulus is correlated with its subjective pleasantness. *Cereb. Cortex* **13**, 1064–1071.

Krugel, U., Schraft, T., Kittner, H., Kiess, W., and Illes, P. (2003). Basal and feeding-evoked dopamine release in the rat nucleus accumbens is depressed by leptin. *Eur. J. Pharmacol.* **482**, 185–187.

Lamonte, N., Ackerman, T. F., and Bodnar, R. J. (2002a). Opioid subtype antagonist modulation of mu opioid-induced feeding: ventral tegmental area and nucleus accumbens interactions in rats. *Soc. Neurosci. Abst.*

Lamonte, N., Echo, J. A., Ackerman, T. F., Christian, G., and Bodnar, R. J. (2002b). Analysis of opioid receptor subtype antagonist effects upon mu opioid

agonist-induced feeding elicited from the ventral tegmental area of rats. *Brain Res.* **929**, 96–100.

Lang, I. M., Strahlendorf, J. C., Strahlendorf, H. K., and Barnes, C. D. (1981). Effects of chronic administration of naltrexone on appetitive behaviors of rats. *Prog. Clin. Biol. Res.* **68**, 197–207.

Lepore, M., Vorel, S. R., Lowinson, J., and Gardner, E. L. (1995). Conditioned place preference induced by delta9-tetra-hydrocannabinol: comparison with cocaine, morphine, and food reward. *Life Sci.* **56**, 2073–2080.

Lerman, C., Berrettini, W., Pinto, A., Patterson, F., Crystal-Mansour, S., Wileyto, E. P., Restine, S., Leonard, D. G. B., Shields, P. G., and Epstein, L. H. (2004). Changes in food reward following smoking cessation; a pharmacogenetic investigation. *Psychopharmacology (Berlin)* **174**, 571–577.

Levine, A. S., Grace, M., Billington, C. J., and Portoghese, P. S. (1990). Nor-binaltorphimine decreases deprivation and opioid-induced feeding. *Brain Res.* **534**, 60–64.

Levine, A. S., Grace, M., and Billington, C. J. (1991). Beta-funaltrexamine (beta-FNA) decreases deprivation and opioid-induced feeding. *Brain Res.* **562**, 281–284.

Levine, A. S., Weldon, D. T., Grace, M., Cleary, J. P., and Billington, C. J. (1995). Naloxone blocks that portion of feeding driven by sweet taste in food-restricted rats. *Am. J. Physiol.* **268**, R248–R252.

Levine, A. S., Kotz, C. M., and Gosnell, B. A. (2003). Sugars and fats: the neurobiology of preference. *J. Nutr.* **133**, 831S–834S.

Lu, L., Shepard, J. D., Scott Hall, F., and Shaham, Y. (2003). Effect of environmental stressors on opiate and psychostimulant reinforcement, reinstatement and discrimination in rats: a review. *Neurosci. Biobehav. Rev.* **27**, 457–491.

Ludwig, D. S., Tritos, N. A., Mastaitis, J. W., Kulkarni, R., Kokkotou, E., Elmquist, J., Lowell, B., Flier, J. S., and Maratos-Flier, E. (2001). Melanin-concentrating hormone overexpression in transgenic mice leads to obesity and insulin resistance. *J. Clin. Invest.* **107**, 379–386.

Maldonado-Irizarry, C. S., Swanson, C. J., and Kelley, A. E. (1995). Glutamate receptors in the nucleus accumbens shell control feeding behavior via the lateral hypothalamus. *J. Neurosci.* **15**, 6779–6788.

Margules, D. L., and Olds, J. (1962). Identical "feeding" and "rewarding" systems in the lateral hypothalamus of rats. *Science* **135**, 374–375.

Marinelli, M., and Piazza, P. V. (2002). Interaction between glucocorticoid hormones, stress, and psychostimulant drugs. *Eur. J. Neurosci.* **16**, 387–394.

Marks-Kaufman, R., Balmagiya, T., and Gross, E. (1984). Modifications in food intake and energy metabolism in rats as a function of chronic naltrexone infusions. *Pharmacol. Biochem. Behav.* **20**, 911–916.

Marks-Kaufman, R., Plager, A., and Kanarek, R. B. (1985). Central and peripheral contributions of endogenous opioid systems to nutrient selection in rats. *Psychopharmacology (Berlin)* **85**, 414–418.

Martin, W. R., Wikler, A., Eades, C. G., and Pescor, F. T. (1963). Tolerance to and physical dependence on morphine in rats. *Psychopharmacologia* **65**, 247–260.

Maurice, N., Deniau, J. M., Glowinski, J., and Thierry, A. M. (1999). Relationships between the prefrontal cortex and the basal ganglia in the rat: physiology of the cortico-nigral circuits. *J. Neurosci.* **19**, 4674–4681.

McBride, W. J., Murphy, J. M., and Ikemoto, S. (1999). Localization of brain reinforcement mechanisms: intracranial self-administration and intracranial place-conditioning studies. *Behav. Brain Res.* **101**, 129–152.

McGregor, I. S., Menendez, J. A., and Atrens, D. M. (1990a). Metabolic effects obtained from excitatory amino acid stimulation of the sulcal prefrontal cortex. *Brain Res.* **529**, 1–6.

McGregor, I. S., Menendez, J. A., and Atrens, D. M. (1990b). Metabolic effects of neuropeptide Y injected into the sulcal prefrontal cortex. *Brain Res. Bull.* **24**, 363–367.

Miller, C. C., Murray, T. F., Freeman, K. G., and Edwards, G. L. (2004). Cannabinoid agonist, CP55,940, facilitates intake of palatable foods when injected into the hindbrain. *Physiol. Behav.* **80**, 611–616.

Mokdad, A., Bowman, B., Ford, E., Vinicor, R., Marks, J., and Koplan, J. (2001). The continuing epidemics of obesity and diabetes in the United States. *JAMA* **286**, 1195–1200.

Monteleone, P., Di Lieto, A., Tortorella, A., Longobardi, N., and Maj, M. (2000). Circulating leptin levels in patients with anorexia nervosa, bulimia nervosa or binge-eating disorder: relationship to body weight, eating patterns, psychopathology and endocrine changes. *Psychiat. Res.* **94**, 121–129.

Monteleone, P., Martiadis, V., Colurcio, B., and Maj, M. (2002). Leptin secretion is related to chronicity and severity of the illness in bulimia nervosa. *Psychosom. Med.* **64**, 874–879.

Naleid, A. M., Grace, M. K., Cummings, D. E., and Levine, A. S. (2005). Ghrelin induces feeding in the mesolimbic reward pathway between the ventral tegmental area and the nucleus accumbens. *Peptides* **26**, 2274–2279.

Noel, M. B., and Wise, R. A. (1995). Ventral tegmental injections of a selective mu or delta opioid enhance feeding in food-deprived rats. *Brain Res.* **673**, 304–312.

Olds, J. (1958). Self-stimulation of the brain. *Science* **127**, 315–324.

Olds, J. (1962). Hypothalamic substrate of reward. *Physiol. Rev.* **42**, 554–604.

Olszewski, P. K., and Levine, A. S. (2004). Minireview: characterization of influence of central nociceptin/orphanin FQ on consummatory behavior. *Endocrinology* **145**, 2627–2632.

Oomura, Y., Ono, T., Ooyama, H., and Wayner, M. J. (1969). Glucose and osmosensitive neurons of the rat hypothalamus. *Nature* **222**, 282–284.

Oomura, Y., Ooyama, H., Sugimori, M., Nakamura, T., and Yamada, Y. (1974). Glucose inhibition of the glucose-sensitive neurone in the rat lateral hypothalamus. *Nature* **247**, 284–286.

Papp, M. (1988). Different effects of short- and long-term treatment with imipramine on the apomorphine- and food-induced place preference conditioning in rats. *Pharmacol. Biochem. Behav.* **30**, 889–893.

Pecina, S., Cagniard, B., Berridge, K. C., Aldridge, J. W., and Zhuang, X. (2003). Hyperdopaminergic mutant mice have higher "wanting" but not "liking" for sweet rewards. *J. Neurosci.* **23**, 9395–9402.

Petrovic, G. D., Setlow, B., Holland, P. C., and Gallagher, M. (2002). Amygdalo-hypothalamic circuit allows learned cues to override satiety and promote eating. *J. Neurosci.* **22**, 8748–8753.

Phillipson, O. T., and Griffiths, A. C. (1985). The topographic order of inputs to nucleus accumbens in the rat. *Neuroscience* **16**, 275–296.

Pi-Sunyer, X. (2003). A clinical view of the obesity problem. *Science* **299**, 859–860.

Pomonis, J. D., Levine, A. S., and Billington, C. J. (1997). Interaction of the hypothalamic paraventricular nucleus and central nucleus of the amygdala in naloxone blockade of neuropeptide Y-induced feeding revealed by c-fos expression. *J. Neurosci.* **17**, 5175–5182.

Presse, F., Sorokovsky, I., Max, J.-P., Nicolaidis, S., and Nahon, J. L. (1996). Melanin-concentrating hormone is a potent anorectic peptide regulated by food-deprivation and glucopenia in the rat. *Neuroscience* **71**, 735–745.

Qu, D., Ludwig, D. S., Gammeltoft, S., Piper, M., Pelleymounter, M. A., Cullen, M. J., Mathes, W. F., Przypek, R., Kanarek, R., and Maratos-Flier, E. (1996). A role for melanin-concentrating hormone in the central regulation of feeding behavior. *Nature* **380**, 243–247.

Ravinet Trillou, C., Arnone, M., Delgorge, C., Gonalons, N., Keane, P., Maffrand, J.-P., and Soubrie, P. (2003). Anti-obesity effect of SR141716, a CB1 receptor antagonist, in diet-induced obese mice. *Am. J. Physiol.* **284**, R345–R353.

Richardson, N. R., and Gratton, A. (1996). Behavior-relevant changes in nucleus accumbens dopamine transmission elicited by food reinforcement: an electrochemical study in rat. *J. Neurosci.* **24**, 8160–8169.

Robbins, T. W., and Koob, G. F. (1980). Selective disruption of displacement behavior by lesions of the mesolimbic dopamine system. *Nature* **285**, 409–412.

Robbins, T. W., and Everitt, B. J. (1996). Neurobehavioural mechanisms of reward and motivation. *Curr. Opin. Neurobiol.* **6**, 228–236.

Robbins, T. W., Cador, M., Taylor, J. R., and Everitt, B. J. (1989). Limbic-striatal interactions in reward-related processes. *Neurosci. Biobehav. Rev.* **13**, 155–162.

Rolls, E. T. (1994). Neural processing related to feeding in primates. *In* "Appetite: Neural and Behavioral Bases" (C. R. Legg and D. A. Booth, eds.), pp. 11–53. Oxford University Press, Oxford.

Rolls, E. T. (2004). The functions of the orbitofrontal cortex. *Brain Cogn.* **55**, 11–29.

Rolls, E. T., Critchley, H. D., Browning, A. S., Hernadi, I., and Lenard, L. (1999). Responses to the sensory properties of fat of neurons in the primate orbitofrontal cortex. *J. Neurosci.* **19**, 1532–1540.

Rudski, J. M., Billington, C. J., and Levine, A. S. (1994). Naloxone's effects on operant responding depend upon level of deprivation. *Pharmacol. Biochem. Behav.* **49**, 377–383.

Saito, Y., Cheng, M., Leslie, F. M., and Civelli, O. (2001). Expression of the melanin-concentrating hormone (MCH) receptor mRNA in the rat brain. *J. Comp. Neurol.* **435**, 26–40.

Salamone, J. D., Correa, M., Mingote, S., and Weber, S. M. (2003). Nucleus accumbens dopamine and the regulation of effort in food-seeking behavior: implications for studies of natural motivation, psychiatry, and drug abuse. *J. Pharmacol. Exp. Ther.* **305**, 1–8.

Sanchis-Segura, C., Cline, B. H., Marsicano, G., Lutz, B., and Spanagel, R. (2004). Reduced sensitivity to reward in CB1 knockout mice. *Psychopharmacology (Berlin)* **176**, 223–232.

Saper, C. B., Chou, T. C., and Elmquist, J. K. (2002). The need to feed: homeostatic and hedonic control of eating. *Neuron* **36**, 199–211.

Schultz, W. (2002). Getting formal with dopamine and reward. *Neuron* **36**, 241–263.

Schultz, W. (2004). Neural coding of basic reward terms of animal learning theory, game theory, microeconomics and behavioural ecology. *Curr. Opin. Neurobiol.* **14**, 139–147.

Sergeyev, V., Broberger, C., Gorbatyuk, O., and Hokfelt, T. (2000). Effect of 2-mercaptoacetate and 2-deoxy-D-glucose administration on the expression of NPY, AGRP, POMC, MCH and hypocretin/orexin in the rat hypothalamus. *Neuroreport* **11**, 117–121.

Shalev, U., Yap, J., and Shaham, Y. (2001). Leptin attenuates food deprivation-induced relapse to heroin seeking. *J. Neurosci.* 21: RC129.

Shalev, U., Grimm, J. W., and Shaham, Y. (2002). Neurobiology of relapse to heroin and cocaine seeking: a review. *Pharmacol. Rev.* **54**, 1–42.

Shimada, M., Tritos, N. A., Lowell, B. B., Flier, L. S., and Maratos-Flier, E. (1998). Mice lacking melanin-concentrating hormone are hypophagic and lean. *Nature* **396**, 670–674.

Shinohara, M., Mizushima, H., Hirano, M., Shioe, K., Nakazawa, M., Hiejima, Y., Ono, Y., and Kanba, S. (2004). Eating disorders with binge-eating behaviour are associated with the s allele of the 3'-UTR VNTR polymorphism of the dopamine transporter gene. *J. Psych. Neurosci.* **29**, 134–137.

Shizgal, P., Fulton, S., and Woodside, B. (2001). Brain reward circuitry and the regulation of energy balance. *Int. J. Obes.* **25**(Suppl. 5), S17–S21.

Sipols, A. J., Bayer, J., Bennett, R., and Figlewicz, D. P. (2002). Intraventricular insulin decreases kappa opioid-mediated sucrose intake in rats. *Peptides* **23**, 2181–2187.

Small, D. M., Jones-Gotman, M., and Dagher, A. (2003). Feeding-induced dopamine release in dorsal striatum correlates with meal pleasantness ratings in healthy human volunteers. *Neuroimage* **19**, 1709–1715.

Smith, S. L., Harrold, J. A., and Williams, G. (2002). Diet-induced obesity increases mu opioid receptor binding in specific regions of the rat brain. *Brain Res.* **953**, 215–222.

Spring, B., Pagoto, S., McChargue, D., Hedeker, D., and Werth, J. (2004). Altered reward value of carbohydrate snacks for female smokers withdrawn from nicotine. *Pharmacol. Biochem. Behav.* **76**, 351–360.

Stanley, B. G., Willett, V. L., Donias, H. W., Ha, L. H., and Spears, L. C. (1993). The lateral hypothalamus: a primary site mediating excitatory amino acid-elicited eating. *Brain Res.* **630**, 41–49.

Stratford, T. R., and Kelley, A. E. (1997). GABA in the nucleus accumbens shell participates in the central regulation of feeding behavior. *J. Neurosci.* **17**, 4434–4440.

Stratford, T. R., and Kelley, A. E. (1999). Evidence of a functional relationship between the nucleus accumbens shell and lateral hypothalamus subserving the control of feeding behavior. *J. Neurosci.* **19**, 11040–11048.

Stunkard, A. J., and Allison, K. C. (2003). Two forms of disordered eating in obesity: binge eating and night eating. *Int. J. Obes.* **27**, 1–12.

Swerdlow, N. R., Van der Kooy, D., Koob, G. F., and Wenger, J. R. (1983). Cholecystokinin produces conditioned place-aversions, not place preferences, in food-deprived rats: evidence against involvement in satiety. *Life Sci.* **32**, 2087–2093.

Tanda, G., and DiChiara, G. (1998). A dopamine-mu 1 opioid link in the rat ventral tegmentum shared by palatable food (Fonzies) and non-psychostimulant drugs of abuse. *Eur. J. Neurosci.* **10**, 1179–1187.

Trojniar, W., and Wise, R. A. (1991). Facilitory effect of delta 9-tetrahydrocannabinol on hypothalamically induced feeding. *Psychopharmacology (Berlin)* **103**, 172–176.

Tucci, S. A., Rogers, E. K., Korbonits, M., and Kirkham, T. C. (2004). The cannabinoid CB1 receptor antagonist SR141716 blocks the orexigenic effects of intrahypothalamic ghrelin. *Br. J. Pharmacol.* **143**, 520–523.

Tzschentke, T. M. (1998). Measuring reward with the conditioned place preference paradigm: a comprehensive review of drug effects, recent progress, and new issues. *Prog. Neurobiol.* **56**, 613–672.

Volkow, N. D., Wang, G.-J., Fowler, J. S., Logan, J., Jayne, M., Franceschi, D., Wong, C., Gatley, S. J., Gifford, A. N., Ding, Y. S., and Pappas, N. (2002). "Nonhedonic" food motivation in humans involves dopamine in the dorsal striatum and methylphenidate amplifies this effect. *Synapse* **44**, 175–180.

Volkow, N. D., Wang, G.-J., Maynard, L., Jayne, M., Fowler, J. S., Zhu, W., Logan, J., Gatley, S. J., Ding, Y.-S., Wong, C., and Pappas, N. (2003). Brain dopamine is associated with eating behaviors in humans. *Int. J. Eat. Disord.* **33**, 136–142.

Wang, G.-J., Volkow, N. D., Logan, J., Pappas, N. R., Wong, C. T., Zhu, W., Netusil, L., and Fowler, J. S. (2001). Brain dopamine and obesity. *Lancet* **357**, 354–357.

Wang, G.-J., Volkow, N. D., and Fowler, J. S. (2002). The role of dopamine in motivation for food in humans: implications for obesity. *Expert Opin. Ther. Targets* **6**, 601–609.

Welch, C. C., Kim, E. M., Grace, M. K., Billington, C. J., and Levine, A. S. (1996). Palatability-induced hyperphagia increases hypothalamic dynorphin peptide and mRNA levels. *Brain Res.* **721**, 126–131.

Weldon, D. T., O'Hare, E., Cleary, J., Billington, C. J., and Levine, A. S. (1996). Effect of naloxone on intake of cornstarch, sucrose, and polycose diets in restricted and non-restricted rats. *Am. J. Physiol.* **270**, R1183–R1188.

Westerhaus, M. J., and Loewy, A. D. (2001). Central representation of the sympathetic nervous system in the cerebral cortex. *Brain Res.* **903**, 117–127.

Williams, G., Bing, C., Cai, X. J., Harrold, J. A., King, P. J., and Liu, X. H. (2001). The hypothalamus and the control of energy homeostasis: different circuits, different purposes. *Physiol. Behav.* **74**, 683–701.

Wilson, C., Nomikos, G. G., Collu, M., and Fibiger, H. C. (1995). Dopaminergic correlates of motivated behavior: importance of drive. *J. Neurosci.* **15**, 5169–5178.

Wise, R. A. (1988). Psychomotor stimulant properties of addictive drugs. *Ann. N.Y. Acad. Sci.* **537**, 228–234.

Wise, R. A. (2002). Brain reward circuitry: insights from unsensed incentives. *Neuron* **36**, 229–240.

Wise, R. A., and Hoffman, D. C. (1992). Localization of drug reward mechanisms by intracranial injections. *Synapse* **10**, 247–263.

Wolinsky, T. D., Carr, K. D., Hiller, J. M., and Simon, E. J. (1994). Effects of chronic food restriction on mu and kappa opioid binding in rat forebrain: a quantitative autoradiographic study. *Brain Res.* **656**, 274–280.

Wolinsky, T. D., Abrahamsen, G. C., and Carr, K. D. (1996a). Diabetes alters mu and kappa opioid binding in rat brain regions: comparison with effects of food restriction. *Brain Res.* **738**, 167–171.

Wolinsky, T. D., Carr, K. D., Hiller, J. M., and Simon, E. J. (1996b). Chronic food restriction alters mu and kappa opioid receptor binding in the parabrachial nucleus of the rat: a quantitative autoradiographic study. *Brain Res.* **706**, 333–336.

Woods, S. C., Lotter, E. C., McKay, L. D., and Porte, D., Jr. (1979). Chronic intracerebroventricular infusion of insulin reduces food intake and body weight of baboons. *Nature* **282**, 503–505.

Yanovski, S. Z., and Yanovski, J. A. (2002). Obesity. *N. Engl. J. Med.* **346**, 591–602.

Yeomans, M. R., Wright, P., Macleod, H. A., and Critchley, J. A. J. H. (1990). Effect of nalmefene on feeding in humans. Dissociation of hunger and palatability. *Psychopharmacology (Berlin)* **100**, 426–432.

Yu, W. Z., Ruegg, H., and Bodnar, R. J. (1997). Delta and kappa opioid receptor subtypes and ingestion: antagonist and glucoprivic effects. *Pharmacol. Biochem. Behav.* **56**, 353–361.

Yu, W. Z., Sclafani, A., Delamater, A. R., and Bodnar, R. J. (1999). Pharmacology of flavor preference conditioning in sham-feeding rats: effects of naltrexone. *Pharmacol. Biochem. Behav.* **64**, 573–584.

Zhang, M., and Kelley, A. E. (2000). Enhanced intake of high-fat food following striatal mu-opioid stimulation: microinjection mapping and fos expression. *Neuroscience* **99**, 267–277.

Zhang, M., Gosnell, B. A., and Kelley, A. E. (1998). Intake of high-fat food is selectively enhanced by mu opioid receptor stimulation within the nucleus accumbens. *J. Pharmacol. Exp. Ther.* **285**, 908–914.

Zheng, H., Corkern, M., Stoyanova, I., Patterson, L. M., Tian, R., and Berthoud, H.-R. (2003). Appetite-inducing accumbens manipulation activates hypothalamic orexin neurons and inhibits POMC neurons. *Am. J. Physiol.* **284**, R1436–R1444.

8

Brain Reward Systems for Food Incentives and Hedonics in Normal Appetite and Eating Disorders

KENT C. BERRIDGE

Department of Psychology (Biopsychology Program),
University of Michigan

I. Introduction
II. Possible Roles of Brain Reward Systems in Eating Disorders
   A. Reward Dysfunction As Cause
   B. Passively Distorted Reward Function As Consequence
   C. Normal Resilience in Brain Reward
III. Understanding Brain Reward Systems for Food "Liking" and "Wanting"
   A. Measuring Pleasure
   B. Brain Systems for Food Pleasure
   C. Pleasure Brain Hierarchy
   D. Forebrain Limbic Hedonic Hot Spot in Nucleus Accumbens
   E. Ventral Pallidum: "Liking" and "Wanting" Pivot Point for Limbic Food Reward Circuits
IV. "Wanting" Without "Liking"
   A. What Is "Wanting"?
   B. Addiction and Incentive Sensitization
   C. Is There a Neural Sensitization Role in Food Addictions?
   D. Cognitive Goals and Ordinary Wanting
V. A Brief History of Appetite: Food Incentives, Not Hunger Drives
   A. Separating Reward and Drive Reduction
VI. Connecting Brain Reward and Regulatory Systems
VII. Conclusion
   Acknowledgments
   References

What brain reward systems mediate motivational "wanting" and hedonic "liking" for food rewards? And what roles might these systems play in eating disorders? This chapter identifies more recent discoveries in hedonic "liking" mechanisms, such as the location of opioid hedonic hotspots in nucleus accumbens and ventral pallidum that paint a pleasure gloss onto sweet sensation. It also considers other incentive motivation systems that mediate only a nonhedonic "wanting" component of reward, such as nearby limbic opioid "wanting" zones, mesolimbic dopamine contributions to incentive salience, and other components of brain limbic systems, and discusses potential roles in eating disorders.

I. INTRODUCTION

Obesity, bulimia, anorexia, and related eating disorders have risen in recent decades, leading to increasing concern. Can improved knowledge about brain reward systems guide us in thinking about eating disorders and normal eating?

First it is important to recognize that brain reward systems are active participants, not just passive conduits, in the act of eating. Sweetness and other food tastes are merely sensation, and their pleasure arises within the brain. Hedonic brain systems must actively paint the pleasure onto sensation to generate a "liking" reaction—as a sort of "pleasure gloss." What brain systems paint a pleasure gloss onto sensation? And what brain systems convert pleasure into a desire to eat? Answers to these questions requires combining information from neuroimaging experiments and studies of eating in humans and information from brain manipulation and pharmacological experiments that ethically can be done only in animals.

II. POSSIBLE ROLES OF BRAIN REWARD SYSTEMS IN EATING DISORDERS

To begin with, we can sketch several alternative possibilities for how brain reward systems might function in any particular eating disorder. Let us set out some alternatives to frame the issue.

A. Reward Dysfunction As Cause

First, it is possible that some aspects of brain reward function may go wrong and actually cause an eating disorder. Foods might become hedonically "liked" too much or too little via reward dysfunction. Or incentive salience "wanting" to eat might detach from normal close association with hedonic "liking," leading to changes in motivated food consumption that are no longer hedonically driven. Or yet again, suppression of positive hedonic reward systems or activation of dysphoric stress systems might prompt persistent attempts to self-medicate by eating palatable food. All of these possibilities have been suggested at one time or another. Each of them deserves consideration because different answers might apply to different disorders.

B. Passively Distorted Reward Function As Consequence

As a second category of possibilities, brain reward systems might remain intrinsically normal and have no essential pathology in eating disorders, but still become distorted in function as a passive secondary consequence of disordered intake. In that case, brain systems of "liking" and "wanting" might well attempt to function normally. The abnormal feedback from physiological signals that are altered by binges of eating or by periods of anorexia might induce reward dysfunction as a consequence of the behavioral disorder that arose from other causes. This would provide a potential red herring to neuroscientists searching for causes of eating disorder, because brain abnormalities might appear as neural markers for a particular disorder, but be mistaken as causes when they were actually consequences. However, it might still provide a window of opportunity for medication treatments that aim to correct eating behavior in part by modulating reward function back to a normal range.

C. Normal Resilience in Brain Reward

Third, it is possible that most aspects of brain reward systems will function even more normally than suggested by the passively distorted consequence model above. Many compensatory changes can take place in response to physiological alterations, to oppose them via homeostatic or negative feedback corrections. The final consequence of those compensations might restore normality to brain reward functions. In such cases, the causes of eating disorder might then be found to lie completely outside brain reward functions. Indeed, brain reward functions will persist largely normally, and may even serve as aids to eventually help spontaneously normalize eating behavior even without treatment.

The answer to which of these alternative possibilities is best may well vary from case to case. Different eating disorders may require different answers. Perhaps even different individuals with the "same" disorder will involve different answers, at least if there are distinct subtypes within the major types of eating disorder.

It is important to strive toward discovering which answers are most correct for particular disorders or subtypes, because those answers carry implications for what treatment strategy might be best. For example, should one try to restore normal eating by reversing brain reward dysfunction via medications to correct the underlying problem? That would be appropriate if reward dysfunction is the underlying cause.

Or should one use drugs instead only as compensating medications, not cures? Such a medication might aim to boost aspects of brain reward function and so correct eating, even though it may not address the original underlying cause? For example, just as aspirin often helps treat pain, even though the original cause of pain was never a deficit in endogenous aspirin, so a medication that altered reward systems might still help to oppose whatever original underlying factors are altering eating, even though it will not reverse those causal factors.

Or instead should treatment be focused entirely on separate brain or peripheral targets that are unrelated to food reward? That might be the best choice if brain reward systems simply remain normal in all cases of eating disorders, and thus perhaps essentially irrelevant to the expression of pathological eating behavior.

Placing these alternatives side by side helps illustrate that there are therapeutic implications that would follow from a better understanding of brain reward systems. Only if we know how food reward is processed normally in the brain will we be able to recognize pathology in brain reward function. And only if we can recognize reward pathology when it occurs will we be able to judge which of the possibilities above best applies to a particular eating disorder.

III. UNDERSTANDING BRAIN REWARD SYSTEMS FOR FOOD "LIKING" AND "WANTING"

This section turns to some issues involved in measuring and understanding components of brain reward function (Berridge and Robinson, 2003; Everitt and Robbins, 2005). At the heart of reward is hedonic impact or pleasure, and so it is fitting to begin with the practical problem of measuring pleasure "liking" for food rewards in affective neuroscience studies.

A. Measuring Pleasure

Fortunately for psychologists and neuroscientists, pleasure is not just a metaphysical will-o'-the-wisp. Pleasure is a psychological process with neural reality, and has objective markers in brain and behavior. These objective aspects give a tremendously useful handle to neuroscientists and psychologists in their efforts to gain a scientific purchase on pleasure.

To identify the brain mechanisms that generate pleasure we must be able to identify when pleasure "liking" occurs or changes in magnitude. A useful "liking" reaction to measure taste pleasure is the affective facial expression elicited by the hedonic impact of sweet tastes in newborn human infants. Fortunately for pleasure causation studies, many animals display "liking–disliking" reactions elicited by sweet/bitter tastes that are similar and homologous to affective facial expressions to the same tastes displayed by human infants (Grill and Norgren, 1978b; Steiner, 1973; Steiner et al., 2001). These affective expressions seem to have developed from the same evolutionary source in humans, orangutans, chimpanzees, monkeys, and even rats and mice (Berridge, 2000; Steiner et al., 2001). Sweet tastes elicit positive facial "liking" expressions (tongue protrusions, etc.), whereas bitter tastes instead elicit facial "disliking" expressions (gapes, etc.).

A particular set of taste "liking–disliking" reactions are remarkably similar across species, and even their apparent differences often reflect a deeper shared identity, such as identical allometric timing laws for expression duration that are scaled to the size of the species. For example, human or gorilla tongue protrusions to sweetness or gapes to bitterness may appear languidly slow, whereas the same reactions by rats or mice seem blinkingly fast, yet, they are actually the "same" reactions durations in what is called an allometric sense; that is, each species is timed proportionally to their evolved sizes and that timing is programmed deep in their brains. Such shared universals further underline the common brain origins of these "liking" and "disliking" reactions in rats and humans. That sets the stage for animal affective neuroscience studies to use these affective expressions to identify brain mechanisms that generate hedonic impact.

B. Brain Systems for Food Pleasure

What brain systems paint a pleasure gloss onto mere sensation? Many brain sites are activated in humans by food pleasures: Cortical sites in the front of the brain implicated in the regulation of emotion, such as orbitofrontal cortex and anterior cingulated cortex, gustatory-visceral-emotion-related zones of cortex such as insular cortex; subcortical forebrain limbic structures such as amygdala, nucleus accumbens, and ventral pallidum; mesolimbic dopamine projections and even deep brainstem sites (Berns et al., 2001; Cardinal et al., 2002; Everitt and Robbins, 2005; Kringelbach, 2004; Kringelbach et al., 2004; Levine et al., 2003; O'Doherty et al., 2002; Pelchat et al., 2004; Rolls, 2005; Schultz, 2006; Small et al., 2001; Volkow et al., 2002; Wang et al., 2004a). All of these code pleasurable foods, in the sense of activating during the experience of seeing, smelling, tasting, or eating palatable foods.

But let us also ask: Which of these many brain structures actually cause the pleasure of foods? Do all generate pleasure "liking" or only some? Some activations might reflect causes of pleasure, whereas other activations might reflect consequences of pleasure that was caused elsewhere. How can causation be identified? Typically only by results of brain manipulation studies: a manipulation of a particular brain system will reveal pleasure causation if it produces an increase or decrease in "liking" reactions to food pleasure.

Recent years have seen progress in identifying brain systems responsible for generating the pleasure gloss that makes palatable foods "liked" (Berridge, 2003; Cardinal et al., 2002; Cooper, 2005; Higgs et al., 2003; Kelley et al., 2005a; Levine and Billington, 2004; Rolls, 2005; Wise, 2004; Yeomans and Gray, 2002). What has emerged most recently is a connected network of forebrain sites that use opioid neurotransmission to increase taste "liking" and "wanting" together to enhance food reward.

Pleasure "liking" appears to be generated by a distributed network of brain islands scattered across sites like an archipelago that trails throughout the limbic forebrain and brainstem (Berridge, 2003; Kelley et al., 2005a; Levine and Billington, 2004; Peciña and Berridge, 2005; Smith and Berridge, 2005). These sites include nucleus accumbens, ventral pallidum, and possibly amygdala and even limbic cortical sites, and also deep brainstem sites including the parabrachial nucleus in the pons. These distributed "liking" sites are all connected together so that they interact as a single integrated "liking" system, which operates by largely hierarchical control rules.

C. Pleasure Brain Hierarchy

Certain elemental reaction circuits within brain affective systems are contained in the brainstem. By themselves, brainstem circuits have a basic autonomy in the sense of functioning as simple reflexes when they have no other signals to modulate them. For example, basic positive or negative facial expressions are still found in human anencephalic infants born with a midbrain and hindbrain, but no cortex, amygdala, or classic limbic system, due to a congenital defect that prevents prenatal development of their forebrain. Yet, sweet tastes still elicit normal positive affective facial expressions from anencephalic infants, and bitter or sour tastes elicit negative expressions (Steiner, 1973). Similarly, a decerebrate rat has an isolated brainstem because a surgical transaction that separates the brainstem's connections from the forebrain, but the decerebrate brainstem also remains able to generate normal taste reactivity expressions to sweet or bitter tastes placed in the decerebrate's mouth (Grill and Norgren, 1978a).

However, brainstem generation of basic reactions does not mean "liking" lives only in the brainstem. Normal "liking" reactions are not brainstem reflexes in a whole-brained individual. This becomes obvious when we consider a related example: anencephalic infants cry and vocalize and even a decerebrate rat squeaks and emits distress cries if its tail gets pinched. But no one would suggest the vocal ability of anencephalic infants to cry and decerebrate rats to squeak means that normal human speech is merely a brainstem reflex. Obviously neither speech nor normal affective expressions generated by an entire brain is merely a brainstem reflex when forebrain systems determine them via hierarchical control.

When a brainstem is connected to the forebrain, the entire affective system operates in a hierarchical, flexible, and complex fashion. In a neural hierarchy, forebrain operations overrule brainstem elements, and dictate the output of the whole system. This means that brainstem reflex aspects of affective reactions are largely an artifact of brainstem isolation in decerebrates and anencephalics. In a fully connected brain, affective "liking" and "disliking" reactions are determined by an extensive forebrain network, and the final behavioral expression of affective taste reactivity reflects forebrain "liking" processes.

D. Forebrain Limbic Hedonic Hot Spot in Nucleus Accumbens

One forebrain hedonic island able to cause "liking" is an opioid hot spot in the nucleus accumbens. The nucleus accumbens contains major subdivisions called core and shell. While the core appears especially important for learning about rewards, the shell is more important for generating actual affective and motivational components of rewards themselves, including the pleasure gloss of "liking" for food rewards (Fig. 1).

The shell of nucleus accumbens is an L-shaped structure: its vertical back (called

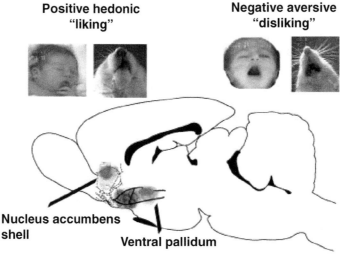

Opioid hedonic hot spots

FIGURE 1   "Liking" reactions and brain hedonic hot spots. Top: Positive hedonic "liking" reactions are elicited by sucrose taste from human infant and adult rat (e.g., rhythmic tongue protrusion). By contrast, negative aversive "disliking" reactions are elicited by bitter quinine taste. Lower: Forebrain hedonic hot spots in limbic structures where μ-opioid activation causes a brighter pleasure gloss to be painted on sweet sensation. Red/yellow shows hot spots in nucleus accumbens and ventral pallidum where opioid microinjections caused the biggest increases in the number of sweet-elicited "liking" reactions. Modified from Peciña and Berridge (2005) and Smith and Berridge (2005). (Color Plate).

medial shell), stands against the middle of the brain in each hemisphere, and its bottom horizontal foot points outward toward the lateral sides of the brain (the core is held in the concave crook formed between vertical back and lateral foot). It is the medial shell, the vertical upright of the L, which has received the most attention in the search for pleasure generators—with success.

The pleasure generator in the medial shell runs in part on opioid neurotransmitters. Opioids are natural brain neurotransmitters, such as enkephalin, endorphin, and dynorphin, that act on the same receptors as opiate drugs such as morphine or heroin. Opioid neurotransmitters that activate the μ type of opioid receptor appear particularly important to causing food reward "wanting" and "liking."

An important fact about opioid neurotransmitters in food reward is that they stimulate eating behavior in nearly the entire nucleus accumbens, and in quite a number of other forebrain limbic structures too (Cooper and Higgs, 1994; Kelley, 2004; Kelley et al., 2005a; Levine and Billington, 2004; Yeomans and Gray, 2002). So if opioid activation causes taste pleasure wherever it stimulates appetite for palatable food, then the widespread brain distribution of appetite-promoting sites means the brain has an extensive opioid pleasure network that stretches throughout much of the forebrain. That happy possibility would give every brain a really large hedonic causation system for generating pleasure.

But alternatively, if opioid circuits of food hedonic ("liking") versus motivational ("wanting") functions are organized somewhat differently from each other, then opioid activation might enhance taste pleasure at only some of the sites where it stimulates appetite. In that case, we must grapple with a complexity in opioid psychology and brain function. In other words, brain opioid activation might contribute to "liking" food and "wanting" food in different ways in different brain places. Understanding precisely how each opioid brain region contributes psychologically is a demanding but also interesting task for affective neuroscientists.

This issue was more recently addressed in a hedonic mapping study by Susana Peciña at the University of Michigan (Peciña and Berridge, 2005). It turns out that only one relatively small site in the medial shell may generate food "liking" as an opioid pleasure island or hedonic hot spot. That hedonic hot spot is in the rostral and dorsal one-quarter of medial shell. Here, µ-opioid activation acts as a hedonic generator to paint a pleasure gloss on sweetness, generating more "liking" for the food it also makes more "wanted."

Peciña's study used a "Fos plume" mapping technique to find where microinjections of a drug that activates opioid circuits causes increased "liking" reactions to the hedonic impact of a pleasant taste. To map pleasure generation, Peciña first made painless microinjections into rats' brains of tiny droplets of a drug known as DAMGO, which stimulates µ opioid receptors (natural neurotransmitter receptors for heroin or morphine). The opioid drug caused nearby neurons with appropriate receptors to begin transcribing genes on chromosomes in their cellular nucleus. One is a rapid-onset or immediate early gene called c-*fos*, which is transcribed to produce a protein called Fos that plays important roles in subsequent neuronal function. Fos protein stains very dark in appearance when postmortem brain slices are appropriately processed later with chemicals and antibodies that bind to the protein, and as a result the neurons with drug-induced Fos could be seen later as forming a dark plume on a slice of brain tissue. The size of each microinjection plume showed how far in the brain its drug had acted.

Some opioid drug microinjections caused sweet taste to carry increased positive hedonic impact. That increase in apparent pleasure-activating quality was reflected by dramatic increases in sucrose-elicited facial "liking" expressions, which often doubled or even tripled in number above control levels. But the "liking" increase was anatomically restricted to a single hedonic hot spot in the brain's medial shell of nucleus accumbens (Peciña and Berridge, 2005). In rats, the hedonic hot spot was roughly just a cubic millimeter in size, contained entirely in the rostral and dorsal quadrant of medial shell. Inside the hedonic hot spot, opioid activation made sweet sucrose taste "liked" even more than normal, and made bitter quinine taste "disliked" less (not as aversive)—shifting both positive/negative dimensions toward a positive pole. Outside the hot spot, positive hedonic impact was no longer increased, even though the "disliking" suppression zone extended more caudally. In fact, in one posterior site outside the hedonic hot spot affective suppression applied to both sweet "liking" and bitter "disliking"—essentially an affective cold spot of general suppression. Thus Peciña's mapping of opioid mechanisms indicates that a localized hedonic island in the medial shell of nucleus accumbens helps generate the pleasure gloss that opioid circuits paint onto sweet sensation to make it positively "liked."

1. Larger Opioid Sea of "Wanting" in Nucleus Accumbens

In addition to causing increased "liking," almost all opioid microinjections in the nucleus accumbens also caused the rats to eat more food soon afterward, increasing "wanting" for food reward too, even outside the hedonic hot spot (and even in the suppressive cold spot). The appetite-increasing zone was much larger than the pleasure hot spot: it was as though a large opioid sea of "wanting" in the shell of nucleus accumbens contains a smaller opioid island of "liking" for the same reward (Peciña and Berridge, 2005).

The large size of the opioid "wanting" zone fits with many earlier results from elegant appetite-mapping studies that have shown that many limbic brain structures support opioid-increased eating of palatable sweet or fatty foods (Gosnell and Levine, 1996; Kelley et al., 2005b; Levine and Billington, 2004; Will et al., 2003, 2004; Zhang and Kelley, 2000). But the opioid "liking" island identified by Peciña indicates that only in the rostral/dorsal hedonic hot spot does opioid activation in medial shell also increase "liking" reactions to food at the same time as increasing "wanting" to eat.

2. Implications for Normal Eating and Disorders

So it seems that the same μ-opioid neurotransmitter stimulation does different psychological things in different spots within the same brain structure. This leads us to outline several speculative possibilities by which eating disorders might relate endogenous opioid neurotransmission in nucleus accumbens shell.

First, one could speculate that pathological overactivation in regions of the opioid hedonic hot spot might cause enhanced "liking" reaction to taste pleasure in some individuals. An endogenously produced increase in opioid tone there could magnify the hedonic impact of foods, making an individual "like" food more than other people, and "want" to eat more.

A second and alternative speculative possibility is that activation in the opioid sea of "wanting" outside the hedonic island could cause people to "want" to overeat palatable, without making them "like" food more. If so, the preceding results would predict that this increased appetite would occur from excessive opioid function in a relatively large region of nucleus accumbens shell. The resulting "wanting" to eat would occur without any concomitant increase in the perceived palatability of food if the major locus of elevated activity

lay outside the anterior and dorsal region of medial shell.

3. Beyond Opioid "Liking": Other Hedonic Neurotransmitters in Nucleus Accumbens?

Opioid activation is only one type of neurochemical signal received by neurons in the shell nucleus accumbens. Are any others able to modulate "liking" reactions to food hedonic impact? The answer appears to be yes.

4. Cannabinoid "Liking"

Neurons in the nucleus accumbens also manufacture an endogenous cannabinoid neurochemical called anandamide, the CB1 receptors for which are activated by active ingredient THC in the drug, marijuana. Anandamide has been suggested to be a reverse neurotransmitter, which would be released by a target neuron in the shell to float back to nearby presynaptic axon terminals.

Marijuana, in addition to its own central rewarding effects, is widely known to cause an appetite-stimulating effect sometimes called the "marijuana munchies," raising intake especially of palatable snack foods. Several investigators have suggested that endogenous cannabinoid receptor activation stimulates appetite in part by enhancing "liking" for the perceived palatability of food (Cooper, 2004; Dallman, 2003; Higgs et al., 2003; Kirkham, 2005; Kirkham and Williams, 2001; Sharkey and Pittman, 2005).

A more recent taste reactivity study by Jarrett and Parker found that THC causes increased "liking" reactions to sugar tastes in rats, just as opiate drugs do (Jarrett et al., 2005). Focusing on natural endogenous brain cannabinoids, a microinjection study by Stephen Mahler and Kyle Smith in our laboratory pinned the neural causation of cannabinoid "liking" on activation of natural anandamide receptors in the nucleus accumbens shell (Mahler et al., 2004). Microinjections of anandamide directly into the medial shell of nucleus accumbens promoted positive "liking" reactions to the pleasure of sucrose taste.

The ability of natural anandamide to magnify the hedonic impact of natural food reward raises interesting questions about whether natural sensory reward functions will be disrupted in people by taking drugs that have been proposed to help dieters or addicts suppress their excessive consumption by blocking the natural CB1 receptors for anandamide and other endogenous cannabinoids (Cooper, 2004; Higgs et al., 2003; Kirkham and Williams, 2004). The answer is currently not known: the fact that a brain event can cause enhanced pleasure (i.e., sufficient cause for hedonic impact) does not necessarily mean that the brain mechanism is needed for normal pleasure (i.e., necessary cause for hedonic impact). The two questions must be separately answered by future research.

In either case, it appears that the cannabinoid pleasure zone overlaps with the opioid hedonic hot spot described previously in medial shell of nucleus accumbens. That suggests that both natural opioids and natural cannabinoids act in overlapping hedonic islands of the nucleus accumbens shell. In those overlapping zones, both neurochemicals act to enhance "liking" for the hedonic impact of natural sensory pleasures.

This raises possibilities for potential interactions between these two neurochemical forms of pleasure gloss, opioid and cannabinoid (Vigano et al., 2005). And there are additional "liking" interactions to consider too. Beyond neurotransmitters related to classic drugs of abuse, other neurotransmitters such as GABA are also used by the same neurons in the medial shell. These neurons both send and receive GABA signals. In addition, these neurons receive further neurochemical inputs, such as glutamate from the neocortex, hippocampus, and basolateral amygdala, and dopamine from mesolimbic neurons in the midbrain ventral tegmental area.

Of these various neurochemical signals, GABA at least can potently alter "liking" reactions to the hedonic impact of sugar tastes. But the positive/negative valence of GABA on "liking" versus "disliking" depends very much on precisely where in the shell the GABA is (Reynolds and Berridge, 2002). For example, GABA stimulation in the anterior subregion of the shell can increase "liking" reactions as well as food intake. GABA in most of the front half of the shell increases food intake without increasing "liking" reactions to hedonic impact, a bit like opioid activation in much of the shell described before. And GABA delivered to the posterior half of the shell dramatically suppresses intake, and makes sweet tastes "disliked," reversing their usual hedonic impact (Reynolds and Berridge, 2002).

It remains to be known whether similar changes in "liking" are evoked by nucleus accumbens glutamate blockade, by microinjections of a drug that blocks AMPA receptor signals, and so which may similarly hyperpolarize neurons in medial shell. However, glutamate AMPA blockade does alter eating behavior and food intake in ways similar to GABA (Reynolds and Berridge, 2003), which increases the plausibility that "liking" might be modulated by glutamate just as "wanting" is. The role of cortico-amygdala-hippocampal glutamate signals to nucleus accumbens in modulating "liking" reactions to hedonic impact of sweetness is a topic of substantial interest.

E. Ventral Pallidum: "Liking" and "Wanting" Pivot Point for Limbic Food Reward Circuits

Leaving the nucleus accumbens, output projections head to several destinations but the most single heavy projections are posteriorly to two nearby neighbors, the ventral pallidum and lateral hypothalamus. Of these two structures, the lateral hypothalamus has long been famous for roles in food intake and food reward. Lesions of the lateral hypothalamus disrupt eating and drinking behaviors, sending food and water intakes to zero (Teitelbaum and Epstein, 1962; Winn, 1995). After electrolytic lesions to lateral hypothalamus, rats would starve to death unless they were given intensive nursing care and artificial intragastric feeding.

Decades ago, lateral hypothalamic lesions were thought not only to abolish food "wanting," but also to abolish food "liking" too. Even sweet tastes were reported to elicit bitter-type disliking reactions (Schallert and Whishaw, 1978; Stellar et al., 1979; Teitelbaum and Epstein, 1962). However, it appears that lateral hypothalamus may have been blamed through a case of mistaken identity for the effects of lesions that stretched beyond it in lateral and anterior directions. Early studies on this topic indicated that sucrose "liking" would be replaced by sucrose "disliking" only if the lesion were in the anterior zone of lateral hypothalamus—and not if the lesion were in the posterior part of lateral hypothalamus, where it would produce loss of eating and drinking, but leave "liking" reactions essentially normal (Schallert and Whishaw, 1978). Further lesion studies by Cromwell mapped more carefully the boundaries of sites where neuron death caused aversion, and found that the "disliking" lesions might actually have to be so far anterior and lateral that they actually were outside the lateral hypothalamus itself—and in another structure, now called the ventral pallidum (Cromwell and Berridge, 1993).

Until about 10 years ago the ventral pallidum was known often as the substantia innominata, or brain structures without a name, and earlier than 20 years ago it was often mistaken for part of the lateral hypothalamus, as we have seen. Today it has a name, actually several names that correspond to different divisions of this intriguing part of the ventral forebrain. The chief names today are "ventral pallidum" for the part known to cause "liking" for sensory pleasure, and "sublenticular extended amygdala" for a bit behind that lies between ventral pallidum and lateral hypothalamus.

Ventral pallidum is the chief target of the heaviest projections emanating from the nucleus accumbens, and so it is the primary output channel through which mesocorticolimbic circuits must work (Zahm, 2000). The ventral pallidum is relatively new on the affective neuroscience scene, but there is reason to believe this chief target of nucleus accumbens is crucial for both normal reward "liking" and for enhanced "liking" caused under some neurochemical conditions.

An astounding fact is that the ventral pallidum is the only brain region known so far where the loss of neurons is capable of abolishing all "liking" for sweetness. It is the only brain site absolutely necessary for normal sucrose "liking" in the sense that damage to it makes "liking" go away (at least for several weeks). Sucrose no longer elicits "liking" reactions from a rat that has an excitotoxin lesion of its ventral pallidum, a type of lesion that selectively destroys the neurons that live in that structure while preserving fibers from neurons elsewhere that are simply passing through. Instead sucrose taste elicits only "disliking" reactions after the ventral pallidal lesion, as though the sweet taste had become bitter quinine (Cromwell and Berridge, 1993).

Ventral pallidum can also generate enhancement of natural pleasure, at least when it is intact. Ventral pallidum contains its own hedonic hot spot where μ opioid activation can increase the pleasure gloss that gets painted on sweetness. In a hedonic mapping study, Kyle Smith discovered that opioid DAMGO microinjections into a hedonic hot spot within the posterior VP caused sucrose to elicit over twice as many "liking" reactions than it normally did (Smith and Berridge, 2005). Opioid activation in the posterior ventral pallidum increased the hedonic impact of the natural taste reward, and also caused rats to eat over twice as much food. The hedonic hot spot was localized quite tightly within only the posterior one-third of ventral pallidum. If the same opioid microinjections were moved anteriorly toward the front of the structure, it actually suppressed hedonic "liking" reactions to sucrose and suppressed food intake too (Smith and Berridge, 2005).

A final reason to suppose ventral pallidum mediates hedonic impact is that the activity of neurons in the posterior hedonic hot spot appears to code "liking" for sweet and other tastes (Tindell et al., 2004, 2005). Recording electrodes can be permanently implanted in the ventral pallidum, and neurons there fire faster when rats eat a sweet taste. The firing of sucrose-triggered neurons appears to reflect hedonic "liking" for the taste. For example, the same neurons will not fire to an intensely salty solution that is unpleasant three times saltier than seawater. However, the neurons suddenly begin to fire to the triple-seawater taste if a physiological state of "salt appetite" is induced in the rats, by administering hormones that causes the body to need more salt, and which increase the perceived "liking" for intensely salty taste (Smith et al., 2004). Thus neurons in the ventral pallidum code taste pleasure in a way that is sensitive to the physiological need of the moment. When a taste becomes more pleasant during a particular physiological hunger, in a hedonic shift called "alliesthesia," the ventral pallidum neurons code the increase in salty pleasure. The observation that those hedonic neurons are in the same hedonic hot spot where opioid activation causes increased "liking" reactions to taste suggests that their firing rate might actually be part of the causal mechanism that paints the pleasure gloss onto taste sensation.

1. Speculative Implications for Eating Disorders

Just as for the hedonic island in nucleus accumbens, it is easy to imagine that if a human pathology caused increased opioid activation effects in the posterior ventral pallidum, that change would cause a person to experience foods as more pleasant and would stimulate appetite accordingly. Conversely, if it were possible to target opioid activation to the opposite anterior end of ventral pallidum, both food pleasure and food intake would be expected to be suppressed. Whether either of these events actually occurs in any human eating disorder is unknown. At the moment, they are simply possibilities to be compared against future observations that may carry the answer.

In any case, it seems likely that the hedonic hot spot in ventral pallidum ordinarily interacts together with its counterpart in the nucleus accumbens shell. There are extensive intercommunications between these two brain structures. A consequence is that the two hedonic islands probably actually function together as part of a single hedonic system, in close synchrony with each other. The details of this interaction remain to be unraveled by experiments but when they are, an integrative network of hedonic "liking" likely will be revealed by which opioid limbic circuits modulate the final hedonic impact and appetite for foods.

IV. "WANTING" WITHOUT "LIKING"

A very different product of affective neuroscience studies of taste "liking" reactions is the revelation of a number of false hedonic brain mechanisms, which turn out to mediate motivational "wanting" to eat without mediating hedonic "liking" for the same food. False hedonic mechanisms do not mediate "liking" for sensory pleasures—even though some were once thought to do so by neuroscientists. Their production of "wanting" without "liking" opens up fascinating possibilities for understanding particular pathologies of appetite and desire, including cases of irrational desire.

Perhaps the most famous false hedonic mechanism is dopamine: the mesolimbic projection of fibers to the nucleus accumbens from dopamine neurons in the midbrain ventral tegmental area. Dopamine release is triggered by pleasant foods and many other pleasant rewards, and dopamine neurons themselves fire more to pleasant food (especially when it is suddenly and unexpectedly received) (Ahn and Phillips, 1999; Di Chiara, 2002; Hajnal and Norgren, 2005; Montague et al., 2004; Roitman et al., 2004; Schultz, 2006; Small et al., 2003). Everyone agrees that dopamine release has important consequences on some aspect of reward, but the question is which aspect (Berridge and Robinson, 2003; Everitt and Robbins, 2005). Beyond correlative activations by rewards, the causal importance of dopamine is seen in the well-known observation that drugs that are rewarding or addictive typically cause dopamine activation—either directly or by acting on other neurochemical systems that in turn cause dopamine activation (Everitt and Robbins, 2005; Koob and Le Moal, 2006). Conversely, dopamine suppression reduces the degree to which animals and people seem to want rewarding foods, or rewards of other types (Berridge and Robinson, 1998; Dickinson et al., 2000; Wise, 2004).

The suppression of reward "wanting" by dopamine blockade or loss gave rise decades ago to the idea that dopamine must also mediate reward "liking" (Wise, 1985). The view of most neuroscientists has shifted subsequently, although some correlative evidence collected in more recent years can still be viewed as consistent with this dopamine-pleasure hypothesis of reward. For example, PET neuroimaging studies have suggested that obese people may have lower levels of dopamine D_2 receptor binding in their brains' striatum than others (Wang et al., 2001, 2004b). At first take, if one supposes that dopamine causes pleasure, reduced dopamine receptors in obese individuals can be viewed as

reducing the pleasure they get from food. By that view, reduced pleasure has been suggested to cause those individuals to eat more food in a quest to regain normal amounts of pleasure. A difficulty may arise for this account in that it also seems to require that the less people like a food the more they will eat it. By contrast, much evidence from psychology and neuroscience indicates that reducing the value of a food reward causes reduced pursuit of it as an incentive, rather than increasing pursuit and consumption (Cooper and Higgs, 1994; Dickinson and Balleine, 2002; Grigson, 2002; Kelley et al., 2005b; Levine and Billington, 2004). One might perhaps rescue this anhedonia account of D_2 decrement by supposing that all other life pleasures are reduced even more by dopamine receptor suppression than food pleasure, so that food remains the only pleasure available for consumption. However, we can see that actually getting increased food consumption from reduced pleasure out of any known psychological-brain system of food reward may prove trickier than first appears. So alternatives are worth entertaining too. A reverse interpretation of reduced dopamine D_2 binding in obese people is that the reduction is a consequence of overeating and obesity, rather than its cause. As a parallel example, overconsumption of drug rewards that provide increased stimulation to dopamine receptors eventually causes the receptors to reduce in number, even if dopamine receptors were normal to begin with—this is a down regulation mechanism of drug tolerance and withdrawal (Koob and Le Moal, 2006). So it is conceivable that similar sustained overactivation of dopamine systems by overeating food rewards in obese individuals perhaps could cause a similar eventual down regulation of their dopamine receptors. In a related vein, other physiological aspects of preexisting obesity states might also send excessive signals to brain systems sensitive to body weight, which indirectly cause reduction of D_2 receptor as a negative feedback consequence or a sort of long-term satiety signal that down regulates incentive systems. These speculative alternatives are enough to illustrate that possibilities exist by which reduced dopamine receptor binding could be a consequence, not the cause, of sustained obesity. Future research will be needed to resolve the fascinating question of whether correlated changes in levels of dopamine receptors is a cause or a consequence of human obesity.

If we turn to evidence from animal studies in which dopamine's causal roles can be manipulated, then dopamine does not appear to be important for "liking" the hedonic impact of food rewards after all. For example, mutant mice that lack any dopamine in their brains remain quite able to register the hedonic impact of sucrose or food rewards (Cannon and Palmiter, 2003; Robinson et al., 2005). Similarly, dopamine suppression or complete lesion in normal rats does not suppress taste "liking" facial expressions elicited by the taste of sucrose (Berridge and Robinson, 1998). Instead, the hedonic impact of sweetness remains robust even in a nearly dopamine-free forebrain (also, still robust is the ability to learn some new reward values for a sweet taste, which indicates that "liking" expressions remain faithful readouts of forebrain "liking" systems after dopamine loss (Berridge and Robinson, 1998)).

And conversely, too much dopamine in the brain, either in mutant mice whose gene mutation causes extra dopamine to remain in synapses or in ordinary rats given amphetamine that causes dopamine release (or that have drug-sensitized dopamine systems), show elevated "wanting" for sweet food rewards, but no elevation in "liking" sweet rewards (Peciña et al., 2003; Tindell et al., 2005; Wyvell and Berridge, 2000). Supporting evidence that dopamine mediates "wanting" but not "liking" also comes from PET neuroimaging studies of humans, which report that dopamine release triggered when people encounter a food or drug reward may better correlate to their subjective ratings of wanting the reward than to their pleasure ratings of liking the same reward (Leyton et al., 2002; Volkow et al., 2002).

Thus, the idea that dopamine is a pleasure neurotransmitter may have faded in neuroscience, with only a few hedonia pockets remaining (though dopamine seems important to "wanting" rewards, even if not to "liking" rewards). Separating true "liking" substrates from false ones is a useful step in identifying the real affective neural circuits for hedonic processes in the brain.

A. What Is "Wanting"?

"Wanting" is a shorthand term for the psychological process of incentive salience attribution, which helps determine the motivational incentive value of a pleasant reward (Berridge and Robinson, 1998; Everitt and Robbins, 2005; Robinson and Berridge, 2003; Salamone and Correa, 2002). But incentive "wanting" is not a sensory pleasure. "Wanting" is purely the incentive motivational value of a stimulus, not its hedonic impact.

Why did brains evolve separate "wanting" and "liking" mechanisms for the same reward? Originally, "wanting" might have evolved separately as an elementary form of goal directedness to pursue particular innate incentives even in advance of experience of their hedonic effects. Later incentive salience became harnessed by evolution to serve learned "wanting" for predictors of "liking," following learned or conditioned incentive motivation rules, guided by Pavlovian associations (Berridge, 2001; Dickinson and Balleine, 2002; Schultz, 2006; Toates, 1986). "Wanting" may also have evolved as distinct from "liking" to provide a common neural currency of incentive salience shared by all rewards, which could compare and decide competing choices for food, sex, or other rewards that might each involve partly distinct neural "liking" circuits (Shizgal, 1997). The important point is that "liking" and "wanting" normally go together, but they can be split apart under certain circumstances, especially by certain brain manipulations.

"Liking" without "wanting" can be produced, and so can "wanting" without "liking." "Liking" without "wanting" happens after brain manipulations that cause mesolimbic dopamine neurotransmission to be suppressed. For example, disruption of mesolimbic dopamine systems, via neurochemical lesions of the dopamine pathway that projects to nucleus accumbens or by receptor-blocking drugs, dramatically reduces incentive salience or "wanting" to eat a tasty reward, but does not reduce affective facial expressions of "liking" for the same reward (Berridge and Robinson, 1998; Peciña et al., 1997).

Dopamine suppression can leave individuals nearly without motivation for any pleasant incentive at all: food, sex, drugs, brain-stimulation reward, etc. (Cannon and Palmiter, 2003; Everitt and Robbins, 2005; Salamone and Correa, 2002; Wise, 2004). Yet, "liking," or the hedonic impact of the same incentives, remains intact after dopamine loss or suppression, at least in many studies where it can be specifically assessed by either facial affective expressions or subjective ratings. Intact "liking" in animal affective neuroscience studies has usually been manifest via normal positive affective facial expressions elicited by the hedonic impact of sweet tastes, or by normal learning about food reward, after dopamine lesion, blockade, or genetic lack (Berridge and Robinson, 1998; Cannon and Palmiter, 2003; Peciña et al., 1997; Robinson et al., 2005). Similarly, in humans, drugs that block dopamine receptors may completely fail to reduce the subjective pleasure ratings that people give to a reward stimulus such as amphetamine (Brauer and De Wit, 1997; Brauer et al., 1997; Wachtel et al., 2002).

Conversely, "wanting" without "liking" can be produced by several brain manipulations in rats (and perhaps by real-life brain sensitization in human drug addicts, see later). For example, electrical stimulation of the lateral hypothalamus in rats, as mentioned before, triggers a number of motivated behaviors such as eating. In normal hunger, increased appetite is accompanied by increased hedonic appreciation of food, a phenomenon named alliesthesia (Cabanac, 1971). But during lateral hypothalamic stimulation that made them eat avidly, rats facial expressions to a sweet taste actually became more aversive, if anything, as though the taste became bitter (Berridge and Valenstein, 1991). In other words, the hypothalamic stimulation did not make them "want" to eat by making them "like" the taste of food more. Instead, it made them "want" to eat more despite making them "dislike" the taste.

Similarly, eating or pursuit of food caused by a variety of dopamine, GABA, or other neurochemical manipulations of nucleus accumbens or ventral pallidum is not accompanied by enhanced hedonic reactions to the taste of food. Mutant mice with brain receptors that receive more dopamine than normal, because of a genetic mutation that causes released dopamine to remain in synapses, also show excessive "wanting" of sweet reward, while nonetheless "liking" sweetness less than normal mice do (Peciña et al., 2003). Similarly increased food intake without increased "liking" occurs in rats that have received accumbens microinjections of amphetamine or a GABA agonist or opioid agonist in certain shell regions, or ventral pallidum microinjections of a GABA antagonist (Peciña et al., 2003; Peciña and Berridge, 2005; Reynolds and Berridge, 2002; Smith and Berridge, 2005; Tindell et al., 2005; Wyvell and Berridge, 2000). All of these brain manipulations make animals "want" to eat food more, though they fail to make the animals "like" food more (and sometimes even make them "like" it less).

What is "wanting" if it is not "liking?" According to the incentive salience concept, "wanting" is a mesolimbic-generated process that can tag certain stimulus representations in the brain that have Pavlovian associations with reward. When incentive salience is attributed to a reward stimulus representation, it makes that stimulus attractive, attention grabbing, and that stimulus and its associated reward suddenly become enhanced motivational targets. Because incentive salience is often triggered by Pavlovian conditioned stimuli or reward cues, it often manifests as cue-triggered "wanting" for reward. When attributed to a specific stimulus, incentive salience may make an autoshaped cue light appear foodlike to the autoshaped pigeon or rat that perceives it, causing the animal to try to eat the cue. In autoshaping, animals sometimes direct behavioral pursuit and consummatory responses toward the Pavlovian cue, literally trying to eat the conditioned stimulus if it is a cue for food reward (Jenkins and Moore, 1973; Tomie, 1996). When attributed to the smell emanating from a bakery, incentive salience can rivet a person's attention and trigger sudden thoughts of lunch—and perhaps it can do so under some circumstances even if the person merely vividly imagines the delicious food.

But "wanting" is not "liking," and both together are necessary for normal reward. "Wanting" without "liking" is merely a sham or partial reward, without sensory pleasure in any sense. However, "wanting" is still an important component of normal reward, especially when combined with "liking." Reward in the full sense cannot happen without incentive salience, even if hedonic "liking" is present. Hedonic "liking" by itself is simply a triggered affective state—there is no object of desire or incentive target, and no motivation for reward. It is the process of incentive salience attribution that makes a specific associated stimulus or action the object of desire, and that tags a specific behavior as the rewarded response. "Liking" and "wanting" are needed together for full

reward. Fortunately, both usually happen together in human life.

B. Addiction and Incentive Sensitization

For some human addicts, however, drugs such as heroin or cocaine may cause real-life "wanting" without "liking" because of long-lasting sensitization changes in brain mesolimbic systems. Addicts sometimes take drugs compulsively even when they do not derive much pleasure from them (Everitt and Robbins, 2005). For example, drugs such as nicotine generally fail to produce great sensory pleasure in most people, but still are infamously addictive.

In the early 1990s, Terry Robinson and I proposed the incentive-sensitization theory of addiction, which combines neural sensitization and incentive salience concepts (Robinson and Berridge, 1993, 2003). The theory does not deny that drug pleasure, withdrawal, or habits are all reasons people sometimes take drugs, but suggests that something else, sensitized "wanting," may be needed in order to understand the compulsive and long-lasting nature of addiction, and especially for why relapse occurs even after weeks or months of abstinence from any drugs, and even when addicts do not expect to gain much pleasure from their relapse.

Many addictive drugs cause neural sensitization in the brain mesocorticolimbic systems (e.g., cocaine, heroin, amphetamine, alcohol, nicotine). Sensitization means that the brain system can be triggered into abnormally high levels of activation by drugs or related stimuli. Sensitization is nearly the opposite of drug tolerance. Different processes within the same brain systems can simultaneously instantiate both sensitization (e.g., via increase in dopamine release) and tolerance (e.g., via decrease in dopamine receptors) (Koob and Le Moal, 2006; Robinson and Berridge, 1993, 2003; Vezina, 2004). However, tolerance mechanisms usually recover within days to weeks once drugs are given up, whereas neural sensitization can last for years. If the incentive-sensitization theory is true for drug addiction, it helps explain why addicts may sometimes even "want" to take drugs that they do not particularly "like." The long-lasting nature of neural sensitization may also help explain why recovered addicts, who have been drug-free and out of withdrawal for months or years, are still sometimes liable to relapse back into addiction.

Sensitization of incentive salience does not mean that addicts "want" all rewards more in a general fashion. "Wanting" increases instead are highly specific, in target object and in time. In target object, only specific conditioned stimuli or cues for reward get attributed with higher incentive salience after sensitization (Tindell et al., 2005; Wyvell and Berridge, 2001). Other unrelated stimuli get no increase at all (e.g., CS-). And when multiple reward targets are available, some targets get much more increase than others (Nocjar and Panksepp, 2002; Tindell et al., 2005). The key to which stimulus becomes "wanted" may lie in Pavlovian learning mechanisms that control the direction of incentive salience attribution, the relative intensities of reward unconditioned stimuli, and in other features that differentially distribute elevated attribution (Berridge and Robinson, 1998; Robinson and Berridge, 1993). This directional specificity may relate to why a drug addict particularly "wants" drugs, whereas someone with an eating disorder might particularly "want" food.

In time, sensitized cue-triggered "wanting" is specific to sudden intense peaks that rapidly fall off when the cue is taken away, only to reappear later when it is reencountered again (Tindell et al., 2005; Wyvell and Berridge, 2001). High incentive salience comes and goes with cues for the particularly "wanted" reward (and in people possibly also with vivid reward imagery). This may relate to why addicts are particularly vulnerable to relapse on occasions when they reencounter drug-associated cues such as drug paraphernalia or places and social contexts where drugs were taken before.

C. Is There a Neural Sensitization Role in Food Addictions?

Could incentive-sensitization apply to food addictions too? A conclusive answer cannot yet be given. In favor of the possibility, once sensitized, brain mesolimbic systems do overrespond to cues for sugar rewards with excessive incentive salience (Tindell et al., 2005; Wyvell and Berridge, 2001), and repeated exposure to drugs can produce sensitization-type increases in food intake (Bakshi and Kelley, 1994; Nocjar and Panksepp, 2002).

But does sensitization occur to food incentives in the absence of drugs? More recently, several investigators have suggested that sensitization-like changes in brain systems are indeed produced by exposure to certain regimens of food and restriction that model oscillations between dieting and binging on palatable foods (Avena and Hoebel, 2003a,b; Bell et al., 1997; Bello et al., 2003; Carr, 2002; Colantuoni et al., 2001; de Vaca and Carr, 1998; Gosnell, 2005). These food-sensitization studies have generally shown that when rats are given a number of brief chances to consume sucrose (sucrose binges) a number of accumulating sensitization-like changes are seen, especially when binges are separated by periods of food restriction. These include an increasing propensity to overconsume when allowed, an enduring enhanced neural response to the presentation of food reward and cues, and an overresponse to the psychostimulant effects of drugs such as amphetamine (a typical behavioral marker of drug-induced neural sensitization, which suggests a common underlying mechanism).

The possibility of food-induced sensitization suggests the speculative scenario that some diet–binge combinations conceivably might be able to induce incentive-sensitization for food rewards. That is, individuals who develop neural sensitization profiles to food might come to "want" to eat, perhaps even with compulsive intensity that ordinary individuals do not usually encounter, that outstrips their "liking" for the foods they eat. This is an intriguing possible mechanism for a "food addiction" that certainly deserve further consideration.

However, several cautions are in order before concluding that sensitized food addictions do indeed exist. Repeated food presentation and repeated food restriction both introduce complications that could masquerade as neural sensitization, and these complications could interact. For example, repeated presentation of a palatable food is likely to induce strong Pavlovian conditioning, creating conditioned incentive stimuli that strongly activate mesolimbic systems for incentive salience. It is no surprise that a strong conditioned food incentive would activate dopamine systems more than in an individual that has not undergone the same conditioning experience. It would be an error to mistake conditioned incentive activation of mesolimbic systems via serial experiences for an underlying neural sensitization.

Hunger itself also may promote mesolimbic activation under some conditions (Ahn and Phillips, 1999; Nader et al., 1997; Wilson et al., 1995). The use of food restriction can be expected to enhance the psychological incentive properties of food directly, with consequences for underlying neurobiological activations too. During hunger, sucrose tastes more pleasant than when one is full (alliesthesia) (Cabanac, 1971). When foods become more potent unconditioned incentives, a variety of unconditioned and conditioned food-related stimuli might be expected to more strongly activate mesolimbic incentive systems too (Toates, 1986).

Further interactions between these potential confounds may ensue. If hunger facilitates the unconditioned activation of mesolimbic systems by food, then it may also promote stronger incentive conditioning, leading to later enhanced mesolimbic responses to food conditioned stimuli. Essentially hunger would make food into a better and stronger hedonic reward, which would function as a stronger unconditioned stimulus in a Pavlovian sense to promote greater learning. Animals that repeatedly consume while hungry, in alternating dieting-binge periods, could essentially develop a stronger learned or conditioned incentive motivation triggered by predictive cues than animals that are not food-restricted between periods of sucrose access. Those more strongly conditioned animals might well be expected to show enhanced mesolimbic activation later whenever related cues are presented later. If these things occurred, the resulting picture might look a bit like sensitization, without actually being it.

This is not to discount the possibility that real sensitization might occur induced by diet–binge cycles or similar exposure regimens, or to deny that the studies mentioned before might be examples of food incentive-sensitization. Sensitization-like states do indeed seem to be implicated by certain types of physiological deprivation states (Dietz et al., 2005; Roitman et al., 2002). However, the previous cautions point to the need to better disentangle direct effects of true food-induced sensitization from masquerading confounds of unconditioned deprivation states and ordinary incentive conditioning on mesolimbic activation. Otherwise these confounds could mimic sensitization patterns of enhanced activation, and mislead us into thinking sensitization has occurred when it has not.

Future studies may be able to better assess these potential confounds, and perhaps separate them from sensitization and set them to rest. If so, it will put food-induced sensitization of incentive salience on a stronger empirical basis, and raise incentive-sensitization as a possible mechanism for food addiction.

D. Cognitive Goals and Ordinary Wanting

Before leaving this topic of incentive salience and "wanting," it is useful to note how the incentive salience meaning of the word "wanting" above differs from what most people mean by the ordinary sense of the word wanting. A subjective feeling of desire meant by the ordinary word wanting implies something both cognitive (involving an explicit goal) and conscious (involving a subjective feeling). When you say you want something, you usually have in mind a cognitive expectation or idea of the something-you-want: a declarative representation of your goal. Your representation is based usually on your experience with that thing in the past. Or, if you have never before experienced that thing, then, the representation is based on your imagination of what it would be like to experience. In other words, in these cases, you know or imagine cognitively what it is you want, you expect to like it, and you may even have some idea of how to get it. This is a very cognitive form of wanting, involving declarative memories of the valued goal, explicit predictions for the potential future based on those memories, and cognitive understanding of causal relationships that exist between your potential actions and future attainment of your goal.

By contrast, none of this cognition need be part of incentive salience "wants" discussed before (Berridge, 2001; Dickinson and Balleine, 2000). Incentive salience attributions do not need to be conscious and are mediated by relatively simple brain mechanisms (Berridge and Winkielman, 2003). Incentive salience "wants" are triggered by relatively basic stimuli and perceptions (not requiring more elaborate cognitive expectations). Cue-triggered "wanting" does not require understanding of causal relations about the hedonic outcome. Sometimes

as a result of this difference, as described for addiction, excessive incentive salience may lead to irrational "wants" for outcomes that are not cognitively wanted, and that are neither liked nor even expected to be liked.

Behavioral neuroscience experiments have indicated that these forms of wanting may depend on different brain structures. For example, incentive salience "wanting" depends highly on subcortical mesolimbic dopamine neurotransmission, whereas cognitive forms of wanting depend instead on cortical brain regions such as orbitofrontal cortex, prelimbic cortex, and insular cortex (Corbit and Balleine, 2003; Dickinson et al., 2000; Rolls, 2005). The implication is that there may be multiple kinds of wanting, with different neural substrates (Berridge, 2001).

If eating disorders involve a pathology in a particular "wanting" system such as incentive salience, it would be possible that such an individual would "want" to eat food at the same time that they cognitively do not want to eat, and when "liking" would not be enhanced. In other words, the sight, smell, or vivid imagination of food could trigger a compulsive urge to eat, even though the person would not expect to find the experience very pleasurable. Neural sensitization of incentive salience systems, if it truly happens in any eating disorder, might be one way by which excessive "wanting" to eat could generate excessive food intake.

V. A BRIEF HISTORY OF APPETITE: FOOD INCENTIVES, NOT HUNGER DRIVES

To place into a better perspective the role of "liking" and "wanting" mechanisms we have discussed, it might be helpful to consider here a brief overview of how psychologists and neuroscientists' view of appetite has evolved in recent decades. For many years it was thought that eating behavior is driven directly by hunger drives, and that food was a reward because it reduces hunger drive. But actually, physiological hunger works primarily by modulating food incentive values, not driving behavior directly through deficit or drive signals.

Any food is rewarding primarily because of its hedonic value in taste, texture, and the act of eating, not by its eventual drive reduction. A vivid early example against pure drive reduction is the anecdotal medical case of a man named Tom, from the timber and mining frontier of upper Michigan a century ago, whose esophagus was permanently damaged in childhood when he accidentally drank scalding soup without knowing it was too hot (Wolf and Wolff, 1943). The burn sealed his esophagus and, thereafter, blocked the passage of food to the stomach. To help him live, a surgical opening or gastrostomy fistula was implanted in his stomach, and he was sustained afterwards by placing food and drink directly through the fistula into his stomach. There was no longer any apparent purpose in putting food in his mouth first because food in the mouth could not descend through the closed esophagus.

Yet, Tom insisted on munching food at meals, when he would chew and then spit out the food before placing it in his stomach. Why? Because "introducing (food) directly into his stomach failed to satisfy his appetite" (p. 8, (Wolf and Wolff, 1943)).

A. Separating Reward and Drive Reduction

The most important evidence against drive reduction concepts came in the 1960s from studies of brain stimulation reward and related studies of motivated behavior elicited by "free" brain stimulation. A single electrode in the lateral hypothalamus could both elicit motivated behavior (if just turned on freely) and have reward or pun-

ishment effects (if given contingent on the animal's response). Many behavioral neuroscientists of the time believed in drive reduction theory. So, at first, they expected to find that the brain sites where stimulation would reduce eating (presumably by reducing drive) would also be the sites where stimulation was rewarding (again, presumably by reducing drive). Conversely, they believed the opposite would be true too. They expected to find that punishing electrodes would sometimes activate drives like hunger.

The best descriptions of these beliefs come from the experimenters' own words. As James Olds (the codiscoverer of brain stimulation reward; p. 89, (Olds, 1973)) put it, he was originally guided by the drive reduction hypothesis that an "electrical simulation which caused the animal to respond as if it were very hungry might have been a drive-inducing stimulus and might therefore have been expected to have aversive properties." In other words, if the drive reduction theory were true, an "eating electrode" should also have been a "punishment electrode."

Conversely, a "satiety electrode" that stopped eating should have been a "reward electrode." As Miller (pp. 54–55, 1973), a major investigator of brain stimulation effects, recounted later in describing how drive reduction concepts guided his research.

If I could find an area of the brain where electrical stimulation had the other properties of normal hunger, would sudden termination of that stimulation function as a reward? If I could find such an area, perhaps recording from it would provide a way of measuring hunger which would allow me to see the effects of a small nibble of food that is large enough to serve as a reward but not large enough to produce complete satiation. Would such a nibble produce a prompt, appreciable reduction in hunger, as demanded by the drive-reduction hypothesis?

Thus, if the drive reduction theory were true, you might actually be able to watch hunger drive shrink by recording the shrinking activity of the drive neuron each time a nibble of food reduced that drive and caused reward.

But the drive-reduction theory was not true, and nearly all of the predictions based on it turned out to be wrong. In many cases, the opposite results were found instead. The brain sites where the stimulation caused eating behavior were almost always the same sites where stimulation was rewarding (Valenstein et al., 1970; Valenstein, 1976). Eating electrodes were not punishing electrodes. Instead, the eating electrodes were reward electrodes. Stimulation-induced reward and stimulation-induced hunger drive appeared identical or, at least, had identical causes in the activation of the same electrode. This meant that the reward could not be due to drive reduction. The reward electrode increased the motivation to eat, it did not reduce that drive. Instead reward must be understood as a motivational phenomenon of its own, involving its own active brain mechanisms. This is a reason why many affective neuroscientists of recent decades have rejected drive-reduction explanations of appetite and food reward, and instead focused on understanding hedonic/incentive brain mechanisms of reward in their own right.

VI. CONNECTING BRAIN REWARD AND REGULATORY SYSTEMS

A fascinating final topic is the interaction between brain systems of mesolimbic reward on the one hand and of hypothalamic hunger and body weight regulation on the other (Baldo et al., 2003; Berthoud, 2002, 2004; Kelley et al., 2005b). How do these brain systems connect and influence each other?

Control signals go back and forth between mesolimbic reward systems and hypothalamic systems for hunger regulation. For example, hypothalamic orexin-

hypocretin neurons send projections to modulate the nucleus accumbens in ways that might allow hunger states to enhance food reward (Scammell and Saper, 2005), and even interact with other rewards such as drugs (Harris et al., 2005). In return, nucleus accumbens influences hypothalamic circuits. For example, manipulations of nucleus accumbens that cause increased food intake and that modulate reward, such as GABA microinjections into the medial shell, send descending signals that activate orexin neurons in the hypothalamus (Baldo et al., 2004; Zheng et al., 2003).

Neuroscientists have only begun to understand the nature and role of interactions between mesolimbic reward systems and hypothalamic hunger systems, but more recent developments show that such interactions are there and of great importance. They undoubtedly play major roles in modulating the pleasure and incentive value of food rewards during normal hunger versus satiety states, possibly also connecting reward modulation to longer term body weight elevation and dieting states, and finally perhaps even in allowing food reward cues to influence the activation of hunger deficit systems. These interactions also provide avenues, at least in principle, by which eating disorders cause distortion in the function of reward systems, so that they become either exaggerated or suppressed in function. Such interactions will be important to try to understand better in the future.

VII. CONCLUSION

For most people, eating patterns and body weights remain within normally prescribed bounds. Perhaps it is the prevalence of normal body weights, rather than obesity, that should be most surprising in modern affluent societies where tasty foods abound. As is often pointed out, brain mechanisms for food reward and appetite evolved under pressures to protect us from scarcity. As a result, overeating in the face of present abundance could be an understandable overshoot inherited from our evolutionary past.

When eating patterns and body weight do diverge from the norm, questions arise concerning the involvement of food reward systems in the brain. All eating patterns are controlled intimately by brain mechanisms of food reward, whether those mechanisms operate in normal mode or abnormal modes. A primary signpost to help guide future thinking is to know whether any pathological patterns of eating are caused by identifiable pathologies in brain reward function. Can distorted patterns of eating be corrected by medications that alter brain reward mechanisms? Or are the causes of eating disorders essentially independent of brain reward systems? These questions remain to guide future research on how brain substrates of food reward relate to eating disorders.

Acknowledgments

This research was supported by grants from the NIH (DA015188 and MH63649). I am also grateful for the kind hospitality during preparation of this manuscript of faculty, staff, and students at the University of Cambridge in the Department of Experimental Psychology and Downing College, and for fellowship support from the John Simon Guggenheim Memorial Foundation.

References

Ahn, S., and Phillips, A. G. (1999). Dopaminergic correlates of sensory-specific satiety in the medial prefrontal cortex and nucleus accumbens of the rat. *J. Neurosci.* **19**, B1–B6.

Avena, N. A., and Hoebel, B. G. (2003a). Amphetamine-sensitized rats show sugar-induced hyperactivity (cross-sensitization) and sugar hyperphagia. *Pharmacol. Biochem. Behav.* **74**, 635–639.

Avena, N. M., and Hoebel, B. G. (2003b). A diet promoting sugar dependency causes behavioral cross-sensitization to a low dose of amphetamine. *Neuroscience* **122**, 17–20.

Bakshi, V. P., and Kelley, A. E. (1994). Sensitization and conditioning of feeding following multiple morphine microinjections into the nucleus accumbens. *Brain Res.* **648**, 342–346.

Baldo, B. A., Daniel, R. A., Berridge, C. W., and Kelley, A. E. (2003). Overlapping distributions of orexin/hypocretin- and dopamine-beta-hydroxylase immunoreactive fibers in rat brain regions mediating arousal, motivation, and stress. *J. Comp. Neurol.* **464**, 220–237.

Baldo, B. A., Gual-Bonilla, L., Sijapati, K., Daniel, R. A., Landry, C. F., and Kelley, A. E. (2004). Activation of a subpopulation of orexin/hypocretin-containing hypothalamic neurons by GABAA receptor-mediated inhibition of the nucleus accumbens shell, but not by exposure to a novel environment. *Eur. J. Neurosci.* **19**, 376–386.

Bell, S. M., Stewart, R. B., Thompson, S. C., and Meisch, R. A. (1997). Food-deprivation increases cocaine-induced conditioned place preference and locomotor activity in rats. *Psychopharmacology (Berlin)* **131**, 1–8.

Bello, N. T., Sweigart, K. L., Lakoski, J. M., Norgren, R., and Hajnal, A. (2003). Restricted feeding with scheduled sucrose access results in an upregulation of the rat dopamine transporter. *Am. J. Physiol. Regul. Integr. Comp. Physiol.* **284**, R1260–R1268.

Berns, G. S., McClure, S. M., Pagnoni, G., and Montague, P. R. (2001). Predictability modulates human brain response to reward. *J. Neurosci.* **21**, 2793–2798.

Berridge, K. C. (2000). Measuring hedonic impact in animals and infants: microstructure of affective taste reactivity patterns. *Neurosci. Biobehav. Rev.* **24**, 173–198.

Berridge, K. C. (2001). Reward learning: Reinforcement, incentives, and expectations. *In* "The Psychology of Learning and Motivation" (D. L. Medin, ed.), Vol. 40, pp. 223–278. Academic Press, New York.

Berridge, K. C. (2003). Pleasures of the brain. *Brain Cogn.* **52**, 106–128.

Berridge, K. C., and Valenstein, E. S. (1991). What psychological process mediates feeding evoked by electrical stimulation of the lateral hypothalamus? *Behav. Neurosci.* **105**, 3–14.

Berridge, K. C., and Robinson, T. E. (1998). What is the role of dopamine in reward: hedonic impact, reward learning, or incentive salience? *Brain Res. Rev.* **28**, 309–369.

Berridge, K. C., and Robinson, T. E. (2003). Parsing reward. *Trends Neurosci.* **26**, 507–513.

Berridge, K. C., and Winkielman, P. (2003). What is an unconscious emotion? (The case for unconscious "liking"). *Cogn. Emot.* **17**, 181–211.

Berthoud, H. R. (2002). Multiple neural systems controlling food intake and body weight. *Neurosci. Biobehav. Rev.* **26**, 393–428.

Berthoud, H.-R. (2004). Mind versus metabolism in the control of food intake and energy balance. *Physiol. Behav.* **81**, 781–793.

Brauer, L. H., and De Wit, H. (1997). High dose pimozide does not block amphetamine-induced euphoria in normal volunteers. *Pharmacol. Biochem. Behav.* **56**, 265–272.

Brauer, L. H., Goudie, A. J., and De Wit, H. (1997). Dopamine ligands and the stimulus effects of amphetamine: animal models versus human laboratory data. *Psychopharmacology (Berlin)* **130**, 2–13.

Cabanac, M. (1971). Physiological role of pleasure. *Science* **173**, 1103–1107.

Cannon, C. M., and Palmiter, R. D. (2003). Reward without dopamine. *J. Neurosci.* **23**, 10827–10831.

Cardinal, R. N., Parkinson, J. A., Hall, J., and Everitt, B. J. (2002). Emotion and motivation: the role of the amygdala, ventral striatum, and prefrontal cortex. *Neurosci. Biobehav. Rev.* **26**, 321–352.

Carr, K. D. (2002). Augmentation of drug reward by chronic food restriction: behavioral evidence and underlying mechanisms. *Physiol. Behav.* **76**, 353–364.

Colantuoni, C., Schwenker, J., McCarthy, J., Rada, P., Ladenheim, B., Cadet, J. L., et al. (2001). Excessive sugar intake alters binding to dopamine and mu-opioid receptors in the brain. *Neuroreport* **12**, 3549–3552.

Cooper, S. J. (2004). Endocannabinoids and food consumption: comparisons with benzodiazepine and opioid palatability-dependent appetite. *Eur. J. Pharmacol.* **500**, 37–49.

Cooper, S. J. (2005). Palatability-dependent appetite and benzodiazepines: new directions from the pharmacology of GABA(A) receptor subtypes. *Appetite* **44**, 133–150.

Cooper, S. J., and Higgs, S. (1994). Neuropharmacology of appetite and taste preferences. *In* "Appetite: Neural and Behavioural Bases" (C. R. Legg, and D. A. Booth, eds.), pp. 212–242. Oxford University Press, New York.

Corbit, L. H., and Balleine, B. W. (2003). The role of prelimbic cortex in instrumental conditioning. *Behav. Brain Res.* **146**, 145–157.

Cromwell, H. C., and Berridge, K. C. (1993). Where does damage lead to enhanced food aversion: the ventral pallidum/substantia innominata or lateral hypothalamus? *Brain Res.* **624**, 1–10.

Dallman, M. F. (2003). Fast glucocorticoid feedback favors "the munchies". *Trends Endocrinol. Metab.* **14**, 394–396.

de Vaca, S. C., and Carr, K. D. (1998). Food restriction enhances the central rewarding effect of abused drugs. *J. Neurosci.* **18**, 7502–7510.

Di Chiara, G. (2002). Nucleus accumbens shell and core dopamine: differential role in behavior and addiction. *Behav. Brain Res.* **137**, 75–114.

Dickinson, A., and Balleine, B. W. (2000). Causal cognition and goal-directed action. *In* "The Evolution of Cognition," pp. 185–204. The MIT Press, Cambridge, MA.

Dickinson, A., and Balleine, B. (2002). The role of learning in the operation of motivational systems. *In*

"Stevens' Handbook of Experimental Psychology: Learning, Motivation, and Emotion" (C. R. Gallistel, ed.), 3rd ed., Vol. 3, pp. 497–534. Wiley & Sons, New York.

Dickinson, A., Smith, J., and Mirenowicz, J. (2000). Dissociation of Pavlovian and instrumental incentive learning under dopamine antagonists. *Behav. Neurosci.* **114**, 468–483.

Dietz, D. M., Curtis, K. S., and Contreras, R. J. (2005). Taste, salience, and increased NaCl ingestion after repeated sodium depletions. *Chem. Senses.*

Everitt, B. J., and Robbins, T. W. (2005). Neural systems of reinforcement for drug addiction: from actions to habits to compulsion. *Nat. Neurosci.* **8**, 1481–1489.

Gosnell, B. A. (2005). Sucrose intake enhances behavioral sensitization produced by cocaine. *Brain Res.* **1031**, 194–201.

Gosnell, B. A., and Levine, A. S. (1996). Stimulation of ingestive behavior by preferential and selective opioid agonists. *In* "Drug Receptor Subtypes and Ingestive Behavior" (S. J. Cooper, and P. G. Clifton, eds.), pp. 147–166. Academic Press, London.

Grigson, P. S. (2002). Like drugs for chocolate: separate rewards modulated by common mechanisms? *Physiol. Behav.* **76**, 389–395.

Grill, H. J., and Norgren, R. (1978a). The taste reactivity test. II. Mimetic responses to gustatory stimuli in chronic thalamic and chronic decerebrate rats. *Brain Res.* **143**, 281–297.

Grill, H. J., and Norgren, R. (1978b). The taste reactivity test. I. Mimetic responses to gustatory stimuli in neurologically normal rats. *Brain Res.* **143**, 263–279.

Hajnal, A., and Norgren, R. (2005). Taste pathways that mediate accumbens dopamine release by sapid sucrose. *Physiol. Behav.* **84**, 363–369.

Harris, G. C., Wimmer, M., and Aston-Jones, G. (2005). A role for lateral hypothalamic orexin neurons in reward seeking. *Nature* **437**, 556–559.

Higgs, S., Williams, C. M., and Kirkham, T. C. (2003). Cannabinoid influences on palatability: microstructural analysis of sucrose drinking after delta(9)-tetrahydrocannabinol, anandamide, 2-arachidonoyl glycerol and SR141716. *Psychopharmacology (Berlin)* **165**, 370–377.

Jarrett, M. M., Limebeer, C. L., and Parker, L. A. (2005). Effect of delta9-tetrahydrocannabinol on sucrose palatability as measured by the taste reactivity test. *Physiol. Behav.* **86**, 475–479.

Jenkins, H. M., and Moore, B. R. (1973). The form of the auto-shaped response with food or water reinforcers. *J. Exp. Anal. Behav.* **20**, 163–181.

Kelley, A. E. (2004). Ventral striatal control of appetitive motivation: role in ingestive behavior and reward-related learning. *Neurosci. Biobehav. Rev.* **27**, 765–776.

Kelley, A. E., Baldo, B. A., and Pratt, W. E. (2005a). A proposed hypothalamic-thalamic-striatal axis for the integration of energy balance, arousal, and food reward. *J. Comp. Neurol.* **493**, 72–85.

Kelley, A. E., Baldo, B. A., Pratt, W. E., and Will, M. J. (2005b). Corticostriatal-hypothalamic circuitry and food motivation: integration of energy, action and reward. *Physiol. Behav.* **86**, 773–795.

Kirkham, T. C. (2005). Endocannabinoids in the regulation of appetite and body weight. *Behav. Pharmacol.* **16**, 297–313.

Kirkham, T. C., and Williams, C. M. (2001). Endogenous cannabinoids and appetite. *Nutr. Res. Rev.* **14**, 65–86.

Kirkham, T. C., and Williams, C. M. (2004). Endocannabinoid receptor antagonists: potential for obesity treatment. *Treat. Endocrinol.* **3**, 345–360.

Koob, G. F., and Le Moal, M. (2006). "Neurobiology of Addiction." Academic Press, New York.

Kringelbach, M. L. (2004). Food for thought: hedonic experience beyond homeostasis in the human brain. *Neuroscience* **126**, 807–819.

Kringelbach, M. L., de Araujo, I. E., and Rolls, E. T. (2004). Taste-related activity in the human dorsolateral prefrontal cortex. *Neuroimage* **21**, 781–788.

Levine, A. S., and Billington, C. J. (2004). Opioids as agents of reward-related feeding: a consideration of the evidence. *Physiol. Behav.* **82**, 57–61.

Levine, A. S., Kotz, C. M., and Gosnell, B. A. (2003). Sugars: hedonic aspects, neuroregulation, and energy balance. *Am. J. Clin. Nutr.* **78**, 834S–842S.

Leyton, M., Boileau, I., Benkelfat, C., Diksic, M., Baker, G., and Dagher, A. (2002). Amphetamine-induced increases in extracellular dopamine, drug wanting, and novelty seeking: a PET/[11C]raclopride study in healthy men. *Neuropsychopharmacology* **27**, 1027–1035.

Mahler, S. V., Smith, K. S., and Berridge, K. C. (2004). What is the "motivational" mechanism for the marijuana munchies? The effects of intra-accumbens anandamide on hedonic taste reactions to sucrose. Paper presented at the Society for Neuroscience Conference, San Diego.

Miller, N. E. (1973). How the project started. *In* "Brain Stimulation and Motivation: Research and Commentary" (E. S. Valenstein, ed.), pp. 53–68. Scott, Foresman and Company, Glenview, IL.

Montague, P. R., Hyman, S. E., and Cohen, J. D. (2004). Computational roles for dopamine in behavioural control. *Nature* **431**, 760–767.

Nader, K., Bechara, A., and van der Kooy, D. (1997). Neurobiological constraints on behavioral models of motivation. *Annu. Rev. Psychol.* **48**, 85–114.

Nocjar, C., and Panksepp, J. (2002). Chronic intermittent amphetamine pretreatment enhances future appetitive behavior for drug- and natural-reward: interaction with environmental variables. *Behav. Brain Res.* **128**, 189–203.

O'Doherty, J. P., Deichmann, R., Critchley, H. D., and Dolan, R. J. (2002). Neural responses during anticipation of a primary taste reward. *Neuron* **33**, 815–826.

Olds, J. (1973). The discovery of reward systems in the brain. *In* "Brain Stimulation and Motivation: Research and Commentary" (E. S. Valenstein, ed.), pp. 81–99. Scott, Foresman and Company, Glenview, IL.

Peciña, S., and Berridge, K. C. (2005). Hedonic hot spot in nucleus accumbens shell: where do mu-opioids cause increased hedonic impact of sweetness? *J. Neurosci.* **25**, 11777–11786.

Peciña, S., Berridge, K. C., and Parker, L. A. (1997). Pimozide does not shift palatability: separation of anhedonia from sensorimotor suppression by taste reactivity. *Pharmacol. Biochem. Behav.* **58**, 801–811.

Peciña, S., Cagniard, B., Berridge, K. C., Aldridge, J. W., and Zhuang, X. (2003). Hyperdopaminergic mutant mice have higher "wanting" but not "liking" for sweet rewards. *J. Neurosci.* **23**, 9395–9402.

Pelchat, M. L., Johnson, A., Chan, R., Valdez, J., and Ragland, J. D. (2004). Images of desire: food-craving activation during fMRI. **23**, 1486–1493.

Reynolds, S. M., and Berridge, K. C. (2002). Positive and negative motivation in nucleus accumbens shell: bivalent rostrocaudal gradients for GABA-elicited eating, taste "liking"/"disliking" reactions, place preference/avoidance, and fear. *J. Neurosci.* **22**, 7308–7320.

Reynolds, S. M., and Berridge, K. C. (2003). Glutamate motivational ensembles in nucleus accumbens: rostrocaudal shell gradients of fear and feeding in the rat. *Eur. J. Neurosci.* **17**, 2187–2200.

Robinson, S., Sandstrom, S. M., Denenberg, V. H., and Palmiter, R. D. (2005). Distinguishing whether dopamine regulates liking, wanting, and/or learning about rewards. *Behav. Neurosci.*

Robinson, T. E., and Berridge, K. C. (1993). The neural basis of drug craving: an incentive-sensitization theory of addiction. *Brain Res. Rev.* **18**, 247–291.

Robinson, T. E., and Berridge, K. C. (2003). Addiction. *Annu. Rev. Psychol.* **54**, 25–53.

Roitman, M. F., Na, E., Anderson, G., Jones, T. A., and Bernstein, I. L. (2002). Induction of a salt appetite alters dendritic morphology in nucleus accumbens and sensitizes rats to amphetamine. *J. Neurosci.*

Roitman, M. F., Stuber, G. D., Phillips, P. E. M., Wightman, R. M., and Carelli, R. M. (2004). Dopamine operates as a subsecond modulator of food seeking. *J. Neurosci.* **24**, 1265–1271.

Rolls, E. T. (2005). "Emotion Explained." Oxford University Press, Oxford and New York.

Salamone, J. D., and Correa, M. (2002). Motivational views of reinforcement: implications for understanding the behavioral functions of nucleus accumbens dopamine. *Behav. Brain Res.* **137**, 3–25.

Scammell, T. E., and Saper, C. B. (2005). Orexin, drugs and motivated behaviors. *Nat. Neurosci.* **8**, 1286–1288.

Schallert, T., and Whishaw, I. Q. (1978). Two types of aphagia and two types of sensorimotor impairment after lateral hypothalamic lesions: observations in normal weight, dieted, and fattened rats. *J. Comp. Physiol. Psychol.* **92**, 720–741.

Schultz, W. (2006). Behavioral theories and the neurophysiology of reward. *Annu. Rev. Psychol.*

Sharkey, K. A., and Pittman, Q. J. (2005). Central and peripheral signaling mechanisms involved in endocannabinoid regulation of feeding: a perspective on the munchies. *Sci. STKE* (277), pe15.

Shizgal, P. (1997). Neural basis of utility estimation. *Curr. Opin. Neurobiol.* **7**, 198–208.

Small, D. M., Zatorre, R. J., Dagher, A., Evans, A. C., and Jones-Gotman, M. (2001). Changes in brain activity related to eating chocolate—From pleasure to aversion. *Brain* **124**, 1720–1733.

Small, D. M., Jones-Gotman, M., and Dagher, A. (2003). Feeding-induced dopamine release in dorsal striatum correlates with meal pleasantness ratings in healthy human volunteers. *Neuroimage* **19**, 1709–1715.

Smith, K. S., and Berridge, K. C. (2005). The ventral pallidum and hedonic reward: neurochemical maps of sucrose "liking" and food intake. *J. Neurosci.* **25**, 8637–8649.

Smith, K. S, Tindell, A. J., Berridge, K. C., and Aldridge, J. W. (2004). Ventral pallidal neurons code the hedonic enhancement of an NaCl taste after sodium depletion. Paper presented at the Society for Neuroscience Conference, San Diego.

Steiner, J. E. (1973). The gustofacial response: observation on normal and anencephalic newborn infants. *Symp. Oral Sensa. Percep.* **4**, 254–278.

Steiner, J. E., Glaser, D., Hawilo, M. E., and Berridge, K. C. (2001). Comparative expression of hedonic impact: affective reactions to taste by human infants and other primates. *Neurosci. Biobehav. Rev.* **25**, 53–74.

Stellar, J. R., Brooks, F. H., and Mills, L. E. (1979). Approach and withdrawal analysis of the effects of hypothalamic stimulation and lesions in rats. *J. Comp. Physiol. Psychol.* **93**, 446–466.

Teitelbaum, P., and Epstein, A. N. (1962). The lateral hypothalamic syndrome: recovery of feeding and drinking after lateral hypothalamic lesions. *Psychol. Rev.* **69**, 74–90.

Tindell, A. J., Berridge, K. C., and Aldridge, J. W. (2004). Ventral pallidal representation of Pavlovian cues and reward: population and rate codes. *J. Neurosci.* **24**, 1058–1069.

Tindell, A. J., Berridge, K. C., Zhang, J., Pecina, S., and Aldridge, J. W. (2005). Ventral pallidal neurons code incentive motivation: amplification by mesolimbic sensitization and amphetamine. *Eur. J. Neurosci.* **22**, 2617–2634.

Toates, F. (1986). "Motivational Systems." Cambridge University Press, Cambridge.

Tomie, A. (1996). Locating reward cue at response manipulandum (CAM) induces symptoms of drug abuse. *Neurosci. Biobehav. Rev.* **20**, 31.

Valenstein, E. S. (1976). The interpretation of behavior evoked by brain stimulation. *In* "Brain-Stimulation Reward" (A. Wauquier, and E. T. Rolls, eds.), pp. 557–575. Elsevier, New York.

Valenstein, E. S., Cox, V. C., and Kakolewski, J. W. (1970). Reexamination of the role of the hypothalamus in motivation. *Psychol. Rev.* **77**, 16–31.

Vezina, P. (2004). Sensitization of midbrain dopamine neuron reactivity and the self-administration of psychomotor stimulant drugs. *Neurosci. Biobehav. Rev.* **27**, 827–839.

Vigano, D., Rubino, T., and Parolaro, D. (2005). Molecular and cellular basis of cannabinoid and opioid interactions. *Pharmacol. Biochem. Behav.* **81**, 360–368.

Volkow, N. D., Wang, G. J., Fowler, J. S., Logan, J., Jayne, M., Franceschi, D., et al. (2002). "Nonhedonic" food motivation in humans involves dopamine in the dorsal striatum and methylphenidate amplifies this effect. *Synapse* **44**, 175–180.

Wachtel, S. R., Ortengren, A., and de Wit, H. (2002). The effects of acute haloperidol or risperidone on subjective responses to methamphetamine in healthy volunteers. *Drug Alcohol Depend.* **68**, 23–33.

Wang, G. J., Volkow, N. D., Logan, J., Pappas, N. R., Wong, C. T., Zhu, W., et al. (2001). Brain dopamine and obesity. *Lancet* **357**, 354–357.

Wang, G. J., Volkow, N. D., Telang, F., Jayne, M., Ma, J., Rao, M., et al. (2004a). Exposure to appetitive food stimuli markedly activates the human brain. *Neuroimage* **21**, 1790–1797.

Wang, G. J., Volkow, N. D., Thanos, P. K., and Fowler, J. S. (2004b). Similarity between obesity and drug addiction as assessed by neurofunctional imaging: a concept review. *J. Addict. Dis.* **23**, 39–53.

Will, M. J., Franzblau, E. B., and Kelley, A. E. (2003). Nucleus accumbens mu-opioids regulate intake of a high-fat diet via activation of a distributed brain network. *J. Neurosci.* **23**, 2882–2888.

Will, M. J., Franzblau, E. B., and Kelley, A. E. (2004). The amygdala is critical for opioid-mediated binge eating of fat. *Neuroreport* **15**, 1857–1860.

Wilson, C., Nomikos, G. G., Collu, M., and Fibiger, H. C. (1995). Dopaminergic correlates of motivated behavior: importance of drive. *J. Neurosci.* **15**(7 Pt. 2), 5169–5178.

Winn, P. (1995). The lateral hypothalamus and motivated behavior: an old syndrome reassessed and a new perspective gained. *Curr. Dir. Psychol.* **4**, 182–187.

Wise, R. A. (1985). The anhedonia hypothesis: Mark III. *Behav. Brain Sci.* **8**, 178–186.

Wise, R. A. (2004). Drive, incentive, and reinforcement: the antecedents and consequences of motivation. *Nebr. Symp. Motiv.* **50**, 159–195.

Wolf, S., and Wolff, H. G. (1943). "Human Gastric Function, an Experimental Study of a Man and His Stomach." Oxford University Press, London, New York.

Wyvell, C. L., and Berridge, K. C. (2000). Intra-accumbens amphetamine increases the conditioned incentive salience of sucrose reward: enhancement of reward "wanting" without enhanced "liking" or response reinforcement. *J. Neurosci.* **20**, 8122–8130.

Wyvell, C. L., and Berridge, K. C. (2001). Incentive-sensitization by previous amphetamine exposure: increased cue-triggered "wanting" for sucrose reward. *J. Neurosci.* **21**, 7831–7840.

Yeomans, M. R., and Gray, R. W. (2002). Opioid peptides and the control of human ingestive behaviour. *Neurosci. Biobehav. Rev.* **26**, 713–728.

Zahm, D. S. (2000). An integrative neuroanatomical perspective on some subcortical substrates of adaptive responding with emphasis on the nucleus accumbens. *Neurosci. Biobehav. Rev.* **24**, 85–105.

Zhang, M., and Kelley, A. E. (2000). Enhanced intake of high-fat food following striatal mu-opioid stimulation: microinjection mapping and fos expression. *Neuroscience* **99**, 267–277.

Zheng, H. Y., Corkern, M., Stoyanova, I., Patterson, L. M., Tian, R., and Berthoud, H. R. (2003). Peptides that regulate food intake—Appetite-inducing accumbens manipulation activates hypothalamic orexin neurons and inhibits POMC neurons. *Am. J. Physiol. -Regul. Integr. Comp. Physiol.* **284**, R1436–R1444.

Pharmacology of Food, Taste, and Learned Flavor Preferences

STEVEN J. COOPER
School of Psychology, University of Liverpool

I. Introduction
II. Pharmacology of Food Preference
  A. Food Preferences of Rats
  B. Benzodiazepines and Food Preference
  C. Opioids and Food Preference
  D. Dopamine and Food Preference
  E. Summary
III. Pharmacology of Unlearned Taste Preference and Reactivity
  A. Measuring Taste Preference and Taste Reactivity
  B. Benzodiazepines, Taste Preference, and Taste Reactivity
  C. Opioids, Taste Preference, and Taste Reactivity
  D. Dopamine and Taste Reactivity
  E. Summary
IV. Pharmacology of Learned Flavor Preference
  A. Flavor Preference Conditioning
  B. "Electronic Esophagus" Preparation
  C. Opioids and Conditioned Flavor Preferences
  D. Dopamine and Conditioned Flavor Preferences
  E. Taste Reactivity and Conditioned Flavor Preferences
V. Concluding Remarks
  Acknowledgments
  References

Not one man in a billion, when taking his dinner, ever thinks of utility. He eats because the food tastes good and makes him want more. If you ask him why *he should want to eat more of what tastes like that, instead of revering you as a philosopher he will probably laugh at you for a fool.*

William James

I. INTRODUCTION

Some physiological functions, respiration and cardiac activity, for example, are not matters of choice, and we refer to involuntary and autonomous patterns of responses. These are literally *vital* functions, the integrity of which is necessary for continued life. At a more psychological level, however, choice becomes all-important whenever circumstances permit. Characteristics of behavior such as selection, preference, and avoidance, favoring one outcome over another, bear the hallmarks of choice. At first sight, these characteristics embody the adaptability of behavior, the ability to exploit resources to the best advantage, to get the best return for the required outlay. This is not necessarily the case, however; a moment's reflection will provide examples of the avoidance of that which may be good for us, and a preference for that which can do us harm. Choice, even if free, does not carry innate good sense.

Among the poorest inhabitants of this world, the question of choice may scarcely arise, especially with regard to the types and amounts of food consumed. Many starve and more go hungry. In wealthier economies, in contrast, amounts and varieties of foodstuffs are often available to excess, beyond what might be described as meeting basic needs for survival. Food is available, it is nutritious, it is tempting, and there is plenty of variety from which to choose. In the midst of such plenty, many people, young and old, are succumbing to overweight and obesity, in an increasing trend that now extends across continents. A central factor in the obesification of people is the overconsumption of food beyond immediate physiological need (and in the absence of periodic famine). It is here we must consider issues of choice, preference, selection, and avoidance in respect to the foods available to us, on a daily, continuing basis. This chapter will deal with some rather simplistic animal models of preference for foods and flavors, which, admittedly, can scarcely do full justice to the range of factors which help determine human appetite for food, the care and effort that goes into the production, provision and preparation of foodstuffs for consumption, and the exquisite sense of preference we all bring to the choice and consumption of our food. Nevertheless, I contend that experi-

mental results derived from these models may help us to understand some of the important characteristics of human ingestive behavior. They may also pave the way to develop therapeutic interventions to mitigate the threatening medical consequences of overweight and obesity.

Expressing preference depends on what is available. We could poll people on their preferences among cabbage, broccoli, or brussels sprouts, for example. However, it is likely that for most people adding a chocolate-based dessert to the choice list would overwhelmingly decide the matter. Some flavors and some foods are so inviting and delicious, that they invariably come at or close to the top of anyone's preference league table. In general terms, this chapter is concerned with such high-palatability items, which have a particularly strong appeal to our taste buds. We can leave the question of preferences among vegetables to another time. Importantly, the palatability and nutrient content of the available diet are critical factors in the development of obesity in animals (Sclafani, 1984). High fat or high sugar diets are both highly palatable to rats and are energy dense. Cafeteria diets, which provide rats with a rich assortment of highly palatable, calorically dense foods, lead to the greatest weight gains (Sclafani, 1984). Palatability, as such, has a major effect on food overconsumption and subsequent obesity (Drewnowski, 1998; Sclafani et al., 1996). Pragmatically, therefore, we should take a particular interest in the role of palatability in respect to food preference and to levels of food consumption.

While we come equipped with some innately determined preferences for certain tastes and flavors, it is a matter of common observation that we also acquire likes and dislikes, food passions and aversions. Learned, or conditioned, flavor preferences can be acquired by rats, and the processes of acquisition and expression of such preferences can be studied experimentally (Sclafani, 1997). An aim of this chapter is, in consequence, a consideration of both unlearned and conditioned preferences: are they two species of the same type, or are they qualitatively dissimilar, at least in certain important respects? An additional aim is to consider the *pharmacology* of food and flavor preferences, that is to focus attention on the actions of certain classes of drugs in respect to behavioral indices of preference and palatability. A central interest in pharmacological research is to expose degrees of specificity in the actions of drugs—in this case, are there drug treatments which, in some degree, have specific effects on preferences, their acquisition, or their expression? If we can answer in the affirmative, then we may have some basis on which to address two critical issues. The first is a basic research question: can the pharmacological data provide important indications to the neural mechanisms (biochemically and/or neuroanatomically defined) which underpin choice, preference, and palatability? The second is an applied question: Can the same data provide important new leads in the development of therapeutically effective drug treatments for obese patients who require to lose weight?

II. PHARMACOLOGY OF FOOD PREFERENCE

A. Food Preferences of Rats

Like us, rats are omnivores, and wild as well as tamed laboratory rats (*Rattus norvegicus*) show great flexibility and adaptability in their choice of foods. In wild rats, novel objects (including potential food items) may initially elicit avoidance and circumspection, but in laboratory rats such reactions are considerably attenuated (Barnett, 1963). While wild rats may evince object *neophobia*, and explicit avoidance, laboratory rats may exhibit "curiosity" and actively approach and investigate new food items (*neophilia*) (Corey, 1978). Once familiarized, rats will show acceptance of a wide range of food items, and also exhibit clear

preferences among available choices. Preferences may be determined by a number of factors. For example, in a study by Barnett and Spencer (1953a), colonies of wild rats preferred wholemeal to whole wheat grains. Adding sucrose or saccharin to the wholemeal induced a preference for the sweetened food over plain wholemeal. Wild and laboratory rats also prefer diets to which fats or oils have been added (Barnett and Spencer, 1953a; Scott and Verney, 1948). Wild rats are likely to display stronger aversions to some flavors or odors than laboratory rats. Thus, laboratory rats were not deterred by adding aniseed oil or butyric acid to their food (Scott and Quint, 1946), whereas wild rats were repelled (Barnett and Spencer, 1953b). The studies we shall consider in this chapter all involve tamed laboratory rats—ideal subjects that express great interest in and liking for a wide variety of foods which we find appetizing too.

In an innovative laboratory experiment, Rolls and Rolls (1973b) gave food-deprived rats a choice between six different food items in a 15-min test. One was familiar, regular maintenance diet food pellets, while the other five items were novel: greens, sultanas, orange peel, potato, and cauliflower. While the rats investigated and sampled the unfamiliar foods, they overwhelmingly settled to eat the pellets to alleviate their hunger. On the first trial, therefore, the rats expressed a marked preference for the familiar maintenance diet. With repeated testing, not only did the total time devoted to eating increase considerably, but the rats spent a smaller proportion of their time eating the regular food pellets, and more time eating the other foods. Under conditions of familiarity therefore (familiar foods, test environment, food container, and animal handling), rats expressed food preferences which are more closely linked to relative liking among the foods on offer. When a highly palatable food, such as chocolate chip cookies, was included in this cafeteria, rats expressed a strong preference for them once they had become familiar choices. This is not surprising, but does illustrate the behavioral differences exhibited by rats under conditions of familiarity compared with novelty.

Now consider the experimental manipulation introduced by Rolls and Rolls (1973b). They compared normal intact rats with animals that had bilateral electrolytic lesions of the basolateral amygdala. In contrast to the control animals, the basolateral amygdala-lesioned animals expressed preference for the unfamiliar food items over the regular maintenance diet on first exposure. Rolls and Rolls (1973b) carried out further tests which ruled out an alteration in fear to explain the basolateral amygdala-lesioned animals' behavior. Instead, they concluded that the basolateral amygdala may be involved in the selection of foods on the basis of earlier experience. In related work, Rolls and Rolls (1973b) showed that basolateral amygdala-lesioned rats also exhibited a major impairment in learned (or conditioned) taste aversions. Learning a taste aversion is, of course, an opposite process to the acquisition of food or taste preferences. Impairment of taste aversion learning following lesions to the basolateral amygdala has been reported many times since the Rolls and Rolls (1973b) study (e.g., Aggleton et al., 1981; Borsini and Rolls, 1984; Fitzgerald and Burton, 1981; Nachman and Ashe, 1974). In a more recent comprehensive review, Reilly and Bornovalova (2005) closely examined the evidence for disruption of conditioned taste aversions as a result of damage to the basolateral region of the amygdala. On balance, they conclude that the basolateral amygdala lesions do not impair animals' ability to form associative connections between taste cues and aversive consequences, but instead affect the animals' responsiveness to novelty and/or familiarization.

B. Benzodiazepines and Food Preference

We shall now turn to some early pharmacological studies on food preferences which follow on from the Rolls and Rolls (1973b) experiments. Classic benzodiazepine receptor agonists, such as chlordiazepoxide, diazepam, or midazolam, increase food consumption in food-deprived and nondeprived rats, increase lever-pressing for food reward, and overcome the suppressive effect of bitter taste on feeding responses (Margules and Stein, 1967; Randall et al., 1960; Wise and Dawson, 1974). Cooper and Crummy (1978) presented food-deprived rats with a choice of six foods available in a "cafeteria" test situation: familiar food pellets, and five novel foods (apple, carrot, cheddar cheese, currants, and milk chocolate wholemeal cookies). Somewhat surprisingly, chlordiazepoxide did not have the same effect as lesions to the basolateral amygdala: there was no increase in the time spent eating the novel food items, even though highly palatable chocolate-coated cookies were available. Instead, chlordiazepoxide selectively enhanced the feeding response to the familiar maintenance food pellets. Hence, chlordiazepoxide, in this situation at least, had no evident effect on food neophobia, or on the discrimination between novelty and familiarity in food items.

As we have seen, though, rats' food preferences in such a test situation can alter quite substantially provided the animals became fully habituated to the test environment and the available foods. Cooper and McClelland (1980) showed that, following test familiarization, the animals' initial preference for the maintenance diet declined quite markedly, whereas the choice of the other foods increased commensurately. Under these new conditions of familiarity, chlordiazepoxide now selectively enhanced the feeding response not to the regular maintenance diet but to newly preferred cafeteria foods. Cooper and McClelland (1980) also demonstrated that the feeding effect of chlordiazepoxide was unaffected by prior handling experience in the rats, or by contextual fear conditioning. They hypothesized that "chlordiazepoxide essentially acts to enhance the response to the preferred foods" (p. 26). Under conditions of test familiarity, more *palatable* foods are preferred to maintenance food pellets, and chlordiazepoxide enhanced the feeding response to the more palatable food items. The anxiolytic and amnesic properties of the benzodiazepines are not relevant to these specific effects on food preferences. Even in animals that have been selectively bred to exhibit high or low emotionality (fear-reactivity) phenotypes, chlordiazepoxide's effects in a cafeteria-style food preference test relate solely to the food choices available (Cooper and Webb, 1984).

Of course, the food items in the cafeteria-style test differ from one another in terms of taste, flavor, texture, caloric density, macronutrient content, and so on. Hence, Cooper (1987a) reported an experiment in which food-deprived animals were given a choice between three foods which were identical except for distinguishing taste cues. Three wet mashes were prepared from powdered standard diet and water; however, in one case a 0.13% sodium saccharin solution was used instead of plain water, and in a second case a 0.002% quinine solution was used instead of plain water. Under control circumstances, the rats sampled each of the food sources, averaging a couple of feeding bouts for one source before switching to sample another. In this test, chlordiazepoxide exerted quite specific effects on feeding responses. First, chlordiazepoxide promoted increased food consumption as expected. However, the increased food intake was due entirely to a selective increase in the consumption of the sweet-tasting mash. The responses to the other two mashes, plain and quinine-adulterated, were unaffected. Importantly, the animals' sampling behavior was not affected in any way by the drug treatment; following chlordiazepoxide, they continued to devote a couple of feeding bouts at one source before switching to eat from another source. A microstructural analysis of the animals' feeding behavior revealed that chlordiazepoxide did not increase the number of eating bouts in the 15-min test, but did increase the mean duration of feeding bouts from about 15 sec on average to about 70 sec (at 10 mg/kg i.p., chlordiazepoxide). This represents an increase in feeding bout duration of 367%! The benzodiazepine selectively enhanced the consumption of the sweet-flavored mash, inducing a marked taste preference, and did so by selectively increasing feeding bout duration.

C. Opioids and Food Preference

Taken together, these results clearly implicate taste and food palatability in the effects of benzodiazepines on food preferences. We shall now consider the equally important role of opioids in relation to rats' food preferences. Following Holtzman's pioneering observations that opioid receptor antagonists, like naloxone and naltrexone, reduced food and fluid consumption in rats and mice (Brown and Holtzman, 1979; Holtzman, 1974, 1975), many investigators quickly confirmed the results (e.g., Brands et al., 1979; Cooper, 1980; Frenk and Rogers, 1979). Treatments with the opioid receptor agonist, morphine, increased food and water consumption (e.g., Cooper, 1981; Jalowiec et al., 1981; Sanger and McCarthy, 1980). These and other data led to the influential theory that the controls of ingestive behavior depend in important ways on the functions of endogenous opioid peptides in the central nervous system (e.g., Baile et al., 1986; Cooper et al., 1988; Levine et al., 1985; Sanger, 1981). Of particular relevance, here, is the subsidiary hypothesis that endogenous opioid mechanisms may be particularly involved in the mediation and behavioral expression of food palatability, and hence of palatability-dependent food preferences (e.g., Apfelbaum and Mandenoff, 1981; Cooper et al., 1985a,b).

Turkish and I conducted a study to determine the effects of naloxone on food preference in food-deprived and nondeprived rats (Cooper et al., 1988). The two groups of rats were given a concurrent choice of familiar maintenance food pellets and chocolate-coated cookies in a 15-min food preference test. In one condition, the chocolate-coated cookies were novel food items, and in a second, the chocolate-coated cookies were familiar. Hence, four groups of rats were tested: (FAM-DEP) where the animals were familiar with both foods as well as being food-deprived before the test; (FAM-NON) where the animals were familiar with both foods, but not food-deprived before the test; (NOV-DEP) where the chocolate-coated cookies were a novel food item and the animals were food-deprived before the test; (NOV-NON) where the chocolate-coated cookies were novel and the animals were not deprived. Naloxone treatments had no effect whatsoever in three of the four groups: FAM-NON; NOV-DEP; NOV-NON; respectively. Most importantly, naloxone had no effect on the marked preference for the regular maintenance food over the chocolate-coated cookies in the NOV-DEP group of animals. Hence, naloxone had no effect on ingestive behavior when preference was based solely on the familiarity of the food item. In the FAM-DEP group, however, naloxone did exhibit significant effects on ingestive responses. Following drug vehicle administration, this group of animals manifested an overwhelming preference for the high-palatability chocolate-coated cookies over the regular maintenance food. Naloxone exerted two contrasting effects: It produced a sharp reduction in the feeding response elicited by the high-palatability food, but also enhanced the response to the regular maintenance diet. In effect, naloxone reduced the preference for the chocolate-coated cookies, a preference based on relative palatability as distinct from a preference based on relative familiarity. Moreover, these data rule against a simple interpretation of naloxone's effects in terms of anorexia or reduction in food consump-

tion. Nothing of the kind was seen in the NOV-DEP group, where feeding behavior was robust.

Hence, the opioid receptor antagonist, naloxone, reduced an expressed food preference determined by palatability (Cooper et al., 1988). Cooper and Turkish (1989) confirmed this result using a second antagonist, naltrexone. Food-deprived rats were given the concurrent choice of familiar maintenance pellets and familiarized chocolate-coated cookies (the FAM-DEP condition) in a 15-min food preference test. Naltrexone (0.05–5.0 mg/kg, s.c.) produced a dose-dependent reduction in the duration of eating the preferred chocolate-coated cookies. At the same time, there was a modest increase in the time devoted to eating maintenance pellets. As a result, the relative preference for the high-palatability chocolate-coated food dropped from about 100% to 67% after naltrexone, based on time spent eating in the test. Cooper and Turkish (1989) interpreted the data to suggest that endogenous opioid peptide activity is involved in determining the palatability of preferred foods. In further support of this idea, morphine (0.1–5.0 mg/kg, s.c.) was shown to enhance significantly the response to the preferred, familiarized chocolate-coated cookies, without affecting the response to maintenance pellets, in a comparable FAM-DEP food preference test (Cooper and Kirkham, 1990). More recent experiments reported by Barbano and Cador (2005), in which naloxone (0.015–1.0 mg/kg, s.c.) was administered to food-deprived and nondeprived rats, also indicate that endogenous opioids may modulate food palatability.

It has been suggested that opioid receptor agonists and antagonists may exert selective effects on macronutrient intakes. For example, morphine was reported to decrease protein and carbohydrate intake, but to increase fat intake over a 6-h feeding period (Marks-Kaufman and Kanarek, 1980). Following naloxone administration, rats consumed less calories from a fat ration over a comparable 6-h test period (Marks-Kaufman and Kanarek, 1981). In another study, systemic administration of morphine produced a preferential increase in protein consumption, in satiated rats, but a preferential increase in fat ingestion in food-restricted rats (Shor-Posner et al., 1986). These and other, related studies (Marks-Kaufman, 1982; Romsos et al., 1987) appear to imply that opioid receptor agonists and antagonists may act selectively with regard to macronutrients in the diet, and that, by extension, endogenous opioid peptides may mediate the selection of, and preference for, types of food rich in one macronutrient or other (e.g., high fat or high sugar foods). In a departure from this view, Gosnell and colleagues (1990a) investigated the effects of morphine on diet-selection in carbohydrate-preferring and in fat-preferring rats. They found that morphine increased carbohydrate intake in the carbohydrate-preferring rats, and fat intake in the fat-preferring rats. These authors concluded that endogenous opioid peptides are involved in the mediation of food preference and palatability, which is not specific to the macronutrient content of the diet. Echoing this view, Glass et al. (1996) showed that naloxone (0.01–0.3 mg/kg, s.c.) selectively reduced intake of the preferred diet but not the intake of the nonpreferred diet, in diet-selection tests which involved high fat and high carbohydrate diets. Similarly, Giraudo et al. (1993) concluded that the apparent anorectic effect of naloxone depended on the palatability of the food. Naloxone had a more potent effect in the case of high-palatability foods. It follows from these results that the palatability of foods, as distinct from their macronutrient content, determines the potency of the effects of opioid drugs, agonists and antagonists. That is not to say, of course, that the macronutrient content is irrelevant, since it may well be that sugar and fat content of a food contribute strongly to the palatability of that food.

Kelley and her colleagues have demonstrated that administration of the selective MOR agonist DAMGO (D-Ala^2, NMe-Phe^4, Glyol^5-enkephalin) into the nucleus accumbens produces strong enhancement of the consumption of highly palatable food (Bakshi and Kelley, 1993; Zhang and Kelley, 1997; Zhang et al., 1998). More recent work indicates that opioid activity in the ventral striatum is closely involved in the control of highly palatable food consumption via more caudal diencephalic and brainstem structures (Will et al., 2003). Hence, a distributed brain network appears to be engaged when energy-dense high palatability food is encountered and consumed. Opioid activity in the nucleus accumbens appears to be a key component in this integrative activity.

D. Dopamine and Food Preference

During ingestive behavior in rats, dopamine release in the nucleus accumbens is enhanced, resulting in higher levels of extrasynaptic dopamine and its metabolites (Hernandez and Hoebel, 1988; Radhakishun et al., 1998). Consuming sucrose solutions is particularly effective in causing elevated dopaminergic activity in the nucleus accumbens (Hajnal and Norgren, 2001; Hajnal et al., 2004; Rada et al., 2005). Injection of the selective dopamine D_2/D_3 receptor agonist, 7-OH-DPAT, into the nucleus accumbens reduced sucrose consumption in the rat (Gilbert and Cooper, 1995). This effect was probably secondary to a reduction in dopamine release in the nucleus accumbens (Gilbert et al., 1995). Hence, consumption of highly palatable sucrose solutions may require mobilization of nucleus accumbens dopamine. Taha and Fields (2005) have more recently identified a population of nucleus accumbens neurones which specifically encode the palatability of sucrose solutions. Dopaminergic activity in the nucleus accumbens is also increased when rats consume high-palatability snack foods

(Bassareo and Di Chiara, 1997, 1999). We should therefore expect dopaminergic mechanisms to be involved, in a critical sense, in food preference measures where food palatability is a determining factor. Martel and Fantino (1996a,b) have examined this possibility. In their experiments, rats were provided with their regular maintenance diet and highly palatable short cakes, containing 20% sucrose. This kind of food preference situation closely resembles some of the food preference studies described previously, in relation to benzodiazepines and opioids. Using a microdialysis approach, Martel and Fantino (1996a,b) found that dopamine levels in the nucleus accumbens were greater when the rats consumed the highly palatable, sweetened food. Following from their work, we should predict that dopaminergic drugs would affect high-palatability food preference in the rat. Quinpirole, a selective dopamine D_2/D_3 receptor agonist, inhibits dopamine release at dopaminergic terminals (Kennedy et al., 1992) and inhibits licking behavior for sucrose solutions (Genn et al., 2003). In more recent work, we have shown that quinpirole selectively reduced the preference for chocolate-coated cookies, the high-palatability item in a food-preference test (Cooper and Al-Naser, 2006). Attenuating the dopamine signal in the nucleus accumbens, therefore, appears to be instrumental in reducing the preference for and consumption of high-palatability foods containing chocolate and sugar.

E. Summary

In this section we have seen that benzodiazepine, opioid, and dopamine mechanisms are involved in the determination of food preferences in the rat. These provide the bases for the principal pharmacological *dramatis personae* of this chapter. This list is by no means exhaustive, of course, but gives access to a reasonably sound database. For other candidate neurochemical systems, the evidence remains sparse and

patchy. There are two limitations which necessarily confront us and which limit whatever conclusions we draw. Pharmacological analysis, to be sufficiently complete, depends on systematic comparisons across multiple drugs in which extensive dose–response data are generated (Cooper, 1987b). Likewise, behavioral analysis requires systematic comparisons across numerous test paradigms under standardized conditions. Nevertheless, this chapter will proceed along paths defined in terms of benzodiazepine, opioid, and dopamine mechanisms, and will seek to reach at least some, if only provisional, conclusions.

III. PHARMACOLOGY OF UNLEARNED TASTE PREFERENCE AND REACTIVITY

A. Measuring Taste Preference and Taste Reactivity

Many factors influence our choice of foods, our preferences, and the amount we consume in meals. Simple taste factors have received a great deal of attention; however, because "tastes" can be characterized in terms of a small number of categories (sweet, salty, bitter, acid, umami), their intensity is easily manipulated, and they can elicit characteristic "liking," on the one hand, or rejection responses on the other. Moreover, manipulating taste factors can exert strong influences over food and fluid consumption, not only in terms of levels of consumption but also in terms of preferences and aversions (Grill et al., 1987). Responses to some of the primary taste categories appear to be present at birth. For example, human neonates respond to the sweet taste of sucrose with facial expressions indicative of enjoyment. In contrast, they respond to the bitter taste of quinine with facial expressions of disgust (Steiner, 1977). Newborn infants will consume more sweetened water than plain water (Desor et al., 1973). Similarly, infant rats exhibit licking and rhythmic mouth movements (comprising ingestive responses) to sweet solutions, but show mouth gapes and head movements (indicating aversive response) to bitter taste (Ganchrow et al., 1986). There are developmental changes which take place with respect to taste preferences and aversions (Bernstein, 1991), but we can assume for present purposes that learning mechanisms need not be invoked to account for preferences exhibited to the basic taste categories.

Typically, experiments on taste stimuli and ingestive behavior employ well-defined taste qualities delivered in aqueous solutions. Stimulus concentration can easily be varied to investigate relationships between stimulus intensity and dependent variables such as intake, preference, or licking responses. Grill and colleagues (1987) distinguished between *voluntary stimulus sampling* and direct *intraoral infusion* as methodologies useful in studying the effects of taste factors on ingestion in the rat. In the former case, ingestive responses to taste stimuli have been determined using either *one-bottle* (acceptance) or *two-bottle* (preference) tests (Falk, 1971). Early experiments tended to use 24-h measurement periods, but short-duration tests are now typically used to reduce postingestive effects. In very brief exposure tests, the measurement of licking responses may provide more accuracy and better discrimination between test conditions than simple intake measures (Davis, 1973). Such tests focus attention on oral stimulation factors in relation to fluid acceptance and preference outcomes.

The *taste reactivity test* is an example of an intraoral fusion method, which requires no voluntary sampling of tastants; taste stimuli are delivered directly within the oral cavity in an obligatory manner. Infusion of acceptable taste stimuli elicits a characteristic set of stereotyped movement patterns called *ingestive responses* (mouth movements, tongue protrusions, lateral tongue movements, paw licking). *Aversive responses*, on the other hand, are elicited by unacceptable taste stimuli and comprise gaping, chin rubbing, head-shaking, and forelimb flailing (Grill and Berridge, 1985; Grill and Norgren, 1978a). Sucrose, sodium chloride, and dilute hydrochloric acid elicit ingestive responses in control and chronic decerebrate rats (Grill and Norgren, 1978b). Quinine, on the other hand, elicits aversive responses.

B. Benzodiazepines, Taste Preference, and Taste Reactivity

Benzodiazepines increase all ingestive responses, and therefore enhance not only food consumption but also water intake and the consumption of palatable, dilute salt solutions (Cooper, 1982a; Cooper and Estall, 1985; Cooper and Francis, 1979; Turkish and Cooper, 1984). There is good evidence, though, that over and above these effects, benzodiazepines have particular effects on the ingestion of tastants, and enhance unlearned taste preferences. The evidence rests on studies of sweet and salt preferences, and aversion to quinine-adulterated solutions (Cooper, 1989).

Cooper and Yerbury (1988), for example, trained thirsty rats in a short-exposure two-bottle test to drink water versus a sodium saccharin solution (0.01% or 0.05% solution). The potent benzodiazepine receptor agonist, clonazepam, had no effect on the marginal preference for the weaker 0.01% solution. In contrast, it selectively enhanced the robust preference for the stronger 0.05% solution. Hence, clonazepam did not *induce* a preference for the weaker sweet solution, but did *enhance* the preference expressed for the stronger solution. In the same study, the selective anxiolytic/hypnotic compound, zolpidem (Depoortere et al., 1986) had no effect on sweet taste preference. This rules out an explanation of clonazepam's positive effect in terms of anxiolytic or hypnotic activity. This result provides evidence for a scheme whereby benzodiazepines can affect taste preferences which, in turn, underlie their effects on food preferences. The result also encouraged further investigation of the effects of a variety of benzodiazepine receptor agonists on taste preferences and aversions.

The β-carboline drug, abecarnil, is a benzodiazepine receptor agonist with anxiolytic and anticonvulsant properties (Stephens et al., 1990). Additionally, abecarnil not only enhances consumption of sucrose or a highly palatable sweetened food in nondeprived rats, it also affects taste preferences in thirsty rats (Cooper and Greenwood, 1992). Thus we showed that abecarnil selectively enhanced the preferences for both saccharin and isotonic salt solutions. Benzodiazepine receptor *partial* agonists have been developed which retain anxiolytic and anticonvulsant effects, but which lack side effects such as sedation and muscle relaxation (Haefely et al., 1992). Two such partial agonists, bretazenil (Ro16-6028) and Ro17-1812, induce a strong hyperphagic effect in nondeprived rats (Yerbury and Cooper, 1987). They also enhance the taste preference for a 0.05% sodium saccharin solution in a two-choice test, while also suppressing an aversion for a 0.005% quinine solution in water-deprived rats (Cooper and Green, 1993). Bretazenil and Ro17-1812 also promote selectively the ingestion of a preferred isotonic salt solution over water in a two-choice test (Cooper and Barber, 1993). Pharmacologically, these benzodiazepine effects on ingestive responses are dissociable from anxiolytic and anticonvulsant effects (from the zolpidem example), and from side effects such as sedation and muscle relaxation (from the bretazenil and Ro17-1812 examples). Behaviorally, there is a close association between effects of benzodiazepines on taste preferences and their effects on sucrose or palatable food consumption. The most parsimonious account is that benzodiazepine receptor agonists act to promote taste palatability which in turn has a determining effect on taste preferences and food preferences (Cooper, 1989).

The central receptors for benzodiazepines are located in synaptic membrane-bound proteins which also contain receptor sites for the major inhibitory neurotransmitter γ-aminobutyric acid (GABA), and are classified as $GABA_A$ receptors (Rudolph et al., 2001; Rudolph and Möhler, 2004). Benzodiazepine receptor agonists enhance central inhibitory effects mediated by synaptic $GABA_A$ receptors. By a quirk of fate, however, there are other benzodiazepine receptor ligands which have the opposite modulatory effect on GABAergic neurotransmission at $GABA_A$ receptors, and these have been called *inverse agonists* (Barnard et al., 1998; Mehta and Ticku, 1999). (For an introduction to the theory of inverse agonists in a general pharmacological context, see Kenakin, 2004a,b.) These benzodiazepine receptor inverse agonists provided us with an unexpected opportunity to test the relationship assumed to exist between modulation of food consumption on the one hand and taste preferences on the other. We should expect that benzodiazepine receptor inverse agonists to reduce palatable food consumption (a hypophagic effect) and to reduce taste preferences determined by taste palatability factors (a hypohedonic effect). We should bear in mind that benzodiazepine receptor inverse agonists have been found in a number of different chemical families, and therefore have a range of these drugs to test for the generality of their effects across families.

In an initial study, the two β-carboline inverse agonists, FG 7142 and DMCM (Braestrup et al., 1982, 1983) were shown to produce dose-dependent reductions in the consumption of a highly palatable sweetened food in nondeprived rats (Cooper et al., 1985c). Their hypophagic effect was counteracted by the selective benzodiazepine receptor antagonist, flumazenil (Ro15-1788) (Hunkeler et al., 1981), demonstrating that their action was mediated by specific benzodiazepine receptors. This was the first evidence that benzodiazepine receptor ligands (agonists and inverse agonists, respectively) could exert *bidirectional* effects on palatable food consumption (Cooper, 1985). The pyrazoloquinoline compound, CGS 8216, is also a benzodiazepine receptor inverse agonist (Jensen et al., 1983). In a series of experiments we demonstrated that CGS 8216 reduced: (i) palatable sweet mash consumption, (ii) palatable oily mash consumption, (iii) sweetened milk intake, (iv) intake of a highly palatable saccharin–glucose mixture, and (v) sucrose sham feeding in the chronic gastric-fistula rat (Estall and Cooper, 1986; Kirkham and Cooper, 1987; Kirkham et al., 1987).

In addition to their evident effects on food consumption, these same compounds can also affect taste preference determined in two-choice tests. Thus, sweet taste preference is suppressed by the β-carboline FG 7142 (Cooper, 1986), the pyrazoloquinoline CGS 8216 (Kirkham and Cooper, 1986; Kirkham et al., 1987), and imidazobenzodiazepine partial inverse agonist Ro15-4513 (Cooper et al., 1989). Just as there is a bidirectional control of palatable food consumption mediated by central benzodiazepine receptors, so too there is bidirectional control of sweet taste preference. In the case of salt taste preference, however, we have not found evidence for bidirectionality in drug effects. Kirkham and Cooper (1986) reported that the pyrazoloquinoline CGS 8216 did not reduce salt taste preference in thirsty rats. Similarly, Cooper and Barber (1993) found no reduction in salt taste preference following administration of the imidazobenzodiazepine Ro15-4513. Moreover, Kirkham and Cooper (1986) found no effect of CGS 8216 on either water consumption or intake of a weak (0.0005%) quinine solution in water-deprived rats. Hence, there is a degree of selectivity in the effects of benzodiazepine receptor inverse agonists. They reliably suppress sweet taste preference, appear to have no effect on salt taste preference, and do not reduce ingestional behavior nonspecifically.

Microstructural analysis of licking behavior in brief-duration tests can provide measures of taste palatability, for example, in terms of the cumulative number of licks when animals consume palatable fluids (Davis, 1973); the initial rate of licking before satiety signals begin to have a major influence (Davis and Smith, 1988); and the duration of bursts or clusters of licking as distinct from the frequency of such licking episodes (Davis and Perez, 1993; Davis and Smith, 1992). While palatable sugar solutions increase each of the measures in a concentration-dependent way, aversive quinine taste reduces the quantity of fluid consumed, the number of licks, and reduces the duration of licking bursts (Spector and St. John, 1998). Benzodiazepine receptor agonists have a positive enhancing effect on each of these measures, not only for sucrose ingestion but also for maltodextrin drinking, licking for a palatable fat emulsion (Higgs and Cooper, 1997, 1998a), and for palatable dilute NaCl solutions (Cooper and Higgs, 2005). In contrast, the imidazobenzodiazepine inverse agonist Ro15-4513 reduced the initial rate of licking for sucrose, and reduced the size of licking bursts (or bouts) for sucrose (Higgs and Cooper, 1996a).

Taste reactivity measures have been developed to assess taste palatability based not only on voluntary fluid ingestion, but on stereotyped patterns of oral motor responses to intraoral infusions of tastants (Grill and Berridge, 1985; Grill et al., 1987). Caudal brainstem mechanisms are critical to these taste reactivity responses to sweet, salty, or quinine-adulterated solutions (Flynn and Grill, 1988; Grill and Norgren, 1978b). Berridge and Treit (1986) were the first to observe that chlordiazepoxide had a selective effect in a taste reactivity test. This benzodiazepine selectively enhanced ingestive responses, but had little or no effect on aversive reactions. Subsequently, other investigators have confirmed that benzodiazepine drugs modulate positive ingestive responses in taste reactivity tests (Cooper and Ridley, 2005; Gray and Cooper, 1995; Parker, 1995; Söderpalm and Hansen, 1998).

Importantly, chlordiazepoxide maintained its effect on taste reactivity measures in the chronic decerebrate rat (Berridge, 1988). This result indicates that the site of action of benzodiazepines may lie in the caudal brainstem. Later work first implicated a structure (or structures) close to the fourth ventricle (Peciña and Berridge, 1996), and then identified the parabrachial nucleus as the key site of action for the benzodiazepine-induced enhancement of positive ingestive responses (Söderpalm and Berridge, 2000). These results indicate that benzodiazepine-modulation of taste palatability may take place at the level of the parabrachial nucleus (Berridge and Peciña, 1995). There is additional evidence, however, to indicate that the benzodiazepine modulation of taste palatability is, neuroanatomically as well as functionally, closely related to the control of food consumption. Thus, microinjection of the water-soluble benzodiazepine, midazolam, into the fourth ventricle or into the parabrachial nucleus elicits a strong hyperphagic effect (Higgs and Cooper, 1996b,c). Benzodiazepine actions at specific receptors in the parabrachial nucleus, located in the caudal brainstem, may therefore both enhance positive taste palatability and promote food consumption (Cooper and Higgs, 1996). The pharmacology of GABA$_A$ receptor subtypes, particularly in relation to the hyperphagic and taste palatability effects of benzodiazepines, is not considered here. The reader is referred to a more recent, comprehensive review (Cooper, 2005).

C. Opioids, Taste Preference, and Taste Reactivity

Le Magnen and his colleagues (1980) were the first to report that the opioid receptor antagonist, naloxone, reduced ingestion of saccharin solutions in both single-choice acceptance tests and in two-choice preference tests. He interpreted these and later data in terms of brain opioids and food/taste palatability (Le Magnen, 1992). Sweet taste as such (with calories) is highly attractive to rats (Ernits and Corbit, 1973), and its attractiveness appears to depend on central opioid systems. Sclafani and colleagues (1982b) reported that naloxone produced a greater reduction in the consumption of a highly palatable saccharin–glucose solution than of water. Several other laboratories confirmed that opioid receptor antagonists reduced or abolished sweet taste preference (measured in terms of preferred intake of saccharin solutions over water) (Cooper, 1983a; Lynch and Libby, 1983; Siviy and Reid, 1983) and suppressed saccharin drinking in an acceptance test (Cooper, 1982b). Later work by Lynch (1986) appeared to show that naloxone may block the acquisition of saccharin preference, and the acquisition of sucrose preference (Lynch and Burns, 1990).

These first studies strongly suggested that opioid receptor antagonists can act to blunt the palatability of sweet-tasting solutions. Consistent with this view, other studies showed that opioid receptor antagonists would also reduce sucrose sham feeding in chronic gastric-fistula rats (Kirkham, 1990; Kirkham and Cooper, 1988a,b, 1989; Rockwood and Reid, 1982). In a pharmacological analysis of this effect, Leventhal and colleagues (1995) investigated the effects of receptor-specific antagonists on sucrose sham feeding. In addition to the inhibitory effect of naltrexone, the selective μ-receptor (MOR) antagonist, β-funaltrexamine, and the selective κ-receptor (KOR) antagonist, norbinaltorphamine, both reduced sucrose sham feeding. Since selective δ-receptor (DOR) antagonists were without effect, Leventhal et al. (1995) proposed that MOR and KOR receptor antagonists block sweet taste palatability.

Conversely, we should expect that systemic administration of the prototypic MOR agonist, morphine, would lead to increased sweet solution consumption and taste preference. While Calcagnetti and Reid (1983) initially provided some confirmatory evidence, other investigators failed to find increases in saccharin drinking following morphine administration (Cooper, 1982b, 1983a; Lynch and Libby, 1983). Therefore, Cooper (1994) investigated the effects of subcutaneously administered morphine (0.1–3.0 mg/kg) on the consumption of sodium saccharin solutions (0.02%, 0.1%, and 1.0% concentrations) in nonedeprived rats. The highest baseline level of intake occurred at 0.1% sodium saccharin. Interestingly, morphine had no effect whatsoever on the consumption of 0.02% or 0.1% solutions, in line with the earlier negative findings. However, it dramatically increased the consumption of the highest concentration (1.0% solution). This positive effect was not due to morphine counteracting any aversive quality of the concentrated saccharin solution. When the highly acceptable 0.1% solution was adulterated with quinine to reduce its baseline intake down to that of the 1.0% solution, morphine had no effect. Thus, morphine did not overcome the aversive quality of the quinine adulteration. Moreover, on the basis of these results, it is difficult to sustain the view that morphine generally enhanced the palatability of the saccharin solutions. Route of administration may be an issue here, and intracerebroventricular (i.c.v.) administration may confer advantages. Gosnell and Majchrazak (1989) found that i.c.v. administration of a selective MOR agonist DAGO ([D-Ala^2.MePhe^4.Gly-ol^5]enkephalin) and a selective DOR agonist DTLET (Tyr-d-Thr-Gly. Phe-Leu-Thr) increased consumption of a 0.15% sodium saccharin solution.

There is also evidence that salt taste palatability may involve endogenous opioid mediation. Opioid receptor antagonists, like naloxone and naltrexone, reduce salt solution consumption in single-choice and two-choice tests (Cooper and Gilbert, 1984; Cooper and Turkish, 1983; Levine

et al., 1982; Ukai et al., 1988). Morphine increased the preference for isotonic saline in thirsty rats (Bertino et al., 1988). Employing the i.c.v. route, Gosnell and Majchrazak (1990) reported that naloxone reduced hypotonic saline ingestion but left water intake unaffected. Conversely, i.c.v. administration of opioid receptor agonists enhanced hypotonic saline drinking, implicating MOR- and DOR-mediated mechanisms (Gosnell and Majchrazak, 1990; Gosnell et al., 1990b). Bodnar and colleagues (1995) undertook a thorough pharmacological analysis of opioid receptor subtype involvement in relation to hypotonic and hypertonic salt drinking in thirsty rats. While water ingestion was reduced by i.c.v. administration of MOR- and KOR-selective antagonists, respectively, salt drinking was unaffected. Their work raised the important issue of whether salt drinking depends on taste palatability or reflects sodium appetite. Either way, they could detect no effect of selective opioid antagonists on salt drinking. It is beyond the scope of this chapter to go into greater detail about the respective contributions of opioid receptor subtypes in the controls of ingestive behavior (cf. Bodnar, 1996; Gosnell and Levine, 1996). However, we can discern important potential differences with regard to sweet taste and salt preferences, respectively. In general terms, though, there remain good reasons for proposing that central opioid systems contribute to the determination of taste preference and palatability (Cooper and Kirkham, 1993).

This hypothesis can also be tested using other behavioral paradigms. Taste reactivity tests are informative, although the available data focuses almost exclusively on ingestive responses to sweet solutions, and salty solutions have not (at least to date) been considered in the same terms. Parker and colleagues (1992) reported that naltrexone reduced positive ingestive reactions to sucrose. On the other hand, a number of studies have found that morphine increased positive reactions to sucrose, while decreasing aversive reactions to quinine (Clark and Parker, 1995; Doyle et al., 1993; Parker et al., 1992; Peciña and Berridge, 1995; Rideout and Parker, 1996). More recent studies have identified relatively localized "hot spots" in forebrain regions where specific MOR activation enhances positive taste reactivity responses to intraoral infusion of sucrose solution. Enhanced hedonic evaluation of the infused sucrose can be elicited by direct microinjection of the MOR agonist, DAMGO, into the medial shell region of the nucleus accumbens (Peciña and Berridge, 2000). Detailed mapping indicates that the increase in taste palatability can be localized to an effect of DAMGO in a small site in a rostrodorsal region of the medial shell (Peciña and Berridge, 2005). Additionally, Smith and Berridge (2005) have identified a separate site in the posterior ventral pallidum (VP) at which DAMGO increased the hedonic response to intraorally infused sucrose solution. The relationships between the circumscribed locus of action within the nucleus accumbens medial shell region and the ventral pallidum await future investigation. Nevertheless, a picture is emerging of a distributed though anatomically circumscribed system in the forebrain which mediates opioid effects on taste hedonics. How this putative system interacts functionally with caudal brainstem structures implicated in the mediation of taste palatability also awaits future inquiry.

There is a complexity to opioid pharmacology (Waldhoer et al., 2004), which should be taken into account. This can be illustrated, in a small way, using a licking microstructure approach in which nondeprived rats had been trained to lick for sucrose solutions (at three concentrations) or for intralipid emulsions (at three concentrations) (Higgs and Cooper, 1998b). Comparisons were drawn between the effects of naloxone, morphine, and the selective KOR agonist, U-50,488H. Both opioid agonists

increased licking responses. However, morphine's effect was due to a selective increase in the number of bursts of licking (cf. Cooper, 1994), while the effect of U-50, 488H was due to a selective increase in the mean duration of bursts of licking. Analyzed microstructurally, in this way, the effects of morphine and U-50,488H were distinctly different, even though they both increased the overall level of licking. This difference underscores the need to investigate the roles of opioid receptor subtypes in behavioral paradigms, including the taste reactivity test. It is interesting that U-50, 488H had effects on licking microstructure which were similar to those of benzodiazepines (Higgs and Cooper, 1997, 1998a). This result implies that U-50,488H, but not morphine, enhances taste palatability.

D. Dopamine and Taste Reactivity

Intact taste reactivity responses can be obtained in the chronic decerebrate rat, lacking a forebrain (Grill and Norgren, 1978b). The neural circuitry required to evaluate taste stimuli and to mediate the appropriate behavioral responses measured in taste reactivity tests must lie, therefore, in the caudal brainstem. It follows from this that any loss of forebrain function should not compromise taste reactivity responses, including the loss of forebrain dopaminergic function. This prediction has been confirmed in studies in which 6-hydroxydopamine (6-OHDA) lesions of forebrain dopaminergic projections leave intact taste reactivity responses (Berridge et al., 1989). Moreover, we should predict that dopamine receptor blockade should not affect taste reactivity responses, even though such antagonists not only affect ingestive responses but also potently suppress instrumental responding for food or water reinforcement (Ljungberg, 1987, 1990; Rusk and Cooper, 1994).

Following on from a suggestion that the dopamine receptor antagonist, pimozide, may reduce the hedonic value of sucrose

(Bailey et al., 1986), some conflicting data were obtained in taste reactivity tests. While Treit and Berridge (1990) found no effect of haloperidol in a taste reactivity test, Leeb and colleagues (1991) found that pimozide reduced the palatability of sucrose in a similar test. A collaborative study between two laboratories was therefore undertaken, with the result that pimozide was deemed to have no effect on either positive or aversive patterns of response in the taste reactivity paradigm (Peciña et al., 1997). Hence, dopamine receptor blockage does not appear to affect taste palatability. Taking an opposite tack, taste reactivity responses have been investigated in "hyperdopaminergic" mutant mice (Peciña et al., 2003). Knockout of the dopamine transporter (DAT) gene leads to highly elevated levels of extracellular dopamine (Giros et al., 1996; Sora et al., 1998). Peciña and her colleagues (2003) found that DAT knockout mice did not show an increased palatability response in a test of taste reactivity to intraoral infusion of sucrose.

Yet we have seen that consumption of sucrose solutions leads to the enhanced release of dopamine in the nucleus accumbens (e.g., Hajnal and Norgren, 2001), and that inhibition of nucleus accumbens dopamine can lead to the suppression of sucrose consumption (Gilbert and Cooper, 1995; Gilbert et al., 1995). At the same time, at least one component of the necessary neural circuitry underlying intact taste reactivity responses lies in the caudal brainstem, and does not require forebrain dopaminergic involvement. Nevertheless, this leaves open the possibility that taste palatability assessment is a function of at least a subset of neurones in the nucleus accumbens (Taha and Fields, 2005), after initial processing in the caudal brainstem. Sucrose consumption is associated with heightened dopaminergic activity in the nucleus accumbens, and therefore we cannot exclude the possibility, at this stage, of dopaminergic *modulation* of sweet taste palatability processing in the ventral forebrain.

E. Summary

From both unconditioned taste preference and taste reactivity data, there is strong evidence for benzodiazepine enhancement of taste palatability. Most probably, the relevant site of action lies in the caudal brainstem, where the parabrachial nucleus is the principal candidate as the target structure. Similarly, there is convincing evidence for opioid modulation of taste hedonics, derived again from taste preference and taste reactivity data. The parallels between the two cases may suggest important interactions governing the two (Cooper, 1983b). In support of this view, studies point to benzodiazepine–opioid interactions in relation to palatability measures derived from licking microstructure (Higgs and Cooper, 1997) and taste reactivity (Richardson et al., 2005) experiments.

The picture for dopaminergic mechanisms is more complex. Whereas the evidence is sound that forebrain dopaminergic projections are not essential to the function of caudal brainstem evaluation of taste palatability, we cannot rule out *any* involvement of forebrain dopamine in the modulation of taste palatability processing. More attention should be directed to this important research issue. There is some evidence for dopaminergic involvement in benzodiazepine mechanisms underlying taste palatability processing (Higgs and Cooper, 2000). We should recognize that caudal brainstem processing of taste hedonics is assessed using an involuntary set of elicited responses. Such processing must be interfaced with voluntary ingestive behavior (food consumption) and with the expression of food and taste preferences. Forebrain dopaminergic mechanisms (perhaps involving only a minor subset of forebrain projections) may be crucial to such forms of behavioral interfacing.

IV. PHARMACOLOGY OF LEARNED FLAVOR PREFERENCE

A. Flavor Preference Conditioning

We do not come innately equipped with a full range of flavor (olfactory combined with taste stimuli) preferences. Within our own culinary culture, we acquire strong and specific flavor preferences, much as we acquire our first-language skills. Preferences can be extended beyond traditional cultural boundaries, and it is a well-known feature of contemporary British eating habits, that a range of national cuisines are enjoyed—Indian, Chinese, Thai, French, Italian, American, and so on. Acquisition of new food and flavor preferences, therefore, greatly enlarges the range of flavor experiences and the number of available choices among food items. This wide degree of choice contributes significantly to the attraction of eating.

We noted earlier in this chapter that rats, too, acquire food preferences. Rat pups come equipped with marked preferences for sucrose, starch-derived polysaccharides, and nutritive oil emulsions (Ackroff et al., 1990; Vigorito and Sclafani, 1988), but rats also have considerable capacity for acquiring new flavor preferences, as befits omnivorous animals. This can be demonstrated experimentally in the following two ways, each of which enlists a classical or Pavlovian conditioning procedure. It will be recalled that in classical conditioning training trials (the acquisition phase), an initially neutral stimulus which elicits no response is presented in association with an *unconditioned stimulus* (US), which reliably elicits a target response. When the conditioning procedure is successful, the formerly neutral stimulus elicits a reliable response when presented alone, and is called a *conditioned stimulus*. As an extra refinement, a paired stimulus which elicits a conditioned response is referred to as CS$^+$. There is a control condition, in which a stimulus is presented but not in association with the US

over a number of training trials. In this condition, we refer to the CS$^-$. In broad terms, following the conditioning procedure, the animal learns that the CS$^+$ signals the occurrence or availability of a reinforcer (the US), whereas the CS$^-$ signals its absence and lack of availability.

The two forms of flavor conditioning which have been distinguished are (i) *flavor-flavor* conditioning, and (ii) *flavor-nutrient* (or calorie) conditioning (Fedorchak and Bolles, 1987; Mehiel and Bolles, 1984). In the former, an initially neutral flavor stimulus is paired with a highly palatable taste or flavor stimulus, which rats innately prefer. Following the conditioning procedure, rats acquire an association between the two flavors, and learn to prefer the formerly neutral flavor (Fanselow and Birk, 1982). In the latter case, an initially neutral flavor is paired with the supply of nutrient into the gastrointestinal tract (usually as a consequence of food ingestion). Here it is the supply of nutrients, detected in the gastrointestinal tract or postabsorptively, which acts as the unconditioned reinforcer, allowing rats to acquire new flavor preferences when the flavors are associated with food delivery to the gut (Booth, 1985). While the distinction between the two forms of conditioning is clear conceptually, in practice the two are easily confounded. For example, if rats are exposed to a novel flavor added to a sucrose solution, and they learn to prefer the new flavor, what mechanism is responsible? Is it flavor-flavor conditioning or flavor-nutrient conditioning that underlies the acquired flavor preference? Special experimental procedures have to be employed to tease apart the respective contributions of these two forms of associative conditioning.

In flavor-nutrient conditioning, various sources of calories are effective in the acquisition of learned flavor preferences: glucose, sucrose, hydrolyzed starch, fat emulsions, soluble forms of protein, and ethanol (Ackroff and Sclafani, 1994, 2001; Drucker et al., 1993; Elizalde and Sclafani, 1990a,b;

Mehiel and Bolles, 1988; Pérez et al., 1995, 1996). The learned flavor preferences are usually determined using a "two-bottle" choice procedure, at completion of the training sequence of conditioning trials. Learned flavor preferences can be highly robust, and, moreover, can be highly resistant to extinction (Sclafani and Nissenbaum, 1988).

B. "Electronic Esophagus" Preparation

Sclafani and his colleagues have devised a method to circumvent flavor-flavor conditioning, allowing direct investigation of the mechanisms and processes involved in flavor-nutrient conditioning. According to their approach, surgery is carried out on rats to produce a so-called electronic esophagus preparation (Elizalde and Sclafani, 1990a; Sclafani and Nissenbaum, 1988). The animals are fitted with a stainless steel gastric cannula; whenever a rat drinks from a sipper spout, an infusion pump is activated which infuses fluid intragastrically. By this means, the experimenters can control the composition of the fluid which is infused intragastrically, independently of the fluid which the animal tastes as it drinks from the sipper spout. For example, the animal may drink flavored water or a flavored saccharin solution, while a nutrient solution (e.g., glucose, hydrolyzed starch) is infused intragastrically. By this means, flavor-flavor conditioning can be avoided, and any learned flavor preference can be attributed to the conditioning effect of the intragastric infusion of nutrient.

The conditioning procedure is usually carried out as follows: two flavored waters are prepared (cherry and grape flavors, respectively); drinking one of the flavors is paired with intragastric infusion of a nutrient source (this flavor becomes the CS[+]), drinking the second flavor is paired with nonnutritive water infusion (this flavor becomes the CS[-]). (In half the rats, cherry is the CS[+], while grape is the CS[-]; in the other half, the significance of each flavor is

reversed.) The animals' preference for the two flavors is then determined in two-bottle choice tests. Following this conditioning procedure, rats can display a near-total preference for the flavor associated with intragastric infusion of a nutrient source (Sclafani and Nissenbaum, 1988).

Before we move to consider some more recent pharmacological data, it is important to recognize that the *reinforcing* effect of intragastric infusions of nutrients is operationally distinct from the *satiating* effect of nutrients in the gastrointestinal tract. Ingested food, in effect, has two distinct functions: it can act as a reinforcer in the conditioning of food and flavor preferences, but it can also provide satiety cues which contribute to meal termination (Sclafani and Ackroff, 2004). Thus, the CCK$_A$ receptor antagonist devazepide attenuates the satiating effect of intraintestinal infusion of hydrolyzed starch, but does not affect flavor preference conditioning (Pérez et al., 1998). Capsaicin treatment, which inhibits the satiety effect of intraintestinal infusions of nutrients, has no effect on flavor preference conditioning (Lucas and Sclafani, 1996). Moreover, an intact vagus nerve is not always necessary for conditioned flavor preferences to be obtained (Sclafani and Lucas, 1996).

C. Opioids and Conditioned Flavor Preferences

The approach taken to date is to determine if opioid receptor antagonists affect either the acquisition (learning) or the expression (performance) of conditioned flavor preferences. We saw before that opioid receptor antagonists reliably attenuate unconditioned sweet and salty taste preferences. Mehiel (1996) trained rats in a flavor-nutrient preference conditioning paradigm, in which a distinctive flavor was paired with a dextrose solution. Administration of the antagonist, naloxone, not only suppressed the conditioned flavor preference but also prevented the acquisition of the flavor preference over conditioning trials. Mehiel (1996) suggested that endogenous opioid activity may be involved in flavor-nutrient conditioning. However, we cautioned before that steps have to be taken to investigate flavor-nutrient conditioning separately from flavor-flavor conditioning.

Thus, Azzara and colleagues (2000) conditioned a flavor preference in rats using intragastric infusions of sucrose as the reinforcing stimulus. Under these conditions of flavor-nutrient conditioning, the opioid receptor antagonist, naltrexone, had no effect on the acquisition or expression of flavor preferences. These authors concluded, therefore, that endogenous opioid systems play no part in flavor preferences conditioned by intragastric nutrient infusions. In a companion study, Yu and colleagues (1999) sought to minimize postingestional effects in flavor preference conditioning by using sham-feeding rats. Flavor preference conditioning in this preparation should depend more on orosensory factors rather than postingestional nutritional effects. Nevertheless, in this preparation too, naltrexone had little or no effect on the acquisition or expression of learned flavor preferences. It did reduce fluid intake, however, in agreement with earlier sham-feeding studies (Kirkham, 1990; Kirkham and Cooper, 1988a,b; Leventhal et al., 1995). Most recently, a study by Baker and colleagues (2004) confirmed that naltrexone did not affect the acquisition or expression of flavor preferences conditioned by fructose. On the basis of these data, therefore, one has to conclude that, despite the earlier report of Mehiel (1996), endogenous opioid systems may play no major part in the acquisition or expression of learned flavor preferences.

To understand why this may be so, we should briefly consider two distinguishable accounts of flavor preference conditioning. According to Young (1961), flavor stimuli may have informational value, in that they signal nutrient content and availability in foodstuffs, that is, nutritive expectancy. In addition, flavor stimuli have affective value, that is, palatability. Flavor preference conditioning could either reflect the development of nutritive expectancy, so that the conditioned flavor stimulus signals the delivery of nutrient into the gastrointestinal tract, or it could reflect a process whereby rats learn to like the new flavor. If we accept the body of evidence, discussed before, that endogenous opioid systems are intimately involved in food and taste palatability, then we may wish to hypothesize that conditioned flavor preferences cannot involve opioid-dependent palatability processes.

D. Dopamine and Conditioned Flavor Preferences

Bodnar, Sclafani and their colleagues have also investigated the effects of a selective dopamine D_1 receptor antagonist, SCH 23390, and a selective dopamine D_2 receptor antagonist, raclopride, on the acquisition and expression of conditioned flavor preferences. In work by Azzara et al. (2001), rats were trained to associate a distinctive flavor with intragastric infusions of sucrose, and tested for flavor preferences in two-bottle tests. Their results indicated that SCH 23390, but not raclopride, affected the acquisition of the flavor preference. However, neither drug selectively affected the expression of the learned flavor preferences. These authors concluded that dopamine D_1 receptors may be involved in the acquisition of a sucrose-conditioned flavor preference. In companion studies, using sham-feeding rats, Yu et al. (2000a) reported that both SCH 23390 and raclopride reduced the expression of flavor preferences conditioned by the sweet taste of sucrose. The outcome was confirmed in additional work (Yu et al., 2000b), although neither drug had much effect on the acquisition of the conditioned flavor preference. Despite these data, however, more recent work from the same laboratories indicates that SCH 23390 and raclopride can impair the acquisition of flavor preferences conditioned by fructose ingestion (Baker et al., 2003).

Taken together, these data do not admit to a single, straightforward interpretation, and any explanation of the data has to be speculative at this early stage. In the case of flavor-nutrient conditioning, the evidence suggests little or no involvement of D_1 or D_2 receptor-mediated mechanisms in the expression of conditioned flavor preferences (Azzara et al., 2001). However, where sweet taste is the reinforcer determining the conditioning of flavor preferences, then dopaminergic mechanisms may play some part in either the acquisition or the expression of learned flavor preferences. Once again, one has to call for more research on these problems, especially since there is considerable evidence for mobilization of central dopamine activity associated with sweet taste and sucrose consumption (Hajnal and Norgren, 2001; Hajnal et al., 2004; Rada et al., 2005).

E. Taste Reactivity and Conditioned Flavor Preferences

Earlier, we referred to Young's (1961) distinction between the informational value of conditioned flavor stimuli, and their affective value or palatability. In the former case, the CS^+ becomes preferred because it signals the nutrient availability of the ingested foodstuff. In the latter case, the CS^+ becomes preferred because rats like it more—positive taste palatability has been enhanced through the conditioning process. One means to evaluate this second possibility is to test the hedonic evaluation of the CS^+ using a taste reactivity procedure. Myers and Sclafani (2001, 2003) have investigated this possibility.

In the first experiments Myers and Sclafani (2001) trained rats to prefer a sweet CS^+ flavor that had been paired with an intragastric infusion of 16% glucose. A strong preference for the CS^+ flavor was demonstrated in a two-bottle intake test. When the rats were then tested in a taste

reactivity paradigm, they exhibited more hedonic TR responses to the CS^+ flavor than to the CS^- flavor. The authors concluded from these results that the flavor-conditioning procedure had produced a learned shift in the palatability of the CS^+ flavor. However, they were careful to point out that any learned palatability shift may be only one of possibly several mechanisms that underlie behavioral responses to nutrient-paired flavors. Moreover, a learned palatability shift may not even be necessary for conditioned flavor preferences to emerge.

In a second set of studies (Myers and Sclafani, 2003), either a sour (citric acid) or a bitter (sucrose octaacetate) CS^+ flavor was used, instead of the sweet stimulus (saccharin solution) used in the previous experiments. The conditioning procedure (pairing with intragastric infusion of 16% glucose) was successful, and the rats exhibited a 95% preference for the CS^+ flavor in the two-bottle intake test. However, in these instances, there were no increases in palatability as determined by subsequent taste reactivity tests. Hence, there was no evidence that the rats had increased their liking for the sour or bitter CS^+ flavors, despite the conditioning outcome which had led to the pronounced flavor preferences.

In some situations, therefore, conditioned flavor preferences may be accompanied by an increase in the liking for the CS^+ (enhanced hedonic evaluation or palatability), but in other situations not. This means that increased palatability is not necessary for conditioned flavor preferences to be acquired and expressed, and there is no clear evidence that increased palatability may be a sufficient condition either. A critical difference in the two reports of Myers and Sclafani (2001, 2003) may be the initial hedonic value of the flavor to be conditioned using intragastric nutrient infusions. In Myers and Sclafani (2001) a sweetened flavor was used for the CS^+, but distasteful sour or bitter flavors were subsequently tested (Myers and Sclafani, 2003). The nutrient conditioning procedure may engender an enhancement of the palatability of a sweetened flavor (i.e., a quantitative shift in hedonic value), but may fail to transform an initially unpalatable flavor into a palatable one (i.e., a qualitative shift in hedonic value). In the later case, there was presumably a nonhedonic process which was responsible for the acquisition and expression of the conditioned flavor preferences.

Do these behavioral analyses throw any additional light on the pharmacological studies discussed before, involving the use of opioid and dopaminergic antagonists? The lack of effect of naltrexone, either on the acquisition or on the expression of conditioned flavor preferences (Azzara et al., 2000; Baker et al., 2004; Yu et al., 1999), implies that opioid-dependent processes are not crucial to the conditioning process. Even when an upward shift in palatability can be demonstrated following a flavor-conditioning procedure (Myers and Sclafani, 2001), opioid receptor blockade does not appear to be sufficient to compromise the learned flavor preference. One would not wish to conclude from this that opioid receptor blockade does not attenuate liking for tastes/flavors (cf. Section IIIC); rather it appears that the attenuation of opioidergic taste palatability is not in itself sufficient to reduce a conditioned taste preference. Perhaps a loss of function in brainstem taste palatability processing (e.g., involving the parabrachial nucleus) may have more impact, but this remains to be investigated. The dopaminergic data remain problematic, in the sense that taste palatability processing (the liking of tastes) appears to lie outside forebrain dopamine mechanisms, at least according to taste reactivity data (cf. Section IIID). Whether changes in taste hedonics are ruled in or ruled out, therefore, does not materially assist in identifying the mechanism responsible for any impact dopamine antagonists might have on conditioned flavor preferences.

V. CONCLUDING REMARKS

This chapter has dealt with food, taste, and learned flavor preference in the rat. Measures of preference are derived from observed behavioral responses when animals are given two concurrent choices (e.g., in a flavor-conditioning test) or multiple choices (e.g., in a food cafeteria test). This represents a positivistic approach, in which the attribution of preference is derived from the measurement of observed behavior. This approach does not help us directly, however, in seeking explanations couched in terms of mechanisms. Pharmacological interventions may help us to get to grips with the nature of the mechanisms which give rise to the behavioral expressions of preference.

What are the characteristics of such preferences, and what, indeed, do they represent in terms of function and purpose? In an important sense, expression of preferences reflects the assignment of priorities in the animal's behavioral economy. Such priorities may be seen in terms of *sequence* (which item to try first, which second, and so on), in terms of *time allocation* (how much time to devote to eating one food source compared with another), in terms of *quantity* (a little of this, but a lot of that), in terms of *effort expended* (working energetically for one food item, while remaining indifferent to another), and in terms of *competition* (if something better comes along, I will go for that). Measures of preference, therefore, reflect behavioral outcomes which are highly determined by food item availability, time and effort costs, and sampling strategies.

Without being exhaustive, we have seen in this chapter at least three forms of preference behavior which may arise from nonoverlapping sources. The first case is preference based on *familiarity*. At the start of this chapter, we saw how a rat's preference could be decided on the basis of the relative familiarity or novelty of the available food items. Repeated exposure to initially novel items can lead to marked shifts in the animal's expressed preferences. The second case is preference based on *palatability*, which has attracted an enormous amount of research attention. The third case is preference based on *flavor-nutrient conditioning*, in which the preferred flavor cue acts as a signal for the arrival of food in the gastrointestinal tract. There is no *a priori* reason why these three forms of preference should reflect similar or even related neural and behavioral mechanisms. This is where the pharmacological data, admittedly limited and resting on relatively few drug examples, may provide some valuable insights.

Opioid antagonists have been used in a great number of experimental situations. On present evidence, opioid antagonists do not appear to affect preference based on familiarity (cf. Cooper et al., 1988), but do appear to reduce preference based on taste palatability. The studies on learned flavor preferences suggest that these drugs affect neither the acquisition nor the expression of such preferences. Benzodiazepines exert a different pattern of effects. They enhance not only taste palatability and taste preferences, but also preference based on familiarity (cf. Cooper and Crummy, 1978). Data are not as yet available for the possible effects of benzodiazepines on learned flavor preferences, but such information would be extremely interesting. Since benzodiazepines appear to interact directly with caudal brainstem mechanisms which initially process taste hedonics, their effect (or lack of effect) on learned flavor preferences could provide important clues to the nature and possible location of relevant neural mechanisms. In the case of dopaminergic systems, their role in taste palatability has come in for severe questioning; dopamine does not seem to be necessary for basic taste hedonics. However, the flavor preference conditioning studies of Sclafani, Bodnar, and their colleagues strongly imply that there is some dopamine involvement. Pulling together these various strands of

evidence suggests that significant pharmacological differences exist in comparisons among the several varieties of food, flavor, and taste preferences.

Can this insight help in the development of effective drug interventions in the therapeutics of obesity? I think it can. The three forms of preference which I have outlined all have a bearing on food selection, levels of overconsumption, and resistance to change to alternative forms of food choice. Sticking with a highly familiar diet is a commonplace; it helps to define one's characteristic food culture. Even if it can be shown to have undesirable or deficient characteristics (too much fat or too few vegetables, e.g.), many people are resistant to change, are reluctant to try new food items, and, quite simply, prefer what they are familiar with and habitually consume. No one doubts that the palatability of food, its attractiveness in terms of a range of interacting sensory characteristics (flavor, texture, smoothness, temperature, and so on), exerts a powerful influence over food choice and food consumption (Drewnowski, 1998). Palatability is, in effect, the inducement to overconsume and overindulge. Finally, the acquisition of new flavor preferences, based on the powerful conditioning effect of food as a reinforcer, is the means to expand the choice of foods and fluids to consume, and to multiply the variety of food items from which to select.

Each of these forms of preference could be legitimate targets for the development of effective antiobesity treatments, where the overall requirement is to reduce levels of energy intake below those of energy expenditure to achieve sustainable weight loss. Nevertheless, we should accept the clear implication of current pharmacological research on preferences. No single drug intervention can possibly interact with all forms of preference in exactly complementary ways to achieve such a desired outcome. We cannot yet specify the necessary characteristics of ultimately effective drug treatments, but we can at least recognize that single drug treatments are unlikely to be more than only marginally effective. Expanded research on the behavioral, physiological, and neural mechanisms underlying the several forms of preference is required, coupled with more extensive pharmacological investigations. Such work may pave the way for improved drug-based interventions in the treatment of human obesity.

Acknowledgments

I should like to thank Dr. A. Sclafani for his helpful comments on this chapter, and Mrs B. A. Halliwell for her expert manuscript preparation.

References

Ackroff, K., and Sclafani, A. (1994). Flavor preferences conditioned by intragastric infusions of dilute Polycose solutions. *Physiol Behav.* **55**, 957–962.

Ackroff, K., and Sclafani, A. (2001). Flavour preferences conditioned by intragastric infusion of ethanol in rats. *Pharmacol. Biochem. Behav.* **68**, 327–338.

Ackroff, K., Vigorito, M., and Sclafani, A. (1990). Fat appetite in rats: the response of infant and adult rats to nutritive and non-nutritive oil emulsions. *Appetite* **15**, 171–188.

Aggleton, J. P., Petrides, M., and Iversen, S. D. (1981). Differential effects of amygdaloid lesions on conditioned taste aversion learning by rats. *Physiol. Behav.* **27**, 397–400.

Apfelbaum, M., and Mandenoff, A. (1981). Naltrexone suppresses hyperphagia induced in the rat by a highly palatable diet. *Pharmacol. Biochem. Behav.* **15**, 89–91.

Azzara, A. V., Bodnar, R. J., Delamater, A. R., and Sclafani, A. (2000). Naltrexone fails to block the acquisition or expression of a flavour preference conditioned by intragastric carbohydrate infusions. *Pharmacol. Biochem. Behav.* **67**, 545–557.

Azzara, A. V., Bodnar, R. J., Delamater, A. R., and Sclafani, A. (2001). D_1 but not D_2 dopamine receptor antagonism blocks the acquisition of a flavor preference conditioned by intragastric carbohydrate infusions. *Pharmacol. Biochem. Behav.* **68**, 709–720.

Baile, C. A., McLaughlin, C. L., and Della-Fera, M. A. (1986). Role of cholecystokinin and opioid peptides in the control of food intake. *Physiol. Rev.* **66**, 172–234.

Bailey, C. S., Hsiao, S., and King, J. E. (1986). Hedonic reactivity to sucrose in rats: modification by pimozide. *Physiol. Behav.* **38**, 447–452.

Baker, R. M., Shah, M. J., Sclafani, A., and Bodnar, R. J. (2003). Dopamine D_1 and D_2 antagonists reduce the acquisition and expression of flavor-preferences conditioned by fructose in rats. *Pharmacol. Biochem. Behav.* **75**, 55–65.

Baker, R. W., Li, Y., Lee, M. G., Sclafani, A., and Bodnar, R. J. (2004). Naltrexone does not prevent acquisition or expression of flavor preferences conditioned by fructose in rats. *Pharmacol. Biochem. Behav.* **78**, 239–246.

Bakshi, V. P., and Kelley, A. E. (1993). Feeding induced by opioid stimulation of the ventral striatum: role of opiate receptor subtypes. *J. Pharmacol. Exp. Ther.* **265**, 1253–1260.

Barbano, M. F., and Cador, M. (2005). Differential regulation of the consummatory, motivational and anticipatory aspects of feeding behavior by dopaminergic and opioidergic drugs. *Neuropsychopharmacology* **31**, 1371–1381.

Barnard, E. A., Skolnick, P., Olsen, R. W., Möhler, H., Sieghart, W., Biggio, G., et al. (1998). International Union of Pharmacology. XV. Subtypes of γ-aminobutyric acid$_A$ receptors: classification on the basis of subunit structure and receptor function. *Pharmacol. Rev.* **50**, 291–313.

Barnett, S. A. (1963). "A Study in Behaviour." Methuen, London.

Barnett, S. A., and Spencer, M. M. (1953a). Experiments on the food preferences of wild rats. *J. Hyg. Camb.* **51**, 16–34.

Barnett, S. A., and Spencer, M. M. (1953b). Responses of wild rats to offensive smells and tastes. *Br. J. Anim. Behav.* **1**, 32–37.

Bassareo, V., and Di Chiara, G. (1997). Differential influence of associative and nonassociative learning mechanisms on the responsiveness of prefrontal and accumbal dopamine transmission to food stimuli in rats fed ad libitum. *J. Neurosci.* **17**, 815–861.

Bassareo, V., and Di Chiara, G. (1999). Modulation of feeding-induced activation of mesolimbic dopamine transmission by appetitive stimuli and its relation to motivational state. *Eur. J. Neurosci.* **11**, 4389–4397.

Bernstein, I. L. (1991). Development of taste preferences. *In* "The Hedonics of Taste" (R. C. Bolles, ed.), pp. 143–157. Erlbaum, Hillsdale, NJ.

Berridge, K. C. (1988). Brainstem systems mediate the enhancement of palatability by chlordiazepoxide. *Brain Res.* **447**, 262–268.

Berridge, K. C., and Treit, D. (1986). Chlordiazepoxide directly enhances positive ingestive reactions in rats. *Pharmacol. Biochem. Behav.* **24**, 217–221.

Berridge, K. C., and Peciña, S. (1995). Benzodiazepines, appetite and taste palatability. *Neurosci. Biobehav. Rev.* **19**, 121–131.

Berridge, K. C., Venier, I. L., and Robinson, T. E. (1989). Taste reactivity analysis of 6-hydroxydopamine-induced aphagia: implications for arousal and hedonia hypotheses of dopamine function. *Behav. Neurosci.* **103**, 36–45.

Bertino, M., Abelson, M. L., Marglin, S. H., Neuman, R., Burkhardt, C. A., and Reid, L. D. (1988). A small dose of morphine increases intake of and preference for isotonic saline among rats. *Pharmacol. Biochem. Behav.* **29**, 617–623.

Bodnar, R. J. (1996). Opioid receptor subtype antagonists and ingestion. *In* "Drug Receptor Subtypes and Ingestive Behaviour" (S. J. Cooper and P. G. Clifton, eds.), pp. 147–166. Academic Press, London.

Bodnar, R. J., Glass, M. J., and Koch, J. E. (1995). Analysis of central opioid receptor subtype antagonism of hypotonic and hypertonic saline intake in water-deprived rats. *Brain Res. Bull.* **36**, 293–300.

Booth, D. A. (1985). Food-conditioned eating preferences and aversions with interoceptive elements: conditioned appetites and satieties. *Ann. N.Y. Acad. Sci.* **443**, 22–41.

Borsini, F., and Rolls, E. T. (1984). Role of noradrenaline and serotonin in the basolateral region of the amygdala in food preferences and learned taste aversions in the rat. *Physiol. Behav.* **33**, 37–43.

Braestrup, C., Schmiechen, R., Neef, G., Nielsen, M., and Petersen, E. N. (1982). Interaction of convulsive ligands with benzodiazepine receptors. *Science* **216**, 1241–1243.

Braestrup, C., Nielsen, M., Honoré, T., Jensen, L. H., and Petersen, E. N. (1983). Benzodiazepine receptor ligands with positive and negative efficacy. *Neuropharmacology* **22**, 1451–1457.

Brands, B., Thornhill, J. A., Hirst, M., and Gowdey, C. W. (1979). Suppression of food intake and body weight gain by naloxone in rats. *Life Sci.* **24**, 1773–1778.

Brown, D. R., and Holtzman, S. G. (1979). Suppression of deprivation-induced food and water intake in rats and mice by naloxone. *Pharmacol. Biochem. Behav.* **11**, 567–573.

Calcagnetti, D. J., and Reid, L. D. (1983). Morphine and acceptability of putative reinforcers. *Pharmacol. Biochem. Behav.* **18**, 567–569.

Clarke, S. N. D. A., and Parker, L. A. (1995). Morphine-induced modification of quinine palatability: effects of multiple morphine-quinine trials. *Pharmacol. Biochem. Behav.* **51**, 505–508.

Cooper, S. J. (1980). Naloxone: effects on food and water consumption in the non-deprived and deprived rat. *Psychopharmacology (Berlin)* **71**, 1–6.

Cooper, S. J. (1981). Behaviourally-specific hyperdipsia in the non-deprived rat following acute morphine treatment. *Neuropharmacology* **20**, 469–472.

Cooper, S. J. (1982a). Benzodiazepine mechanisms and drinking in the water-deprived rat. *Neuropharmacology* **21**, 775–801.

Cooper, S. J. (1982b). Palatability-induced drinking after administration of morphine, naltrexone and diazepam in the non-deprived rat. *Subst. Alcohol. Actions/Misuse* **3**, 259–265.

Cooper, S. J. (1983a). Effects of opiate agonists and antagonists on fluid intake and saccharin choice in the rat. *Neuropharmacology* **22**, 323–328.

Cooper, S. J. (1983b). Benzodiazepine-opiate antagonist interactions in relation to feeding and drinking reward. *Life Sci.* **32**, 1043–1051.

Cooper, S. J. (1985). Bidirectional control of palatable food consumption through a common benzodiazepine receptor: theory and evidence. *Brain Res. Bull.* **15**, 397–410.

Cooper, S. J. (1986). Effects of the beta-carboline FG 7142 on saccharin preference and quinine aversion in water-deprived rats. *Neuropharmacology* **25**, 213–216.

Cooper, S. J. (1987a). Chlordiazepoxide-induced selection of saccharin-flavored food in the food-deprived rat. *Physiol. Behav.* **41**, 539–542.

Cooper, S. J. (1987b). Drugs and hormones: effects on ingestion. *In* "Feeding and Drinking" (N. E. Rowland and F. M. Toates, eds.), pp. 231–269. Elsevier Biomedical Press, Amsterdam.

Cooper, S. J. (1989). Benzodiazepine receptor-mediated enhancement and inhibition of taste reactivity, food choice, and intake. *Ann. N.Y. Acad. Sci.* **575**, 321–337.

Cooper, S. J. (1994). Palatability and endogenous opioids. *Regul. Pept.* **54**, 67–68.

Cooper, S. J. (2005). Palatability-dependent appetite and benzodiazepines: new directions from the pharmacology of GABA$_A$ receptor subtypes. *Appetite* **44**, 133–150.

Cooper, S. J., and Crummy, Y. M. T. (1978). Enhanced choice of familiar food in a food preference test after chlordiazepoxide administration. *Psychopharmacology (Berlin)* **59**, 51–56.

Cooper, S. J., and Francis, R. L. (1979). Water intake and time course of drinking after single or repeated chlordiazepoxide injections. *Psychopharmacology (Berlin)* **65**, 191–195.

Cooper, S. J., and McClelland, A. (1980). Effects of chlordiazepoxide, food familiarization, and prior shock experience on food choice in rats. *Pharmacol. Biochem. Behav.* **12**, 23–28.

Cooper, S. J., and Turkish, S. (1983). Effects of naloxone and its quaternary analogue on fluid consumption in water-deprived rats. *Neuropharmacology* **22**, 797–800.

Cooper, S. J., and Gilbert, D. B. (1984). Naloxone suppresses fluid consumption in tests of choice between sodium chloride solutions and water in male and female water-deprived rats. *Psychopharmacology (Berlin)* **84**, 362–367.

Cooper, S. J., and Webb, Z. M. (1984). Microstructural analysis of chlordiazepoxide's effects on food preference behaviour in Roman high-, control- and low-avoidance rats. *Physiol. Behav.* **32**, 581–588.

Cooper, S. J., and Estall, L. B. (1985). Behavioural pharmacology of food, water and salt intake in relation to drug actions at benzodiazepine receptors. *Neurosci. Biobehav. Rev.* **9**, 5–19.

Cooper, S. J., and Yerbury, R. E. (1988). Clonazepam selectively increases saccharin ingestion in a two-choice test. *Brain Res.* **456**, 173–176.

Cooper, S. J., and Turkish, S. (1989). Effects of naltrexone on food preferences and concurrent behavioural responses in food-deprived rats. *Pharmacol. Biochem. Behav.* **33**, 17–20.

Cooper, S. J., and Kirkham, T. C. (1990). Basic mechanisms of opioids' effects on eating and drinking. *In* "Opioids, Bulimia, and Alcohol Abuse & Alcoholism" (L. D. Reid, ed.), pp. 91–110. Springer-Verlag, New York.

Cooper, S. J., and Greenwood, S. E. (1992). The β-carboline abecarnil, a novel agonist at central benzodiazepine receptors, influences saccharin and salt taste preferences in the rat. *Brain Res.* **599**, 144–147.

Cooper, S. J., and Barber, D. J. (1993). The benzodiazepine receptor partial agonist bretazenil and the partial inverse agonist Ro15-4513: effects on salt preference and aversion in the rat. *Brain Res.* **612**, 313–318.

Cooper, S. J., and Green, A. E. (1993). The benzodiazepine receptor partial agonists, bretazenil (Ro16-6028) and Ro17-1812, affect saccharin preference and quinine aversion in the rat. *Behav. Pharmacol.* **4**, 81–85.

Cooper, S. J., and Kirkham, T. C. (1993). Opioid mechanisms in the control of food consumption and taste preferences. *In* "Handbook of Experimental Pharmacology, Volume 104/II. Opioids II" (A. Herz, ed.), pp. 239–262. Springer-Verlag, Berlin.

Cooper, S. J., and Higgs, S. (1996). Benzodiazepine receptors and the determination of palatability. *In* "Drug Receptor Subtypes and Ingestive Behaviour" (S. J. Cooper and P. G. Clifton, eds.), pp. 347–368. Academic Press, London.

Cooper, S. J., and Higgs, S. (2005). Benzodiazepine effects on licking responses for sodium chloride solutions in water-deprived male rats. *Physiol. Behav.* **85**, 252–258.

Cooper, S. J., and Ridley, E. T. (2005). Abecarnil and palatability: taste reactivity in normal ingestion in male rats. *Pharmacol. Biochem. Behav.* **81**, 517–523.

Cooper, S. J., and Al-Naser, H. A. (2006). Dopaminergic control of food choice: contrasting effects of SKF 38393 and quinpirole on high-palatability food preference in the rat. *Neuropharmacology* **50**, 953–963.

Cooper, S. J., Jackson, A., Morgan, R., and Carter, R. (1985a). Evidence for opiate receptor involvement in the consumption of a high palatability diet in non-deprived rats. *Neuropeptides* **5**, 345–348.

Cooper, S. J., Jackson, A., and Kirkham, T. C. (1985b). Endorphins and food intake: *kappa* opioid receptor agonists and hyperphagia. *Pharmacol. Biochem. Behav.* **23**, 889–901.

Cooper, S. J., Barber, D. J., Gilbert, D. B., and Moores, W. R. (1985c). Benzodiazepine receptor ligands and the consumption of a highly palatable diet in nondeprived male rats. *Psychopharmacology (Berlin)* **86**, 348–355.

Cooper, S. J., Jackson, A., Kirkham, T. C., and Turkish, S. (1988). Endorphins, opiates and food intake. *In* "Endorphins, Opiates and Behavioural Processes" (R. J. Rodgers and S. J. Cooper, eds.), pp. 143–186. John Wiley, Chichester.

Cooper, S. J., Bowyer, D. M., and van der Hoek, G. (1989). Effects of the imidazobenzodiazepine Ro15-4513 on saccharin choice and acceptance, and on food intake, in the rat. *Brain Res.* **494**, 172–176.

Corey, D. T. (1978). The determinants of exploration and neophobia. *Neurosci. Biobehav. Rev.* **2**, 235–253.

Davis, J. D. (1973). The effectiveness of some sugars in stimulating licking behavior in the rat. *Physiol Behav.* **11**, 39–45.

Davis, J. D., and Smith, G. P. (1988). Analysis of lick rate measures the positive and negative feedback effects of carbohydrates on eating. *Appetite* **11**, 229–238.

Davis, J. D., and Smith, G. P. (1992). Analysis of the microstructure of the rhythmic tongue movements of rats ingesting maltrose and sucrose solutions. *Behav. Neurosci.* **106**, 217–228.

Davis, J. D., and Perez, M. C. (1993). Food deprivation- and palatability-induced microstructural changes in ingestive behavior. *Am. J. Physiol.* **264**, R97–R103.

Depoortere, H., Zivkovic, B., Lloyd, K. G., Sanger, D. J., Derrault, G., Langer, S. Z., and Bartholini, G. (1986). Zolpidem, a novel non-benzodiazepine hypnotic. I. Neuropharmacological and behavioural effects. *J. Pharmacol. Exp. Ther.* **237**, 649–658.

Desor, J. A., Mallor, O., and Turner, R. (1973). Taste in acceptance of sugars by human infants. *J. Comp. Physiol. Psychol.* **19**, 163–174.

Doyle, T. G., Berridge, K. C., and Gosnell, B. A. (1993). Morphine enhances hedonic taste palatability in rats. *Pharmacol. Biochem. Behav.* **46**, 745–749.

Drewnowski, A. (1998). Energy density, palatability and satiety: implications for weight control. *Nutr. Rev.* **56**, 347–353.

Drucker, D. B., Ackroff, K., and Sclafani, A. (1993). Flavor preference produced by intragastric Polycose infusions using a concurrent conditioning procedure. *Physiol. Behav.* **54**, 351–355.

Elizalde, G., and Sclafani, A. (1990a). Flavor preferences conditioned by intragastric Polycose: a detailed analysis using an electronic esophagus preparation. *Physiol. Behav.* **47**, 63–77.

Elizalde, G., and Sclafani, A. (1990b). Fat appetite in rats: flavour preferences conditioned by nutritive and non-nutritive oil emulsions. *Appetite* **15**, 189–197.

Ernits, T., and Corbit, J. D. (1973). Taste as a dipsogenic stimulus. *J. Comp. Physiol. Psychol.* **83**, 27–31.

Estall, L. B., and Cooper, S. J. (1986). Benzodiazepine receptor-mediated effect of CGS 8216 on milk consumption in the non-deprived rat. *Psychopharmacology (Berlin)* **89**, 477–479.

Falk, J. L. (1971). Determining changes in vital functions: ingestion. *In* "Methods in Psychobiology,

Volume 1" (R. D. Myers, ed.), pp. 301–329. Academic Press, New York.

Fanselow, M., and Birk, J. (1982). Flavor-flavor associations induce hedonic shifts in taste preference. *Anim. Learn. Behav.* **10**, 223–228.

Fedorchak, P. M., and Bolles, R. C. (1987). Hunger enhances the expression of calorie- but not taste-mediated conditioned flavour preferences. *J. Exp. Physiol.: Anim. Behav. Proc.* **13**, 73–79.

Fitzgerald, R. E., and Burton, M. J. (1981). Effects of small basolateral amygdala lesions on ingestion in the rat. *Physiol. Behav.* **27**, 431–437.

Flynn, F. W., and Grill, H. J. (1988). Intraoral intake and taste reactivity responses elicited by sucrose and sodium chloride in chronic decerebrate rats. *Behav. Neurosci.* **102**, 934–941.

Frenk, H., and Rogers, G. H. (1979). Suppressant effects of naloxone on food and water intake in the rat. *Behav. Neural Biol.* **26**, 23–40.

Ganchrow, J. R., Steiner, J. E., and Canetto, S. (1986). Behavioral displays to gustatory stimuli in newborn rat pups. *Dev. Psychobiol.* **19**, 163–174.

Genn, R. F., Higgs, S., and Cooper, S. J. (2003). The effects of 7-OH-DPAT, quinpirole and raclopride on licking for sucrose solutions in the non-deprived rat. *Behav. Pharmacol.* **14**, 609–617.

Gilbert, D. B., and Cooper, S. J. (1995). 7-OH-DPAT injected into the accumbens reduces locomotion and sucrose ingestion: D$_3$ autoreceptor-medicated effects? *Pharmacol. Biochem. Behav.* **52**, 274–280.

Gilbert, D. B., Millar, J., and Cooper, S. J. (1995). The putative dopamine D$_3$ agonist, 7-OH-DPAT, reduces dopamine release in the nucleus accumbens and electrical self-stimulation to the ventral tegmentum. *Brain Res.* **68**, 1–7.

Giraudo, S. Q., Grace, M. K., Welch, C. C., Billington, C. J., and Levine, A. S. (1993). Naloxone's anorectic effect is dependant upon the relative palatability of food. *Pharmacol. Biochem. Behav.* **46**, 917–921.

Giros, B., Jaber, M., Jones, S. R., Wightman, R. M., and Caron, M. G. (1996). Hyperlocomotion and indifference to cocaine and amphetamine in mice lacking the dopamine transporter. *Nature* **379**, 606–612.

Glass, M. J., Grace, M., Cleary, J. P., Billington, C. J., and Levine, A. S. (1996). Potency of naloxone's anorectic effect in rats is dependent on diet preference. *Am. J. Physiol.* **271**, R217–R221.

Gosnell, B. A., and Majchrzak, M. J. (1989). Centrally administered opioid peptides stimulate saccharin intake in non-deprived rats. *Pharmacol. Biochem. Behav.* **33**, 805–810.

Gosnell, B. A., and Majchrzak, M. J. (1990). Effects of a selective mu opioid receptor agonist and naloxone on the intake of sodium chloride solutions. *Psychopharmacology (Berlin)* **100**, 66–71.

Gosnell, B. A., and Levine, A. S. (1996). Stimulation of ingestive behaviour by preferential and selective opioid agonists. *In* "Drug Receptor Subtypes and Ingestive Behaviour" (S. J. Cooper and P. G. Clifton, eds.), pp. 147–166. Academic Press, London.

Gosnell, B. A., Krahn, D. D., and Majchrzak, M. J. (1990a). The effects of morphine on diet selection are dependent upon baseline diet preferences. *Pharmacol. Biochem. Behav.* **37**, 207–212.

Gosnell, B. A., Majchrzak, M. J., and Krahn, D. D. (1990b). Effects of preferential delta and kappa opioid receptor agonists on the intake of hypotonic saline. *Physiol. Behav.* **47**, 601–603.

Gray, R. W., and Cooper, S. J. (1995). Benzodiazepines and palatability: taste reactivity in normal ingestion. *Physiol. Behav.* **58**, 853–859.

Grill, H. J., and Norgren, R. (1978a). The taste reactivity test. I. Mimetic responses to gustatory stimuli in neurologically normal rats. *Brain Res.* **143**, 263–279.

Grill, H. J., and Norgren, R. (1978b). The taste reactivity test. II. Mimetic responses to gustatory stimuli in chronic thalamic and chronic decerebrate rats. *Brain Res.* **143**, 281–297.

Grill, H. J., and Berridge, K. C. (1985). Taste reactivity as a measure of the neural control of palatability. *In* "Progress in Psychobiology and Physiological Psychology, Volume 11" (J. M. Sprague and A. N. Epstein, eds.), pp. 1–61. Academic Press, Orlando, FL.

Grill, H. J., Spector, A. C., Schwartz, G. J., Kaplan, J. M., and Flynn, F. W. (1987). Evaluating taste effects on ingestive behavior. *In* "Feeding and Drinking" (F. M. Toates and N. E. Rowland, eds.), pp. 151–188. Elsevier Science, Amsterdam.

Haefely, W., Facklam, M., Schoch, P., Martin, J. R., Bonetti, E. P., Moreau, J. -L., et al. (1992). Partial agonists of benzodiazepine receptors for the treatment of epilepsy, sleep, and anxiety disorders. *In* "GABAergic Synaptic Transmission" (G. Biggio, A. Concas, and E. Costa, eds.), pp. 379–394. Raven Press, New York.

Hajnal, A., and Norgren, R. (2001). Accumbens dopamine mechanisms in sucrose intake. *Brain Res.* **904**, 76–84.

Hajnal, A., Smith, G. P., and Norgren, R. (2004). Oral sucrose stimulation increases accumbens dopamine in the rat. *Am. J. Physiol.* **286**, R31–R37.

Hernandez, L., and Hoebel, B. G. (1988). Feeding and hypothalamic stimulation increases dopamine turnover in the accumbens. *Physiol. Behav.* **44**, 599–606.

Higgs, S., and Cooper, S. J. (1996a). Effects of the benzodiazepine receptor inverse agonist Ro15-4513 on the ingestion of sucrose and sodium saccharin solutions: a microstructural analysis of licking behavior. *Behav. Neurosci.* **110**, 559–566.

Higgs, S., and Cooper, S. J. (1996b). Increased food intake following injection of the benzodiazepine receptor agonist midazolam into the IVth ventricle. *Pharmacol. Biochem. Behav.* **55**, 81–86.

Higgs, S., and Cooper, S. J. (1996c). Hyperphagia induced by direct administration of midazolam into the parabrachial nucleus of the rat. *Eur. J. Pharmacol.* **313**, 1–9.

Higgs, S., and Cooper, S. J. (1997). Midazolam-induced rapid changes in licking behaviour: evidence for involvement of endogenous opioid peptides. *Psychopharmacology (Berlin)* **131**, 278–286.

Higgs, S., and Cooper, S. J. (1998a). Effects of benzodiazepine receptor ligands on the ingestion of sucrose, intralipid and maltodextrin: an investigation using a microstructural analysis of licking behavior in a brief contact test. *Behav. Neurosci.* **112**, 447–457.

Higgs, S., and Cooper, S. J. (1998b). Evidence for early opioid modulation of licking responses to sucrose and intralipid: a microstructural analysis in the rat. *Psychopharmacology (Berlin)* **139**, 342–355.

Higgs, S., and Cooper, S. J. (2000). The effect of dopamine D_2 receptor antagonist raclopride on the pattern of licking microstructure induced by midazolam in the rat. *Eur. J. Pharmacol.* **409**, 73–80.

Holtzman, S. G. (1974). Behavioral effects of separate and combined administration of naloxone and *d*-amphetamine. *J. Pharmacol. Exp. Ther.* **189**, 51–60.

Holtzman, S. G. (1975). Effects of narcotic antagonists on fluid intake in the rat. *Life Sci.* **16**, 1465–1470.

Hunkeler, W., Möhler, J., Pieri, L., Polc, P., Bonetti, E. P., Cumin, R., Schaffner, R., and Haefely, W. (1981). Selective antagonists of benzodiazepines. *Nature* **290**, 514–516.

Jalowiec, J. E., Panksepp, J., Zolovick, A. J., Najam, N., and Herman, B. H. (1981). Opioid modulation of ingestive behavior. *Pharmacol. Biochem. Behav.* **15**, 477–484.

Jensen, L. H., Petersen, E. N., and Braestrup, C. (1983). Audiogenic seizures in DBA/2 mice discriminate sensitively between low efficacy benzodiazepine receptor agonists and inverse agonists. *Life Sci.* **33**, 393–399.

Kenakin, T. (2004a). Principles: receptor theory in pharmacology. *Trends Pharmacol. Sci.* **25**, 186–192.

Kenakin, T. (2004b). Efficacy as a vector: the relative prevalence and paucity of inverse agonism. *Mol. Pharmacol.* **65**, 2–11.

Kennedy, R. T., Jones, S. R., and Wightman, R. M. (1992). Dynamic observation of dopamine autoreceptor effects in rat striatal slices. *J. Neurochem.* **59**, 449–455.

Kirkham, T. C. (1990). Enhanced anorectic potency of naloxone in rats sham feeding 30% sucrose: reversal by repeated naloxone administration. *Physiol. Behav.* **47**, 419–426.

Kirkham, T. C., and Cooper, S. J. (1986). CGS 8216, a novel anorectic agent, selectively reduces saccharin solution consumption in the rat. *Pharmacol. Biochem. Behav.* **25**, 341–345.

Kirkham, T. C., and Cooper, S. J. (1987). The pyrazoloquinoline, CGS 8216, reduces sham feeding in the rat. *Pharmacol. Biochem. Behav.* **26**, 497–501.

Kirkham, T. C., and Cooper, S. J. (1988a). Attenuation of sham feeding by naloxone is stereospecific: evidence for opioid mediation of orosensory reward. *Physiol. Behav.* **43**, 845–847.

Kirkham, T. C., and Cooper, S. J. (1988b). Naloxone attenuation of sham feeding is modified by manipulation of sucrose concentration. *Physiol. Behav.* **44**, 419–494.

Kirkham, T. C., and Cooper, S. J. (1989). Interactions between sucrose concentration and the time course of naloxone suppression of sham feeding: evidence for opioid mediation of food palatability. *Adv. Biosci.* **75**, 655–658.

Kirkham, T. C., Barber, D. J., Heath, R. W., and Cooper, S. J. (1987). Differential effects of CGS 8216 and naltrexone on ingestional behaviour. *Pharmacol. Biochem. Behav.* **26**, 145–151.

Leeb, K., Parker, L., and Eikelboom, R. (1991). Effects of pimozide on the hedonic properties of sucrose: analysis by the taste reactivity test. *Pharmacol. Biochem. Behav.* **39**, 895–901.

Le Magnen, J. (1992). "Neurobiology of Feeding and Nutrition." Academic Press, San Diego, CA.

Le Magnen, J., Marfaing-Jallat, P., Miceli, D., and Devos, M. (1980). Pain modulating and reward systems: a single brain mechanism? *Pharmacol. Biochem. Behav.* **12**, 729–733.

Leventhal, L., Kirkham, T. C., Cole, J. L., and Bodnar, R. J. (1995). Selective actions of central μ and κ opioid antagonists upon sucrose intake in sham-fed rats. *Brain Res.* **685**, 205–210.

Levine, A. S., Morley, J. E., Gosnell, B. A., Billington, C. J., and Bartness, T. J. (1985). Opioids and consummatory behaviour. *Brain Res. Bull.* **14**, 663–672.

Levine, S. A., Murray, D. S., Kneip, J., Grace, M., and Morley, J. E. (1982). Flavour enhances the antidipsogenic effect of naloxone. *Physiol. Behav.* **28**, 23–25.

Ljungberg, T. (1987). Blockade by neuroleptics of water intake and operant responding for water in the rat: anhedonia, motor deficit or both? *Pharmacol. Biochem. Behav.* **27**, 341–350.

Ljungberg, T. (1990). Differential attenuation of water intake and water rewarded operant responding by repeated administration of haloperidol and SCH 23390 in the rat. *Pharmacol. Biochem. Behav.* **35**, 111–115.

Lucas, F., and Sclafani, A. (1996). Capsaicin attenuates feeding suppression but not reinforcement by intestinal nutrients. *Am. J. Physiol.* **270**, R1059–R1064.

Lynch, W. C. (1986). Opiate blockade inhibits saccharin intake and blocks normal preference acquisition. *Pharmacol. Biochem. Behav.* **24**, 833–836.

Lynch, W. C., and Libby, L. (1983). Naloxone suppresses intake of highly preferred saccharin solutions in food deprived and sated rats. *Life Sci.* **33**, 1909–1914.

Lynch, W. C., and Burns, G. (1990). Opioid effects on intake of sweet solutions depend both on prior drug experience and on prior ingestive experience. *Appetite* **15**, 23–32.

Margules, D. L., and Stein, L. (1987). Neuroleptics vs. tranquilizers: evidence from animal studies of mode and site of action. *In* "Neuropsychopharmacology" (H. Brill, J. O. Cole, P. Deniker, H. Hippius, and P. B. Bradley, eds.), pp. 108–120. Excerpta Medical Foundation, Amsterdam.

Marks-Kaufman, R. (1982). Increased fat consumption induced by morphine administration in rats. *Pharmacol. Biochem. Behav.* **16**, 949–955.

Marks-Kaufman, R., and Kanarek, R. (1980). Morphine selectively influences macronutrient intake in the rat. *Pharmacol. Biochem. Behav.* **12**, 427–430.

Marks-Kaufman, R., and Kanarek, R. (1981). Modifications of nutrient selection by naloxone in rats. *Psychopharmacology* **74**, 321–324.

Martel, P., and Fantino, M. (1996a). Mesolimbic dopaminergic system activity as a function of food reward: a microdialysis study. *Pharmacol. Biochem. Behav.* **53**, 221–226.

Martel, P., and Fantino, M. (1996b). Influence of the amount of food ingested on mesolimbic dopaminergic system activity: a microdialysis study. *Pharmacol. Biochem. Behav.* **55**, 297–302.

Mehiel, R. (1996). The effects of naloxone on flavor-calorie preference learning indicate involvement of opioid reward systems. *Psychol. Rec.* **46**, 435–450.

Mehiel, R., and Bolles, R. C. (1984). Learned flavour preferences based on caloric outcome. *Anim. Learn. Behav.* **12**, 421–427.

Mehiel, R., and Bolles, R. C. (1988). Learned flavour preferences based on calories are independent of initial hedonic value. *Anim. Learn. Behav.* **16**, 383–387.

Mehta, A. K., and Ticku, M. K. (1999). An update on GABA$_A$ receptors. *Brain Res. Rev.* **29**, 196–217.

Myers, K. P., and Sclafani, A. (2001). Conditioned enhancement of flavor evaluation reinforced by intragastric glucose. II. Taste reactivity analysis. *Physiol. Behav.* **74**, 495–505.

Myers, K. P., and Sclafani, A. (2003). Conditioned acceptance and preference but not altered taste reactivity responses to bitter and sour flavors paired with intragastric glucose infusion. *Physiol. Behav.* **78**, 173–183.

Nachman, M., and Ashe, J. H. (1974). Effects of basolateral amygdala lesions on neophobia, learned taste aversions, and sodium appetite in rats. *J. Comp. Physiol. Psychol.* **87**, 622–643.

Parker, L. A. (1995). Chlordiazepoxide enhances the palatability of lithium-, amphetamine-, and saline-paired saccharin solution. *Pharmacol. Biochem. Behav.* **50**, 345–349.

Parker, L. A., Maier, S., Rennie, M., and Creboider, J. (1992). Morphine- and naltrexone-induced modification of palatability: analysis by the taste reactivity test. *Behav. Neurosci.* **106**, 999–1010.

Peciña, S., and Berridge, K. C. (1995). Central enhancement of taste pleasure by intraventricular morphine. *Neurobiology* **3**, 269–280.

Peciña, S., and Berridge, K. C. (1996). Brainstem mediates enhancement of palatability and feeding: microinjections into fourth ventricle versus lateral ventricle. *Brain Res.* **727**, 22–30.

Peciña, S., and Berridge, K. C. (2000). Opioid eating site in nucleus accumbens shell mediates food intake and hedonic 'liking': map based on microinjection FoS plumes. *Brain Res.* **863**, 71–86.

Peciña, S., and Berridge, K. C. (2005). Hedonic hot spot in nucleus accumbens shell: where do μ-opioids cause increased hedonic impact of sweetness? *J. Neurosci.* **25**, 11777–11786.

Peciña, S., Berridge, K. C., and Parker, L. A. (1997). Pimozide does not shift palatability: separation of anhedonia from sensorimotor suppression by taste reactivity. *Pharmacol. Biochem. Behav.* **58**, 801–811.

Peciña, S., Cagniard, B., Berridge, K. C., Aldridge, J. W., and Zhuang, X. (2003). Hyperdopaminergic mutant mice have higher "wanting" but not "liking" for sweet rewards. *J. Neurosci.* **23**, 9395–9402.

Pérez, C., Lucas, F., and Sclafani, A. (1995). Carbohydrate, fat and protein condition similar flavour preferences in rats using an oral-delay procedure. *Physiol. Behav.* **57**, 549–554.

Pérez, C., Ackroff, K., and Sclafani, A. (1996). Carbohydrate- and protein-conditioned flavor preferences: effects of nutrient preloads. *Physiol. Behav.* **59**, 467–474.

Pérez, C., Ackroff, K., and Sclafani, A. (1998). Devazepide, a CCK$_A$ antagonist, attenuates the satiating but not the preference and conditioning effects of intestinal carbohydrate infusions in rats. *Pharmacol. Biochem. Behav.* **59**, 451–457.

Rada, P., Avena, N. M., and Hoebel, B. G. (2005). Daily bingeing on sugar repeatedly releases dopamine in the accumbens shell. *Neuroscience* **134**, 737–744.

Radhakishun, F. S., van Ree, J. M., and Westerink, B. H. C. (1988). Scheduled eating increases dopamine release in the nucleus accumbens of food-deprived rats as assessed with on-line brain dialysis. *Neurosci. Lett.* **85**, 351–356.

Randall, L. O., Schallek, W. G. A., Keith, E. F., and Bagdon, R. E. (1960). The psychosedative properties of methaminodiazepoxide. *J. Pharmacol. Exp. Ther.* **129**, 163–171.

Reilly, S., and Bornovalova, M. A. (2005). Conditioned taste aversion and amygdala lesions in the rat: a critical review. *Neurosci. Biobehav. Rev.* **29**, 1067–1088.

Richardson, D. K., Reynolds, S. M., Cooper, S. J., and Berridge, K. C. (2005). Endogenous opioids are necessary for benzodiazepine palatability enhancement: naltrexone blocks diazepam-induced increase of sucrose-'liking'. *Pharmacol. Biochem. Behav.* **81**, 657–663.

Rideout, H. J., and Parker, L. A. (1996). Morphine enhancement of sucrose palatability: analysis by the taste reactivity test. *Pharmacol. Biochem. Behav.* **53**, 731–734.

Rockwood, G. A., and Reid, L. D. (1982). Naloxone modifies sugar-water intake in rats drinking with open gastric fistulas. *Physiol. Behav.* **29**, 1175–1178.

Rolls, B. J., and Rolls, E. T. (1973a). Effects of lesions in the basolateral amygdala on fluid intake in the rat. *J. Comp. Physiol. Psychol.* **83**, 240–247.

Rolls, E. T., and Rolls, B. J. (1973b). Altered food preferences after lesions in the basolateral region of the amygdala in the rat. *J. Comp. Physiol. Psychol.* **83**, 248–259.

Romsos, D. R., Gosnell, B. A., Morley, J. E., and Levine, A. S. (1987). Effects of kappa opioid agonists, cholecystokinin and bombesin on intake of diets varying in carbohydrate-to-fat ratio in rats. *J. Nutr.* **117**, 976–985.

Rudolph, U., and Möhler, H. (2004). Analysis of GABA_A receptor function and dissection of the pharmacology of benzodiazepines and general anesthetics through mouse genetics. *Ann. Rev. Pharmacol. Toxicol.* **44**, 475–498.

Rudolph, U., Crestani, F., and Möhler, H. (2001). GABA_A receptor subtypes: dissecting their pharmacological functions. *Trends Pharmacol. Sci.* **22**, 188–194.

Rusk, I. N., and Cooper, S. J. (1994). Parametric studies of selective D_1 or D_2 antagonists: effects on appetitive and feeding behaviour. *Behav. Pharmacol.* **5**, 615–622.

Sanger, D. J. (1981). Endorphinergic mechanisms in the control of food and water intake. *Appetite* **2**, 193–208.

Sanger, D. J., and McCarthy, P. S. (1980). Differential effects of morphine on food and water intake in food deprived and freely-feeding rats. *Psychopharmacology (Berlin)* **72**, 103–106.

Sclafani, A. (1984). Animal models of obesity: classification and characterization. *Int. J. Obes.* **8**, 491–508.

Sclafani, A. (1997). Learned controls of ingestive behaviour. *Appetite* **29**, 153–158.

Sclafani, A., and Nissenbaum, J. W. (1988). Robust conditioned flavour preference produced by intragastric starch infusions in rats. *Am. J. Physiol.* **255**, R672–R675.

Sclafani, A., and Lucas, F. (1996). Abdominal vagotomy does not block carbohydrate-conditioned flavour preferences in rats. *Physiol. Behav.* **60**, 447–453.

Sclafani, A., and Ackroff, K. (2004). The relationship between food reward and satiation revisited. *Physiol. Behav.* **82**, 89–95.

Sclafani, A., Aravich, P. F., and Xenakis, S. (1982). Dopaminergic and endorphinergic mediation of a sweet reward. *In* "The Neural Basis of Feeding and Reward" (B. G. Hoebel and D. Novin, eds.), pp. 507–515. Haer Institute for Electrophysiological Research, Brunswick ME.

Sclafani, A., Lucas, F., and Ackroff, K. (1996). The importance of taste and palatability in carbohydrate-induced overeating in rats. *Am. J. Physiol.* **270**, R1197–R1202.

Scott, E. M., and Quint, E. (1946). Self selection of diet. II. The effect of flavor. *J. Nutr.* **32**, 113–120.

Scott, E. M., and Verney, E. L. (1948). Self selection of diet. VIII. Appetite for fats. *J. Nutr.* **36**, 91–98.

Shor-Posner, G., Azar, A. P., Filart, R., Tempel, D., and Leibowitz, S. F. (1986). Morphine-stimulated feeding: analysis of macronutrient selection and paraventricular nucleus lesions. *Pharmacol. Biochem. Behav.* **24**, 931–939.

Siviy, S. M., and Reid, R. D. (1983). Endorphinergic modulation of acceptability of putative reinforcers. *Appetite* **4**, 249–257.

Smith, K. S., and Berridge, K. C. (2005). The ventral pallidum and hedonic reward: neurochemical maps of sucrose "liking" and food intake. *J. Neurosci.* **25**, 8637–8649.

Söderpalm, A. H. V., and Hansen, S. (1998). Benzodiazepines enhance the consumption and palatability of alcohol in the rat. *Psychopharmacology (Berlin)* **137**, 215–222.

Söderpalm, A. H. V., and Berridge, K. C. (2000). The hedonic impact and intake of food are increased by midazolam microinjection in the parabrachial nucleus. *Brain Res.* **877**, 288–297.

Sora, I., Wichens, C., Takahashi, N., Li, X. F., Zeng, Z., Revay, R., Lesch, K. P., Murphy, D. L., and Uhl, G. R. (1998). Cocaine reward models: conditioned place preference can be established in dopamine- and in serotonin-transporter knockout mice. *Proc. Natl. Acad. Sci. U.S.A.* **95**, 7699–7704.

Spector, A. C., and St. John, S. J. (1998). Role of taste in the microstructure of quinine ingestion by rats. *Am. J. Physiol.* **274**, R1687–R17103.

Steiner, J. E. (1977). Facial expressions of the neonate infant indicating the hedonics of food-related chemical stimuli. *In* "Taste and Development: the Genesis of Sweet Preference" (J. M. Weiffenbach, ed.), pp. 173–189. DHEW Publication No. NIH 77-1068. U.S. Government Printing Office, Washington, D.C.

Stephens, D. N., Schneider, H. H., Kehr, W., Andrews, J. S., Rettig, K.-J., Turski, L., Schmiechen, R., Turner, J. D., Jensen, L. H., and Petersen, E. N. (1990). Abecarnil, a metabolically stable, anxioselective beta-carbolinc acting at benzodiazepine receptors. *J. Pharmacol. Exp. Ther.* **253**, 334–343.

Taha, S. A., and Fields, H. L. (2005). Encoding of palatability and appetitive behaviours by distinct neuronal populations in the nucleus accumbens. *J. Neurosci.* **25**, 1193–1202.

Treit, D., and Berridge, K. C. (1990). A comparison of benzodiazepine, serotonin, and dopamine agents in the taste-reactivity paradigm. *Pharmacol. Biochem. Behav.* **37**, 451–456.

Turkish, S., and Cooper, S. J. (1984). Enhancement of saline consumption by chlordiazepoxide in thirsty rats: antagonism by Ro15-1788. *Pharmacol. Biochem. Behav.* **20**, 869–873.

Ukai, M., Nakayama, S., and Kameyama, T. (1988). The opioid antagonist, Mr2266, specifically decreases saline intake in the mouse. *Neuropharmacology* **27**, 1027–1031.

Vigorito, M., and Sclafani, A. (1988). Ontogeny of Polycose and sucrose appetite in neonatal rats. *Dev. Psychobiol.* **21**, 457–465.

Waldhoer, M., Bartlett, S. E., and Whistler, J. L. (2004). Opioid receptors. *Annu. Rev. Biochem.* **73**, 953–990.

Will, M. J., Franzblau, E. B., and Kelley, A. E. (2003). Nucleus accumbens μ-opioids regulate intake of a high-fat diet via activation of a distributed brain network. *J. Neurosci.* **23**, 2882–2888.

Wise, R. A., and Dawson, V. (1974). Diazepam-induced eating and lever pressing for food in sated rats. *J. Comp. Physiol. Psychol.* **86**, 930–941.

Yerbury, R. E., and Cooper, S. J. (1987). The benzodiazepine partial agonists, Ro16-6028 and Ro17-1812, increase palatable food consumption in nondeprived rats. *Pharmacol. Biochem. Behav.* **28**, 427–431.

Young, P. T. (1961). "Motivation and Emotion." John Wiley, New York.

Yu, W.-Z., Sclafani, A., Delamater, A. R., and Bodnar, R. J. (1999). Pharmacology of flavour preference conditioning in sham-feeding rats: effects of naltrexone. *Pharmacol. Biochem. Behav.* **64**, 573–584.

Yu, W.-Z., Silva, R. M., Sclafani, A., Delamater, A. R., and Bodnar, R. J. (2000a). Pharmacology of flavour preference conditioning in sham-feeding rats: effects of dopamine receptor antagonists. *Pharmacol. Biochem. Behav.* **65**, 635–647.

Yu, W.-Z., Silva, R. M., Sclafani, A., Delamater, A. R., and Bodnar, R. J. (2000b). Role of D_1 and D_2 dopamine receptors in the acquisition and expression of flavour-preference conditioning in sham-feeding rats. *Pharmacol. Biochem. Behav.* **67**, 537–544.

Zhang, M., and Kelley, A. E. (1997). Opiate agonists microinjected into the nucleus accumbens enhance sucrose drinking in rats. *Psychopharmacology (Berlin)* **132**, 350–360.

Zhang, M., Gosnell, B. A., and Kelley, A. E. (1998). Intake of high-fat food is selectively enhanced by mu opioid receptor stimulation within the nucleus accumbens. *J. Pharmacol. Exp. Ther.* **285**, 908–914.

The Role of Palatability in Control of Human Appetite: Implications for Understanding and Treating Obesity

MARTIN R. YEOMANS

Department of Psychology, University of Sussex

I. Introduction
II. Assessing the Effects of Palatability on Appetite
III. Palatability and the Control of Normal Appetite
    A. Effects of Manipulated Palatability on Short-Term Appetite
    B. Neurochemical Basis of Palatability Effects
    C. Neural Basis of Sensory Hedonics
    D. Palatability and Appetite Control
IV. Palatability and Obesity
    A. Oversensitivity to Palatability in Obesity
    B. Palatability and the Development of Therapeutic Approaches to Obesity
V. Conclusion
    References

Overconsumption of palatable foods has been implicated as one of the factors contributing to the current worldwide increase in the incidence of obesity. In order to summarize current understanding of the role of palatability in appetite, this chapter reviews the nature of palatability effects and associated physiological controls. It is also important to consider how palatability relates to other components of appetite control, and particularly how sensory cues may counteract satiety. The conclusion is that palatability effects reflect stimulation of central reward systems by the sensory qualities of foods, and that these effects interact with satiety mechanisms to determine meal size. Detailed analysis of past studies also concludes that appetite stimulation is a direct function of perceived palatability, with incremental increases in intake as foods become more liked. A brief discussion of the nature of palatability concludes that most food likes are acquired responses. Obese people appear to overrespond to palatability, and the cause and consequence of this oversensitivity is discussed. The overall conclusion is that enhanced responsivity to palatability may underlie the failure to respond to obesity treatment, and that future treatments are likely to benefit by the inclusion of strategies or specific drug treatments which ameliorate the role of palatability in subsequent overeating.

I. INTRODUCTION

When energy intake exceeds expenditure, the excess energy is stored, principally in our long-term fat stores (adipose tissue). If the positive energy balance is maintained, the inevitable consequence is weight gain and ultimately obesity. The recognition that the current obesity crisis arises from long-term overeating relative to energy needs (Doucet and Tremblay, 1997; Jequier and Tappy, 1999) has resulted in increased interest in the factors which promote short-term overeating. Overconsumption of palatable foods is recognized as a factor in generating positive energy balance (Blundell and Cooling, 2000), thus making resolution of the nature of palatability an urgent issue. In this chapter, it is suggested that the relationship between palatability and controls of nutrient intake are central to our understanding of how sensory factors are involved in appetite regulation. Consequently, if the current obesity crisis can be explained by the generation of a positive energy balance through overconsumption of palatable foods, understanding the role of palatability in appetite control will be critical to identifying target behaviors and neural systems to facilitate development of targeted treatments for obesity. However, despite many reviews (Young, 1967; Kissileff, 1976; Le Magnen, 1987; Naim and Kare, 1991; Drewnowski, 1998; Yeomans, 1998; Yeomans et al., 2004a), theories (Davis and Levine, 1977; Cabanac, 1989; Berridge, 1996; Berridge and Robinson, 1998), and debates (Kissileff, 1990; Ramirez, 1990;

Rogers, 1990), there is still a lack of consensus on the role of palatability in appetite control.

The overall aim of this chapter is to review our current understanding of the role of palatability in appetite control in humans. The initial focus is on the phenomenology of palatability effects: what is the short-term effect of palatability on appetite, and to what extent do we understand the neural and physiological bases of these effects? The focus then switches to theoretical interpretations of palatability effects, and the relationship between palatability and homeostatic controls of eating. The conclusion from that review is that the impact of palatability is best understood in terms of positive feedback modulation of appetite independent of satiety. This then raises the obvious but difficult question of what makes a food palatable, and the review briefly contrasts acquired relative to innate influences on palatability. Finally, this chapter focuses specifically on obesity, and explores first whether individual differences in response to palatability might explain phenotypical variation in susceptibility to weight gain, and second how increased understanding of the role of palatability in appetite control might inform the future treatment and prevention of obesity.

II. ASSESSING THE EFFECTS OF PALATABILITY ON APPETITE

The simplest methodology for examining palatability effects on appetite control in a laboratory context is to selectively manipulate the sensory characteristics of a food using selective flavorings and record the impact on appetite in the absence of changes in nutrient content. In laboratory studies, this approach has been used to examine palatability effects with a wide range of foods, including yogurt (Bobroff and Kissileff, 1986; Vickers et al., 2001), ice cream (Lahteenmaki and Tuorila, 1995), milkshake (Kauffman et al., 1995), sandwiches (Bellisle et al., 1984; Hill et al., 1984; Helleman and Tuorila, 1991; Zandstra et al., 2000), pasta (Yeomans, 1996; Yeomans et al., 1997, 2001a, 2004b; Yeomans and Symes, 1999), and porridge (Yeomans et al., 2005). Other researchers have taken the same basic design of contrasting versions varying in palatability, but have used alternative methods to alter the sensory characteristics of the food, for example, by allowing participants to choose their preferred food from a wide selection, and then contrasting appetite when the food is served at the optimal serving temperature or frozen (Melchior et al., 1994). The artificiality of this manipulation could, however, have adversely effected the outcome and so limit conclusions about the true role of palatability in appetite control. An alternative approach was to compose a multi-item palatable meal, and then blend all of the ingredients and freeze-dry these into an unpalatable biscuit form (Sawaya et al., 2001). Although this method is ingenious, a possible disadvantage is that it confounds differences in palatability with possible effects of variety, which is well-known to increase intake (Rolls et al., 1983; Treit et al., 1983; Spiegel and Stellar, 1990; Wansink, 2004). At present, flavor manipulations remain the simplest method for assessing palatability effects in the absence of confounding effects of differences in nutritional content provided that flavorings do not alter energy content, the proportions of macronutrients, or other factors which might impact on short-term eating, such as enhanced osmotic content (osmotic satiety: Mook and Yoo, 1991).

A second key element in laboratory studies of effects of palatability on appetite control is the appropriate selection of measures of ingestion. Ultimately, the key variable is overall energy intake. However, because of the subtlety of the effects of palatability, measurement of intake alone is inadequate if the studies are to help identify the basis of the effects of manipulated

palatability, since altered intake can arise from a multitude of mechanisms. To get around this, studies have used a variety of methods to measure different aspects of appetite in response to manipulated palatability. Most of these involve some form of microstructural evaluation of eating. Variables which appear sensitive to palatability manipulations include eating rate (Bellisle and Le Magnen, 1980; Spiegel et al., 1989, 1993; Yeomans, 1996), meal duration (Hill, 1974; Bellisle and Le Magnen, 1980; Yeomans, 1996), and chewing rate (Pierson and Le Magnen, 1969; Bellisle et al., 2000; Spiegel, 2000). These additional measures are invaluable in interpreting the nature of palatability effects, as we will discuss later.

As well as examining objective measures of ingestion, studies have examined how palatability manipulations alter the subjective experience of appetite alongside objective measures of food intake and ingestive behavior (Yeomans, 2000). Pioneering work by Hill et al. (1984) contrasted ratings of hunger when people consumed their preferred sandwich lunch relative to a less preferred meal. Their finding that rated hunger was increased in the preferred condition paved the way for more detailed studies where rated hunger was recorded at fixed intervals throughout test meals with palatable and bland versions of the same foods (Yeomans, 1996; Yeomans and Symes, 1997; Yeomans et al., 2001a, 2004b; Robinson et al., 2005). To facilitate such studies, specialized software (the Sussex Ingestion Pattern Monitor system) was developed which combined automated recording of food intake throughout each test meal using an adaptation of the Universal Eating Monitor first described by Kissileff et al. (1980) with automatic recording of computerized ratings of appetite at set intervals throughout the test meals. This methodology allowed detailed analysis of the relationship between the experience of food palatability and both subjective and objective measures of ingestion, thereby allowing mathematical modeling of the interrelationships between these variables (Yeomans, 2000). The use of curve fitting to simplify the large quantity of data generated by this methodology facilitated the evaluation of the effects of palatability (Yeomans, 1996; Yeomans et al., 1997, 2001a, 2004b; Yeomans and Symes, 1999), satiety (Yeomans et al., 1998), the interaction between palatability and satiety (Yeomans et al., 2001a; Robinson et al., 2005), the pharmacological manipulation of palatability (Yeomans and Gray, 1997), and most recently the effects of learning on palatability and satiation (Yeomans et al., 2005).

Since studies of the effects of palatability rely on differences in hedonic evaluations of foods, most, but not all, studies also include some form of hedonic rating of the foods to be consumed to confirm that the sensory manipulations achieved the expected effect on palatability. These studies all use some form of simple rating scale to assess hedonic evaluation of the consumed food. Clear interpretation of effects of palatability are contingent on clear demonstrations that the test foods used to examine palatability effects were rated significantly different on hedonic measures, and at the same time that the palatability contrast was not confounded by differences in nutritional composition (either in terms of overall energy or macronutrient composition). Consequently, studies which failed to include measures confirming that putative manipulations of palatability resulted in clear differences in hedonic evaluations of the test foods between conditions, or which contrasted foods which differed in terms of nutritional content and hedonic contrast, have been excluded from further analysis in this chapter.

III. PALATABILITY AND THE CONTROL OF NORMAL APPETITE

Before the potential role of palatability as a cause of overeating in obesity can be assessed (Section IV), it is important to establish the nature of palatability effects on normal eating, and discuss current knowledge of the mechanisms underlying the influence of palatability on appetite.

A. Effects of Manipulated Palatability on Short-Term Appetite

The consistent, and unsurprising, finding from studies using flavor manipulation to explore effects of palatability is that volunteers consume more in the palatable condition (Bellisle et al., 1984; Bobroff and Kissileff, 1986; Guy-Grand et al., 1989; Helleman and Tuorila, 1991; Kauffman et al., 1995; Yeomans, 1996; Yeomans et al., 1997, 2004b, 2005; Zandstra et al., 2000; Sawaya et al., 2001; Vickers et al., 2001; Robinson et al., 2005). Detailed analysis of the outcome of these studies allows some examination of the actual relationship between reported differences in rated palatability and consequent differences in intake. A complication in this analysis is that studies vary in the scales used to assess hedonic qualities; however, if these are rescaled to a hypothetical 100-pt scale, it is possible to estimate the relationship between differences in palatability of test conditions and the associated differences in food intake. Moreover, this analysis can be used to assess whether palatability manipulation alters mass or energy consumed as an indication of the mechanism underlying palatability-induced eating. Most studies have also only examined acute effects of flavor manipulation, and only more recently have studies started to examine how subsequent learning about postingestive consequences might moderate the effects of palatability (Yeomans et al., 2005). Thus the present analysis is limited to the immediate impact of hedonic differences, and only includes studies where palatability effects were examined in a singe contrast of a palatable relative to a less-palatable intake condition. To simplify the analysis, where palatability contrasts were combined with other manipulations, such as manipulated satiety, data were averaged across conditions.

The outcome of the reanalysis of the relationship between manipulated palatability and overall food intake (Fig. 1) shows two noteworthy features: first, that there is a clear, linear relationship between differences in hedonic evaluation of two foods and consequent intake, and second that this effect is stronger when intake is expressed in terms of mass consumed ($r = 0.85$, $p < 0.001$) than in terms of energy intake ($r = 0.58$, $p < 0.01$). Given that some of the studies in Figure 1 tested intake at a meal (lunch or breakfast), and others snack intake, and that the studies used a wide variety of foods, the consistency in the rela-

tionship between altered palatability and consequent change in the mass of food consumed is remarkably consistent. This latter point has particular relevance to the idea that obesity may be due to excess energy intake stimulated by palatability (Blundell and Cooling, 2000), and is discussed further later. The idea that palatability alters mass consumed fits well with the idea that palatability effects are due to orosensory stimulation of appetite: it is the amount of oral stimulation that is important, and mass is a better approximation of the amount of orosensory information the body will receive from food than is energy intake, which is more likely to generate different postingestive rather than orosensory signals. It may be that volume ingested

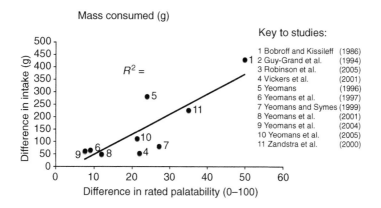

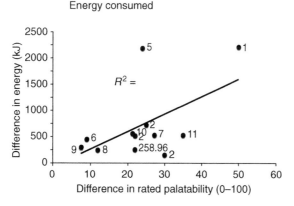

FIGURE 1 The relationship between increased food pleasantness and consequent increases in food intake from studies using flavor manipulation to explore the impact of palatability on appetite.

would be a better measure still of the amount of orosensory stimulation, but it is not possible to calculate volume from most published studies of manipulated palatability.

A criticism of studies that utilize simple flavor manipulations to explore effects of palatability is that they risk circularity: palatability is defined as the sensory stimulation of appetite, and increased intake is then the evidence for effects of palatability. This circularity has to some extent detracted from our understanding of the true role of sensory factors in short-term appetite control, and remains a major limitation of attempts to study palatability effects in animal studies. However, this circularity can be avoided in human studies either if differences in palatability (in terms of hedonic evaluation, for example) are established prior to the intake test, or if a specific mechanism underlying the short-term enhancement of food intake through palatability is evident. The analysis across studies of the impact of manipulated palatability on intake in Figure 1 argues for a clear mechanism underlying palatability effects: changes in palatability have a predictable and consistent effect on intake. At a behavioral level, these changes in flavor also produce measurable differences in the pattern of change of rated appetite within a meal (Yeomans, 1996; Yeomans et al., 2001a), with hunger tending to increase in the early stages of meals which are rated above neutral in terms of palatability (Fig. 2). Rated appetite can also be enhanced merely by the sight of a more-preferred food (Hill et al., 1984), suggesting immediate modulation of appetite by palatability. Manipulations which increase food pleasantness also enhance eating rate (Bellisle and Le Magnen, 1980; Spiegel et al., 1989; Yeomans, 1996), as well as overall length of meals (Spiegel et al., 1989; Yeomans, 1996).

The overall conclusion from analysis of laboratory studies of effects of manipulated palatability is that incremental increases in palatability have consistent and predictable

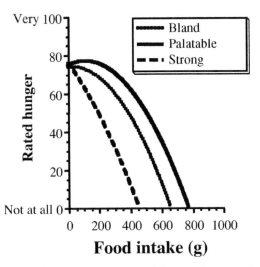

FIGURE 2 Changes in rated hunger for normal weight men eating a palatable, bland, or overly strong flavored test meal. Redrawn from Yeomans (1996), with permission.

effects on food intake and rated appetite. More naturalistic studies of effects of variation in the palatability of foods people select in their own diet confirm the relevance of laboratory studies to everyday eating behavior. Based on detailed analysis of self-reported diary records where participants completed an overall evaluation of the palatability of the meal alongside details of how much they ate, a clear effect of palatability on intake emerged (Fig. 3) even though people rarely voluntarily selected foods they rated as less palatable (De Castro et al., 2000a,b). These data cannot be explained as an artefact of different levels of hunger at the time when food was consumed, since participants also completed hunger ratings and analysis established that rated hunger and palatability acted independently on intake. The issue of how palatability effects relate to hunger and satiety is one we return to later. The conclusion from this section of this review is that palatability has predictable effects on intake consistent with the stimulation of appetite through sensory stimulation. The next question is what are the underly-

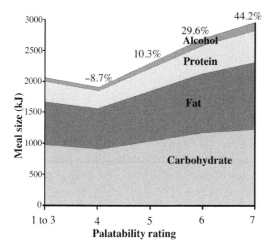

FIGURE 3 Meal size as a function of palatability based on 7-day diary reports by 564 U.S. adults. Reprinted from De Castro et al. (2000b) with permission from Elsevier.

ing controls of this hedonic influence on eating.

B. Neurochemical Basis of Palatability Effects

The previous section suggests that palatability acts to stimulate intake in an orderly manner, with mass consumed increasing as a linear function of the hedonic evaluation of the food being consumed. These data could be interpreted in terms of orosensory reward processes modulating appetite and so acting through a positive feedback system to stimulate further ingestion. Many theories of appetite control recognize hedonically driven positive feedback mechanisms as important components of the control of meal size (Wiepkema, 1971; Davis and Levine, 1977; Toates, 1986; Le Magnen, 1987). Today it is generally accepted that meal size involves the integration of two interacting feedback systems: a positive feedback system relating to orosensory stimulation of appetite, and associated with neural reward systems, and a negative feedback control system associated with satiation (Smith, 2000). The

neural basis of these two control systems is becoming clearer. In particular, the use of psychopharmacological interventions to explore the neurochemical basis of hedonic evaluations of foods in human participants has extended findings in other species to characterize the key neurochemical systems involved in determining food palatability.

If the stimulatory effect of palatability was the consequence of positive feedback mechanisms generated by orosensory reward mechanisms, then disruption of these pathways should modify the response to palatability. Current models of reward implicate three putative neurotransmitter systems as the main chemicals involved in the neural circuits underlying food-related reward: dopamine (Berridge, 1996), endogenous opioid peptides (Cooper and Kirkham, 1990; Kelley et al., 2002; Yeomans and Gray, 2002), and most recently endocannabinoids (Kirkham and Williams, 2001a). Of these, opioid systems have been the subject of the most clear investigations using pharmacological modulation in humans to explore the relationship between palatability and appetite control, and the outcome of these studies supports the idea that palatability reflects stimulation of central reward pathways by orosensory cues (reviewed by Yeomans and Gray, 2002). The most common experimental approach has been to examine how placebo-controlled administration of opiate receptor antagonists modify the hedonic evaluation of food flavors (Yeomans et al., 1990; Bertino et al., 1991; Yeomans and Wright, 1991; Drewnowski et al., 1992; Yeomans and Gray, 1996; Arbisi et al., 1999). In all published studies to date, foods were rated as tasting less pleasant following the opioid antagonist relative to placebo control. It then logically follows that if foods taste less pleasant when endogenous opioids are blocked, then the normal hedonic experience must relate to some action of endogenous opioids. While these types of study clearly implicate opioids in the response to palatability, since the opioid antagonists are administered either orally or by intravenous injection, it is not possible to determine whether the role of opioids reflects some form of peripheral or central effect. This issue could be addressed using modern brain-scanning techniques to examine the relationship between hedonic experience of foods, opioid antagonism, and activation of brain sites implicated in food-related reward processes in animals.

Research with animals has made significant progress in elucidating the likely neural basis of opioid effects, thereby paving the way for groundbreaking human studies in the future. Whereas the primary homeostatic control of appetite is regulated by discrete structures in the hypothalamus, opioid effects relate primarily to actions in precortical areas associated with reward and emotion. A full review of the evidence from animal studies is beyond the scope of this chapter, and has been the subject of excellent more recent reviews (Bodnar, 2004; Kelley, 2004). In brief, most evidence suggests that opioid modulation of areas in the ventral striatum, and particularly the nucleus accumbens (Bodnar et al., 1995; Ragnauth et al., 2000; Kim et al., 2001; Lamonte et al., 2002; MacDonald et al., 2003, 2004; Will et al., 2003; Zhang et al., 2003), in combination with the amygdala (Smith et al., 2002; Kim et al., 2004; Levine et al., 2004; Will et al., 2004), are critical to the role of opioids in hedonic components of feeding. The involvement of the nucleus accumbens and amygdala in opioid-mediated hedonic components of feeding fits well with evidence that the nucleus accumbens has a general role in hedonic and rewarding aspects of behavior (Grigson, 2002; Pelchat, 2002; Berridge, 2003), while the amygdala is implicated in processing of emotional information including the experience of positive affect (Zald, 2003; Burgdorf and Panksepp, 2005; Panksepp, 2005). It may be that these brain areas evolved partly to deal with the complexities of feeding, and that the palatability

response evolved in response to the need to ensure that animals made maximal use of available energy sources at a time when food was scarce.

The data reviewed so far provide clear evidence for opioid involvement in the hedonic experience of eating. An important next step is to determine the relationship between this hedonic experience and short-term modulation of appetite by palatability. To test this, the effects of the opioid antagonist naltrexone on hedonic stimulation of appetite within a meal (the appetizer effect: Yeomans, 1996) was investigated (Yeomans and Gray, 1997). Naltrexone was found to abolish the stimulatory effects of palatability on subjective appetite within a meal (Fig. 4) without altering the normal rate of decline of appetite as the meal progressed (satiation). These data provide strong evidence that opioids mediate the stimulatory effects of palatability on short-term food intake in humans.

Dopamine is strongly implicated in control of feeding in animals (Wellman, 2005) and in general reward processes in addiction (Wise, 1981; Berridge, 1996; Berridge and Robinson, 1998; Di Chiara, 2005). An obvious question is then whether drugs which modify dopaminergic neurotransmission have effects on eating in humans consistent with a critical role of dopamine in food-related reward. More recent studies suggest that this is the case. First, consumption of a favorite meal by healthy lean human volunteers was associated with reduction in binding of the dopamine D_2 receptor antagonist raclopride in the dorsal putamen and caudate, although surprisingly not the ventral striatum. Notably, the change in raclopride binding correlated with the rated pleasantness ratings of the food, consistent with a role of dopamine in food-related reward (Small et al., 2003). Second, the degree of binding of raclopride in the brain is negatively correlated with body mass index, with obese individuals showing reduced binding (Wang et al., 2001; Volkow et al.,

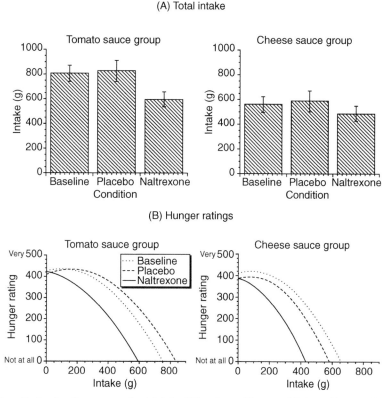

FIGURE 4 The effects of naltrexone on food intake (A) and rated hunger (B) for subjects eating a pasta lunch with a cheese or tomato sauce. Reprinted from Yeomans and Gray (1997) with permission from Elsevier.

2003), which could be interpreted either in terms of obese having reduced numbers of striatal dopamine D_2 receptors predisposing them to overeat (consistent with the broader dopamine-deficiency model: Blum et al. (1996)) or that weight gain leads to down regulation of dopamine D_2. Both these pioneering studies clearly suggest that dopamine is involved in food-related reward, and possibly hedonic experience, in humans.

Finally, the relatively recent discovery of brain cannabis-like molecules (endocannabinoids), and the finding that endocannabinoids modify feeding in animals (Kirkham, 2005; Vickers and Kennett, 2005) suggest a potential role of endocannabinoids in hedonic components of eating in humans. Indeed, there are strong parallels and interactions between the actions of endocannabinoids and opioids (Kirkham and Williams, 2001b; Pietras and Rowland, 2002; Solinas and Goldberg, 2005). However, evidence for cannabis altering human appetite remains largely anecdotal (Kirkham and Williams, 2001a), and although the cannabinoid antagonist rimonabant is being developed as an anti-obesity agent, to date there are no controlled studies of the effects of rimonabant on eating in humans. Thus the idea that endocannabinoids are involved in hedonic aspects of eating remains plausible, it is also largely untested.

C. Neural Basis of Sensory Hedonics

The advent of modern brain-imaging techniques has facilitated investigation of the neural basis of palatability in humans (de Araujo et al., 2003), as part of a broader interest in the basis of affect in general (Kringelbach, 2004), giving rise to affective neuroscience as a new discipline in psychological investigation. To be able to understand how the neural systems are involved in the experience of hedonic aspects of eating, it is first necessary to understand the way the brain encodes the sensory stimuli underlying our perception of food flavor. Although consumers describe the sensory qualities of food in terms of "taste," in reality food flavor is a complex percept resulting from multimodal integration of sensory inputs (Prescott, 2004). Although most of our experience of flavor is based on integration of olfactory and gustatory cues (Small and Prescott, 2005), other senses including temperature, oral irritation, touch, hearing, and visual cues all contribute to our overall experience of a food. It is thus arguable whether any other experience integrates as many different sensory components as does the experience of food in our mouth, perhaps explaining our fascination with the sensory characteristics of foods.

A full evaluation of the neural encoding of flavor is beyond the scope of this chapter, and has been reviewed more recently elsewhere (Small and Prescott, 2005). Since most of our experience of flavor is based on integration of gustatory and olfactory cues, it is first important to summarize current knowledge of the primary gustatory and olfactory pathways. The five basic tastes (sweet, sour, bitter, salty, and umami) are detected by specific receptors (taste papillae), primarily on the tongue. These gustatory cues, along with information about temperature, touch, and pain, are transmitted to the brain by several cranial nerves, and the taste information converges at the nucleus of the solitary tract. The taste pathway then projects via the thalamus to the primary taste cortex in the anterior insula and frontal operculum. Crucially, none of these structures are sensitive to olfactory stimuli, and the first site where

taste and smell information converge on the orbitofrontal cortex (OFC), which is interpreted as secondary taste cortex. Olfactory stimuli activate olfactory receptors in the olfactory bulb, and this activation is transmitted to the primary olfactory cortex (the piriform cortex, along with anterior olfactory nucleus and the anterior cortical nucleus of the amygdala). As with the primary taste correct, primary olfactory cortex projects to the OFC. In line with the idea that the OFC is crucial to flavor integration, simultaneous presentation of a sweet taste and food-related odor experienced retronasally (which is the route odors are experienced when flavors are sensed) resulted in superadditive activation of the OFC (Small et al., 2004).

Given the importance of the OFC in odor–taste integration, the OFC is of clear interest as the neural area underlying hedonic evaluation of flavors. There is now a large body of evidence in demonstrating activation of the OFC in relation to affective judgments of gustatory or olfactory stimuli either alone or together (Zald et al., 1998, 2002; O'Doherty et al., 2001; Small et al., 2001; Anderson et al., 2003; Royet et al., 2003). The more recent finding that the pleasantness but not intensity of binary mixtures of a sweet taste with a food-related odor were correlated with the level of activity in the OFC (de Araujo et al., 2003) is consistent with the view that the OFC is crucial to hedonic evaluation of flavors. This study also distinguishes hedonic experience of the integrated odor–taste stimulus from activation by sensory stimulation alone (indexed by the flavor intensity judgment).

Earlier, the discussion of the neurochemical basis of palatability effects concluded that the ventral striatum was critical to the expression of palatability. In contrast, the human brain-imaging data implicate OFC as critical. In an attempt to resolve this apparent contradiction, Berridge (2003) suggests that the OFC may be a useful neural marker of positive affective response, but may not be causal. There are major projections from OFC to the nucleus accumbens (Zahm, 2000), and as discussed earlier the nucleus accumbens is strongly implicated in opioid-mediated hedonic components of feeding. Thus the OFC may be the area where flavor information is integrated, but areas in the striatum then translate flavor perception into hedonic modulation of appetite. However, until studies combine opioid pharmacology with brain scanning in humans, this hypothesis remains hard to test.

D. Palatability and Appetite Control

The preceding discussions characterize the short-term effects of manipulated palatability on appetite, and highlight neural and neurochemical substrates of these effects. However, ultimately a real understanding of the role of palatability in appetite control can only be achieved when the relationship between palatability effects and homeostatic controls of eating is clarified, and the next section examines this by reviewing the relationship between palatability and factors involved in initiation and cessation of eating.

Theories about the relationship between palatability and homeostatic controls (hunger and satiety) fall roughly into two categories. The traditional (homeostatic) view of palatability was that it reflected underlying biological need for the nutrient predicted by the sensory properties of the available food. The classic example is with sweet tastes, where liking for sweet tastes when hungry has been interpreted in terms of expression of the current need for energy (Cabanac, 1971, 1989). The second class of theories offer a radically different view of the fundamental nature of palatability, relating palatability to reward processes which may operate, at least to some extent, independently of homeostatic controls of eating. More recently these types of theories have been discussed as part of a broader interest into "nonhomeostatic" controls of

eating (Berthoud, 2004; Corwin and Hajnal, 2005), and the approach draws heavily on theories arising from the study of drug abuse (Di Chiara, 2005; Kelley et al., 2005). However, the term "nonhomeostatic" is problematic in this context since it is difficult to conceive how an eating episode can be interpreted either as a homeostatic event (i.e., one driven by nutritional needs) or as an event driven by hedonic factors alone (putative nonhomeostatic controls). Even if hedonic factors lead to short-term over-consumption, subsequent compensatory reductions in food intake may mitigate against a positive energy balance, and so be interpreted in terms of energy homeostasis. For example, manipulating palatability at lunch resulted in increased lunchtime energy intake, but subsequent reductions in voluntary intake so that overall daily energy intake did not differ between palatable and bland lunch conditions (Yeomans et al., 2001a). The problem is determining the duration over which energy intake is regulated. Analysis of detailed diary records of food intake suggest that short-term overeating may be compensated for over 2 days (De Castro, 2000). Thus even if palatability enhances short-term intake, there will be a subsequent inhibition of appetite through greater stimulation of postmeal satiety by the ingested nutrients.

The previous discussion makes a strong case for short-term overconsumption through hedonic stimulation of appetite, and suggests that palatability cannot be explained adequately by homeostatic accounts of appetite control alone. However, once ingested, food must impact on the physiological systems underlying energy homeostasis. The next sections therefore explore the relationship between hedonic stimulation of appetite and the satiety processes which have been the traditional targets for drug-based interventions in obesity treatment. The aim here is to develop a more integrated model of appetite control combining hedonic and need-state driven components.

1. Interactions with Hunger and Satiety

The classic view of palatability expressed in Cabanac's interpretation of the physiological role of sensory pleasure (Cabanac, 1971) interpreted hedonic experience as an indicator of biological needs: when we are hungry, we need energy and consequently sweet tastes are more pleasant, while pleasantness of sweet tastes has been reported to decrease when sated. However, although this model is elegant, it does not offer an explanation for differential modulation of appetite by flavor, implying instead that hedonic evaluation should reflect the interpretation of the nutritional significance of the flavor to current motivational needs.

If we accept that the interpretation of palatability as a hedonic expression of homeostatic needs is an inadequate explanation for the substantial body of evidence demonstrating effects of palatability independently of nutritional needs, the next question is how do palatability and satiety interact. Most models of meal control accept that meal size reflects the interaction of palatability as a positive feedback control, and negative feedbacks associated with the development of satiation and generation of postmeal satiety (Smith, 2000). Theoretically, these two components could interact in three ways. First, palatability and satiety may operate independently, and therefore manipulations of these two systems should have additive effects. Alternatively, homeostatic needs (low levels of satiety) might magnify the effects of orosensory reward and thereby enhance the effects of palatability. Third, and in line with the early ideas about palatability as a reflection of nutritional needs, palatability effects may be most apparent in the absence of satiety. There are numerous strategies for examining these possibilities in human laboratory studies.

Ultimately, investigations of palatability–satiety interactions must rely on simultaneous manipulations of palatability effects with a concurrent manipulation of satiety. The classic method for achieving this is to examine how manipulations of pretest energy consumption modulate both the pleasantness of a subsequent test meal and the relationship between pleasantness and intake within that meal. This approach is clearest in preload studies where a fixed energy load is followed by an ad libitum test meal. Preload studies can be modified to test appetite in a variety of ways, depending on the timing of the preload, the macronutrient composition of preloads, whether preloads are matched in sensory and hedonic qualities, and whether the preload is eaten or infused postorally. The preload literature is large, and complex, however, more recent preload studies have provided strong evidence that palatability effects counter the satiating effects of disguised energy preloads.

Preload studies have allowed evaluation of two separate questions about palatability–satiety interactions. First, whether disguised ingestion of nutrients prior to a meal alters the pleasantness of the subsequent test meal, and second whether the magnitude to which meal size is adjusted in response to disguised energy preloads depends on the palatability of the subsequent test meals. If food pleasantness reflects the current level of need for energy, as suggested in homeostatic accounts of palatability effects, pleasantness of a test meal should be lower after a high energy than low energy preload. However, while there have been some studies where test meal pleasantness was reduced after high energy preloads (Booth et al., 1982; Johnson and Vickers, 1993; Kim and Kissileff, 1996), many other studies report no effect of preload energy on the rated pleasantness of a test meal despite compensatory reductions in subsequent energy intake (Birch and Deysher, 1986; Vandewater and Vickers, 1996; Yeomans et al., 1998, 2001a,b; Raynor and Epstein, 2000). Thus, enhanced satiety does not reliably produce reductions in subsequent food pleasantness, suggesting that any adjustment in food intake generated by disguised energy preloads is not a consequence of changes in liking for the flavor of test foods. These data also discount homeostatic accounts of the expression of palatability.

How then do manipulations of palatability and satiety interact? The use of combinations of disguised energy preloads and the flavor of the test lunch meal has allowed direct assessment of the palatability–satiety interaction (Yeomans et al., 2001a; Robinson et al., 2005). The results suggest that orosensory stimulation decreases the ability of short-term satiety cues generated by moderate energy preloads to reduce intake, resulting in an increase in overall energy intake in conditions where moderate energy fat or carbohydrate preloads were combined with a test meal with enhanced flavor (Fig. 5). These data not only confirm that satiety and orosensory stimulation have opposing effects on short-term food intake, but also suggest that palatability has a greater influence in conditions where satiety is enhanced, contradicting ideas that satiety and orosensory reward have either additive or positively interacting effects. The implication is that palatability may lead to overconsumption particularly when sated. This conclusion is further supported by the observation that when intake of a preload was enhanced through a palatability manipulation, adequate compensatory reductions in intake were not seen at a subsequent test, resulting in overconsumption following high-palatability preloads (de Graaf et al., 1999).

An alternative to preloading as a manipulation of need state is simply to increase the time since the previous meal (deprivation state). As with satiety, the effects of deprivation state on palatability are ambiguous. Some studies report enhanced increased palatability of food under conditions of deprivation (Spiegel et al., 1989). However, when the reinforcing value (i.e., ability to reduce hunger) and hedonic evaluation of food were separated using a novel operant task, the outcome suggested that deprivation enhanced the reinforcing properties of foods without altering hedonic evaluations (Epstein et al., 2003). This distinction between need-state based and hedonic influences fits well with theories of motivation originating in the drug-abuse literature (Berridge and Robinson, 1998), and particularly the distinction between "wanting" and "liking" as independent influences on motivated behavior in general. Motivation to take drugs and consume food share the same neural substrates (Berridge, 1996; Carr, 1996; Grigson, 2002), with the effectiveness of addictive drugs arising from their ability to artificially

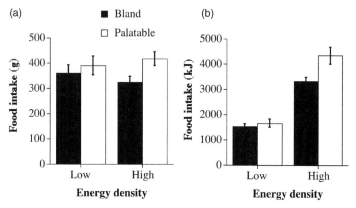

FIGURE 5 Effects of high (360 kJ) and low (60 kJ) soup preloads on mass (a) and energy (b) of a lunch presented in bland or palatable forms. Data modified from Yeomans et al. (2001b), and reprinted with permission from Nature Publishing Group.

stimulate neural systems which evolved to deal with the problems of motivating key behaviors such as eating, drinking, and reproduction. The increasing recognition that the desire to consume a food and the hedonic experience of that food are dissociable fits with the earlier discussion of palatability as a purely hedonic response to food.

The discussion so far has contrasted stimulation of appetite by palatability with the modulation of appetite by systems underlying hunger and satiety. However, the cessation of eating might also be explained by a second sensory influence, often referred to as sensory-specific satiety (Rolls et al., 1981, 1988; Rolls and Rolls, 1997). Sensory-specific satiety (SSS) refers to the decline in rated pleasantness of the consumed food relative to other, uneaten items, and is a robust component of short-term intake. SSS appears to operate through oral habituation to the hedonic qualities of the ingested food (Raynor and Epstein, 2000), and may act to counter the stimulatory effects of palatability, as well as promoting variety. Thus, orosensory stimulation may be modulated by orosensory habituation to determine short-term intake, while postingestive effects may then

modify subsequent intake in order to promote energy balance.

2. Palatability and Energy Density

The energy density (ED) of foods, defined as the energy per mass of food, has been the subject of considerable interest in more recent years since there is evidence that ED of the typical Western diet has increased, and that the increased availability of palatable, high ED foods may be a significant component of the worldwide increase in the incidence of obesity. In practice, effects of differences in ED on energy intake are most evident in the short term, with greater variation in energy intake when high ED foods are first consumed (Westerterp-Plantenga, 2004) but better control of energy intake once high ED foods are established as dietary components. This difference may be attributable to the role of learning in development of food preferences and control of meal size (Stubbs et al., 1998; Stubbs and Whybrow, 2004), and is discussed further later. In the context of this review, the critical observation is that ED and palatability are positively correlated (Holt et al., 1995; Drewnowski, 1998; Yeomans, 2003). This effect is evident across a wide range of foods (Fig. 6), but is still

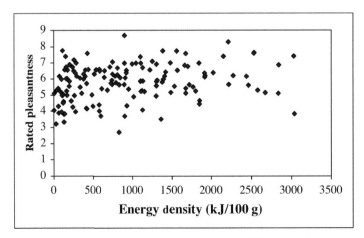

FIGURE 6   Self-reported ratings of the pleasantness of the flavor of 150 representative foods and beverages as a function of energy density. Ratings are average responses from a sample of 78 U.K. consumers (Yeomans, unpublished data).

evident in a subclass of foods such as children's evaluation of fruit and vegetables (Gibson and Wardle, 2003). As discussed later, the reason why ED and palatability are correlated probably reflects the role of learning in development of palatability. However, if we consider that palatability promotes short-term overconsumption, and that foods with high ED are rated as more palatable, then it is clear that palatability effects have the potential to have a major impact on short-term energy intake. Indeed, the reason why it is mass rather than energy content that was more closely related to the short-term impact of manipulated palatability (see Fig. 1) is attributable to differences in ED between the foods in these tests.

3. Palatability and Learned Satiety

The concept of learned satiety was introduced by Booth (1972) when discussing the impact of changes from controls of meal size in rats. According to this idea, how much of a food we consume is in part a learned response in anticipation of the level of postingestive satiety we may experience afterwards. The prediction of learned satiety is that foods with high ED may be overconsumed in the short term, but that with experience energy intake becomes controlled. This describes well the effects of ED or overall energy balance, as we already discussed (Westerterp-Plantenga, 2004). However, whether learned satiety is the mechanism underlying the short-term nature of the effects of ED on energy intake is still to be established, and there are very few published studies with unequivocal demonstrations of learned satiety in humans (Booth et al., 1976, 1982; Yeomans et al., 2005). A further complication is that according to the concept of conditioned flavor preferences, flavors which predict positive postingestive consequences should be rated as more palatable (Brunstrom, 2005; Yeomans, 2006). Thus, foods with high ED should be particularly effective at reducing appetite, and so should become

more palatable. However, foods with higher palatability should promote intake, contrary to the prediction of learned satiety. To date, no study has been able to adequately resolve this issue, although one study has been able to demonstrate changes in relative intake of high and low ED foods as a consequence of experience (Yeomans et al., 2005). In that study, volunteers consumed low and high ED versions of flavored porridge. Initially, they consumed the same mass of both versions (Fig. 7a), however, intake of the low ED version increased significantly once they had the opportunity to experience the relative satiating effects of the two foods, although never sufficiently to attach equal energy intakes in the two conditions (Fig. 7b). In this case, increased intake of the low ED food was accompanied by increased liking for that version, contrary to predictions. Until further research has been conducted, the role of learned satiety in meal control, and its relationship with palatability, remains uncertain.

4. What Makes a Food Palatable: Innate and Acquired Influences?

The focus of this review is how palatability effects are expressed. However, a brief summary of the basis of palatability is warranted. The only innate preferences in humans appear to be an innate preference for sweet tastes (Steiner et al., 2001), and an innate dislike of bitter tastes. Although olfactory information provides most of our experience of flavor (Small and Prescott, 2005), there is no evidence of any innate preference for food-related odors. Thus most of our hedonic responses to food stimuli have been acquired, and the most widely cited mechanisms for flavor preference acquisition involve conditioned associations (Rozin and Vollmecke, 1986; Sclafani et al., 1999; Yeomans, 2006) either between flavors and either postingestive consequences (flavor-consequence learning) or with already liked or disliked flavors (flavor-flavor learning). A detailed

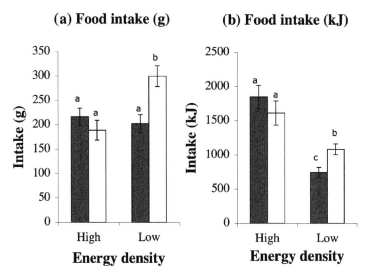

FIGURE 7 Total mass (a) and energy (b) consumed of distinctively flavored high and low energy density porridge breakfasts before (■) and after (□) 4 training days where participants consumed fixed quantities of each breakfast on alternate days to facilitate the development of associations between flavor and consequence. Adapted from Yeomans et al. (2005) with permission from Elsevier.

evaluation of these learning mechanisms is provided elsewhere in this volume. However, as yet there have been few attempts to explore the relationship between acquired food likes and the effects of palatability on appetite. Thus studies which have used flavor modulation to manipulate hedonic response to foods are probably making use of preferences acquired prior to testing. However, in order to fully understand learned influences on appetite, future studies need to examine this relationship more explicitly.

IV. PALATABILITY AND OBESITY

A. Oversensitivity to Palatability in Obesity

A classic finding in laboratory studies contrasting intake of palatable foods between obese and lean subjects was that obese overconsume palatable foods relative to lean controls (Nisbett, 1968; Price and Grinker, 1973; Hill and McCutcheon, 1975; Rodin, 1975; Kissileff and Thornton, 1982;

Spiegel et al., 1989). This apparent heightened sensitivity to food palatability was originally interpreted in support of the broader externality theory of obesity (Schachter, 1968), whereby obese people were seen to be relatively insensitive to internal (homeostatic) influences on eating but overresponsive to external food cues, including palatability. An influential interpretation of these ideas was that obese people were actually currently below their body-weight "set point," and so overresponded to palatability as a consequence of their chronic dieting habit (Nisbett, 1972), in line with the broader idea discussed earlier that palatability reflected an internal need for energy (Cabanac, 1971). An alternative interpretation is that the overresponsivity to palatability associated with palatability reflects individual differences in hedonic response as part of a broader attempt to characterize the obese phenotype (Blundell and Finlayson, 2004; Blundell, et al., 2005). The worldwide increase in prevalence of obesity could be interpreted either way, with the increased affordability and availability of palatable foods allowing people to respond to external food cues where they were restricted from doing so in the past. However, the idea that overresponsiveness to externality requires self-restriction is not consistent with current data. First, sufferers from obesity associated with binge-eating disorder, where hedonic stimulation of eating is clearly an issue (Mitchell et al., 1999), score low on measures of dietary restraint (Yanovski et al., 1992; Peterson et al., 1998), suggesting that dieting and self-restriction does not precede their sensitivity to palatability. Second, when responses of normal weight women to manipulated palatability were assessed, it was women with low restraint scores but who scored on a self-report measure of tendency to overeat (the disinhibition scale from the three-factor eating questionnaire: Stunkard and Messick, 1985) who overresponded to palatability (Yeomans et al., 2004b). Thus the notion that sensitivity to palatability is a consequence of weight loss associated with dieting appears incorrect, although desire for palatable foods is clearly a factor in the breakdown of dieting (Stirling and Yeomans, 2004; Lowe and Levine, 2005).

Overall, evaluation of current research on the responsivity to palatability by obese humans strongly suggests that individual differences in hedonic response may be a factor in overeating. However, whether this reflects differences in predisposition (i.e., genetic influences) or is a consequence of repeated overeating is unclear. Evidence suggestive of a genetic influence includes the observation that twins with high levels of body fat have a higher preference for fatty foods (Rissanen et al., 2002), since foods high in fat are typically rated as palatable. More recently, studies have started to explore the possibility that obese participants show differences in brain function in areas implicated in rewarding aspect of eating. To date, no such study has examined changes in the opioid system clearly implicated in our response to palatability. However, studies have found evidence of differences in dopamine D_2 receptor func-

tion between obese and normal participants (Wang et al., 2004), consistent with the idea either that repeated overeating leads to down regulation of dopamine D_2 receptors or that people prone to becoming obese have a premorbid abnormality in their dopamine reward system. Either way, these data are consistent with the idea that obesity is associated with abnormal function in a brain neurochemical system heavily implicated in the rewarding experience of eating.

B. Palatability and the Development of Therapeutic Approaches to Obesity

Does our current knowledge of the role of palatability in overeating and obesity allow us to predict what types of obesity treatment may be successful? The final section of this chapter examines the extent to which the desire to consume palatable foods may have hindered past attempts to treat overeating in relation to obesity, and how future potential treatments might try and get around these problems.

Given the importance of opioids in mediating the effects of palatability on appetite, early attempts to treat obesity using opioid receptor antagonists had mixed success (Yeomans and Gray, 2002). Most studies examining effects of opioid blockade on food intake in obese patients reported a significant decrease in food consumed. However, this did not translate into consistent weight loss, and coupled with concerns about side effects, the potential for use of opioid antagonists as putative antiobesity agents was rejected. Today the same drugs are being investigated as a way of reducing relapse in alcoholism (Oswald and Wand, 2004), and many of the arguments for a role of opioids in alcohol use parallel the arguments underlying the opioid–palatability relationship. Should more specific opioid antagonists emerge with greater selectively for the μ opioid receptor, there is the clear possibility of reevaluating the value of pharmacological manipulation of endogenous opioids as an adjunct to behavioral treatments for obesity since these drugs may reduce the risk of relapse to overconsumption of palatable foods, thereby aiding the development of a healthy eating style.

While opioid-based drugs have not proved successful to date, there is considerable interest in drugs affecting the endogenous cannabinoid system (Cooper, 2004; Kirkham, 2005). As discussed earlier, as yet the evidence for therapeutic benefit of cannabinoid-related drugs is based primarily on preclinical data. Moreover, although there are parallels in the animal literature between the effects of manipulation of endocannbinoids and other drugs which modulate food palatability, at present there is no controlled exploration of the effects of cannabinoid antagonists on palatability-related aspects of appetite in humans. There is clear potential for a therapeutic role of cannabinoid antagonists in obesity, but considerable research is needed to evaluate whether such drugs do result in consistent sustained weight loss, and test whether this is due to altered hedonic experience while eating.

If obesity is mediated, even in part, by overconsumption of palatable foods, would drugs that reduce the experience of pleasure during eating really work? There are obvious concerns that people who have used food as a source of sensory pleasure would find it hard to comply with a drug regime which reduced their experience of pleasure, and this may prove problematic in therapies based purely on drug-based treatments. However, if reduced pleasure during eating was an acute effect of a drug, a novel way of using such a drug would be to use it not as a drug which was taken daily, but one which was taken by people who are voluntarily restricting food intake on occasions when they felt the urge to relapse to overeating. The reported ability of opioid antagonists to reduce binge size in bulimics (Jonas and Gold, 1988) fits well with this argument. Thus, the issue of drugs as "crutches" to support the lifestyle changes needed by obese people to sustain a reduced body weight may be a more successful approach to treatment than simple drug administration alone. This approach parallels the effects of orlistat, which impairs fat absorption so that overconsumption of it leads to unpleasant GI side effects: if relapse to overeating is driven by anticipation of the sensory pleasure of consuming food, acutely removing that sensory pleasure at the point of relapse should act to reduce the hedonic experience, and so reduce the likelihood of future relapse.

V. CONCLUSION

Overall, we now know a great deal about the role of palatability in short-term overeating in general, including understanding at the behavioral, neurochemical, and neural level. While significant gaps in our understanding remain, particularly in relation to acquisition of palatability, the current level of knowledge and technical expertise is sufficient to offer significant progress in the evaluation of novel drug-based therapies for obesity. However, considerable research effort is now needed to translate these basic research findings into effective treatments for obesity.

References

Anderson, A. K., Christoff, K., Stappen, I., Panitz, D., Ghahremani, D. G., Glover, G., Gabrieli, J. D., and Sobel, N. (2003). Dissociated neural representations of intensity and valence in human olfaction. *Nat. Neurosci.* **6**, 196–202.

Arbisi, P. A., Billington, C. J., and Levine, A. S. (1999). The effect of naltrexone on taste detection and recognition threshold. *Appetite* **32**, 241–249.

Bellisle, F., and Le Magnen, J. (1980). The analysis of human feeding patterns: the Edogram. *Appetite* **1**, 141–150.

Bellisle, F., Lucas, F., Amrani, R., and Le Magnen, J. (1984). Deprivation, palatability and the microstructure of meals in human subjects. *Appetite* **5**, 85–94.

Bellisle, F., Guy-Grand, B., and Le Magnen, J. (2000). Chewing and swallowing as indices of the stimulation to eat during meals in humans: effects revealed by the Edogram method and video recordings. *Neurosci. Biobehav. Rev.* **24**, 223–228.

Berridge, K. C. (1996). Food reward: brain substrates of wanting and liking. *Neurosci. Biobehav. Rev.* **20**, 1–25.

Berridge, K. C. (2003). Pleasures of the brain. *Brain Cogn.* **52**, 106–128.

Berridge, K. C., and Robinson, T. E. (1998). What is the role of dopamine in reward: hedonic impact, reward learning, or incentive salience? *Brain Res. Bull.* **28**, 309–369.

Berthoud, H. R. (2004) Neural control of appetite: cross-talk between homeostatic and non-homeostatic systems. *Appetite* **43**, 315–317.

Bertino, M., Beauchamp, G. K., and Engelman, K. (1991). Naltrexone, an opioid blocker, alters taste perception and nutrient intake in humans. *Am. J. Physiol.* **261**, R59–R63.

Birch, L. L., and Deysher, M. (1986). Caloric compensation and sensory-specific satiety: evidence for self-regulation of food intake by young children. *Appetite* **7**, 323–331.

Blum, K., Cull, J. G., Braverman, E. R., and Comings, D. E. (1996). Reward deficiency syndrome. *Am. Scient.* **84**, 132–145.

Blundell, J. E., and Cooling, J. (2000). Routes to obesity: phenotypes, food choices and activity. *Br. J. Nutr.* **83** (Suppl. 1), S33–S38.

Blundell, J. E., and Finlayson, G. (2004). Is susceptibility to weight gain characterized by homeostatic or hedonic risk factors for overconsumption? *Physiol. Behav.* **82**, 21–25.

Blundell, J. E., Stubbs, R. J., Golding, C., Croden, F., Alam, R., Whybrow, S., Le Noury, J., and Lawton, C. L. (2005). Resistance and susceptibility to weight gain: Individual variability in response to a high-fat diet. *Physiol. Behav.*

Bobroff, E. M., and Kissileff, H. (1986). Effects of changes in palatability on food intake and the cumulative food intake curve of man. *Appetite* **7**, 85–96.

Bodnar, R. J. (2004). Endogenous opioids and feeding behavior: a 30-year historical perspective. *Peptides* **25**, 697–725.

Bodnar, R. J., Glass, M. J., Ragnauth, A., and Cooper, M. L. (1995). General, mu and kappa opioid antagonists in the nucleus accumbens alter food intake under deprivation, glucoprivic and palatable conditions. *Brain Res.* **700**, 205–212.

Booth, D. A. (1972). Conditioned satiety in the rat. *J. Comp. Physiol. Psychol.* **81**, 457–471.

Booth, D. A., Lee, M., and McAleavey, C. (1976). Acquired sensory control of satiation in man. *Br. J. Psychol.* **67**, 137–147.

Booth, D. A., Mather, P., and Fuller, J. (1982). Starch content of ordinary foods associatively conditions human appetite and satiation, indexed by intake and pleasantness of starch-paired flavors. *Appetite* **3**, 163–184.

Brunstrom, J. M. (2005). Dietary learning in humans: directions for future research. *Physiol. Behav.* **85**, 57–65.

Burgdorf, J., and Panksepp, J. (2005). The neurobiology of positive emotions. *Neurosci. Biobehavior. Rev.*

Cabanac, M. (1971). Physiological role of pleasure. *Science* **173**, 1103–1107.

Cabanac, M. (1989). Palatability of food and the ponderostat. *Ann. N.Y. Acad. Sci.* **575**, 340–352.

Carr, K. D. (1996). Feeding, drug abuse and the sensitization of reward by metabolic need. *Neurochem. Res.* **21**, 1455–1467.

Cooper, S. J. (2004). Endocannabinoids and food consumption: comparisons with benzodiazepine and opioid palatability-dependent appetite. *Eur. J. Pharmacol.* **500**, 37–49.

Cooper, S. J., and Kirkham, T. C. (1990). Basic mechanisms of opioids' effects on eating and drinking. *In* "Opioids, Bulimia and Alcohol Abuse and Alcoholism" (L. D. Reid, ed.), pp. 91–110. Springer-Verlag, New York.

Corwin, R. L., and Hajnal, A. (2005). Too much of a good thing: neurobiology of non-homeostatic eating and drug abuse. *Physiol. Behav.* **86**, 5–8.

Davis, J. D., and Levine, M. W. (1977). A model for the control of ingestion. *Psychol. Rev.* **84**, 379–412.

de Araujo, I. E., Rolls, E. T., Kringelbach, M. L., McGlone, F., and Phillips, N. (2003). Taste–olfactory convergence, and the representation of the pleasantness of flavor, in the human brain. *Eur. J. Neurosci.* **18**, 2059–2068.

De Castro, J. M. (2000). Eating behaviour: lessons from the real world of humans. *Nutrition* **16**, 800–813.

De Castro, J. M., Bellisle, F., and Dalix, A.-M. (2000a). Palatability and intake relationships in free-living humans: measurement and characterisation in the French. *Physiol. Behav.* **68**, 271–277.

De Castro, J. M., Bellisle, F., Dalix, A.-M., and Pearcey, S. M. (2000b). Palatability and intake relationships in free-living humans: characterization and independence of influence in North Americans. *Physiol. Behav.* **70**, 343–350.

de Graaf, C., de Jong, L. S., and Lambers, A. C. (1999). Palatability affects satiation but not satiety. *Physiol. Behav.* **66**, 681–688.

Di Chiara, G. (2005). Dopamine in disturbances of food and drug motivated behavior: a case of homology? *Physiol. Behav.* **86**, 9–10.

Doucet, E., and Tremblay, A. (1997). Food intake, energy balance and body weight control. *Eur. J. Clin. Nutr.* **51**, 846–855.

Drewnowski, A. (1998). Energy density, palatability and satiety: implications for weight control. *Nutr. Rev.* **56**, 347–353.

Drewnowski, A., Krahn, D. D., Demitrack, M. A., Nairn, K., and Gosnell, B. A. (1992). Taste responses and preferences for sweet high-fat foods: evidence for opioid involvement. *Physiol. Behav.* **51**, 371–379.

Epstein, L. H., Truesdale, R., Wojcik, A., Paluch, R. A., and Raynor, H. A. (2003). Effects of deprivation on hedonics and reinforcing value of food. *Physiol. Behav.* **78**, 221–227.

Gibson, E. L., and Wardle, J. (2003). Energy density predicts preferences for fruit and vegetables in 4-year-old children. *Appetite* **41**, 97–98.

Grigson, P. S. (2002). Like drugs for chocolate: separate rewards modulated by common mechanisms. *Physiol. Behav.* **76**, 389–395.

Guy-Grand, B., Lehner, V., and Doassans, M. (1989). Effects of palatability and meal type on food intake in normal weight males. *Appetite* **12**, 213–214.

Helleman, U., and Tuorila, H. (1991). Pleasantness ratings and consumption of open sandwiches with varying NaCl and acid contents. *Appetite* **17**, 229–238.

Hill, A. J., Magson, L. D., and Blundell, J. E. (1984). Hunger and palatability: tracking ratings of subjective experience before, during and after the consumption of preferred and less preferred food. *Appetite* **5**, 361–371.

Hill, S. W. (1974). Eating responses of humans during dinner meals. *J. Comp. Physiol. Psychol.* **86**, 652–657.

Hill, S. W., and McCutcheon, N. B. (1975). Eating responses of obese and non-obese humans during dinner meals. *Psychosom. Med.* **37**, 395–401.

Holt, S. H., Miller, J. C., Petocz, P., and Farmakalidis, E. (1995). A satiety index of common foods. *Eur. J. Clin. Nutr.* **49**, 675–690.

Jequier, E., and Tappy, L. (1999). Regulation of body weight in humans. *Physiol. Rev.* **79**, 451–480.

Johnson, J., and Vickers, Z. (1993). Effects of flavor and macronutrient composition of food servings on liking, hunger and subsequent intake. *Appetite* **21**, 25–39.

Jonas, J. M., and Gold, M. S. (1988). Naltrexone treatment of bulimia: clinical and theoretical findings linking eating disorders and substance abuse. *Adv. Alcohol Substance Abuse* **7**, 29–37.

Kauffman, N. A., Herman, C. P., and Polivy, J. (1995). Hunger-induced finickiness in humans. *Appetite* **24**, 203–218.

Kelley, A. E. (2004). Ventral striatal control of appetitive motivation: role in ingestive behavior and reward-related learning. *Neurosci. Biobehav. Rev.* **27**, 765–776.

Kelley, A. E., Bakshi, V. P., Haber, S. N., Steininger, T. L., Will, M. J., and Zhang, M. (2002). Opioid modulation of taste hedonics within the ventral striatum. *Physiol. Behav.* **76**, 365–377.

Kelley, A. E., Schiltz, C. A., and Landry, C. F. (2005). Neural systems recruited by drug- and food-related cues: studies of gene activation in corticolimbic regions. *Physiol. Behav.* **86**, 11–14.

Kim, E.-M., Shi, Q., Olszewski, P. K., Grace, M. K., O'Hare, E. O., Billington, C. J., and Levine, A. S. (2001). Identification of central sites involved in butorphanol-induced feeding in rats. *Brain Res.* **907**, 125–129.

Kim, E. M., Quinn, J. G., Levine, A. S., and O'Hare, E. (2004). A bi-directional mu-opioid-opioid connection between the nucleus of the accumbens shell and the central nucleus of the amygdala in the rat. *Brain Res.* **1029**, 135–139.

Kim, J. Y., and Kissileff, H. R. (1996). The effect of social setting on response to a preloading manipulation in nonobese women and men. *Appetite* **27**, 25–40.

Kirkham, T. C. (2005). Endocannabinoids in the regulation of appetite and body weight. *Behav. Pharmacol.* **16**, 297–313.

Kirkham, T. C., and Williams, C. M. (2001a). Endogenous cannabinoids and appetite. *Nutr. Res. Rev.* **14**, 65–86.

Kirkham, T. C., and Williams, C. M. (2001b). Synergistic effects of opioid and cannabinoid antagonists on food intake. *Psychopharmacology (Berlin)* **153**, 267–270.

Kissileff, H. R. (1976). Palatability. *In* "International Encyclopedia of Psychiatry, Psychology, Psychoanalysis and Neurology" 10th ed. p. 172. Academic Press, New York.

Kissileff, H. R. (1990). Some suggestions on dealing with palatability—response to Ramirez. *Appetite* **14**, 162–166.

Kissileff, H. R., and Thornton, J. (1982). Facilitation and inhibition in the cumulative food intake curve in man. *In* "Changing Concepts of the Nervous System" (A. R. Morrison and P. L. Strick, eds.), pp. 585–607. Academic Press, New York.

Kissileff, H. R., Kilngsberg, G., and Van Italie, T. B. (1980). Universal eating monitor for continuous recording of solid or liquid consumption in man. *Am. J. Physiol.* **238**, R14–R22.

Kringelbach, M. L. (2004). Food for thought: hedonic experience beyond homeostasis in the human brain. *Neuroscience* **126**, 807–819.

Lahteenmaki, L., and Tuorila, H. (1995). Consistency of liking and appropriateness ratings and their relation to consumption in a product test of ice cream. *Appetite* **25**, 189–197.

Lamonte, N., Echo, J. A., Ackerman, T. F., Christian, G., and Bodnar, R. J. (2002). Analysis of opioid receptor subtype antagonist effects upon mu opioid agonist-induced feeding elicited from the ventral tegmental area of rats. *Brain Res.* **929**, 96–100.

Le Magnen, J. (1987). Palatability: Concept, Terminology and Mechanisms. *In* "Eating Habits: Food, Physiology and Learned Behaviour" (R. A. Boakes, D. A. Popplewell, and M. J. Burton, eds.). pp. 131–154. Wiley, Chichester.

Levine, A. S., Olszewski, P. K., Mullett, M. A., Pomonis, J. D., Grace, M. K., Kotz, C. M., and Billington, C. J. (2004). Intra-amygdalar injection of DAMGO: effects on c-Fos levels in brain sites associated with feeding behavior. *Brain Res.* **1015**, 9–14.

Lowe, M. R., and Levine, A. S. (2005). Eating motives and the controversy over dieting: eating less than needed versus less than wanted. *Obes. Res.* **13**, 797–806.

MacDonald, A. F., Billington, C. J., and Levine, A. S. (2003). Effects of the opioid antagonist naltrexone on feeding induced by DAMGO in the ventral tegmental area and in the nucleus accumbens shell region in the rat. *Am. J. Physiol.—Regul. Integr. Comp. Physiol.* **285**, R999–R1004.

MacDonald, A. F., Billington, C. J., and Levine, A. S. (2004). Alterations in food intake by opioid and dopamine signaling pathways between the ventral tegmental area and the shell of the nucleus accumbens. *Brain Res.* **1018**, 78–85.

Melchior, J. C., Rigaud, D., Chayvialle, J. A., Colas-Linhart, N. C., Laforest, M. D., Petiet, A., Comoy, E., and Apfelbaum, M. (1994). Palatability of a meal influences release of beta-endorphin, and of potential regulators of food intake in healthy human subjects. *Appetite* **22**, 233–244.

Mitchell, J. E., Mussell, M. P., Peterson, C. B., Crow, S., Wonderlich, S., Crosby, R. D., Davis, T., and Weller, C. (1999). Hedonics of binge eating in women with bulimia nervosa and binge eating disorder. *Int. J. Eat. Disord.* **26**, 165–170.

Mook, D. G., and Yoo, D. K. (1991). Inhibition of sham feeding by hyperosmotic gastric infusions in rat. *Psychobiology* **19**, 359–364.

Naim, M., and Kare, M. R. (1991). Sensory and postingestional components of palatability in dietary obesity: an overview. *In* "Appetite and Nutrition" (M. I. Friedman, M. G. Tordoff, and M. R. Kare, eds.). 4th ed., pp. 109–126. Dekker, New York.

Nisbett, R. E. (1968). Taste, deprivation and weight determinants of eating behavior. *J. Pers. Soc. Psychol.* **10**, 107–116.

Nisbett, R. E. (1972). Hunger, obesity and the ventromedial hypothalamus. *Psychol. Rev.* **79**, 433–453.

O'Doherty, J., Rolls, E. T., Francis, S., Bowtell, R., and McGlone, F. (2001). Representation of pleasant and aversive taste in the human brain. *J. Neurophysiol.* **85**, 1315–1321.

Oswald, L. M., and Wand, G. S. (2004). Opioids and alcoholism. *Physiol. Behav.* **81**, 339–358.

Panksepp, J. (2005). Affective consciousness: core emotional feelings in animals and humans. *Conscious. Cogn.* **14**, 30–80.

Pelchat, M. L. (2002). Of human bondage: food craving, obsession, compulsion, and addiction. *Physiol. Behav.* **76**, 347–352.

Peterson, C. B., Mitchell, J. E., Engbloom, S., Nugent, S., Mussell, M. P., Crow, S. J., and Miller, J. P. (1998). Binge eating disorder with and without a history of purging symptoms. *Int. J. Eat. Disord.* **24**, 251–257.

Pierson, A., and Le Magnen, J. (1969). Étude quantificaiton du processus de regulation des reponses alimentaires chez hommes. *Physiol. Behav.* **4**, 61–67.

Pietras, T. A., and Rowland, N. E. (2002). Effect of opioid and cannabinoid receptor antagonism on orphanin FQ-induced hyperphagia in rats. *Eur. J. Pharmacol.* **442**, 237–239.

Prescott, J. (2004). Psychological processes in flavor perception. *In* "Flavor Perception" (A. J. Taylor and D. Roberts, eds.). pp. 256–277. Blackwell, London.

Price, J. M., and Grinker, J. (1973). Effects of degree of obesity, food deprivation and palatability on eating behavior of humans. *J. Comp. Physiol. Psychol.* **85**, 265–271.

Ragnauth, A., Moroz, M., and Bodnar, R. J. (2000). Multiple opioid receptors mediate feeding elicited by mu and delta opioid receptor subtype agonists in the nucleus accumbens shell in rats. *Brain Res.* **876**, 76–87.

Ramirez, I. (1990). What do we mean when we say "palatable food"? *Appetite* **14**, 159–161.

Raynor, H. A., and Epstein, L. H. (2000). Effects of sensory stimulation and post-ingestive consequences on satiation. *Physiol. Behav.* **70**, 465–470.

Rissanen, A., Hakala, P., Lissner, L., Mattlar, C. E., Koskenvuo, M., and Ronnemaa, T. (2002). Acquired preference especially for dietary fat and obesity: a study of weight-discordant monozygotic twin pairs. *Int. J. Obes.* **26**, 973–977.

Robinson, T. M., Gray, R. W., Yeomans, M. R., and French, S. J. (2005). Test-meal palatability alters the effects of intragastric fat but not carbohydrate preloads on intake and rated appetite in healthy volunteers. *Physiol. Behav.* **84**, 193–203.

Rodin, J. (1975). Effects of obesity and set point on taste responsiveness and ingestion in humans. *J. Comp. Physiol. Psychol.* **89**, 1003–1009.

Rogers, P. J. (1990). Why a palatability construct is needed. *Appetite* **14**, 167–170.

Rolls, E. T., and Rolls, J. H. (1997). Olfactory sensory-specific satiety in humans. *Physiol. Behav.* **61**, 461–473.

Rolls, B. J., Rolls, E. T., Rowe, E. A., and Sweeney, K. (1981). Sensory-specific satiety in man. *Physiol. Behav.* **27**, 137–142.

Rolls, B. J., Van Duijvenvoorde, P. M., and Rowe, E. A. (1983). Variety in the diet enhances intake in a meal and contributes to the development of obesity in the rat. *Physiol. Behav.* **31**, 21–27.

Rolls, B. J., Hetherington, M., and Burley, V. J. (1988). The specificity of satiety: the influence of foods of different macronutrient content on the development of satiety. *Physiol. Behav.* **43**, 145–153.

Royet, J. P., Plailly, J., Delon-Martin, C., Kareken, D. A., and Segebarth, C. (2003). fMRI of emotional responses to odors: influence of hedonic valence and judgment, handedness, and gender. *Neuroimage* **20**, 713–728.

Rozin, P., and Vollmecke, T. A. (1986). Food likes and dislikes. *Annu. Rev. Nutr.* **6**, 433–456.

Sawaya, A. L., Fuss, P. J., Dallal, G. E., Tsay, R., McCrory, M. A., Young, V., and Roberts, S. B. (2001).

Meal palatability, substrate oxidation and blood glucose in young and older men. *Physiol. Behav.* **72**, 5–12.

Schachter, S. (1968). Obesity and eating. *Science* **161**, 751–756.

Sclafani, A., Fanizza, L. J., and Azzara, A. V. (1999). Conditioned flavor avoidance, preference, and indifference produced by intragastric infusions of galactose, glucose and fructose in rats. *Physiol. Behav.* **67**, 227–234.

Small, D. M., and Prescott, J. (2005). Odor/taste integration and the perception of flavor. *Exp. Brain Res.*

Small, D. M., Zatorre, R. J., Dagher, A., Evans, A. C., and Jones-Gotman, M. (2001). Changes in brain activity related to eating chocolate: from pleasure to aversion. *Brain* **124**, 1720–1733.

Small, D. M., Jones-Gotman, M., and Dagher, A. (2003). Feeding-induced dopamine release in dorsal striatum correlates with meal pleasantness ratings in healthy human volunteers. *Neuroimage* **19**, 1709–1715.

Small, D. M., Voss, J., Mak, Y. E., Simmons, K. B., Parrish, T., and Gitelman, D. (2004). Experience-dependent neural integration of taste and smell in the human brain. *J. Neurophysiol.* **92**, 1892–1903.

Smith, G. P. (2000). The controls of eating: a shift from nutritional homeostasis to behavioral neuroscience. *Nutrition* **16**, 814–820.

Smith, S. L., Harrold, J. A., and Williams, G. (2002). Diet-induced obesity increases mu opioid receptor binding in specific regions of the rat brain. *Brain Res.* **953**, 215–222.

Solinas, M., and Goldberg, S. R. (2005). Motivational effects of cannabinoids and opioids on food reinforcement depend on simultaneous activation of cannabinoid and opioid systems. *Neuropsychopharmacology* **30**, 2035–2045.

Spiegel, T. A. (2000). Rate of intake, bites and chews: the interpretation of obese-lean differences. *Neurosci. Biobehav. Rev.* **24**, 229–237.

Spiegel, T. A., and Stellar, E. (1990). Effects of variety on food intake of underweight, normal-weight and overweight women. *Appetite* **15**, 47–61.

Spiegel, T. A., Shrager, E. E., and Stellar, E. (1989). Responses of lean and obese subjects to preloads, deprivation and palatability. *Appetite* **13**, 46–69.

Spiegel, T. A., Kaplan, J. M., Tomassini, A., and Stella, E. (1993). Bite size, ingestion rate and meal size in lean and obese women. *Appetite* **21**, 131–145.

Steiner, J. E., Glaser, D., Hawilo, M. E., and Berridge, K. C. (2001). Comparative expression of hedonic impact: affective reactions to taste by human infants and other primates. *Neurosci. Biobehav. Rev.* **25**, 53–74.

Stirling, L. J., and Yeomans, M. R. (2004). Effects of exposure to a forbidden food on eating in restrained and unrestrained women. *Int. J. Eat. Disord.* **35**, 59–68.

Stubbs, R. J., and Whybrow, S. (2004). Energy density, diet composition and palatability: influences on overall food energy intake in humans. *Physiol. Behav.* **81**, 755–764.

Stubbs, R. J., Johnstone, A. M., O'Reilly, L. M., Barton, K., and Reid, C. (1998). The effect of covertly manipulating the energy density of mixed diets on ad libitum food intake in 'pseudo free-living' humans. *Int. J. Obes.* **22**, 980–987.

Stunkard, A. J., and Messick, S. (1985). The three-factor eating questionnaire to measure dietary restraint, disinhibition and hunger. *J. Psychosom. Res.* **29**, 71–83.

Toates, F. M. (1986). "Motivational Systems." Cambridge University Press, Cambridge.

Treit, D., Spetch, M. L., and Deutsch, J. A. (1983). Variety in the flavor of food enhances eating in the rat: a controlled demonstration. *Physiol. Behav.* **30**, 207–211.

Vandewater, K., and Vickers, Z. (1996). Higher-protein foods produce greater sensory-specific satiety. *Physiol. Behav.* **59**, 579–583.

Vickers, S. P., and Kennett, G. A. (2005). Cannabinoids and the regulation of ingestive behaviour. *Curr. Drug Targets* **6**, 215–223.

Vickers, Z., Holton, E., and Wang, J. (2001). Effect of ideal-relative sweetness on yoghurt consumption. *Food Qual. Prefer.* **12**, 521–526.

Volkow, N. D., Wang, G. J., Maynard, L., Jayne, M., Fowler, J. S., Zhu, W., Logan, J., Gatley, S. J., Ding, Y.-S., Wong, C., and Pappas, N. (2003). Brain dopamine is associated with eating behaviors in humans. *Int. J. Eat. Disord.* **33**, 136–142.

Wang, G. J., Volkow, N. D., Logan, J., Pappas, N. R., Wong, C. T., Zhu, W., Netusil, N., and Fowler, J. S. (2001). Brain dopamine and obesity. *Lancet* **357**, 354–357.

Wang, G. J., Volkow, N. D., Thanos, P. K., and Fowler, J. S. (2004). Similarity between obesity and drug addiction as assessed by neurofunctional imaging: a concept review. *J. Addict. Dis.* **23**, 39–53.

Wansink, B. (2004). Environmental factors that increase the food intake and consumption volume of unknowing consumers. *Annu. Rev. Nutr.* **24**, 455–479.

Wellman, P. J. (2005). Modulation of eating by central catecholamine systems. *Curr. Drug Targets* **6**, 191–199.

Westerterp-Plantenga, M. S. (2004). Effects of energy density of daily food intake on long-term energy intake. *Physiol. Behav.* **81**, 765–771.

Wiepkema, P. R. (1971). Positive feedbacks at work during feeding. *Behaviour* **39**, 266–273.

Will, M. J., Franzblau, E. B., and Kelley, A. E. (2003). Nucleus accumbens mu-opioids regulate intake of a high-fat diet via activation of a distributed brain network. *J. Neurosci.* **23**, 2882–2888.

Will, M. J., Franzblau, E. B., and Kelley, A. E. (2004). The amygdala is critical for opioid-mediated binge eating of fat. *Neuroreport* **15**, 1857–1860.

Wise, R. A. (1981). Brain dopamine and reward. *In* "Theory in Psychopharmacology" (S. J. Cooper, ed.). Vol. 1, pp. 103–122. Academic Press, London.

Yanovski, S. Z., Leet, M., Yanovski, J. A., Flood, M., Gold, P. W., Kissileff, H. R., and Walsh, B. T. (1992). Food selection and intake of obese women with binge-eating disorder. *Am. J. Clin. Nutr.* **56**, 975–980.

Yeomans, M. R. (1996). Palatability and the microstructure of eating in humans: the appetiser effect. *Appetite* **27**, 119–133.

Yeomans, M. R. (1998). Taste, palatability and the control of appetite. *Proc. Nutr. Soc.* **57**, 609–615.

Yeomans, M. R. (2000). Rating changes over the course of meals: what do they tell us about motivation to eat? *Neurosci. Biobehav. Rev.* **24**, 249–259.

Yeomans, M. R. (2003). Effects of manipulated palatability and energy density on appetite. *Appetite* **40**, 371.

Yeomans, M. R. (2006). The role of learning in development of food preferences. *In* "Psychology of Food Choice" (R. Shepherd and M. Raats, eds.) (pp. in press). CABI.

Yeomans, M. R., and Wright, P. (1991). Lower pleasantness of palatable foods in nalmefene-treated human volunteers. *Appetite* **16**, 249–259.

Yeomans, M. R., and Gray, R. W. (1996). Selective effects of naltrexone on food pleasantness and intake. *Physiol. Behav.* **60**, 439–446.

Yeomans, M. R., and Gray, R. W. (1997). Effects of naltrexone on food intake and changes in subjective appetite during eating: evidence for opioid involvement in the appetiser effect. *Physiol. Behav.* **62**, 15–21.

Yeomans, M. R., and Symes, T. (1999). Individual differences in the use of palatability and pleasantness ratings. *Appetite* **32**, 383–394.

Yeomans, M. R., and Gray, R. W. (2002). Opioids and human ingestive behaviour. *Neurosci. Biobehav. Rev.* **26**, 713–728.

Yeomans, M. R., Wright, P., Macleod, H. A., and Critchley, J. A. J. H. (1990). Effects of nalmefene on feeding in humans: dissociation of hunger and palatability. *Psychopharmacology (Berlin)* **100**, 426–432.

Yeomans, M. R., Gray, R. W., Mitchell, C. J., and True, S. (1997). Independent effects of palatability and within-meal pauses on intake and subjective appetite in human volunteers. *Appetite* **29**, 61–76.

Yeomans, M. R., Gray, R. W., and Conyers, T. (1998). Maltodextrin preloads reduce intake without altering the appetiser effect. *Physiol. Behav.* **64**, 501–506.

Yeomans, M. R., Lee, M. D., Gray, R. W., and French, S. J. (2001a). Effects of test-meal palatability on compensatory eating following disguised fat and carbohydrate preloads. *Int. J. Obes.* **25**, 1215–1224.

Yeomans, M. R., Lartamo, S., Procter, E. L., Lee, M. D., and Gray, R. W. (2001b). The actual, but not labelled, fat content of a soup preload alters short-term appetite in healthy men. *Physiol. Behav.* **73**, 533–540.

Yeomans, M. R., Blundell, J. E., and Lesham, M. (2004a). Palatability: response to nutritional need or need-free stimulation of appetite? *Br. J. Nutr.* **92**, S3-S14.

Yeomans, M. R., Tovey, H. M., Tinley, E. M., and Haynes, C. L. (2004b). Effects of manipulated palatability on appetite depend on restraint and disinhibition scores from the Three Factor Eating Questionnaire. *Int. J. Obes.* **28**, 144–151.

Yeomans, M. R., Weinberg, L., and James, S. (2005). Effects of palatability and learned satiety on energy density influences on breakfast intake in humans. *Physiol. Behav.* **86**, 487–499.

Young, P. T. (1967). Palatability: the hedonic response to foodstuffs. *In* "Handbook of Physiology, Section 6: Alimentary Canal, Volume 1. Control of Food and Water Intake" (C. F. Code, ed.). pp. 353–366. American Physiology Society, Washington, D.C.

Zahm, D. S. (2000). An integrative neuroanatomical perspective on some subcortical substrates of adaptive responding with emphasis on the nucleus accumbens. *Neurosci. Biobehav. Rev.* **24**, 85–105.

Zald, D. H. (2003). The human amygdala and the emotional evaluation of sensory stimuli. *Brain Res. Rev.* **41**, 88–123.

Zald, D. H., Lee, J. T., Fluegel, K. W., and Pardo, J. V. (1998). Aversive gustatory stimulation activates limbic circuits in humans. *Brain* **121**(Pt 6), 1143–1154.

Zald, D. H., Hagen, M. C., and Pardo, J. V. (2002). Neural correlates of tasting concentrated quinine and sugar solutions. *J. Neurophysiol.* **87**, 1068–1075.

Zandstra, E. H., de Graaf, C., Mela, D. J., and Van Staveren, W. (2000). Short- and long-term effects of changes in pleasantness on food intake. *Appetite* **34**, 253–260.

Zhang, M., Balmadrid, C., and Kelley, A. E. (2003). Nucleus accumbens opioid, GABAergic, and dopaminergic modulation of palatable food motivation: contrasting effects revealed by a progressive ratio study in the rat. *Behav. Neurosci.* **117**, 202–211.

Learned Influences on Appetite, Food Choice, and Intake: Evidence in Human Beings

E. L. GIBSON
School of Human and Life Sciences, Roehampton University,
Whitelands College
J. M. BRUNSTROM
Department of Experimental Psychology, Bristol University

I. Introduction
II. Innate Influences on Human Eating
III. Types of Learning
    A. Habituation, "Mere" Exposure and Learned Safety
    B. Associative Learning
IV. The Learned Appetite for Energy
V. Learned Modulation of Appetite and Meal Size by Associated States
    A. Learned Control of Meal Size
    B. Regulation of Food Choice by Association with Internal State
VI. Nutrient-Specific Learned Appetites
    A. Sodium Appetite
    B. Human Appetite for Protein
VII. Flavor-Flavor Learning
    A. Evidence for Flavor-Flavor Learning
    B. What Do We Know (and Need to Know) about Flavor-Flavor Learning in Human Beings?
    C. Flavor-Flavor Learning and Everyday Dietary Behavior—Some Suggestions for Future Research
VIII. Awareness and Dietary Learning
    A. What Do We Know about Awareness and Dietary Learning?
    B. Awareness and Its Relevance to Understanding Dietary Control
IX. Summary
    References

Learning underlies the development and regulation of habitual eating, including our likes and dislikes, choosing foods most appropriate to our current motivational state, and controlling how much is eaten. In young children, mere exposure to the flavor of a food, such as a novel vegetable, increases acceptance of that food. Similarly, exposure to flavors in amniotic fluid and breast milk might link maternal dietary choice with preference development in children. Children's preferences are strongly correlated with the energy density of foods due to the reinforcing effects of energy eaten when hungry, that is, flavor-consequence learning. Carbohydrate, fat, and protein have all proved effective in reinforcing flavor preferences. Flavors associated with higher energy consumption are preferred when hungry, but conversely less liked when full than lower energy-paired flavors, and they suppress subsequent intake. Sensitivity to postingestive energy differences may weaken with age and externalization of eating control. Frequent eating of high fat energy-dense foods may impair neural inhibition of learned appetite, creating a vicious circle leading to obesity. Flavor-flavor learning occurs when a neutral flavor is eaten together with a flavor that already has strong positive or, more robustly, aversive properties. This could form a shortcut for transferring important information from one sensory property to another. The necessity for explicit awareness of flavor-consequence or flavor-flavor associations for learned control of eating is discussed. This is important, because it has implications as to who should be held accountable for eating behavior, and so for public health strategies to control obesity and dietary-related disease.

I. INTRODUCTION

Experience plays an important part in determining dietary preferences in general, and it will certainly have to be taken into account in future work on that hitherto neglected subject, the psychology of appetite.

Harris et al. (1933)

Broadly defined, learning is a change in behavior as a result of experience. However, the simplicity of that definition belies both the importance and the ubiquitous nature of learning, whether for eating or other behaviors: it is the principal means by which an animal can adapt to a changing environment. One might suppose from the preceding comment, from the seminal work on need-dependent flavor preference learning in rats (Harris et al., 1933), that a substantial knowledge base should have developed steadily over the last 80 years or so. In fact, evidence in human beings for a role for learning in appetite and food choice has by and large only begun to appear in the last two decades. Prior to that, even in rats, remarkably little attention had been paid to this area; which is ironic considering that psychologists studying learning theory had been using hunger, food reinforcement, and taste acceptance as standard tools of the trade for most of the last century. Nevertheless, there has more recently been a growth of interest in this area (e.g., Holland and Petrovich, 2005), which this chapter will summarize.

Eating and drinking will be the product of both learned and unlearned responses to stimuli both within and without the body. As we shall see, there is evidence for only a few unlearned reactions to food stimuli, and although reactions to internal stimuli may not initially depend on learning, it is probable that responses to most such stimuli, both internal and external, are modifiable by learning. This is essential to allow an animal to adapt to changes in the environment, and to encourage a sufficiently varied diet, particularly for omnivores like human beings (Rozin and Zellner, 1985). The influence of learning on eating behavior is not just about developing likes and dislikes, but also about motivating us to find food when needed, making choices that are appropriate for our current nutritional status, and controlling the amount that is eaten. Nevertheless, as will be apparent elsewhere in this book, as the human food environment shifts allegiance from the forces of natural toward unnatural selection, adaptation through learning will not necessarily lead to better health. Instead, we suffer now from having evolved in less comfortable times. It has even been proposed that our propensity to choose fat-rich foods promotes obesity by interfering with neural control of learned adaptive eating (Davidson et al., 2005).

II. INNATE INFLUENCES ON HUMAN EATING

Adaptive innate reactions will have evolved where a specific stimulus or cue, such as the taste of a plant, consistently predicts a consequence of eating the plant that has an impact on the animal's survival or reproductive success. Thus, many animals instinctively reject bitter (e.g., quinine) and astringent (e.g., tannin) tastes, presumably because these are associated with the presence of poisonous plant alkaloids. Conversely, animals commonly show an innate propensity to ingest sweet tasting food or fluids: these innate acceptance and rejection reactions have both been demonstrated in newborn human babies, including those born anencephalic, that is, lacking intact cerebral hemispheres (Steiner et al., 2001). Sweetness might be a consistent cue for ripeness (and so high sugar content) of nontoxic fruit and roots, although ripeness also means loss of acidity, astringency, and inedible toughness. Primate species that eat a wide-ranging diet have a low threshold for detection of sweetness, which encourages consumption even of low sugar foods, whereas those with feeding strategies limited to high energy foods have higher sweetness thresholds (Hladik et al., 2002). However, Booth and Thib (2000) have suggested that a more convincing evolutionary role for sweet liking in mammals may be as an antibitterness device in nitrogen-rich mother's milk. That is, "sweet" receptors respond to chemical groups on both sugars and amino acids, so there would be strong

selection pressure to develop an ingestive reflex when such receptors are stimulated, ensuring that the protein-rich milk is not rejected.

This trade-off between sweetness and bitterness seems to be supported by more recent findings for children with differing alleles for a gene for detection of the bitter taste of 6-n-propylthiouracil (PROP), that is, the TAS2R38 gene (Mennella et al., 2005). Children who were homo- or heterozygous for the bitter-tasting allele preferred higher levels of sucrose in foods and drinks than those having the nonbitter-tasting allele. Intriguingly, mothers who were nontasters of bitter PROP rated children who were bitter tasters were more emotional than children who did not have bitter-sensitive alleles. Although ratings of child food neophobia did not differ here, it is notable that another study concluded that low-threshold PROP tasters may be inherently more neophobic, or at least more emotionally reactive to food (Pasquet et al., 2002). However, these different TAS2R38 genotypes did not show correspondence to sweet preferences in adults, suggesting that such genetic influences on children's eating behavior are modifiable by experience (see also Section VII.C).

In general, the complex sensory experience that results from the combination of tastes, smells, textures, and latent heat of foods, which we and other animals eat, cannot provide sufficiently consistent cues to postingestive consequences for responses to be hardwired by inheritance. Furthermore, omnivores such as rats and ourselves require too many nutrients for optimal health to depend on innate appetites (Rozin and Schulkin, 1990; Gibson, 2001). In other words, adaptive food choice cannot rely solely on a battery of innate reactions of acceptance or rejection to a world of varying sensory stimuli. Instead, we depend on a nervous system designed to detect, record, and respond to reliable relationships between two or more events, that is, to learn.

The precise forms of the types of learning likely to be involved in eating remain a matter for debate (e.g., Tapper, 2005). This is not the place to deal with this topic in detail here: nevertheless, later in this chapter we discuss areas where theoretical distinctions and evidence for certain processes involved in dietary learning may have important implications (Sections VII and VIII). Moreover, the likelihood is that the learning involved in responding to tastes, odors, textures, and other oral sensations, and to internal sensations related to nutritional status, will have different properties from that which underlies responses to visual and auditory stimuli, which arise from more distal sources. It is these latter stimuli that underlie the majority of human cognitive research—hence the current debate. However, we believe the descriptions used in the following sections have a broad consensus.

III. TYPES OF LEARNING

A. Habituation, "Mere" Exposure and Learned Safety

Novel sensations or stimuli normally elicit some forms of orienting or defensive response, whether they be urgent physical orientation or simply briefly focused attention. With repeated exposure, but no reinforcement, the response to the stimulus will weaken or even disappear. This simple form of nonassociative learning is known as habituation, which presumably underlies the specific loss of interest in, or of appetite for, a repeatedly tasted food, even when the food is merely chewed for several minutes but not eaten—a phenomenon termed sensory-specific satiety (Le Magnen, 1967). It is unlikely to be the result of sensory adaptation, because whereas ratings of pleasantness decrease, ratings of intensity of the taste or smell of the food hardly change (Rolls and Rolls, 1997). A related example of habituation is the reduction in

salivation found on repeated tasting of food, followed by "dishabituation," that is, increased salivation, after a different food is tasted. It has been shown that distraction of attention from eating weakens this salivary habituation (Epstein et al., 1997). Moreover, obese women show weaker habituation of salivation (Epstein et al., 1996), which might reflect a learned or unlearned difference in the reinforcing value of the food, or a weakened inhibitory control of learned sensory salience (cf. Davidson et al., 2005; Section V.B).

Sensory-specific satiety occurs after many repetitions of exposure over a time frame equivalent to a meal. By comparison, when exposure to a novel food occurs with intervals of several hours or days between repetitions, the acceptability and intake of that food often increases (Pliner, 1982). This may well reflect the decline in "neophobia" (fear of novelty), that is, a tendency to avoid or treat novel foods cautiously, which many animals, especially omnivores, express—notwithstanding the argument that it is adaptive to explore novel stimuli (Kyriazakis et al., 1999; Rolls, 1999), at least to the point of ingestion. One implication of this learned acceptance with exposure is that familiar foods would be less likely to acquire aversive connotations, for example, by association with illness, than more novel foods. This inhibition of learned aversion is known as learned safety and is well evidenced in animals (Kalat and Rozin, 1973; Siegel, 1974), and probably true for human beings (Bernstein, 1994).

There are many claims for a role for this seemingly nonassociative form of learning due to "mere" exposure (i.e., without reinforcement) in acquired food acceptance in human beings (see later). However, it may often be the case that there has been some form of reinforcement, particularly social or emotional, contingent with the exposure (Birch, 1987), so that in fact the learning may be at least partly associative, for example, a form of evaluative conditioning (Section VII).

A significant advance in understanding the contribution of this form of learning was the finding that maternal diet affected the flavor of the mother's milk, which in turn altered flavor acceptance in their babies. However, the results were not entirely positive. For garlic and carrot flavor, prior exposure to the flavor in breast milk seemed, respectively, to suppress feeding on garlic-flavored breast milk (Mennella and Beauchamp, 1993), or suppress eating of cereal mixed with carrot juice relative to infants exposed for the first time to those flavors (Mennella and Beauchamp, 1999). This was suggested to be a form of sensory-specific satiety to the recently exposed flavors. In a subsequent study, which included exposing fetuses, via their amniotic fluid, to carrot flavor from the pregnant mother's diet, infant acceptance of cereal mixed with carrot juice was judged to be enhanced relative to water (Mennella et al., 2001). Another finding from this group was that early exposure to a particular infant formula milk determined subsequent liking for not only the same and similar formulas, but for related tastes in other drinks, such as sourness acceptance in infants fed on sour-tasting protein hydrolysate formula (Mennella and Beauchamp, 2002).

In infants just being weaned and naïve to vegetables, Sullivan and Birch (1994) found that 10 opportunities to eat a small amount of pureed peas or green beans led to increased intake of this vegetable, but this was only significant in breast-fed babies. It was suggested that breast-fed babies might more readily accept a novel vegetable, having experienced vegetable flavors in breast milk previously. Nevertheless, Gerrish and Mennella (2001) showed later that formula-fed infants given nine exposures to several different vegetables subsequently ate not only more carrot than a group fed only potato, but also more pureed chicken than a group fed only carrot. They also found that prior experience with fruit in these infants

led to greater intake of carrot on first exposure.

Birch and colleagues have provided several demonstrations in preschool children of increased acceptance of foods following repeated tasting (Birch and Marlin, 1982; Birch, 1987)—usually requiring at least 10 exposures. Birch et al. (1987a) also showed that tasting was necessary to enhance rated taste preferences, not just visual exposure. This simple way of encouraging liking for foods has been used effectively in short-term interventions to enhance acceptance of vegetables in young children (Loewen and Pliner, 1999; Wardle et al., 2003a,b), although again contingent social reinforcement was likely. Furthermore, in 4- to 7-month-old infants, substantial increases in intake of an exposed (eaten) food, including fruit, occur after just one feed, and this generalizes to similar foods, such as other fruit (Birch et al., 1998). This suggests that liking for healthy foods can be enhanced much more quickly in infants than in young children aged 2 or more. However, it may also reflect a short-lived phase of easy acceptance, since 2- to 5-year-olds show varying degrees of neophobia for many classes of food, especially meat, fruit, and vegetables, though not sweet and starchy staples (Cooke et al., 2003). This cautious eating is thought to be an adaptation to reduce risk of poisoning in toddlers starting to explore the world (Wright, 1991; Cashdan, 1998).

In adults, Mela et al. (1993) investigated whether exposure to consumption of reduced fat cheese or potato chips by extended home use would increase liking and intake of those foods: in fact, there was no evidence for an improvement due to such exposure. It is possible that some negative effect of a mismatch between expectations of high energy delivery normally obtained from full fat versions and the lower energy actually received may have counteracted any benefit of exposure. There is also a considerable risk with such studies that the monotony of having to eat the same food repeatedly will reduce liking for the food (Zandstra et al., 2000b): in the latter study, involving eating a meat sauce once a week for 10 weeks, effects of monotony were mitigated by allowing a choice of three flavors of the sauce, compared to no choice. Encouragingly, one study found an improvement in liking and consumption of a bland low salt bread, following repeated intake over 5 days (Zandstra et al., 2000a)—in this case, there would be no conflict with energy expectations, and salt-reduced bread is healthier.

B. Associative Learning

Associative learning is a powerful mechanism by which an organism acquires knowledge about the relationship between events, and is thought to be the principal form of learning that influences eating (Booth, 1977b; Rozin and Zellner, 1985). In its simplest form, it is the memory that a perceived "cue" has a particular "consequence." A cue might be an event perceived in the external or internal environment (a stimulus). If the association between the cue and the consequence is remembered, then the cue has been "conditioned" (a conditioned stimulus, CS) or reinforced to the association. The reinforcing consequence is the unconditioned stimulus (US). In some cases, "stimulus substitution" may occur, when the unconditioned response (UR) that would normally follow the US (e.g., saliva secretion in response to food) is now elicited by the CS (e.g., a sound previously paired with food presentation), so that salivation is now a conditioned response (CR). However, the CS–US contingency can also elicit an anticipatory CR that is quite different from any UR induced by the US. These are instances of classical or Pavlovian conditioning. Where the CS or US does not involve ingestion, tight temporal contiguity between CS and US is normally required to see such learning; however, for ingestive behavior, associative learning has unique properties that allow much longer delays

between, for example, a taste and its consequence, whether an aversion or a preference is being learned (Rozin and Zellner, 1985; Chambers, 1990) (but see discussion of flavor-flavor and evaluative conditioning in Section VII).

For ingestive behavior, learning typically involves a two-stage process (Rolls, 1999). The first is the form of stimulus-reinforcement association discussed before that teaches the animal the value of food-related stimuli. The second is learning what appropriate action is required to obtain the food reinforcer. In this type of action–outcome learning, known as instrumental or operant conditioning, internal or external stimuli can act as discriminative stimuli (S^D), predicting that the reinforcing outcome will follow the action, and strengthening the cue-consequence memory (Dickinson and Balleine, 1994; Davidson and Benoit, 1996). Also, classically conditioned CSs previously paired with a food can enhance instrumental responding for that food during food deprivation ("Pavlovian-instrumental transfer"; Holland and Petrovich, 2005). Reinforcement should not be confused with "reward": it is the memory of the association between events that is reinforced, so that if a CS is perceived it will evoke a memory of the properties of the food-related US, and a CR is likely to follow. The reinforcing consequence may not necessarily be pleasurable, and could involve aversive motivational states.

Evidence that learning has occurred depends on measuring a change in behavior in appropriate test conditions. In general, this involves demonstrating a CR to a CS, or an operant response to a S^D, in the absence of the reinforcing US. This is known as testing in *extinction*. The learned response will usually be *extinguished* on repeated testing in this way, since the association of CS to US is no longer reinforced. However, some paradigms produce learned responses that appear to be very resistant to such extinction (Sclafani, 1997).

IV. THE LEARNED APPETITE FOR ENERGY

The physiological need for energy, and especially for sufficient availability of glucose to ensure a constant supply to the brain, probably provides the primary motivation to eat. Energy can be derived from any of the three macronutrients, carbohydrate, protein, and fat, available in a multitude of physical forms. The extent to which a given macronutrient in a food is metabolized to energy, rather than stored or used for growth or other physiological function, depends on the current nutritional status of the eater, as well as the nutrient content of the food. Thus, human beings and other animals cannot rely on some invariant sensory cue to decide how much to eat of what food, to satisfy their immediate needs.

Instead, we have to know about the consequences of eating a particular food in our current state, and how that state will change on eating. So, we must learn to associate the taste, texture, and/or smell experienced while eating the food, with postingestive sensations or changes in state. Removal of an energy deficit as the food is absorbed will be a positive consequence (US) that will reinforce an appetite, or preference (CR), for flavors (CSs) in the eaten food. Thus, we do not strictly have an appetite for energy, but rather a learned appetite for food providing that energy.

There is now considerable evidence that just this sort of cue-consequence associative conditioning of flavor preferences (also called flavor-consequence learning; FCL) occurs in many animals (Booth, 1985; Provenza, 1995; Sclafani, 1997), including people (Booth et al., 1994). In general, a hungry animal will prefer flavors associated with rich sources of energy, and the hungrier the animal, the more sensitive to the caloric consequences of the food it becomes (Gibson and Wardle, 2001). Moreover, rats can learn to select flavors that provide them not just with energy but with the specific macronutrient currently in

deficit (Baker et al., 1987; Davidson et al., 1997). So far, there is little evidence for metabolic selectivity in human eating behavior, other than for a learned protein-need dependent flavor preference and need-dependent sodium appetite (see Section VI).

There was early evidence in adults that adding energy as starch to soup could increase liking for the flavors of that food, at least when tested hungry (see Section V.A.; Booth et al., 1982). Subsequently, evidence for flavor preferences conditioned by pairing with increased postingestive energy came from studies in young children. Birch et al. (1990) showed that, after 8 trials, children learned to prefer flavors of drinks with extra energy from carbohydrate, compared to flavors in low energy drinks (150-kcal difference). This was followed by evidence from the same group that 2- to 5-year-old children learned to like novel flavors paired with high fat energy-rich yogurts after 8 trials, but not flavors in low fat yogurts having 110 kcal less energy (Johnson et al., 1991). Similarly, in 3- to 4-year-old children given 12 conditioning trials, high fat yogurt flavors came to be preferred over fat-free yogurt flavors (energy difference, 162 kcal) (Kern et al., 1993). In this study, one group of children just tasted the yogurts at each exposure (mere exposure control), and this resulted in an increase in liking for both high and low energy flavors.

Children are probably more sensitive to quite small differences in energy delivery than adults, but even among children, increasing age seems to be associated with a weakening of this sensitivity (Cecil et al., 2005). One reason may be increasing dominance of cognitive strategies based on attitudes, and also cultural habits, that develop with aging (Westenhoefer, 2002). This energy sensitivity in young children is illustrated by the finding that 4- to 5-year-old children's preferences for fruit and vegetables were strongly correlated ($r = 0.65$) to the foods' energy densities (Gibson and Wardle, 2003) (see Fig. 1). This is despite the

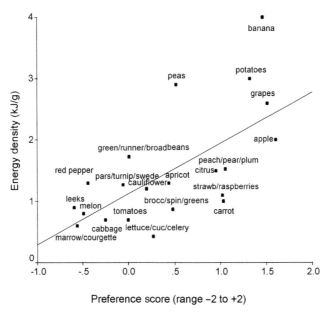

FIGURE 1   Energy density versus 4 to 5-year-old children's preference scores for fruit and vegetables ($R^2 = 0.42$; $r(21) = 0.65$, $p < 0.001$, one tail). Preference scores were averaged from mother's ratings for their children ($N = 228$–416, depending on the number eating the food). Controlling for protein or sugar content (vegetables only) did not remove the correlation. With permission from Gibson and Wardle (2003).

fact that the most energy-dense food in this study was the banana (4 kJ/g), which is one-fifth as energy dense as chocolate. The correlation was shown not to be due to sugar (for vegetables) or protein content. A similar finding across a range of foods was reported more recently for 2- to 3-year-old children choosing their own food in a nursery canteen (Nicklaus et al., 2005): Children most often chose foods with greater energy per portion (see Fig. 2). It is most likely that these findings reflect learned preferences conditioned by postingestive energy absorption.

By contrast to these successful studies in children, clear evidence for learned flavor preferences in adults conditioned by pairing with absorption of energy is scarce. One of the difficulties with studying learned eating behavior in adults is the extent of prior learning and experience that each participant brings to the study. It is known that prior associations with stimuli, or even exposure to a stimulus without

reinforcement, can inhibit new learning (Holland and Petrovich, 2005). Thus, even if novel flavors are used as the experimental CSs, other cues from the foods or drinks carrying these flavors and disguised energy differences may be predictors of postingestive outcomes from prior learning, making new flavor-energy associations harder to learn (Yeomans, 2006). Another issue is that any learned preferences conditioned by energy are more likely to be expressed when energy is needed, that is, when hungry (Booth et al., 1994). Evidence for this is considered in the next section.

V. LEARNED MODULATION OF APPETITE AND MEAL SIZE BY ASSOCIATED STATES

A. Learned Control of Meal Size

It is one thing to acquire a liking for food flavors predicting a good source of energy,

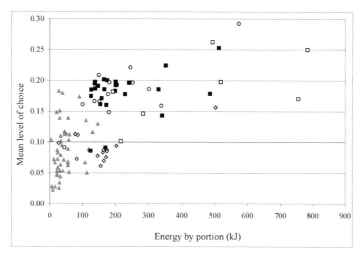

FIGURE 2　Mean level of choice for a food item versus energy per portion, for foods freely selected by 2 to 3-year-old children ($r(106) = 0.65$, $p < 0.001$). (Ratio of portions of food chosen relative to total portions eaten, averaged over all children for each food.) ▲: vegetable; ○: starchy food; ◇: dairy product; ■: animal product; □: combination food. With permission from Nicklaus et al. (2005).

as described before. It is another to know how much of such food to eat, in order to satisfy our current energy needs. It might be supposed that we could rely on a physiological signal, such as a gut hormone, that is released, or reaches a critical level, when enough energy has been absorbed. However, the system needs to be more adaptable than that, and the reason is simple. Consider first that a typical meal (say eaten at a work canteen rather than the occasional luxurious feast) is eaten within 20 min. Yet, the constraints of digestion, and physiological adaptation, mean that food is emptied from the stomach into the upper gut for absorption at a rate of only about 2–3 kcal/min (for a mainly solid meal) (Carbonnel et al., 1994). In other words, a 600-kcal meal will take 3.5 to 5 h to empty from the stomach. For most animals, it would not be adaptive to linger for hours over a meal, so the solution is to learn to anticipate how much energy a particular amount of a food will provide, so that a meal can be completed quickly and well in advance of absorption of all the energy. This principal was first proposed half a century ago by Le Magnen (published in English,

1999), who realized that ending a meal was in part a response to cues predicting later energy absorption consequent on what had just been eaten.

What might these cues be? At the time the eater decides to end a meal, at least three changes in state are signaled to the brain, other than arrival of nutrients into the circulation. One is the oral stimulation from food in the mouth, or the recent memory of it, and perhaps the time taken, allowing some memory of amount swallowed. Another is distension of the stomach by the large volume of food eaten but not yet emptied (Hoad et al., 2004). A third is the detection, by neural receptors in the gut wall, of the rate of passage of nutrients from the stomach into the duodenum (Powley et al., 2005). Through a lifetime's experience, particular levels of these stimuli come to predict particular amounts and effects of subsequently absorbed nutrients.

Therefore, a large part of what ends a meal is a learned rejection of food in the presence of particular internal states generated by eating food, such as gastric distension. Booth called this learned control of meal size "conditioned satiety," and he and colleagues have provided substantial evidence for such learning in rats and primates (Booth, 1972, 1985; Booth and Grinker, 1993). More recently, Gibson and Booth (2000) showed that rats would reject an odor previously paired with concentrated carbohydrate solution in preference for a novel odor, but only in the late stages of a meal. The design also excluded the possibility that the meal-end odor rejection was due to habituation, familiarization, or preference alternation. They argued that this conditioned satiety provides the only known mechanism by which the volume of food consumed can be adjusted to provide a needed amount of energy.

Booth et al. (1976) provided the first evidence for learned satiation in man. In their study, participants drank 100 ml of starch solution prior to eating a lunch which included a particular flavor of yogurt-based dessert. On one occasion, the drink was 65% starch, which was paired with one flavor of the dessert, and on another the drink was 5% starch, which was paired with a different flavor of dessert. Initially, lunch intakes after each type of drink did not differ, but with a few repeated pairings, participants ate less after the high energy drink than after the low energy drink. Furthermore, when the energy of the drinks was subsequently equated at 35% (extinction testing), the participants still ate less of the meal with the flavor of dessert previously associated with the high energy drink, and this difference was due to changes in the later stages of the meal.

In a later study, after 2–4 pairings of soup flavors and dessert flavors with starch-augmented high or low energy content (differences of 190–280 kcal), learned starch-paired flavor-dependent suppression of intake was seen for energy-rich soup flavors, but not for dessert flavors (Booth et al., 1982). In a second experiment, using these same foods, rated appetite for the energy-rich flavors increased when eaten hungry, but decreased when eaten without hunger (Booth et al., 1982). A similar result

was found for preschool children, in whom flavor preferences conditioned by pairing with high fat yogurt were diminished when the children were satiated (Kern et al., 1993): by contrast, increased liking due to mere exposure did not vary with hunger state. Thus, like rats, human appetite for a food can depend on a learned association between a food's flavor, its aftereffects, and the internal state in which it is eaten.

Booth (1985) interprets these state-dependent effects as evidence that the internal state cues and food sensory cues are configured into a Gestalt stimulus complex governing the conditioned response. That is, the learned appetite CR is not fully expressed unless all components of the associated internal and external cues are present. An alternative model proposed by Davidson (2000) holds that the internal state, during which sensory cues come to predict postingestional effects, acts as an "occasion setter," that is, a contiguous or contextual stimulus that modulates associative strength between a CS and US. Thus, the eater has learned that a particular meal has a particular reinforcing consequence when eaten in a particular state, for example, a level of hunger, and any deviation from those circumstances alters the learned response. This is similar to the control of appetite that environmental contexts can acquire after repeated association with a particular eating experience (Weingarten, 1983; Holland and Petrovich, 2005), so that desire for a food or drink might be elicited by a particular room or social gathering. Such cue-potentiated eating has been shown in preschool children, who were repeatedly asked to eat snacks in the presence of one set of visual and auditory stimuli, or even a location (CS+), and did not eat in the presence of another set of such stimuli (CS–) (Birch et al., 1989). To test for learned control of appetite by the paired stimuli, children were first sated with a snack and then exposed to either CS+ or CS– stimuli, on different days. Intake increased, and latency to eat decreased during exposure to CS+ versus CS−, at least for children who were explicitly aware of the CS–snack pairing (see Section VIII). These models allow learning of adaptive context-specific food choice and meal size control, so that appetite and satiety largely reflect acquired "metabolic expectancies" (Booth, 1977a).

Related evidence for learned control of meal size comes from two studies in preschool children. In the first study, children were given repeated experience of consuming preloads with disguised differences in energy paired with particular flavors, followed by ad libitum snack intake (Birch and Deysher, 1985). In extinction testing, with no energy difference in the preloads, children consumed more of the snack foods following the low-energy-paired flavor than the high-energy-paired flavor. In the second study, there was similar evidence for differential intake following preloads with flavors previously paired with high or low energy content. However, in this case the learning trials occurred in one of two conditions: one focusing the children's attention on their internal cues for hunger and satiety, and the other focusing on external cues, including rewards for eating. These latter children showed no evidence of learned responsiveness to flavor cues predicting caloric differences (Birch et al., 1987b). This effect of external, including social, contexts distracting from learned control of eating might underlie the decline with age in ability to compensate for recent energy consumed by reducing intake in a subsequent meal (Birch and Deysher, 1986; Cecil et al., 2005).

There has been a surprising lack of direct evidence for learned control of meal size in adult human beings, perhaps for want of trying, or misunderstanding of the theory (Booth, 1990). Zandstra et al. (2002) provided one of the few studies in adults to investigate effects of covert manipulation of energy intake on learned responses to associated flavor cues. Participants had 20 exposures to differently flavored high and low energy yogurt drinks (206-kcal difference).

There was no change in compensation to the energy content over the exposure trials, but when the flavor–energy pairing was covertly switched, participants increased their energy intake after the high energy yogurt drink containing the flavor that was previously coupled with the low energy yogurt drink. However, switching to the high-energy-paired flavor did not reduce subsequent intake. So there was some evidence for learned control of intake, but this was biased toward "desatiation," rather than satiation.

A more recent study found evidence suggestive of such learning for intake of a breakfast food: Yeomans et al. (2005b) compared ad libitum consumption of high and low energy versions of porridge before and after two training experiences of fixed intake of these versions with distinct arbitrary flavors. On the first occasion, participants ate the same weight of both versions on average. After training, intake and pleasantness of the low energy version increased significantly (learned desatiation), whereas pleasantness of the high energy version decreased. This latter result may have been induced by aversive oversatiation due to eating a large fixed amount of the energy-rich breakfast during training.

The results of that study illustrate an issue that has confused interpretation of much research in this area; that is, short-term intake studies find that people consistently eat meals based on weight or volume of foods, despite considerable, usually disguised, variation in energy density (Stubbs and Whybrow, 2004; Westerterp-Plantenga, 2004). Since eating is habitual, the amount eaten (and associated sensorimotor cues) is usually an approximate (and learned) predictor of energy intake, on which, in the novel context of a laboratory study, participants may be happy to rely. So energy intake will follow disguised manipulations of energy density, as is found (Yeomans et al., 2005b). Gibson and Booth (2000) argued that it is more meaningful to explain this as "active" regulation of amount eaten, through learning, rather than "passive" overconsumption of energy (e.g., Blundell et al., 1995). The point is that with repeated experience, new learning linking the dietary cues with postingestive metabolic changes will allow adaptation of portion size, so that energy intake is constrained again. This idea is now more widely accepted, even if the mechanisms have not been fully appreciated (Stubbs and Whybrow, 2004). Of course, sensory and nutritional properties of the foods, and genetic and psychological characteristics of the eaters, will add variability to this regulation (De Castro, 2001; Westerterp-Plantenga, 2004; Yeomans et al., 2005b).

B. Regulation of Food Choice by Association with Internal State

This ability to form associations between sensory properties of foods and their postingestive consequences in concurrent states should have implications for the way in which particular eating habits influence our desires for particular foods. Certainly, there is substantial evidence in animals that more energy-dense foods become preferred when hungry, and vice versa (Booth, 1985). Similarly in women, Booth et al. (1994) reported a study of the effects of one-trial learning of a preference for a yogurt flavor paired with high or low energy versions. The women ate each version once on separate days either before lunch or after lunch, and preference change was measured before and after those days as rated pleasantness to eat a spoonful "right now" of mid energy mixtures of the versions, but with either the high-energy-paired or low-energy-paired flavor. As with rats (cf. Gibson and Booth, 1989), desire to eat the high energy flavor increased when hungry in women who had eaten the yogurts when hungry but not if they ate them when full. And vice versa, women learned to prefer the low energy flavor when full, only if they had previously experienced that flavor when full. Interestingly, women classified

as "unsuccessful dieters" did not show this learned FCL discrimination (see evidence for lack of flavor-flavor learning in dieters, Section VII.B).

Inspired by these findings, Gibson and Desmond (1999) tested the hypothesis that craving (or a strong appetite) for chocolate could be reduced if chocolate was eaten for a period only when satiated, and conversely could be intensified by eating chocolate exclusively when hungry. They asked people to rate their craving for milk chocolate, while looking at but prior to eating a piece, before and after 2 weeks of eating chocolate twice per day either exclusively when hungry or full. Also, they made these test ratings on separate days when actually hungry or full. As predicted, the "hungry" group showed increased craving after 2 weeks, when rating during hunger but not when full. Moreover, the "full" group showed a substantial reduction in craving, which was apparent, for self-confessed chocolate cravers, even when now currently hungry. Thus, the hunger state in which an energy-rich food such as chocolate is habitually eaten will determine the strength of appetite for that food, and craving can be moderated by avoiding eating chocolate when hungry.

Such findings have relevance to therapeutic interventions for eating disorders and weight management: for example, binge-eating disorder has been characterized as a disorder of learned control of eating (Jansen, 1998), and both binge eaters and patients with bulimia nervosa show exaggerated learned appetitive cephalic phase responses to favorite food stimuli (Vögele and Florin, 1997; Legenbauer et al., 2004). Moreover, food cravings in binge eaters can be substantially reduced following exposure to food cues without eating (Jansen, 1998). Conversely, binging and comfort eating of sweet fatty foods may be learned stress-reducing coping strategies (Hagan et al., 2002; Dallman et al., 2005).

A question relevant to healthy eating interventions and dietary management of weight is whether appetite for healthier low energy foods such as fruit can also be enhanced by habitually eating such food when hungry. This was addressed in a study similar in design to the chocolate-craving study, except that the food was dried fruit bars and participants ate them either when hungry or full for just 1 week (Gibson and Wardle, 2001). In contrast to chocolate, for this low energy food, eating it only when hungry actually reduced the desire to eat the fruit bar. However, in this "hungry" group, intake of the bar increased after the week, but only when currently full. This suggests learned desatiation, that is, participants have learned that they can easily eat a little more of these novel low energy fruit bars when full. Happily, there was at least one benefit from eating fruit when hungry: the preference for the flavor of fruit bar eaten increased relative to others not eaten. Interestingly, this was only true for participants who already regularly ate fruit as a snack, which may limit the usefulness of this approach for encouraging consumption of low energy foods in those without this habit. However, there is some evidence to suggest that habitual fruit snacking may help extinguish the learned hunger-dependent increase in appetite for energy-dense foods described before (Gibson and Wardle, 2004), which could be of considerable benefit in limiting excess energy intake.

The converse explanation is that habitual snacking on more energy-dense higher fat foods, instead of fruit, increases the ability of hunger to elicit a strong appetite for such foods. This seems compatible with a more recent paper describing an important new explanation for the current obesity epidemic (Davidson et al., 2005). The theory proposes that the failure to regulate energy intake reflects an impairment of learning, and in particular involves dysfunction of the hippocampus. It is argued that the hippocampus normally inhibits learned appetite for rewarding foods when appropriate for regulation, such as when satiated

(as in the occasion-setting model described earlier). It is further proposed that high intake of saturated fats, as often found in energy-dense foods, leads to impairment of the hippocampus. Thus, a vicious circle is created, whereby high fat energy-rich foods promote overeating through neurological disruption of dietary learning, which leads to obesity.

VI. NUTRIENT-SPECIFIC LEARNED APPETITES

In a wide variety of species, including both vertebrates and invertebrates, there is evidence that animals learn to choose diets that correct specific nutritional deficiencies, including for vitamins, minerals, essential amino acids, or protein (Rozin and Schulkin, 1990; Kyriazakis et al., 1999; Gibson, 2001; Mayntz et al., 2005). However, evidence for nutrient-specific appetites in humans is scarce, and essentially limited to appetites for salt (sodium) and protein.

A. Sodium Appetite

The crucial point about sodium is that the salty taste of a food is closely related to its sodium content; that is, for sodium, there is a sensory cue in the food eaten that predicts absorption of the nutrient. In that case, it is reasonable to expect that an unlearned appetite for saltiness (the taste of sodium chloride) may have evolved where deficiency can occur, so that a sodium-deficient animal would select and consume saltier foods. Retention of sodium depends on hormones released from the adrenal glands. Indeed, it appears that rats express a rapid and unlearned appetite for sodium when lacking that micronutrient (Denton, 1982), nevertheless, learning may still have a role to play in adaptive appetite for sodium in the rat (Dickinson, 1986).

It would seem reasonable to ask whether human beings have an innate appetite for sodium, as we are also omnivores in whom sodium is an essential nutrient. It is hard to answer this question definitively in people. Evidence from twin studies, which help to separate environment from inheritance, do not support genetic control of sodium appetite, although the studies were methodologically weak (Beauchamp, 1987). The oft-cited classic case of craving for salt by a child with adrenocortical insufficiency reported by Wilkins and Richter (1940) could be a case of innate appetite for sodium when physiological need is pathologically high. However, it is not clear whether the appetite was innate or rapidly learned from the benefits to his health of eating salty food.

By contrast to a lack of clear evidence for an innate sodium appetite in humans, evidence for the influence of experience is strong. Newborn infants do not seem to like salted water, and infants aged 3 to 4 months appear quite indifferent to salt solutions over plain water, unlike their clear preference for sugar solutions at that age, or the preference for both sugar and salt solutions over water seen in older children (Beauchamp, 1987; Bartoshuk and Beauchamp, 1994). There is probably a maturational change in infants' abilities to taste salt. Adults typically enjoy salty foods but not salty water, and this preference is seen to develop in children by about 3 years of age (Beauchamp, 1987), likely reflecting contextual experience of salty tastes. Also, 6-month-old infants previously fed high or low sodium cereal preferred the salt level to which they had been exposed (Harris and Booth, 1987). In adults, if dietary sodium is either raised or lowered for several months, the preferred salt level shifts in the direction of the manipulation (Bertino et al., 1983; Blais et al., 1986; Beauchamp, 1987). These are examples of learning through nonassociative exposure (see Section III.A).

Clearly then, our rather excessive preferred salt level is influenced by the levels to which we are exposed in our diet. Nevertheless, there is more recent evidence that adult human salt appetite can be determined by neonatal or even fetal physiological experience. That is, if the developing baby is exposed to a loss of minerals including sodium and chloride, whether through vomiting and diarrhea (Leshem, 1998), maternal vomiting during pregnancy (Crystal et al., 1999), or deficient infant diet (Bertino et al., 1983; Blais et al., 1986; Stein et al., 1996), salt appetite may be enhanced into adulthood. Presumably, some permanent change in a neurohormonal system has occurred as a result of the mineral insufficiency, to enhance intake of sodium. However, this "fetal learning," or early imprinting, does not appear to adapt again once the insufficiency is removed. This contrasts with some more recent evidence for adaptive changes in sodium appetite in adults: in particular, salt loss through exercise increases appetite for sodium, which also aids hydration, and this sodium need can act as a reinforcing US to condition associated drink flavor preferences (Wald and Leshem, 2003). Thus, human sodium appetite may have innate, physiological and learned attributes (Kochli et al., 2005).

Interestingly, appetite for salt, in rats at least, is also enhanced by other mineral deficiencies, especially calcium (Tordoff, 1996). It has been suggested that salty taste might be a marker for the presence of a number of essential minerals (Schulkin, 1991). Yet, salt appetite seems to be enhanced by protein deficiency too (Torii et al., 1998). Perhaps any state of nutritional deficiency to some extent arouses the only innate appetite the brain can call on. This may be of relevance to the current obesity epidemic, given the increasing prevalence of salty, energy-dense foods, together with a possible dilution of intake of essential nutrients, including protein (see later).

B. Human Appetite for Protein

Unlike sodium, protein content of foods does not provide a consistent sensory indicator that could allow expression of innate protein appetites adapted to need. Although it has been argued that the so-called savory taste, umami, might provide such a "protein taste" (Rolls, 1999), behavioral evidence in both rats and humans does not support this notion (reviewed by Gibson, 2001), although such a flavor could become a learned cue to protein. Instead, in many species, there is clear evidence that animals can learn to select diets that supply them with essential amino acids, or protein, when these nutrients are lacked (Booth, 1974; Gietzen, 2000). Regarding human beings, we will address two questions here: first, what evidence is there to suggest that people can, in certain circumstances, make adaptive changes in their food preferences to provide them with protein when that nutrient is lacking, that is, are human beings capable of acquiring and expressing a need-dependent appetite for protein-rich foods; second, could such an ability have any relevance to eating patterns and weight regulation outside of the laboratory?

There are several studies supporting expression of a need-dependent protein appetite. People with lower blood indices of protein status preferred the flavor of greater concentrations of casein hydrolysate added to an amino acid deficient soup, while more protein-replete subjects preferred casein-free soup (Murphy and Withee, 1987). Similarly, human infants recovering from protein–energy malnutrition, but not healthy controls, preferred a soup to which casein hydrolysate (which has a strong sour-savory flavor) had been added (Vazquez et al., 1982). The question of whether such a protein appetite can be learned was addressed experimentally. When people ate one meal of adequate protein, following a low protein breakfast and overnight fast, they preferred the (arbitrary) flavors associated with that meal (particularly the desserts) compared to flavors associated with a low protein meal (Gibson et al., 1995). Furthermore, on a second test day, this protein-paired flavor preference was abolished specifically by a high protein preload, just as it had been in rats (Gibson and Booth, 1986; Baker et al., 1987). Therefore, in appropriate conditions, human beings can learn a protein-specific appetite for a food flavor, which depends on the presence of a mild protein need.

It is not known whether or how we consciously perceive the reinforcing consequences of eating protein when deprived of it. However, one possibility is suggested by a more recent finding from a study of hormonal and mood responses to protein-rich lunches (Gibson, 2003). When people who have been eating a low protein diet for 5 days were given a protein-rich lunch, a substantial increase in positive arousal occurred 2–3 h later. The change in mood was not immediate or even 1 h after eating, and so is not likely to be merely satisfaction from the sensory properties of the meal. Rather, induction by some internal cue arising from amino acid absorption seems more likely. Indeed, the extent of increase in arousal correlated with the increased secretion of the glucocorticoid hormone, cortisol, which occurred immediately after the meal. This rise in cortisol is a recognized effect of meals, particularly at midday, when the normal circadian decline in cortisol is rapid (Ishizuka et al., 1983). It is probably related to control of nitrogen metabolism, gluconeogenesis and proteolysis (i.e., catabolism of protein), and is not seen after very low protein meals (Gibson et al., 1999). In fact, the lower the amount of protein consumed in the 5 days prior to the protein-rich lunch, the greater the rise in cortisol (Gibson, 2006). However, cortisol has profound effects on neuronal function in limbic areas of the brain (De Kloet et al., 1993), linked to changes in mood state and memory (Lupien et al., 1999).

Does this have any relevance to regulation of appetite and body weight in free-living people? There are several observations that suggest it does. Fifty years ago, it was observed that changes in appetite during the day correlated with levels of amino acids in the blood, but not with blood glucose levels (Mellinkoff et al., 1956). More recently, it was found that people chose to eat more protein in the 4 h following a high energy but protein-free drink than after a very low energy placebo (Stockley et al., 1984). Moreover, it is well-established that, calorie for calorie, protein-rich meals suppress subsequent intake more effectively than high carbohydrate or fat-rich meals (e.g., Weigle et al., 2005). Collectively, these findings suggest that human protein intake is, as for other animals, defended over and above that of carbohydrate or fat. Indeed, more recent analyses of a range of experimental and epidemiological data support this conclusion (Simpson et al., 2003), and Simpson and Raubenheimer (2005) further argue that a significant cause of the current obesity epidemic is that defense of protein intake drives a feed-forward cycle underpinned by increasing carbohydrate and fat intake, adiposity, and hepatic gluconeogenesis.

VII. FLAVOR-FLAVOR LEARNING

Consider a typical meal; it might contain a meat serving, several types of vegetable, and perhaps a dessert. One possibility is that the meal forms a perceptual "Gestalt." That is, an association is formed between its combined sensory properties and its combined postingestive effect. However, more often than not we are able to offer an evaluative opinion about the individual components of a meal, not just the collective whole. Indeed, this appears to be the case even when these components are consistently presented in combination with other foods (e.g., ketchup!). Clearly, these preferences play an important role in everyday food choices. But how do they arise?

The most likely answer to this question is flavor-flavor learning. The underlying logic is simple. Certain flavors convey important information about "biological usefulness," as exemplified by newborn babies' liking for sweet tastes and aversion to intense bitter and sour tastes (Berridge,

2000) (Section II). Flavor-flavor learning offers a way to use already established (innate or learned) likes and dislikes to develop appropriate affective judgments about newly encountered flavors. Rather than forming a relationship with the action of food in the gut, flavor-flavor learning represents a kind of "shortcut" (Brunstrom et al., 2005). The basic idea is couched within an understanding of classical conditioning. When a novel flavor (CS) is presented repeatedly in combination with an already liked flavor (US), then the CS becomes liked. Conversely, when a CS is paired with an unpleasant US, then it adopts the affective quality of the US, and becomes disliked.

A. Evidence for Flavor-Flavor Learning

For many individuals, food valence is the primary determinant of food consumption (Drewnowski, 1997). Therefore, to understand individual differences in food choice it is critical that we appreciate how flavor preferences are acquired in the first place. Unfortunately, as with other forms of dietary learning, researchers have become bewildered and frustrated by a failure to produce reliable examples of the phenomenon in the laboratory. Consequently, many important questions remain unresolved. However, as we shall see, more recent developments might hold the key to understanding why these problems have occurred. They also highlight reasons why researchers should pay greater attention to this particular form of dietary learning.

1. Studies in Animals

Several studies report flavor-flavor learning in animals (e.g., Holman, 1975; Lavin, 1976; Fanselow and Birk, 1982). In particular, this work has taught us quite a lot about how flavor-flavor learning differs from FCL. For example, unlike in FCL, flavor-flavor associations are not formed when there is a delay between the cue flavor and a palatable reinforcer (Sclafani and Ackroff, 1994). FCL is primed by food deprivation whereas flavor-flavor learning is not (Fedorchak and Bolles, 1987; Capaldi et al., 1994). Finally, FCL and flavor-flavor learning have a different underlying neurochemistry, because dopamine antagonists have differential effects on the acquisition and the expression of associations (Yu et al., 2000; Azzara et al., 2001).

2. Studies in Human Beings

Zellner et al. (1983) are credited with the first demonstration of flavor-flavor learning in humans. Over three experiments, they found that preferences for flavored teas could be conditioned using sucrose as a sweet-tasting US. After an initial set of hedonic ratings, two teas were selected to be differentially reinforced. One was presented 24 times with the US added (CS+) and the other was presented 24 times on its own (CS–). After this training, and in the absence of the US, the subjects provided a second set of hedonic ratings. Zellner et al. analyzed the change in hedonic response that occurred and found a greater increase in liking for the CS+ relative to the CS–. Indeed, this effect was present when the teas were reevaluated 1 week later. Unfortunately, since the publication of their work, only one study has come close to replicating this basic effect (Baeyens et al., 1990) and another has reported a failure (Rozin et al., 1998). By contrast, the development of learned dislikes appears to be more reliable. In several studies, participants have been exposed to a novel flavor (CS+) combined with an unpleasant tasting US (Tween 20—a highly unpleasant "soapy" taste). After this training, participants tend to show a relatively greater decrease in liking for the CS+ compared with an unpaired control (CS–) (e.g., Baeyens et al., 1990, 1995a, 1998).

B. What Do We Know (and Need to Know) about Flavor-Flavor Learning in Human Beings?

Most researchers (e.g., Baeyens et al., 1990; Rozin et al., 1998) regard flavor-flavor

learning as an example of "evaluative conditioning" (Martin and Levey, 1978). This term refers to a general phenomenon whereby the valence of a novel target (CS) comes to be altered by the valence (positive or negative) of stimuli or events (US) that are presented contingently or spatio-contingently with the US. Although there has been considerable debate around the underlying mechanism (De Houwer et al., 2001; Field, 2001; Fulcher and Hammerl, 2001), evaluative conditioning has been demonstrated successfully using a range of CSs and USs, including odors (Todrank et al., 1995), faces (Baeyens et al., 1992), and tactile stimuli (Hammerl and Grabitz, 2000).

Evaluative conditioning is also quite unusual. In a typical Pavlovian paradigm a sounding bell (CS) might be paired with a mild electric shock (US). Repeated CS–US presentations will establish an associative link between the CS and the US. When the CS is presented subsequently, it elicits a memory for the US, which in turn leads to an appropriate and observable conditioned response (CR; e.g., a startle reflex). In this context, the CS activates an *expectation* of the physical occurrence of the US. By contrast, evaluative conditioning has been described as *referential* (Baeyens et al., 1995b; Diaz et al., 2005), because the CS activates a representation of the US but without the expectation that something is going to happen.

Evaluative conditioning also differs in other important respects (for a review see De Houwer et al., 2001). For example, both flavor (Harris et al., 2004) and nonflavor (Diaz et al., 2005) conditioning appear to be resistant to extinction. This claim contrasts a fundamental prediction from associative learning theory (e.g., Rescorla and Wagner, 1972). Similarly in most forms of associative learning it is normal for learning to occur at least in conjunction with awareness of CS–US relationships (Lovibond and Shanks, 2002; Shanks and St. John, 1994). However, in both flavor (Baeyens et al., 1990) and nonflavor evaluative conditioning (e.g., Hammerl and Fulcher, 2005) this is not necessarily the case (for a more general discussion on the role of CS–US contingency awareness in dietary learning see Section VIII). More recently, there has been considerable debate around these issues. However, there is a general agreement that paradigms lack reliability, and that this is probably not due to lack of statistical power or to obvious design problems (e.g., Rozin et al., 1998). Instead, problems with reliability are more likely explained by the fact that there are other idiosyncratic features that modulate valence learning (see De Houwer et al., 2005). In relation to flavor-flavor learning, a few possibilities have been rejected.

First, one explanation for the difficulty in finding evidence for learned likes is that sweet-tasting USs are less potent, possibly because sweetened solutions are not liked as much as an unpleasant USs are disliked (Baeyens et al., 1990). It also seems that people differ in their liking for sweet tastes (Looy and Weingarten, 1991; Looy et al., 1992). Some people prefer a highly sweetened solution while others find it aversive. Clearly, this variability could contribute toward a failure to replicate earlier studies (Zellner et al., 1983). Indeed, Yeomans et al. (2004), have more recently reported evidence that supports this proposition. In a closely related form of learning (involving odor–flavor pairings) they found that the resulting direction of affective change (positive or negative) is predicted by general differences in liking for sweet tastes (Yeomans et al., 2004).

A second suggestion is that the outcome of CS–US pairings is determined by factors other than the basic valence characteristics of the CS and the US. In response to the problem of individual differences in sweet liking, Brunstrom et al. (2001), paired novel flavored fruit teas (CSs) with small portions of pleasant-tasting food. In their paradigm, each CS was reinforced either 10%, 50%, or 90% of the time. After conditioning, liking for the CSs correlated with the amount of reinforcement during training. However, this was only the case in unrestrained eaters. Learning was not found in participants who routinely restricted their dietary intake. In a similar study, fruit juices were paired with small portions of chocolate (Brunstrom et al., 2005). The pattern of results was broadly consistent with their earlier work. However, in this study, the restrained eaters ended the conditioning period with a relatively greater preference for the 10% paired CS. This was the case even though restrained and unrestrained eaters indicated a similar liking for the US at the beginning of the experiment. A key implication here is that aspects of our everyday dietary behavior might influence the outcome of flavor-flavor pairings. Based on earlier work (Francis et al., 1997), the authors of this study argued that restrained eaters might be more likely to regard sweet-tasting USs as forbidden. Thus, it would seem possible that the valence of the US was influenced by beliefs and attitudes about the food as well as by its basic hedonic tone. At present, this hypothesis remains to be tested. However, it is consistent with several related findings. For example, in other instances of evaluative conditioning there have been reports that the outcome of CS–US pairings is not determined solely by the simple hedonic qualities of the evaluative stimuli. Instead, it might also be influenced by the extent to which they exhibit a conceptual "belongingness" (Lascelles et al., 2003). There is also evidence that learning is dependent on underlying beliefs about the likely efficacy of the US (Baeyens et al., 2001). Finally, other examples of individual differences have been reported elsewhere, and these have been attributed to particular personality traits that may direct attention to particular aspects of emotional stimuli (Schienle et al., 2001; Sonuga-Barke et al., 2004).

A final suggestion is that the tendency to acquire new flavor preferences might be associated with age (Brunstrom, 2005). A "critical developmental period" has been proposed in relation to obesity in adulthood (Dietz, 1994). Perhaps flavor-flavor associations are formed during a similar period of plasticity? In relation to this idea, it may be relevant that many studies of flavor preference learning have been reported in children (see Sections IV and V). A further possibility is that learning is limited because prior experience with a particular CS–US combination means that behavioral and/or affective responding to a CS is already firmly established, and so the extent to which a CS can form a new association is diminished. Thus, while pictures, music, and so on can all be represented in a wide variety of novel formats, most flavors will remind us of other foods, and this limits new learning. For example, the myriad prior associations of sweetness may inhibit further flavor-flavor associations, or at least their clear expression. By contrast, the bitter soapy taste of Tween 20 would be novel and so easily able to condition changes in reactions to associated flavors. In adults, it may normally be very difficult to create flavors that are both highly novel and sensorially distinct. Notwithstanding this point, evidence for the acquisition of learned dislikes, together with Brunstrom et al.'s examples of learned likes in unrestrained eaters, suggests that learning still remains possible in adulthood. Indeed, in other forms of flavor conditioning (e.g., based on caffeine reinforcement) highly reliable effects are reported (e.g., Yeomans et al., 2005a). On this basis, it would seem that problems with novelty might limit but not exclude all learning in adulthood. Currently, we know very little about the way in which flavor components contribute toward the novelty of a CS, and this represents an interesting area for future research.

C. Flavor-Flavor Learning and Everyday Dietary Behavior—Some Suggestions for Future Research

We have a seemingly limitless capacity to make judgments about our liking for foods.

However, FCL necessarily relies on a systematic pairing of a single food with a single postingestive outcome. In many cases foods are not, and have never been, consumed in this context. Moreover, when one considers the number of food combinations that are consumed at different meals and courses, the opportunity for FCL to take place could be quite limited. When viewed in this context, it would seem that flavor-flavor learning probably plays a critical role in shaping human dietary preferences.

What happens when individuals differ in their liking for basic taste properties? Over a lifetime, could these simple differences have a general effect on the kinds of foods that a person comes to like and choose? In relation to this idea, there is some indication that obese individuals have a greater liking for the taste of high fat and/or high sugar solutions (Drewnowski et al., 1985; Salbe et al., 2004), and this affection is also reflected in their consumption of high fat foods (Macdiarmid et al., 1998). A further possibility is that individual differences are mediated by differences in sensitivity to tastes. An excellent example here is work relating to genetic differences in sensitivity to the compounds PROP and phenylthiocarbamide (PTC) (see Section II). Around 70% of whites in North America and Western Europe taste PROP/PTC as bitter and roughly 30% are identified as nontasters (Bartoshuk et al., 1994). This difference is probably important, because nontasters tend to have a higher BMI (body mass index) (Tepper and Ullrich, 2002; Goldstein et al., 2005). Importantly, nontasters also show a higher preference for high fat foods, and this is likely to be reflected in their diet (for a review, see Duffy, 2004). Although the reason for this relationship is unclear, it might be linked to the fact that nontasters are less adept at discriminating between the fattiness of foods. However, not all evidence suggests that PROP tasters eat more healthily (see Section II): a more recent finding was that adult

PROP bitter tasters found vegetables to be more bitter, less sweet, and ate fewer of them (Dinehart et al., 2006).

In relation to the preceding discussion, the work of Yeomans et al. (2004) is especially relevant, because it represents the first of its kind to show that shifts in valence are predicted by individual differences in liking for sweet tastes. Given evidence for a link between PROP bitter taste sensitivity in children and sweet preference (Section II), in the future it would be interesting to develop this research by comparing evidence for flavor-flavor learning in PCT/PROP tasters and nontasters, or in individuals who choose high or low fat diets. At the same time, it would be useful to extend the work of Brunstrom et al. (2001, 2005) by incorporating measures of beliefs and attitudes about the US. Together, these innovations hold the promise of arriving at a reliable learning paradigm that could offer a useful perspective on the development of particular taste preferences that promote obesity.

VIII. AWARENESS AND DIETARY LEARNING

It has been estimated that around 95% of all dietary learning experiments are conducted using animals (Brunstrom, 2004). This may frustrate researchers who are interested to observe learning in humans. However, to elucidate underlying processes it is important to be able to observe clear and reliable instances of the phenomenon under scrutiny. Researchers of animal behavior are also able to take advantage of pharmacological interventions (Azzara et al., 2001) and brain-lesion procedures (Touzani and Sclafani, 2002) that are regarded as unethical in human beings.

Despite these advantages, one question can only be addressed using human participants. This relates to the role of "awareness" in learning. Specifically, for learning to occur, is it necessary that a person develops an explicit appreciation that a particular CS is, or has been, paired with a particular US? As we shall see, asking this kind of question could be critically important. For one thing, its resolution leads to one of two diametrically opposite conclusions about how learning takes place. Moreover, asking questions about free will and the nature of human agency has implications for an ongoing debate around whom we might blame for associations that promote behaviors leading to obesity.

A. What Do We Know about Awareness and Dietary Learning?

It is self-evident that when we learn something new we are often aware that something has happened. But, is this always the case? Formally, "implicit learning" is identified when it can be demonstrated that new information has been acquired despite having neither the intention to do so, nor the ability to recall what knowledge has been acquired. Over the years, claims about implicit learning have been made in a variety of contexts (Cleeremans et al., 1998). However, with respect to associative learning, the evidence has been less convincing. Instead, associative learning is usually observed only after explicit CS–US contingency awareness has been acquired (Shanks and St. John, 1994; Lovibond and Shanks, 2002), and this appears to be the case across a range of phenomena including autonomic conditioning (Dawson et al., 1985), conditioning in amnesic patients (Kitchell et al., 1986), conditioning under anesthesia (Ghoneim et al., 1992), and conditioning with subliminal stimuli (Shanks and St. John, 1994).

Strangely, it is against this background that researchers of dietary learning have tended to ignore the importance of awareness. Moreover, many procedures are designed deliberately to reduce the possibility that explicit information can be acquired. In part, this represents a sensible attempt to reduce demand characteristics. However, it also reflects an emerging body of evidence that forms of dietary learning might be quite special. In relation to this idea, we briefly review what we know about the role of awareness in separate forms of dietary learning (for a more comprehensive account see Brunstrom, 2004). We then consider ways in which this issue impacts more generally on our understanding of human dietary control.

1. Flavor-Preference Learning and Learned Satiety

In an early study of learned satiety (Booth et al., 1976), participants completed a questionnaire to probe subjective opinion on the taste of CSs and other aspects of the experiment. The authors concluded that there was little evidence that participants knew that intake difference was an important measure. However, the precise details of the participants' responses were not reported. The authors acknowledged the potential limitations of their approach and concluded that the role of awareness in learning remained unresolved.

In most other studies, awareness has tended not to be considered (Booth and Toase, 1983; Birch et al., 1990; Gibson et al., 1995; Johnson et al., 1991; Kern et al., 1993; Tepper and Farkas, 1994) (however, see Booth, 1994, p. 63; Mela, 2001). In part, this probably reflects the fact that many studies have involved testing young children (Birch and Marlin, 1982; Birch et al., 1990, 1987a, b; Kern et al., 1993). An exception here is a more recent experiment in which participants were asked to identify the correct flavored yogurt that had been presented in a high or a low energy condition during a previous training period (Zandstra et al., 2002). Unfortunately, this study found only minimal evidence for learned satiety, and so the role of awareness in learning is difficult to assess. Moreover, if explicit information *is* acquired, then this might remain undetected by a question that only probes views on "high or low energy." Instead, and perhaps more likely, an explicit association will take the form of an expectation based

around the relationship between the taste and textural quality of a CS and the visceral sensations associated with its ingestion.

2. Caffeine-Based Associations

When caffeine is combined with a novel flavor then this produces the same outcome as a flavor–nutrient pairing—the novel flavor becomes liked. However, because learning is attributed to its psychoactive rather than its nutrient properties, strictly speaking, the process is not a form of dietary learning. Nevertheless, at a functional level important similarities exist— preference shifts occur after an association is formed between the sensory properties of a substance and its postingestive effects. For this reason, caffeine-conditioning paradigms offer a useful context within which to explore general aspects of "flavor-postingestive consequences" learning (Brunstrom, 2004). Importantly, caffeine-based research has been studied more extensively than any form of dietary learning. And, evidence to date suggests that CS–US relationships do indeed configure outside awareness (Richardson et al., 1996; Yeomans et al., 1998, 2000a, 2000b). Very recently, implicit learning has been demonstrated using a repeated-measures design (Yeomans et al., 2005a). This is important, because it helps to address a more general concern that results from these studies might represent an artefact, derived from the otherwise exclusive use of between-subjects designs (Brunstrom, 2004, 2005).

3. Flavor-Flavor Conditioning

As noted elsewhere (Lovibond and Shanks, 2002), flavor-flavor conditioning represents another promising candidate for learning outside conscious control. However, because there is a dearth of empirical studies in this area the evidence is not conclusive. In the first published account, a measure of awareness was included in only one of the three studies reported (Zellner et al., 1983), and details of the measure are not discussed. More recently, studies involving both liked (Brunstrom et al., 2001) and disliked USs (Baeyens et al., 1990, 1996, 1998) have reported evidence that learning occurs outside contingency awareness. However, in one study the authors concede that their measure of awareness lacked sufficient rigor and other studies have since been criticized on this basis (see Brunstrom, 2004).

B. Awareness and Its Relevance to Understanding Dietary Control

As we have seen, when we consider general characteristics of associative learning in humans, it seems that dietary learning might need to be treated as a "special case." Indeed, FCL would seem particularly unusual, because most other forms of learning require a temporal contiguity in the range of 500–2000 msec. Of course, these differences are bewildering if we suggest that a single associative mechanism exists. Instead, it might make more sense to assume that separate subsystems have evolved in response to particular needs (for a related point see Garcia et al., 1989). Indeed, one suggestion is that dietary learning is so important that it makes sense to learn in a spontaneous way and this may best be achieved by learning without conscious control (Field, 2000). In relation to this possibility, relevant clues can be found in the literature on food-aversion learning. For example, in animals, aversion learning does not appear to require contingent "experience" of the US (i.e., experience of nausea). Thus, rats will acquire aversions even when they are completely anesthetized, both during, and after a US is administered (Roll and Smith, 1972; although see Pelchat and Rozin, 1982). And, perhaps even more striking, aversions are formed following chemotherapy, and this appears to be the case even when a person is aware that a target food is not the cause of their illness (Bernstein, 1978; Logue et al., 1981).

In our ancient past, dietary learning served us well because it enabled us to orientate toward foods that offer protection against periods when food is scarce. Unfortunately, in societies where inexpensive high energy foods are readily available, this tendency is no longer adaptive. Instead, we acquire preferences that represent risk factors for obesity and weight gain. Set in this context, it is surprising that relatively little attention has been paid to understanding how these "unhealthy associations" are formed in the first place. In particular, we might want to know whether preference changes are an inevitable consequence of food exposure, or whether the underlying process is more complex (see our earlier discussion on flavor-flavor learning). Clearly, asking questions about awareness can be helpful in this respect. Moreover, by developing an understanding of this kind, we might be better placed to develop strategies and health advice aimed at limiting the formation of unhealthy preferences, especially in children.

The question of awareness might also have implications for future claims about who is to blame for unhealthy dietary habits. As we have already noted, when associations are formed outside awareness then learning probably occurs outside voluntary control. This is important, because evidence of this kind could shift the responsibility for overeating and obesity away from the individual and toward those agencies (e.g., parents, food manufacturers/outlets) that expose the individual to potentially damaging CS–US relationships. In a more recent publicized case (*Pelman vs. McDonalds*, 2003) the plaintiffs (who were obese or overweight children) alleged that McDonalds Corporation knew or should have known that its actions would exacerbate obesity and that this kind of information was withheld from its customers. Around that time, similar litigation provoked a generally negative response from the media and this particular lawsuit was dismissed. The judge's ruling referred to issues relating to free consumer choice, "If customers know (or reasonably should know) the potential ill health effects of eating McDonald's, they cannot blame McDonald's if they, nonetheless, choose to satiate their appetite with a surfeit of 'supersized' McDonald's products" (Mello et al., 2003). Of course, free will is not so easy to demonstrate if it is found that preferences for these kinds of food products are acquired outside awareness. In this case, to avoid future litigation, the fast-food industry might be advised to issue a health warning along the lines, "Repeated exposure to our products can bring about an automatic and involuntary change in your liking for their taste." Clearly, we are some way from this position. However, at the very least, this example highlights the possibility that encounters that we have with food during our formative years may have consequences for our dietary preferences later in life. And, importantly, we may have little or no control over their acquisition, especially in cases where we receive exposure on a regular basis, such as in school cafeterias, and so on.

A related issue here is the nature of the information that is acquired following CS–US exposures. If associations are formed outside awareness then they are unlikely to be encoded as logical propositions or expectations that are organized in working memory (e.g., "I know that I like this food because I have found it satisfying on previous occasions"). Rather, dietary associations could be relatively "primitive." In relation to this prospect, parallels can be drawn with observations in the literature on odor memory. For example, associations that are formed between specific odors and particular locations have previously been described as implicit and "presemantic" (Köster, 2005). In these cases a reliable interference effect appears to exist, such that implicit associations are effaced when explicit memory odor representations are also accessed successfully (Degel and Köster, 1999; Degel et al., 2001). Specifically,

it would seem that odor–place representations are degraded in individuals who are able correctly to identify (i.e., name, e.g., "jasmine") odors from a stimuli set. If the same kind of interference generalizes to food-based associations then merely thinking about the consequences of consuming food might undermine previous learning. At present this proposition remains untested. However, it is consistent with studies indicating that flavor-flavor learning is disrupted in individuals who are normally preoccupied with food and body weight (Brunstrom et al., 2001, 2005). This issue also merits attention because most organized weight-loss programs lead to elaboration of this kind.

Finally, the converse proposition should also be considered. If learning necessarily involves awareness of contingency relationships then we might expect it to interact with other concurrent cognitive activity. In this case, we might want to explore how beliefs and attitudes (possibly generated by advertising) about food influence the way that associations are formed and expressed. Furthermore, if learning does require cognizance of CS–US relationships, then we can also expect the process to vie with demands imposed by other everyday nonfood-related tasks, activities, and day-to-day stresses (for a further discussion see Brunstrom and Higgs, 2002). On occasions when we fail to attend to CS–US cooccurrences, we should expect learning to be impaired. This possibility should be taken seriously, because it might explain why human dietary learning paradigms are noted for their unreliability (Rozin et al., 1998; Zandstra et al., 2002). Indeed, in this context, it would seem somewhat ironic that most (human) learning paradigms are designed such that awareness is minimized.

IX. SUMMARY

Evidence is reviewed for the role of learning in human appetite, food choice, and intake. It is argued that learning is fundamental to the development and regulation of habitual eating behavior, including acquisition of likes and dislikes, choosing foods most appropriate to our current motivational state, and controlling how much is eaten. Both nonassociative (e.g., exposure) and associative learning (e.g., flavor-consequence and flavor-flavor learning) models are considered, and are contrasted with the few innate influences on eating. Despite a wealth of evidence in other animals, evidence for the role of learning in human eating is comparatively limited, but continues to accumulate. In young children, mere exposure to the flavor of a food, such as a novel vegetable, increases acceptance of that food. Indeed, even fetuses exposed to maternal dietary flavors *in utero* via amniotic fluid, or infants exposed to flavors in breast milk, show subsequent changes in acceptance of the exposed flavor. This exposure learning might link maternal dietary choice with preference development in children. Even so, children's food preferences are strongly correlated to the energy density of foods. Experimental evidence indicates that this results from the reinforcing effects of energy eaten when hungry, that is, flavor-consequence learning. Carbohydrate, fat, and protein have all proved effective reinforcers of flavor preferences in adults and children, although for protein, the reinforcing stimulus during acute protein shortage was not energy but presumably some amino acid derived signal. Given sufficient opportunity to learn, flavor cues associated with higher energy consumption are preferred when hungry, but conversely these flavors are less liked when full than lower energy-paired flavors, and they suppress subsequent intake. This evidence for learned satiety has been seen in both adults and children, although sensitivity to postingestive energy differences may weaken with age and with externalization of eating control. Eating energy-rich foods when hungry will establish strong appetites for those foods, but this may not

be the case for low energy foods. One proposal is that frequent eating of high fat energy-dense foods impairs neural inhibition of this learned appetite, creating a vicious circle leading to obesity. Another kind of learning, flavor-flavor learning, occurs when a relatively neutral flavor is eaten together with a flavor that already has strong positive or aversive properties. It is thought such learning could form a shortcut for transferring important information already learned about food from one sensory property to another. In humans, the strongest evidence for this learning is for acquired dislikes of flavors paired with unpleasant tastes, perhaps because of the relative novelty in repeated eating of distasteful flavors. In all these forms of learning about food, there is debate about the need for explicit awareness of the associations that have been learned. This is an important issue, because it has implications for public health strategies to control obesity and dietary-related disease. If people learn appetites for fat-rich energy-dense foods without awareness, then who should be held accountable?

References

Azzara, A. V., Bodnar, R. J., Delamater, A. R., and Sclafani, A. (2001). D-1 but not D-2 dopamine receptor antagonism blocks the acquisition of a flavor preference conditioned by intragastric carbohydrate infusions. *Pharmacol. Biochem. Behav.* **68**, 709–720.

Baeyens, F., Eelen, P., Vandenbergh, O., and Crombez, G. (1990). Flavor-flavor and color-flavor conditioning in humans. *Learn. Motiv.* **21**, 434–455.

Baeyens, F., Eelen, P., Vandenbergh, O., and Crombez, G. (1992). The content of learning in human evaluative conditioning—acquired valence is sensitive to US-revaluation. *Learn. Motiv.* **23**, 200–224.

Baeyens, F., Crombez, G., Hendrickx, H., and Eelen, P. (1995a). Parameters of human evaluative flavor-flavor conditioning. *Learn. Motiv.* **26**, 141–160.

Baeyens, F., Eelen, P., and Crombez, G. (1995b). Pavlovian associations are forever—on classical conditioning and extinction. *J. Psychophysiol.* **9**, 127–141.

Baeyens, F., Crombez, G., De Houwer, J., and Eelen, P. (1996). No evidence for modulation of evaluative flavor-flavor associations in humans. *Learn. Motiv.* **27**, 200–241.

Baeyens, F., Hendrickx, H., Crombez, G., and Hermans, D. (1998). Neither extended sequential nor simultaneous feature positive training result in modulation of evaluative flavor-flavor conditioning in humans. *Appetite* **31**, 185–204.

Baeyens, F., Eelen, P., Crombez, G., and De Houwer, J. (2001). On the role of beliefs in observational flavor conditioning. *Curr. Psychol.* **20**, 183–203.

Baker, B. J., Booth, D. A., Duggan, J. P., and Gibson, E. L. (1987). Protein appetite demonstrated: learned specificity of protein-cue preference to protein need in adult rats. *Nutr. Res.* **7**, 481–487.

Bartoshuk, L. M., and Beauchamp, G. K. (1994). Chemical senses. *Annu. Rev. Psychol.* **45**, 419–449.

Bartoshuk, L. M., Duffy, V. B., and Miller, I. J. (1994). PTC/PROP tasting—anatomy, psychophysics, and sex effects. *Physiol. Behav.* **56**, 1165–1171.

Beauchamp, G. K. (1987). The human preference for excess salt. *Am. Scient.* **75**, 27–33.

Bernstein, I. L. (1978). Learned taste aversions in children receiving chemotherapy. *Science* **200**, 1302–1303.

Bernstein, I. L. (1994). Development of food aversions during illness. *Proc. Nutr. Soc.* **53**, 131–137.

Berridge, K. C. (2000). Measuring hedonic impact in animals and infants: microstructure of affective taste reactivity patterns. *Neurosci. Biobehav. Rev.* **24**, 173–198.

Bertino, M., Beauchamp, G. K., and Engelman, K. (1983). Long-term reduction in dietary sodium alters the taste of salt. *Am. J. Clin. Nutr.* **36**, 1134–1144.

Birch, L. L. (1987). The role of experience in children's food acceptance patterns. *J. Am. Diet. Assoc.* **87**, S36–S40.

Birch, L. L., and Marlin, D. W. (1982). I don't like it; I never tried it: effects of exposure on two-year-old children's food preferences. *Appetite* **3**, 353–360.

Birch, L. L., and Deysher, M. (1985). Conditioned and unconditioned caloric compensation: evidence for self-regulation of food intake in young children. *Learn. Motiv.* **16**, 341–355.

Birch, L. L., and Deysher, M. (1986). Caloric compensation and sensory specific satiety: evidence for self regulation of food intake by young children. *Appetite* **7**, 323–331.

Birch, L. L., McPhee, L., Shoba, B. C., Pirok, E., et al. (1987a). What kind of exposure reduces children's food neophobia? Looking vs. tasting. *Appetite* **9**, 171–178.

Birch, L. L., McPhee, L., Shoba, B. C., Steinberg, L., and Krehbiel, R. (1987b). "Clean up your plate": effects of child feeding practices on the conditioning of meal size. *Learn. Motiv.* **18**, 301–317.

Birch, L. L., McPhee, L., Sullivan, S., and Johnson, S. (1989). Conditioned meal initiation in young children. *Appetite* **13**, 105–113.

Birch, L. L., McPhee, L., Steinberg, L., and Sullivan, S. (1990). Conditioned flavor preferences in young children. *Physiol. Behav.* **47**, 501–505.

Birch, L. L., Gunder, L., Grimm-Thomas, K., and Laing, D. G. (1998). Infants' consumption of a new food enhances acceptance of similar foods. *Appetite* **30**, 283–295.

Blais, C., Pangborn, R. M., Borhani, N. O., Ferrell, M. F., Prineas, R. J., and Laing, B. (1986). Effect of dietary sodium restriction on taste responses to sodium chloride: a longitudinal study. *Am. J. Clin. Nutr.* **44**, 323–343.

Blundell, J. E., Cotton, J. R., Delargy, H., Green, S., Greenough, A., King, N. A., and Lawton, C. L. (1995). The fat paradox: fat-induced satiety signals versus high fat overconsumption. *Int. J. Obes.* **19**, 832–835.

Booth, D. A. (1972). Conditioned satiety in the rat. *J. Comp. Physiol. Psychol.* **81**, 457–471.

Booth, D. A. (1974). Acquired sensory preference for protein in diabetic and normal rats. *Physiol. Psychol.* **2**, 344–348.

Booth, D. A. (1977a). Appetite and satiety as metabolic expectancies. *In* "Food Intake and Chemical Senses" (Y. Katsuki, M. Sato, S. F. Takagi, and Y. Oomura, eds.), pp. 317–330. University of Tokyo Press, Tokyo.

Booth, D. A. (1977b). Satiety and appetite are conditioned reactions. *Psychosom. Med.* **39**, 76–81.

Booth, D. A. (1985). Food-conditioned eating preferences and aversions with interoceptive elements: conditioned appetites and satieties. *Ann. N.Y. Acad. Sci.* **443**, 22–41.

Booth, D. A. (1990). How not to think about immediate dietary and postingestional influences on appetites and satieties. *Appetite* **14**, 171–179.

Booth, D. A. (1994). "Psychology of Nutrition." Taylor & Francis, London.

Booth, D. A., and Toase, A. M. (1983). Conditioning of hunger satiety signals as well as flavor cues in dieters. *Appetite* **4**, 235–236.

Booth, D. A., and Grinker, J. A. (1993). Learned control of meal size in spontaneously obese and nonobese bonnet macaque monkeys. *Physiol. Behav.* **53**, 51–57.

Booth, D. A., and Thibault, L. (2000). Macronutrient-specific hungers and satieties and their neural bases, learnt from pre- and postingestional effects of eating particular foodstuffs. *In* "Neural and Metabolic Control of Macronutrient Intake" (H. R. Berthoud and R. J. Seeley, eds.), pp. 61–92. CRC Press, Boca Raton, FL.

Booth, D. A., Lee, M., and McAleavey, C. (1976). Acquired sensory control of satiation in man. *Br. J. Psychol.* **67**, 137–147.

Booth, D. A., Mather, P., and Fuller, J. (1982). Starch content of ordinary foods associatively conditions human appetite and satiation, indexed by intake and eating pleasantness of starch-paired flavours. *Appetite* **3**, 163–184.

Booth, D. A., Gibson, E. L., Toase, A.-M., and Freeman, R. P. J. (1994). Small objects of desire: the recognition of appropriate foods and drinks and its neural mechanisms. *In* "Appetite: Neural and Behavioural

Bases" (C. R. Legg and D. A. Booth, eds.), pp. 98–126. Oxford University Press, Oxford.

Brunstrom, J. M. (2004). Does dietary learning occur outside awareness? *Conscious. Cogn.* **13**, 453–470.

Brunstrom, J. M. (2005). Dietary learning in humans: Directions for future research. *Physiol. Behav.* **85**, 57–65.

Brunstrom, J. M., and Higgs, S. (2002). Exploring evaluative conditioning using a working memory task. *Learn. Motiv.* **33**, 433–455.

Brunstrom, J. M., Downes, C. R., and Higgs, S. (2001). Effects of dietary restraint on flavour-flavour learning. *Appetite* **37**, 197–206.

Brunstrom, J. M., Higgs, S., and Mitchell, G. L. (2005). Dietary restraint and US devaluation predict evaluative learning. *Physiol. Behav.* **85**, 524–535.

Capaldi, E. D., Owens, J., and Palmer, K. A. (1994). Effects of food-deprivation on learning and expression of flavor preferences conditioned by saccharin or sucrose. *Anim. Learn. Behav.* **22**, 173–180.

Carbonnel, F., Lemann, M., Rambaud, J. C., Mundler, O., and Jian, R. (1994). Effect of the energy density of a solid-liquid meal on gastric emptying and satiety. *Am. J. Clin. Nutr.* **60**, 307–311.

Cashdan, E. (1998). Adaptiveness of food learning and food aversions in children. *Soc. Sci. Info.* **37**, 613–632.

Cecil, J. E., Palmer, C. N., Wrieden, W., et al. (2005). Energy intakes of children after preloads: adjustment, not compensation. *Am. J. Clin. Nutr.* **82**, 302–308.

Chambers, K. C. (1990). A neural model for conditioned taste aversions. *Annu. Rev. Neurosci.* **13**, 373–385.

Cleeremans, A., Destrebecqz, A., and Boyer, M. (1998). Implicit learning: News from the front. *Trends Cogn. Sci.* **2**, 406–416.

Cooke, L., Wardle, J., and Gibson, E. L. (2003). Relationship between parental report of food neophobia and everyday food consumption in 2–6 year-old children. *Appetite*, **41**, 205–206.

Crystal, S. R., Bowen, D. J., and Bernstein, I. L. (1999). Morning sickness and salt intake, food cravings, and food aversions. *Physiol. Behav.* **67**, 181–187.

Dallman, M. F., Pecoraro, N. C., and la Fleur, S. E. (2005). Chronic stress and comfort foods: self-medication and abdominal obesity. *Brain Behav. Immun.* **19**, 275–280.

Davidson, T. L. (2000). Pavlovian occasion setting: a link between physiological change and appetitive behavior. *Appetite* **35**, 271–272.

Davidson, T. L., and Benoit, S. C. (1996). The learned function of food-deprivation cues: a role for conditioned modulation. *Anim. Learn. Behav.* **24**, 46–56.

Davidson, T. L., Altizer, A. M., Benoit, S. C., Walls, E. K., and Powley, T. L. (1997). Encoding and selective activation of "metabolic memories" in the rat. *Behav. Neurosci.* **111**, 1014–1030.

Davidson, T. L., Kanoski, S. E., Walls, E. K., and Jarrard, L. E. (2005). Memory inhibition and energy regulation. *Physiol. Behav.* **86**, 731–746.

Dawson, M. E., Schell, A. M., Ackles, P. K., Jennings, J. R., and Coles, M. G. H. (1985). Information processing and human autonomic classical conditioning. *In* "Advances in Psychophysiology" (K. Ackles, J. R. Jennings, and M. G. H. Coles, eds.), pp. 89–165. JAI Press.

De Castro, J. M. (2001). Palatability and intake relationships in free-living humans: the influence of heredity. *Nutr. Res.* **21**, 935–945.

Degel, J., and Köster, E. P. (1999). Odors: Implicit memory and performance effects. *Chem. Senses* **24**, 317–325.

Degel, J., Piper, D., and Köster, E. P. (2001). Implicit learning and implicit memory for odors: The influence of odor identification and retention time. *Chem. Senses* **26**, 267–280.

De Houwer, J., Thomas, S., and Baeyens, F. (2001). Associative learning of likes and dislikes: A review of 25 years of research on human evaluative conditioning. *Psychol. Bull.* **127**, 853–869.

De Houwer, J., Baeyens, F., and Field, A. P. (2005). Associative learning of likes and dislikes: Some current controversies and possible ways forward. *Cogn. Emotion* **19**, 161–174.

De Kloet, E. R., Oitzl, M. S., and Joels, M. (1993). Functional implications of brain corticosteroid receptor diversity. *Cell. Mol. Neurobiol.* **13**, 433–455.

Denton, D. A. (1982). "The Hunger for Salt." Springer-Verlag, New York.

Diaz, E., Ruiz, G., and Baeyens, F. (2005). Resistance to extinction of human evaluative conditioning using a between-subjects design. *Cogn. Emotion* **19**, 245–268.

Dickinson, A. (1986). Re-examination of the role of the instrumental contingency in the sodium-appetite irrelevant incentive effect. *Q. J. Exp. Psychol: B* **38B**, 161–172.

Dickinson, A., and Balleine, B. (1994). Motivational control of goal-directed action. *Anim. Learn. Behav.* **22**, 1–18.

Dietz, W. H. (1994). Critical periods in childhood for the development of obesity. *Am. J. Clin. Nutr.* **95**, 955–959.

Dinehart, M. E., Hayes, J. E., Bartoshuk, L. M., Lanier, S. L., and Duffy, V. B. (2006). Bitter taste markers explain variability in vegetable sweetness, bitterness, and intake. *Physiol. Behav.* **87**, 304–313.

Drewnowski, A. (1997). Taste preferences and food intake. *Ann. Rev. Nutr.* **17**, 237–253.

Drewnowski, A., Brunzell, J. D., Sande, K., Iverius, P. H., and Greenwood, M. R. C. (1985). Sweet tooth reconsidered—taste responsiveness in human obesity. *Physiol. Behav.* **35**, 617–622.

Duffy, V. B. (2004). Associations between oral sensation, dietary behaviors and risk of cardiovascular disease (CVD). *Appetite*, **43**, 5–9.

Epstein, L. H., Paluch, R., and Coleman, K. J. (1996). Differences in salivation to repeated food cues in obese and nonobese women. *Psychosom. Med.* **58**, 160–164.

Epstein, L. H., Paluch, R., Smith, J. D., and Sayette, M. (1997). Allocation of attentional resources during habituation to food cues. *Psychophysiology* **34**, 59–64.

Fanselow, M. S., and Birk, J. (1982). Flavor-flavor associations induce hedonic shifts in taste preference. *Anim. Learn. Behav.* **10**, 223–228.

Fedorchak, P. M., and Bolles, R. C. (1987). Hunger enhances the expression of calorie-mediated but not taste-mediated conditioned flavor preferences. *J. Exp. Psychol. Anim. Behav. Proc.* **13**, 73–79.

Field, A. P. (2000). I like it, but I'm not sure why: can evaluative conditioning occur without conscious awareness? *Conscious. Cogn.* **9**, 13–36.

Field, A. P. (2001). When all is still concealed: are we closer to understanding the mechanisms underlying evaluative conditioning? *Conscious. Cogn.* **10**, 559–566.

Francis, J. A., Stewart, S. H., and Hounsell, S. (1997). Dietary restraint and the selective processing of forbidden and nonforbidden food words. *Cogn. Ther. Res.* **21**, 633–646.

Fulcher, E. P., and Hammerl, M. (2001). When all is revealed: a dissociation between evaluative learning and contingency awareness. *Conscious. Cogn.* **10**, 524–549.

Garcia, J., Brett, L. P., and Rusiniak, K. W. (1989). Limits of Darwinian conditioning. *In* "Contemporary Learning Theories: Instrumental Conditioning Theory and the Impact of Biological Constraints on Learning " (S. B. Klein and R. R. Mowrer, eds.), Erlbaum, Hillsdale, NJ.

Gerrish, C. J., and Mennella, J. A. (2001). Flavor variety enhances food acceptance in formula-fed infants. *Am. J. Clin. Nutr.* **73**, 1080–1085.

Ghoneim, M. M., Block, R. I., and Fowles, D. C. (1992). No evidence of classical-conditioning of electrodermal responses during anesthesia. *Anesthesiology* **76**, 682–688.

Gibson, E. L. (2001). Learning in the development of food cravings. *In* "Food Cravings and Addiction" (M. M. Hetherington, ed.), pp. 193–234. Leatherhead Publishing, Leatherhead.

Gibson, E. L. (2003). Learnt protein appetite in human beings: involvement of cortisol in postingestive reinforcement by protein intake. *Appetite* **40**, 334.

Gibson, E. L. (2006). Mood, emotions and food choice. *In* "Psychology of Food Choice" (R. Shepherd and M. Raats, eds.), pp. 113–140. CAB International, Wallingford CT.

Gibson, E. L., and Booth, D. A. (1986). Acquired protein appetite in rats: dependence on a protein-specific need state. *Experientia* **42**, 1003–1004.

Gibson, E. L., and Booth, D. A. (1989). Dependence of carbohydrate-conditioned flavor preference on internal state in rats. *Learn. Motiv.* **20**, 36–47.

Gibson, E. L., and Desmond, E. (1999). Chocolate craving and hunger state: implications for the acquisition and expression of appetite and food choice. *Appetite* **32**, 219–240.

Gibson, E. L., and Booth, D. A. (2000). Food-conditioned odour rejection in the late stages of the meal, mediating learnt control of meal volume by aftereffects of food consumption. *Appetite* **34**, 295–303.

Gibson, E. L., and Wardle, J. (2001). Effect of contingent hunger state on development of appetite for a novel fruit snack. *Appetite* **37**, 91–101.

Gibson, E. L., and Wardle, J. (2003). Energy density predicts preferences for fruit and vegetables in 4-year-old children. *Appetite* **41**, 97–98.

Gibson, E. L., and Wardle, J. (2004). Resistance to hunger-dependent increased intake of an energy-dense snack in frequent fruit eaters. *Appetite* **43**, 113.

Gibson, E. L., Wainwright, C. J., and Booth, D. A. (1995). Disguised protein in lunch after low-protein breakfast conditions food-flavor preferences dependent on recent lack of protein intake. *Physiol. Behav.* **58**, 363–371.

Gibson, E. L., Checkley, S., Papadopoulos, A., Poon, L., Daley, S., and Wardle, J. (1999). Increased salivary cortisol reliably induced by a protein-rich midday meal. *Psychosom. Med.* **61**, 214–224.

Gietzen, D. W. (2000). Amino acid recognition in the central nervous system. *In* "Neural and Metabolic Control of Macronutrient Intake" (H. R. Berthoud and R. J. Seeley, eds.), pp. 339–357. CRC Press, Boca Raton, FL.

Goldstein, G. L., Daun, H., and Tepper, B. J. (2005). Adiposity in middle-aged women is associated with genetic taste blindness to 6-*n*-propylthiouracil. *Obes. Res.* **13**, 1017–1023.

Hagan, M. M., Wauford, P. K., Chandler, P. C., Jarrett, L. A., Rybak, R. J., and Blackburn, K. (2002). A new animal model of binge eating: key synergistic role of past caloric restriction and stress. *Physiol. Behav.* **77**, 45–54.

Hammerl, M., and Grabitz, H. J. (2000). Affective-evaluative learning in humans: a form of associative learning or only an artifact? *Learn. Motiv.* **31**, 345–363.

Hammerl, M., and Fulcher, E. P. (2005). Reactance in affective-evaluative learning: outside of conscious control? *Cogn. Emotion* **19**, 197–216.

Harris, G., and Booth, D. A. (1987). Infants' preference for salt in food: its dependence upon recent dietary experience. *J. Reprod. Infant Psychol.* **5**, 97–104.

Harris, J. A., Shand, F. L., Carroll, L. Q., and Westbrook, R. F. (2004). Persistence of preference for a flavor presented in simultaneous compound with sucrose. *J. Exp. Psychol. Anim. Behav. Proc.* **30**, 177–189.

Harris, L. J., Clay, J., Hargreaves, F., and Ward, A. (1933). Appetite and choice of diet. The ability of vitamin B deficient rats to discriminate between diets containing and lacking the vitamin. *Proc. R. Soc. London B* **113**, 161–190.

Hladik, C. M., Pasquet, P., and Simmen, B. (2002). New perspectives on taste and primate evolution: the dichotomy in gustatory coding for perception of beneficent versus noxious substances as supported by correlations among human thresholds. *Am. J. Phys. Anthropol.* **117**, 342–348.

Hoad, C. L., Rayment, P., Spiller, R. C., et al. (2004). In vivo imaging of intragastric gelation and its effect on satiety in humans. *J. Nutr.* **134**, 2293–2300.

Holland, P. C., and Petrovich, G. D. (2005). A neural systems analysis of the potentiation of feeding by conditioned stimuli. *Physiol. Behav.* **86**, 747–761.

Holman, E. W. (1975). Immediate and delayed reinforcers for flavor preferences in rats. *Learn. Motiv.* **6**, 91–100.

Ishizuka, B., Quigley, M. E., and Yen, S. S. C. (1983). Pituitary hormone release in response to food ingestion: evidence for neuroendocrine signals from gut to brain. *J. Clin. Endocrinol. Metab.* **57**, 1111–1115.

Jansen, A. (1998). A learning model of binge eating: cue reactivity and cue exposure. *Behav. Res. Ther.* **36**, 257–272.

Johnson, S. L., McPhee, L., and Birch, L. L. (1991). Conditioned preferences: young children prefer flavors associated with high dietary fat. *Physiol. Behav.* **50**, 1245–1251.

Kalat, J. W., and Rozin, P. (1973). "Learned safety" as a mechanism in long-delay taste-aversion learning in rats. *J. Comp. Physiol. Psychol.* **83**, 198–207.

Kern, D. L., McPhee, L., Fisher, J., Johnson, S., and Birch, L. L. (1993). The postingestive consequences of fat condition preferences for flavors associated with high dietary fat. *Physiol. Behav.* **54**, 71–76.

Kitchell, J. A., Clark, D. L., and Gombos, A. M. (1986). Biological selectivity of extinction: a link between background and mass extinction. *Ann. Paleontol.* **1**, 3–23.

Kochli, A., Tenenbaum-Rakover, Y., and Leshem, M. (2005). Increased salt appetite in patients with congenital adrenal hyperplasia 21-hydroxylase deficiency. *Am. J. Physiol. Regul. Integr. Comp. Physiol.* **288**, R1673–R1681.

Köster, E. P. (2005). Does olfactory memory depend on remembering odors? *Chem. Senses* **30**, i236–i237.

Kyriazakis, I., Tolkamp, B. J., and Emmans, G. (1999). Diet selection and animal state: an integrative framework. *Proc. Nutr. Soc.* **58**, 765–771.

Lascelles, K. R. R., Field, A. P., and Davey, G. C. L. (2003). Using foods as CSs and body shapes as UCSs: a putative role for associative learning in the development of eating disorders. *Behav. Ther.* **34**, 213–235.

Lavin, M. J. (1976). Establishment of flavor-flavor associations using a sensory preconditioning training procedure. *Learn. Motiv.* **7**, 173–183.

Legenbauer, T., Vögele, C., and Ruddel, H. (2004). Anticipatory effects of food exposure in women diagnosed with bulimia nervosa. *Appetite* **42**, 33–40.

Le Magnen, J. (1967). Habits and food intake. *In* "Handbook of Physiology, Section 6: Alimentary Canal" (C. F. Code, ed.), Vol. 1, pp. 11–30. Americal Physiology Society, Washington, D.C.

Le Magnen, J. (1999). The state of research into the mechanisms of appetites for energy (first published in French in 1956). *Appetite* **33**, 2–7.

Leshem, M. (1998). Salt preference in adolescence is predicted by common prenatal and infantile mineralofluid loss. *Physiol. Behav.* **63**, 699–704.

Loewen, R., and Pliner, P. (1999). Effects of prior exposure to palatable and unpalatable novel foods on children's willingness to taste other novel foods. *Appetite* **32**, 351–366.

Logue, A. W., Ophir, I., and Strauss, K. E. (1981). The acquisition of taste-aversions in humans. *Behav. Res. Ther.* **19**, 319–333.

Looy, H., and Weingarten, H. P. (1991). Effects of metabolic state on sweet taste reactivity in humans depend on underlying hedonic response profile. *Chem. Senses* **16**, 123–130.

Looy, H., Callaghan, S., and Weingarten, H. P. (1992). Hedonic response of sucrose likers and dislikers to other gustatory stimuli. *Physiol. Behav.* **52**, 219–225.

Lovibond, P. F., and Shanks, D. R. (2002). The role of awareness in Pavlovian conditioning: empirical evidence and theoretical implications. *J. Exp Psychol. Anim. Behav. Proc.* **28**, 3–26.

Lupien, S. J., Nair, N. P., Briere, S., et al. (1999). Increased cortisol levels and impaired cognition in human aging: implication for depression and dementia in later life. *Rev. Neurosci.* **10**, 117–139.

Macdiarmid, J. I., Vail, A., Cade, J. E., and Blundell, J. E. (1998). The sugar-fat relationship revisited: differences in consumption between men and women of varying BMI. *Int. J. Obes.* **22**, 1053–1061.

Martin, I., and Levey, A. B. (1978). Evaluative conditioning. *Adv. Behav. Res. Ther.* **1**, 57–102.

Mayntz, D., Raubenheimer, D., Salomon, M., Toft, S., and Simpson, S. J. (2005). Nutrient-specific foraging in invertebrate predators. *Science* **307**, 111–113.

Mela, D. J. (2001). Why do we like what we like? *J. Sci. Food. Agric.* **81**, 10–16.

Mela, D. J., Trunck, F., and Aaron, J. I. (1993). No effect of extended home use on liking for sensory characteristics of reduced-fat foods. *Appetite* **21**, 117–129.

Mellinkoff, S. M., Frankland, M., Boyle, D., and Greipel, M. (1956). Relation between serum amino acid concentration and fluctuations in appetite. *Am. J. Physiol.* **8**, 535–538.

Mello, M. M., Rimm, E. B., and Studdert, D. M. (2003). Trends—the MacLawsuit: the fast-food industry and legal accountability for obesity. *Health Aff. (Millwood)* **22**, 207–216.

Mennella, J. A., and Beauchamp, G. K. (1993). The effects of repeated exposure to garlic-flavored milk on the nursling's behavior. *Pediatr. Res.* **34**, 805–808.

Mennella, J. A., and Beauchamp, G. K. (1999). Experience with a flavor in mother's milk modifies the infant's acceptance of flavored cereal. *Dev. Psychobiol.* **35**, 197–203.

Mennella, J. A., and Beauchamp, G. K. (2002). Flavor experiences during formula feeding are related to preferences during childhood. *Early Hum. Dev.* **68**, 71–82.

Mennella, J. A., Jagnow, C. P., and Beauchamp, G. K. (2001). Prenatal and postnatal flavor learning by human infants. *Pediatrics* **107**, E88.

Mennella, J. A., Pepino, M. Y., and Reed, D. R. (2005). Genetic and environmental determinants of bitter perception and sweet preferences. *Pediatrics* **115**, e216–e222.

Murphy, C., and Withee, J. (1987). Age and biochemical status predict preference for casein hydrolysate. *J. Gerontol.* **42**, 73–77.

Nicklaus, S., Boggio, V., and Issanchou, S. (2005). Food choices at lunch during the third year of life: high selection of animal and starchy foods but avoidance of vegetables. *Acta Paediatr.* **94**, 943–951.

Pasquet, P., Oberti, B., El, A. J., and Hladik, C. M. (2002). Relationships between threshold-based PROP sensitivity and food preferences of Tunisians. *Appetite* **39**, 167–173.

Pelchat, M. L., and Rozin, P. (1982). The special role of nausea in the acquisition of food dislikes by humans. *Appetite*, **3**, 341–351.

Pliner, P. (1982). The effects of mere exposure on liking for edible substances. *Appetite* **3**, 283–290.

Powley, T. L., Chi, M. M., Schier, L. A., and Phillips, R. J. (2005). Obesity: should treatments target visceral afferents? *Physiol. Behav.* **86**, 698–708.

Provenza, F. D. (1995). Postingestive feedback as an elementary determinant of food preference and intake in ruminants. *J. Range Man.* **48**, 2–17.

Rescorla, R. A., and Wagner, A. R. (1972). A theory of Pavlovian conditioning: variations in the effectiveness of reinforcement and non-reinforcement. *In* "Classical Conditioning II: Current Research and Theory" (A. H. Blake and W. F. Prokasy, eds.), pp. 64–199. Appleton-Century-Crofts, New York.

Richardson, N. J., Rogers, P. J., and Elliman, N. A. (1996). Conditioned flavor preferences reinforced by caffeine consumed after lunch. *Physiol. Behav.* **60**, 257–263.

Roll, D. L., and Smith, J. C. (1972). Conditioned taste aversions in anesthetized rats. *In* "Biological Boundaries of Learning" (M. E. P. Seligman and J. L. Hager, eds.), pp. 98–102. Appleton-Century-Crofts, New York.

Rolls, E. T. (1999). "The Brain and Emotion." Oxford University Press, Oxford.

Rolls, E. T., and Rolls, J. H. (1997). Olfactory sensory-specific satiety in humans. *Physiol. Behav.* **61**, 461–473.

Rozin, P., and Zellner, D. (1985). The role of Pavlovian conditioning in the acquisition of food likes and dislikes. *Ann. N.Y. Acad. Sci.* **443**, 189–202.

Rozin, P., and Schulkin, J. (1990). Food selection. *In* "Neurobiology of Food and Fluid Intake: Handbook of Behavioral Neurobiology, Volume 10" (E. M. Stricker, ed.), pp. 297–328. Plenum Press, New York.

Rozin, P., Wrzesniewski, A., and Byrnes, D. (1998). The elusiveness of evaluative conditioning. *Learn. Motiv.* **29**, 397–415.

Salbe, A. D., Del Parigi, A., Pratley, R. E., Drewnowski, A., and Tataranni, P. A. (2004). Taste preferences and body weight changes in an obesity-prone population. *Am. J. Clin. Nutr.* **79**, 372–378.

Schienle, A., Stark, R., and Vaitl, D. (2001). Evaluative conditioning needs contingency verbalization. *Psychophysiology* **38**, S86–S86.

Schulkin, J. (1991). The allure of salt. *Psychobiology* **19**, 116–121.

Sclafani, A. (1997). Learned controls of ingestive behaviour. *Appetite* **29**, 153–158.

Sclafani, A., and Ackroff, K. (1994). Glucose-conditioned and fructose-conditioned flavor preferences in rats—taste versus postingestive conditioning. *Physiol. Behav.* **56**, 399–405.

Shanks, D. R., and St. John, M. F. (1994). Characteristics of dissociable human learning systems. *Behav. Brain Sci.* **17**, 367–395.

Siegel, S. (1974). Flavor preexposure and "learned safety." *J. Comp. Physiol. Psychol.* **87**, 1073–1082.

Simpson, S. J., and Raubenheimer, D. (2005). Obesity: the protein leverage hypothesis. *Obes. Rev.* **6**, 133–142.

Simpson, S. J., Batley, R., and Raubenheimer, D. (2003). Geometric analysis of macronutrient intake in humans: the power of protein? *Appetite* **41**, 123–140.

Sonuga-Barke, E. J. S., De Houwer, J., De Ruiter, K., Ajzenstzen, M., and Holland, S. (2004). AD/HD and the capture of attention by briefly exposed delay-related cues: Evidence from a conditioning paradigm. *J. Child Psychol. Psychiat.* **45**, 274–283.

Stein, L. J., Cowart, B. J., Epstein, A. N., Pilot, L. J., Laskin, C. R., and Beauchamp, G. K. (1996). Increased liking for salty foods in adolescents exposed during infancy to a chloride-deficient feeding formula. *Appetite* **27**, 65–77.

Steiner, J. E., Glaser, D., Hawilo, M. E., and Berridge, K. C. (2001). Comparative expression of hedonic impact: affective reactions to taste by human infants and other primates. *Neurosci. Biobehav. Rev.* **25**, 53–74.

Stockley, L., Jones, F. A., and Broadhurst, A. J. (1984). The effects of moderate protein or energy supplements on subsequent nutrient intake in man. *Appetite* **5**, 209–219.

Stubbs, R. J., and Whybrow, S. (2004). Energy density, diet composition and palatability: influences on overall food energy intake in humans. *Physiol. Behav.* **81**, 755–764.

Sullivan, S. A., and Birch, L. L. (1994). Infant dietary experience and acceptance of solid foods. *Pediatrics* **93**, 271–277.

Tapper, K. (2005). Motivating operations in appetite research. *Appetite* **45**, 95–107.

Tepper, B. J., and Farkas, B. K. (1994). Reliability of the sensory responder classification to learned flavor cues—a test-retest study. *Physiol. Behav.* **56**, 819–824.

Tepper, B. J., and Ullrich, N. V. (2002). Influence of genetic taste sensitivity to 6-*n*-propylthiouracil (PROP), dietary restraint and disinhibition on body mass index in middle-aged women. *Physiol. Behav.* **75**, 305–312.

Todrank, J., Byrnes, D., Wrzesniewski, A., and Rozin, P. (1995). Odors can change preferences for people in photographs—a cross-modal evaluative conditioning study with olfactory USs and visual CSs. *Learn. Motiv.* **26**, 116–140.

Tordoff, M. G. (1996). The importance of calcium in the control of salt intake. *Neurosci. Biobehav. Rev.* **20**, 89–99.

Torii, K., Kondoh, T., Mori, K., and Ono, T. (1998). Hypothalamic control of amino acid appetite. *Ann. N.Y. Acad. Sci.* **855**, 417–425.

Touzani, K., and Sclafani, A. (2002). Area postrema lesions impair flavor-toxin aversion learning but not flavor-nutrient preference learning. *Behav. Neurosci.* **116**, 256–266.

Vazquez, M., Pearson, P. B., and Beauchamp, G. K. (1982). Flavor preferences in malnourished Mexican infants. *Physiol. Behav.* **28**, 513–519.

Vögele, C., and Florin, I. (1997). Psychophysiological responses to food exposure: an experimental study in binge eaters. *Int. J. Eat. Disord.* **21**, 147–157.

Wald, N., and Leshem, M. (2003). Salt conditions: a flavor preference or aversion after exercise depending on NaCl dose and sweat loss. *Appetite* **40**, 277–284.

Wardle, J., Cooke, L. J., Gibson, E. L., Sapochnik, M., Sheiham, A., and Lawson, M. (2003a). Increasing children's acceptance of vegetables; a randomized trial of parent-led exposure. *Appetite* **40**, 155–162.

Wardle, J., Herrera, M. L., Cooke, L., and Gibson, E. L. (2003b). Modifying children's food preferences: the effects of exposure and reward on acceptance of an unfamiliar vegetable. *Eur. J. Clin. Nutr.* **57**, 341–348.

Weigle, D. S., Breen, P. A., Matthys, C. C., Callahan, H. S., Meeuws, K. E., Burden, V. R., and Purnell, J. Q. (2005). A high-protein diet induces sustained reductions in appetite, ad libitum caloric intake, and body weight despite compensatory changes in diurnal plasma leptin and ghrelin concentrations. *Am. J. Clin. Nutr.* **82**, 41–48.

Weingarten, H. P. (1983). Conditioned cues elicit feeding in sated rats: a role for learning in meal initiation. *Science* **220**, 431–433.

Westenhoefer, J. (2002). Establishing dietary habits during childhood for long-term weight control. *Ann. Nutr. Metab.* **46** (Suppl. 1), 18–23.

Westerterp-Plantenga, M. S. (2004). Effects of energy density of daily food intake on long-term energy intake. *Physiol. Behav.* **81**, 765–771.

Wilkins, L., and Richter, C. P. (1940). A great craving for salt by a child with corticoadrenal insufficiency. *JAMA* **114**, 866–868.

Wright, P. (1991). Development of food choice during infancy. *Proc. Nutr. Soc.* **50**, 107–113.

Yeomans, M. R. (2006). The role of learning in development of food preferences. *In* "Psychology of Food Choice" (R. Shepherd and M. Raats, eds.), pp. 93–112. CAB International, Wallingford, CT.

Yeomans, M. R., Spetch, H., and Rogers, P. J. (1998). Conditioned flavor preference negatively reinforced by caffeine in human volunteers. *Psychopharmacology (Berlin)* **137**, 401–409.

Yeomans, M. R., Jackson, A., Lee, M. D., Nesic, J., and Durlach, P. J. (2000a). Expression of flavor preferences conditioned by caffeine is dependent on caffeine deprivation state. *Psychopharmacology (Berlin)* **150**, 208–215.

Yeomans, M. R., Jackson, A., Lee, M. D., Steer, B., Tinley, E., Durlach, P., and Rogers, P. J. (2000b). Acquisition and extinction of flavour preferences conditioned by caffeine in humans. *Appetite* **35**, 131–141.

Yeomans, M. R., Mobini, S., Elliman, T. D., and Walker, H. C. (2004). Changes in the sensory and hedonic characteristics of odours conditioned by association with tastants. *Appetite* **42**, 415.

Yeomans, M. R., Durlach, P. J., and Tinley, E. M. (2005a). Flavor liking and preference conditioned by caffeine in humans. *Q. J. Exp. Psychol. B.* **58**, 47–58.

Yeomans, M. R., Weinberg, L., and James, S. (2005b). Effects of palatability and learned satiety on energy density influences on breakfast intake in humans. *Physiol. Behav.* **86**, 487–499.

Yu, W. Z., Silva, R. M., Sclafani, A., Delamater, A. R., and Bodnar, R. J. (2000). Role of D-1 and D-2 dopamine receptors in the acquisition and expression of flavor-preference conditioning in sham-feeding rats. *Pharmacol. Biochem. Behav.* **67**, 537–544.

Zandstra, E. H., De, G. C., Mela, D. J., and van Staveren, W. A. (2000a). Short- and long-term effects of changes in pleasantness on food intake. *Appetite* **34**, 253–260.

Zandstra, E. H., De, G. C., and van Trijp, H. C. (2000b). Effects of variety and repeated in-home consumption on product acceptance. *Appetite* **35**, 113–119.

Zandstra, E. H., Stubenitsky, K., De, G. C., and Mela, D. J. (2002). Effects of learned flavour cues on short-term regulation of food intake in a realistic setting. *Physiol. Behav.* **75**, 83–90.

Zellner, D. A., Rozin, P., Aron, M., and Kulish, C. (1983). Conditioned enhancement of human's liking for flavor by pairing with sweetness. *Learn. Motiv.* **14**, 338–350.

12

Gene Environment Interactions and the Origin of the Modern Obesity Epidemic: A Novel "Nonadaptive Drift" Scenario

JOHN R. SPEAKMAN
School of Biological Sciences, University of Aberdeen

I. Introduction
II. Evidence Supporting the Famine Hypothesis
III. Evidence Against the Famine Hypothesis
IV. The Challenge Facing Evolutionary Scenarios for the Genetic Predisposition to Obesity in Modern Societies
V. An Alternative Model for the Evolution of the Genetic Basis for Obesity
   A. Weight Regulation in Small Mammals
   B. The Set-Point Hypothesis
   C. The Evolution of Set-Point Control Systems
   D. Problems with the Set-Point Idea
   E. Application of Set-Point Ideas to the Evolution of Human Body Weight
   F. Quantification of the Nonadaptive Drift Model
VI. Time Trends in Obesity Prevalence: The Interaction of Physiology and Social Factors
VII. Implications
   Acknowledgments
   References

Since the 1950s there has been a major epidemic of obesity and associated comorbidities, throughout the Western world. Twin and genetic-mapping studies reveal that obesity has a strong genetic component, yet the rapidity of its increase cannot be a consequence of population genetic changes. Consequently, the most accepted model is that modern obesity occurs because of a gene–environment interaction. We appear to have a genetic predisposition to deposit fat that is particularly strongly expressed in the modern environment. There is a consensus that over evolutionary time we have been exposed to regular periods of famine, during which fatter individuals would have had lower mortality. Hence, individuals with genes promoting the efficient deposition of fat during periods between famines (so-called "thrifty genes") would be favored. In the modern environment where food is continuously available this genetic predisposition prepares us for a famine that never comes, resulting in obesity. This chapter has two objectives. I present some fundamental difficulties with the "famine hypothesis." Data on mortality during famine and its frequency of occurrence show that famines provide insufficient selective advantage and have occurred over an insufficient time period for thrifty genes to spread in the population. I then present a novel alternative model for the evolutionary background of the epidemic. The biggest challenge for any evolutionary model is to explain not why so many people in modern society get fat, but rather to explain why so many remain slim. In the novel model it is argued that ancient man like other animals would have been subjected to stabilizing selection for body weight/fatness—constrained at the lower end by the risks of disease, and possibly starvation, and at the upper end by the risk of predation. This selection results in a set point around which body weight and fatness is regulated. The major change in our ancient history that altered this pattern was the effective removal of predation by the development of social behavior, weapons, and fire around 1.8 million years ago. The absence of a strong selective disadvantage at the upper end of the body fatness range would lead to a change in the distribution of set points due to random mutations and drift. The important aspect of this model is that the increased frequency of high set points under the absence of selection is *not* argued to be adaptive. A simple model of random mutations in the set points over 1.8 million years, with selection only acting on the lower end of the distribution, yields a predicted distribution

of obesity phenotypes compatible with the modern obesity epidemic. During the Paleolithic period 50,000–10,000 years ago humans probably did not achieve their drifted set points because food was not highly abundant. The development of agriculture-based societies in the Neolithic, however, made food abundantly available. Relatively few people, however, became obese. Only relatively few members of the population had high set points, and only members of the highest social classes could achieve these high set points because only they had access to unlimited quantities of food. The biggest change in modern societies is not that food has become abundant in supply but that it has become abundantly available, so that everyone can attain their drifted set points. However, because the model depends on random drift rather than directed selection it explains why even after 1.8 million years of evolution and 40 years of freely available food supplies, most people are still not obese.

I. INTRODUCTION

Western societies have experienced an epidemic of obesity over the past 50 or so years that has been described by the World Health Organization (WHO) as the greatest health threat facing the Western world (WHO, 1998). Throughout the United States by 2000, the prevalence of obesity had already exceeded 25–30% in many states, depending on the database used (Flegal et al., 2002; Flegal 2005; Yun et al., 2006), with an additional 25–30% being classed as overweight. The American pattern is being repeated throughout the rest of the world (Walker et al., 2001). The rapidity with which levels of obesity have risen indicates that the epidemic cannot be a consequence of genetic restructuring of the population by standard demographic processes, and must therefore have an environmental cause. Yet, when studies have investigated the contribution of genetic and shared environmental factors on individual susceptibility to obesity, the dominant effect is always genetic (Perusse et al., 1998; Barsh et al., 2000; Hebebrand, 2005). The unavoidable conclusion is that obesity results from a gene by environment interaction (e.g., de Castro, 2004; Wilkin and Voss, 2004). Some individuals have a genetic predisposition to become obese and this genetic predisposition is particularly expressed in the modern environment.

Faced with this scenario, several previous researchers have speculated about the evolutionary processes that may have genetically predisposed us to develop these problems in modern society. These evolutionary scenarios are all fundamentally similar. They postulate that the modern epidemic results from the natural selection of traits that in our ancient history were advantageous, but when the resultant adaptive genome is immersed in modern society, it confers disadvantages. Probably the first exposition of these ideas was by Neel (1962), who suggested that diabetes and obesity stemmed from natural selection on our ancient ancestors favoring a "thrifty genotype" that enabled highly efficient storage of fat during periods of food abundance. Neel (1962) argued that such a genotype would be extremely advantageous for primitive man who was exposed to periods of food shortage, because it would allow them to efficiently deposit fat stores and thus survive any subsequent period of food shortage. In modern society, however, where food supply is always available, this thrifty genotype proves deleterious because it promotes efficient storage of fat, in preparation for a period of shortage that never arrives. Neel (1962) viewed the development of insulin resistance as part of this adaptive "thrifty genotype" assisting the process of efficient fat deposition.

Since Neel (1962) there have been many papers broadly reiterating the general theme that obesity and its sequelae are fundamentally a consequence of a genotype that was at some historical stage ideally adjusted to an ancient environment,

characterized primarily by unpredictable energy resources, that fails to cope in a modern environment, characterized by constant and highly available supplies of energy. In the last decade, this original idea has been heavily promoted as a plausible evolutionary scenario underpinning the epidemics of obesity and other chronic diseases (Eaton et al., 1988; Widmaier, 1998; Lev-Ran, 1999, 2001; Campbell and Cajigal, 2001; Prentice, 2001, 2005a,b; Ravussin, 2002; Diamond, 2003; Chakravarthy and Booth, 2004; Wilkin and Voss, 2004; Hebebrand, 2005; Prentice et al., 2005). In particular, the critical role that has been played by historical periods of famine in the process of selection is strongly reiterated.

I have previously presented evidence showing the hypothesis that periods of famine have played a central role in the evolution of our genetic predisposition to obesity is fundamentally flawed (Speakman, 2006a,b). In this chapter I have two aims. First, I will summarize the arguments in favor of, and against, the famine hypothesis, hopefully demonstrating the inadequacies of the famine arguments. My second aim is to introduce a novel evolutionary scenario for the modern epidemic that does not hinge on selection of genes during periods of famine but rather emphasizes the absence of certain selective forces. This builds on a previous paper in which this model was briefly introduced (Speakman, 2004).

II. EVIDENCE SUPPORTING THE FAMINE HYPOTHESIS

Famine has been a common feature of human history (Keys et al., 1950; McCance, 1975; Elia, 2000; Prentice, 2005b), stretching back over at least the last 5000 years (Eaton et al., 1988; Chakravarthy and Booth, 2004; Prentice et al., 2005). During famines individuals face tremendous difficulties, increased mortality, and reduced fertility (Keys et al., 1950; Watkins and Menken,

1985). The spread of genes conferring different traits is a function of the intensity of selection (i.e., how much the genes affect mortality and fecundity) and the frequency at which such selective events takes place. The frequency of famines depends on how a famine is defined, but estimates suggest that they may occur every 10 years or so, and hence human populations would have experienced between 600 and 5000 famines—depending on whether one considers famines have their origin during the Neolithic (Prentice, 2005b) or Paleolithic (Chakravarthy and Booth, 2004) periods.

Estimates of the extent of mortality in early famines suggest that it can be very high including up to 60% of the population. This high risk of mortality even leads to people cannibalizing their own children (Prentice, 2005b; Prentice et al., 2005). Moreover this increased mortality is combined with a decline in fertility. It is argued this combination must have had a profound effect on the evolution of "thrifty genes" predisposing us to efficiently deposit fat whenever famines were not present. A foundation of the "famine hypothesis" is the idea that individuals who have large deposits of fat in their bodies survive periods of famine, while their lean counterparts do not. Consequently, the genes that predispose to efficient deposition of fat during postfamine good times are favored. There is theoretical and empirical data to support this idea. The relationship between energy expenditure and body size has a much shallower gradient than the relationship between energy storage and body size. Although fatter people burn energy up at a faster rate than lean people, they store disproportionately more, so that the time period that a fatter person can survive in the total absence of food is in theory longer. Direct measurements of the time that people of different body weights survive under conditions of total food absence (such as political hunger strikers) support this suggestion.

I have previously (Speakman, 2006a) modeled the expected penetration of thrifty

genes that confers a protection against such famine-induced mortality and have shown that there are easily sufficient famine events for such a gene to spread in the population if it is postulated that famines stretched back into the Paleolithic period.

III. EVIDENCE AGAINST THE FAMINE HYPOTHESIS

If the argument is correct that in hunter-gatherer and subsistence agriculture communities there is strong selection for "thrifty genes," individuals should become obese in periods when their communities do not experience famine. However, hunter-gatherers and subsistence agriculturalists have body mass indices (BMI) that range from 17.5 to 21.0—all at the lean end of normal (Odea, 1991; Alemu and Lindtjorn, 1995; Kesteloot et al., 1997; Kirchengast, 1998; Bribiescas, 2001; Campbell et al., 2003) and nowhere near the cutoff point for being overweight (BMI = 25), never mind obese (BMI = 30). Clearly, these individuals have not inherited the supposed thrifty genes that predispose them to weight gain and obesity during periods between famines.

While details of the deprivations that people undergo during famine are impressive, and stories of eating one's own children are indications of the desperation that famine victims endure, such accounts do not provide quantitative estimates of famine mortality necessary to evaluate their evolutionary significance. Although reports for ancient periods of famine indicate tremendously high mortality rates, evidence supporting these estimates is at best limited, and is normally nonexistent. Moreover, most estimates confound mortality with emigration that can be significant. When reliable estimates can be made that are corrected for the background rate of mortality in the absence of famine, the rates of mortality only exceed 10% per annum very exceptionally. More frequently, the impact is around 2–5%.

There is a positive relationship between how long famines last and how much mortality they cause. Although Keys et al. (1950) identified 190 "famines" that occurred in Britain over 2000 years, giving a rough occurrence rate of one famine per 10 years, most of these were of short duration (Wrigley and Schofield, 1981) with virtually no mortality consequences. In fact estimates of the frequency of severe mortality crises in Europe (defined as crises leading to significant increases in mortality in a given year relative to the baseline over the previous 10 years) were an order of magnitude less frequent, at one per century (Dupaquier, 1979). Similar data from China indicate famines on average affect an area about once every 150 years (Ho, 1959). When this is combined with the suggestion that famine is a modern phenomenon only dating back to the Neolithic, then most human populations would not have encountered 5000 severe famine events but more likely only 60–100 events.

Whether this frequency of these severe events is enough to drive the natural selection of thrifty genes depends on the extent to which the mortality is distributed between lean and obese subjects. The argument is generally made that lean subjects would die first in famines because they would be the first to run out of energy, but this assumes that the mortality during famine is due to starvation. In fact, only a small proportion of total mortality in famine conditions is attributable to starving to death (normally between 5 and 25%: reviewed in Speakman, 2006a). Most people during famines die of infectious disease and diarrhea. This occurs because people in famines facing food starvation become less choosy in their selection of food. People routinely eat weeds, tree bark, various other noxious plants, decomposing carrion, and even corpses. This lack of discrimination in food selection leads many people to die directly of poisoning (Addis et al., 2005), while others acquire food poisoning and lethal diarrhea. Moreover, the decline in the nutritional quality of food may lead to a decrease in immunocompetence and hence the ability to fight off infectious disease, compounded by a general breakdown in sanitary conditions (Carmichael, 1983; Dirks, 1993).

Although we have no direct information about the variation in mortality during famines the reasons why people die suggest that body fatness would not be a significant factor. This is strongly supported by reports of the age distribution of mortality. The burden of mortality in all the famines where adequate data exist indicate that mortality falls disproportionately on two groups: the very young (<5 years old) and the old (>60 years old) (Murray et al., 1976; Watkins and Menken, 1985; Toole and Waldman, 1988; Lindtjorn, 1990). This variation in mortality reflects the fact that most people during famine conditions die of infectious diseases and it is the very young and the old that are most susceptible to these causes of mortality. This pattern of mortality is a serious problem for the famine hypothesis since any mortality in the elderly is completely unimportant in terms of its impact on selection, while mortality in the very young is unlikely to be biased with respect to obesity simply because until recently obesity in this age class was virtually unheard of.

In summary, severe famines involving significant mortality are rare events, probably occurring less than once every 100 years, and are a phenomenon of advanced agriculture-based societies. Hence, most human populations have experienced such events probably less than 60–100 times in their entire history. Even in the worst famines mortality seldom exceeds 10% per annum, and most of that mortality is not starvation, which routinely comprises only one-fifth to one-twentieth of the total. Consequently, excess mortality attributable to starvation during famines rarely exceeds 2% per annum, and more commonly is less than 1%. Moreover, even this mortality predominantly falls on the very young and old, an age distribution that is probably incompatible with a distribution of mortality favoring obese over lean subjects.

IV. THE CHALLENGE FACING EVOLUTIONARY SCENARIOS FOR THE GENETIC PREDISPOSITION TO OBESITY IN MODERN SOCIETIES

If the famine hypothesis fails to adequately explain the genetic underpinning of the modern obesity epidemic, then what are the alternatives? The major challenge facing any evolutionary explanation of the genetic predisposition behind the obesity epidemic is not to explain why we get obese, but rather to explain why only a relatively small fraction of the population gets that way. Even in the United States where there has been unlimited access to energy resources for at least the past 40 years, 35% of the population still has a BMI in the "normal" range of 17.5 to 25 (Flegal et al., 2002; Helmchen and Henderson, 2004; Flegal, 2005). This section of the population is just as exposed as the remainder to the so-called obesogenic environment, yet they do not gain weight. Any scenario that postulates a selective advantage for the obese trait hinging on obese or thrifty genes must explain why 35% of the population apparently did not inherit these genes in spite of their evolutionary advantage. This is a major problem for adaptive scenarios because genes that confer even small advantages spread in the gene pool given sufficient time to propagate.

V. AN ALTERNATIVE MODEL FOR THE EVOLUTION OF THE GENETIC BASIS FOR OBESITY

A. Weight Regulation in Small Mammals

One way that we may gain insight into the processes that underlie body weight regulation in early hominids is to examine the regulation of body weight in other wild animals. Over the past 20 years there has been a considerable amount of work in this field and significant progress has been made. Studies of small animals indicate that they have a very strong regulatory system for body weight that is highly resistant to perturbations that are brought about by, for example, modifying their diets. An example in Figure 1 are data for the bank vole

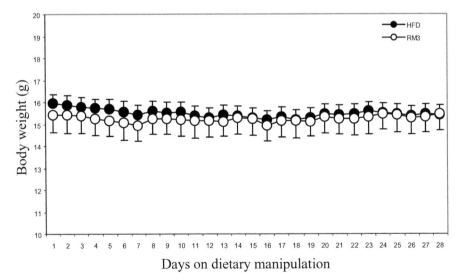

FIGURE 1   Body weights of bank voles (*Clethrionomys glareolus*) fed either a high fat diet (HFD 29% energy from fat closed symbols) or standard rodent chow (RM3 pellets 8% fat open symbols) for a period of 28 days. The voles show complete resistance to weight gain when fed the high fat diet (redrawn after Peacock and Speakman, 2001).

(*Clethrionomys glareolus*) (Peacock and Speakman, 2001). When bank voles are exposed to a high fat diet they do not gain weight. Rather the animals modulate their energy intake and elevate their levels of physical activity so that their weight remains stable.

It is possible to make small rodents lose weight by placing them onto caloric restriction. When this is done the animals oppose the restriction by modulating their energy budgets. The main way that they achieve this is to reduce their locomotor activity although there are also contributions to the total saving by the reduction in resting metabolic rate, which occurs because of the decline in overall body mass as the animals lose weight (Fig. 2). Consequently after a period of losing weight the animals come again into energy balance and remain weight stable (Boyle et al., 1981; Hill et al., 1984, 1985; Keesey and Corbet, 1990; Even and Nicoliadis, 1993; Gonzales-Pachecho et al., 1993; Hambly and Speakman, 2005). If the animals are then given free access to food again they exhibit a profound hyperphagia, which results in rapid weight gain

until their body masses return to the level of control animals that had never undergone the period of restriction. The existence of postrestriction hyperphagic responses in rodents is important for two reasons. The first is that it demonstrates that the animals could eat more food in the control conditions if they wished to. That is to say their intake in captivity is not constrained by some quirk of the housing conditions (such as the dynamics of getting food out of the food hopper). Consequently their actual intake must be internally regulated. They do not simply eat as much as they can and their body mass comes to a dynamic equilibrium where the expenditure of the expanding tissue mass balances this maximal intake. The second aspect of the hyperphagia is that it indicates when the animals come off restriction they perceive themselves to be underweight and overeat relative to controls to redress this imbalance. Taken together the observations that small rodents oppose changes in their body weight—either when fed high fat diets or when placed under caloric restriction, as well as exhibiting postrestriction hyperphagia, suggest that they have a target body weight that they are attempting to attain by varying their food intake levels and levels of energy expenditure (Keesey and Corbett, 1984; Keesey, 1995; Kalra, 1997; Keesey and Hirvonen, 1997; Levin and Keesey, 1998; Hirvonen and Keesey, 2000; Morgan and Mercer, 2001).

B. The Set-Point Hypothesis

The idea of a target body weight or fatness is in fact over 50 years old and was first proposed by Kennedy (1953), who suggested that animals have a lipostatic regulation system and Mellinkoff et al. (1956) who proposed a system based around regulation of lean tissue (protein stores) (see also Samec et al., 1998 for models that include signaling between these body components). The lipostatic system was based on the idea of a signal emanating from body

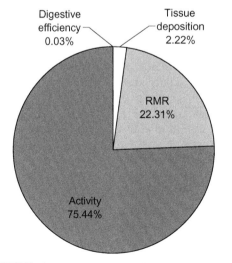

FIGURE 2 Percentage contribution of different effects to energy compensation during food restriction in the mouse (redrawn from data in Hambly and Speakman, 2005). Reductions in physical activity account for most of the energy saving.

fat stores that indicates the size of the stores. Animals read the signal and then compare their implied adiposity from this signal with a reference point or set point encoded in their brains. If their actual body stores exceed the reference point then feedback systems come into play that attempt to reduce body mass, either by reducing food intake or by increasing expenditure. In contrast, if the signal from the body fat stores suggests the animal's adiposity is too low this results in an increase in food intake and/or a decrease in expenditure to elevate body fatness until the two are again in balance. Experimental studies where fat is surgically removed also show that there is compensatory regrowth of the fat tissue supporting the idea that fat levels are centrally regulated (Mauer et al., 2001). The discovery of leptin, now over 10 years ago (Zhang et al., 1994), provided a potential molecular basis for this lipostatic system (Friedman, 2002). Leptin is a cytokine produced by fat tissue, circulating levels of which are directly linked to the levels of adiposity (Considine et al., 1996). It is detected by leptin receptors in the hypothalamus (Tartaglia et al., 1995), which are linked into a complex neural network involving multiple neuroendocrine secretions that ultimately is connected to feedback systems that regulate both appetite and energy expenditure (Schwarz et al., 1997, 1999; Vergoni and Bertolini, 2000; Williams et al., 2000; Mercer and Speakman, 2001; Abdel-Hamid, 2002; Broberger, 2005; Erlanson-Albertsson, 2005; Levin, 2005; Park and Bloom, 2005). A mutant mouse lacking the capacity to generate functional leptin (the *ob/ob* mouse) and a similar mouse incapable of reading the leptin signal (the *db/db* mouse) are both continuously hyperphagic and increase to enormous body weights and fatnesses. This response is consistent with the lipostatic model, because if the body fat produces no signal, or the brain cannot read the signal, then the perceived body fatness will always lie below the reference point of the system.

No matter how fat the animal becomes it will always be perceived by its brain as being thin and needing to eat to put on weight. Treatment of *ob/ob* mice with exogenous leptin produces an immediate reduction in food intake and normalization of their body weight (Pelleymounter, 1995; Romsos et al., 1996; Halaas et al., 1997; Mistry et al., 1997; Pelleymounter et al., 1998). This response is consistent with the prediction from the lipostatic model, while as anticipated treating *db/db* mice in the same way has no effect. In mice with targeted transgenic overexpression of leptin production there is a dramatic fall in body fatness (Masuzaki et al., 1997).

C. The Evolution of Set-Point Control Systems

Why do mice and other small wild rodents have a lipostatic system regulating their body fatness? Small animals probably store fat as an insurance against periods when they have no access to food supplies. For these animals having a fat store can make a big difference. We have previously measured the level of energy demands of free-living field voles using the doubly-labeled water technique. During winter these animals routinely live in environments where the ambient temperature can be 30–40°C below their lower critical temperature. Daily energy demands in these conditions amount to about 120 kJ/day (Speakman et al., 2003). Body fat has an energy content of 39 kJ/g (Schmidt-Nielsen, 1997). If a 25-g vole was to store 0.5 g of fat, then it would have enough energy available in that fat store only to survive for about 4 h in the absence of food. Any food supply crisis that exceeded this period would be fatal. There is a strong selective pressure therefore to store more fat in their bodies to avoid the potential risks of starvation. In fact, voles store about 3 g of fat which is sufficient for them to survive without food for about 24 h. However, 24 h is still a quite tight starvation window. If the animals were to store 6 or 10 g of fat they would obviously increase their probabilities of surviving longer periods without food. There comes a trade-off, however. Longer periods of interruption of food supply become less and less likely to occur, but carrying around 10 g of fat, when you only weigh 25 g in the first place brings other problems. The key one is reduced mobility. Mobility is important for voles because they are the prey of lots of predatory animals. In fact in the United Kingdom alone it is estimated that every second 29 field voles are killed by predators, and the life expectancy of a typical bank vole is under 4 months (Bobek, 1969). The risks of mortality from predation are very high. Consequently carrying around a large fat store may enhance the probability of surviving a crisis of food supply, but also increases the probability of being killed by a predator. This is probably why these animals have very tight regulatory systems that cannot be perturbed by changing their source of food (Fig. 1). Animals have a set-point regulatory system because they need such a system to regulate their mass at a critical point that balances the risks of being too fat with the risks of being too thin. Considerable research suggests that this fundamental balance of risks of starvation pushing body mass up and risks of predation keeping body masses down is a key component of body mass regulation in many wild animals, including small mammals and birds (Kullberg et al., 1996; Fransson and Weber, 1997; Adriaensen et al., 1998; Cresswell, 1998; Pravosudov and Grubb, 1998; van der Veen, 1999; van der Veen and Sivars, 2000; Brodin, 2000, 2001; Cuthill et al., 2000; Pravosudov and Lucas, 2001; Gentle and Gosler, 2001; Covas et al., 2002).

D. Problems with the Set-Point Idea

The idea that there is a set-point regulation system for body weight (or fatness) has attracted considerable criticism (Wirtshafter and Davies, 1977; Harris, 1990;

Weinsier et al., 2000; de Castro and Plunkett, 2002; Levitsky, 2002). Some time ago I was sceptical myself about set points as a control mechanism (Speakman et al., 2002). One criticism is that apparent regulation of body weight around a set point may be an illusion generated by other control systems that do not rely on the set-point concept. An example of such a system is a "settling point" system (Girardier, 1994; Dulloo, 1997; Payne and Dugdale, 1977; Wirtshafter and Davies, 1977; Millward and Wijesinghe, 1998; Levitsky, 2002; Speakman et al., 2002; Speakman, 2004—model 1). A settling point system contrasts with a set-point system in that there is postulated to be no feedback control from existing weights on levels of appetite and expenditure. Such a system can, however, generate an illusion of feedback control. Consider a situation where an animal (or human) is in energy balance eating x calories each day and expending y calories each day. If the animal finds itself in a situation where food supply is limited and it can only obtain y calories each day (where $y < x$) then it will need to supply the deficit from its reserves. In this condition its body mass will fall because it is withdrawing energy to fuel the shortfall. However, because energy expenditure is related to body mass, as the mass falls the expenditure will also fall, until at some point the expenditure will equal y. At this point the animal will be in balance again as intake will equal y and expenditure will equal y and weight loss will stop. It will appear that the animal has opposed the restriction in supply by enabling mechanisms for energy conservation, whereas in fact it has simply reached a new equilibrium body weight—a so-called "settling point" (Wirtshafter and Davies, 1977; Levitsky, 2002).

Once an animal in this condition is given free access to food again, it will go back to its old habits of intake, it will be in energy surplus, and its body mass will rise until it settles back to its original level (the original settling point). One might erroneously infer that the animal was regulating around an internal set point using a feedback system when in fact the animal was simply varying the input parameters of intake and expenditure and the body mass was a simple outcome variable from these parameters—rising under conditions of elevated intake and falling under the converse.

The critical difference separating a set-point and a settling point system is the existence of feedback from the body fat stores driving the level of food intake (Speakman, 2004, models 2 and 3). In this context the observations of postrestriction hyperphagia and reductions in activity when animals are under restriction are crucial. Under the set-point system the hyperphagia is predicted to occur because the animal has a body fatness (or lean tissue levels) lower than its set point. The settling point model, however, postulates that the level of intake is only governed by the environmental conditions. Following a period of restriction therefore the settling point model predicts that intake should only return to the prerestriction levels, if the prerestriction conditions are re-created, and that body mass will return eventually to the same level. The presence of postrestriction hyperphagia strongly indicates that there is an active regulation system based around a set point rather than a passive system based around settling points. Similarly, the settling point system postulates only that under food restriction energy demands will fall consequentially to lowered body mass. However, direct measurements suggest that this element of the response to restriction accounts for only 25% of the total opposition to weight loss and that reductions in locomotor activity are far more important (see Fig. 2). Such activity reductions are more consistent with an active feedback system centered around a set point.

Small animals are not unique in their exhibition of postrestriction hyperphagia and reductions in activity under food restriction. Humans also exhibit these phenomena. This is evident in the many anecdotes of people's phenomenal appetites under the conditions where they are emerging from enforced food restriction. Experimentally, however, we have detailed measurements of profound postrestriction hyperphagia from the experiments of Keys et al. (1950). Many studies have shown that during food deprivation subjects tend to reduce their expenditure more than can be accounted for by the loss of metabolizing tissue (Nair et al., 1981; Munch et al., 1993; Ballor and Poehlman, 1995; Leibel et al., 1995; Dulloo, 2005). The implication is that humans also regulate their body weights using similar control systems based around deviations from centrally encoded set points. A nice experimental study of homeostatic regulation of body weight in humans involved the demonstration that individuals fed isoenergetic diets either gained or lost weight depending on their previous weight-loss history (Macias, 2004).

One objection to the existence of a set-point system in humans, based on leptin as a signal of the level of adiposity, is that if such system exists then it clearly is not working in those individuals that become obese. Except in a few rare instances (Montague et al., 1997) the obese have intact leptin production systems (Maffei et al., 1995; Considine et al., 1996; Haffner et al., 1996; Hinney et al., 1997; Adami et al., 1998). Moreover, clinical trials using exogenous leptin (e.g., Heymsfield et al., 1999) did not live up to their promise of delivering reductions in body weight by tricking the postulated system into reacting as if individuals were fatter than they actually were. I will address later why these arguments are probably erroneous. In the meantime let us suppose that the existence of postrestriction hyperphagia, and reductions in activity during restriction, confirm the existence of a set-point rather than a settling point system, the setting of which is governed by a trade-off between the existing selection pressures that act against the set point being too low (starvation risk) or being set too high (predation risk).

E. Application of Set-Point Ideas to the Evolution of Human Body Weight

Early humans and our hominid ancestors were very likely subject to the same constraints of stabilizing selection as wild animals are today. If an individual were to store virtually no body fat (BMI around 14) they would be at risk of mortality because of the increased risk of starvation during any period of food shortage. I am not referring here to periods of food shortage at the level of a famine, but just a period of a few days when they failed to find or catch food. A second and probably much more important factor working against very low fat storage would be elevated risk of mortality during contraction of infectious diseases. This could work in two ways. The low body mass might cause poor immunocompetence, and hence individuals with very low set points and fat contents may have been more susceptible to contracting disease in the first place and less able to fight it off once contracted. Additionally, when infected and unable to forage, these individuals would have a narrower window during which they could draw on stored reserves. In contrast, individuals with high levels of fat storage might never starve during disease episodes, but would also be selected against because these individuals would be less able to avoid predators. During the early period of human evolution, between 6 and 2 million years ago (Pliocene), large predatory animals were far more abundant than they are today. Studies of the fossil record of early hominids (*Australopithecus afraensis*) suggest that 6–10% of the individual fossils show signs of predation as the cause of death (Hart and Sussman, 2005). Most bones of other *Australopithecines* come from ancient hyena kills (Stearns, 2001), or in assemblages that reflect predator activity (Pickering et al., 2004), consistent with the idea that early hominids suffered high predation risks (Brain, 1981).

Several major events happened in our evolutionary history around the period from 2 million to 1.8 million years ago. The first was the evolution of social behavior. This allowed several individuals to band together to enhance their ability to detect predators and protect each other from their attacks. Modern primates, for example, vervet monkeys have evolved complex signaling systems to warn other members of their social groups about the approach of potential predators (Cheney and Seyfarth, 1985; Owren and Bernacki, 1988; Brown et al., 1992; Baldellou and Henzi, 1992). A second important factor was the discovery of fire (Platek et al., 2002), and the use of tools that could be used as weapons. *Australopithecine* bones found in caves do not have tools or other artefacts associated with them. It seems tool use probably evolved with *Homo habilis* around 2.5 to 2 million years ago (Stearns, 2001). Together fire and weapons would have been very powerful mechanisms for our ancestors to protect themselves against predation, and social structures would have greatly augmented these capacities by enabling more rapid predator detection and effective group protection systems. Modern nonhominid apes, such as chimpanzees, also use weapons such as sticks to protect themselves against predators such as large snakes, and it has been concluded that bands of early hominids with even quite primitive tools could easily succeed in confrontations with predators (Treves and Naughton-Treves, 1999).

The effective removal of predation as an evolutionarily significant force is suggested here to be a key event in the regulation of our body mass because it generates an asymmetry in the selective pressures on the body fatness set point (see also Speakman, 2004). After the evolution of social groups, and the discovery of fire and weapons there would have still been strong disease-related selection against lowering of the set point, but no selective pressure constraining the top end of the distribution. Under this scenario any mutation leading to an increase in the set point would not be removed by selection, but mutations leading to reductions in the set point would still be strongly selected against. Consequently, over time, there would be a shift in the distribution of set points to include individuals regulating their body weights at higher levels.

The key aspect of this novel "drift" scenario is that the genetic predisposition to obesity that underpins the modern epidemic is not interpreted to be an advantageous characteristic favored by the process of natural selection—as in the famine hypothesis, but is rather seen as a neutral drift in set points in response to the absence of selection. As such it is a "nonadaptive" scenario. In this model, obesity is not being interpreted to have ever given us an advantage. Moreover, because the upward drift in set points is presumed to have occurred at random, this explains why many individuals still regulate their body weights in the BMI range from 17.5 to 25, overcoming the major challenge facing "adaptive" scenarios. This is not the first suggestion that obesity results from a heritable variation in body weight set points (e.g., Keesey, 1995; Leibel, 1990; Keesey and Hirvonen, 1997), but it provides an evolutionary reason why such heritable variations in set points exist.

F. Quantification of the Nonadaptive Drift Model

How quantitatively realistic is this drift model? To explore the likely change in the distribution of set points under an absence of selection at the upper end of the distribution, I have modeled the pattern making the following assumptions. I have assumed that the set point is a polygenic trait that is influenced by a large number of genes, each having an independent additive contribution to its level. Since these genes are presumed to be independent and additive we can simplify the model by considering the situation for a single gene with large effects. Hence, if we assume that there is a single gene governing the set point and that mutations in this gene result in increases or

decreases in the set point by 8 BMI units, this is numerically equivalent to the set point being defined by 40 independent and additive genes—each having an impact of 0.2 BMI units. (For a 1.73-m tall person 0.2 BMI units is equivalent to 600 g of body weight.) We will take as a starting point the BMI of modern hunter-gathering communities as an indication of the BMIs of hominids 1.8 million years ago when the predation transition occurred. These communities have BMIs centered on a mean of around 20 (Odea, 1991; Alemu and Lindtjorn, 1995, reviewed in Speakman, 2006b; Bribiescas, 2001; Kesteloot et al., 1997; Kirchengast, 1998; Campbell et al., 2003). Random gene mutations occur at a rate of about 3 per generation (Eyre-Walker and Keightley, 1999; Crow, 1999) and generations last about 25 years. Consequently, in 1.8 million years there have been 72,000 generations, resulting in a total of 216,000 mutations. Given the genome is currently estimated to consist of 20,000–25,000 genes (Venter et al., 2001), and if we assume the mutations occur at random across the genome (this assumption is known to be dubious as there are mutational hot spots, but since we do not know if these hot spots coincide with areas that define the set point we cannot include this into the model), then each gene has on average experienced 3.08 random mutations since the predation transition. We will assume that mutations occurring at random are equally likely to result in an increase or a decrease in the set point (i.e., on average 1.54 mutations result in positive movements and 1.54 mutations result in negative movements) and also that in a single gene model that each mutation results in an 8 point shift in BMI (either up or down).

Because mutations are discontinuous events (i.e., there is no such thing as 1.54 mutations) we will assume that the actual number of mutations in any particular lineage follows a Poisson distribution with a mean intensity of the Poisson process equal to 1.54 mutations (up and down).

Given mutational events are presumed to occur at random we can estimate the probability of any particular combination of numbers of mutations leading to increases and decreases in the set point by combining the respective Poisson probabilities. For example, the probability of a lineage experiencing 5 mutations increasing the set point is 0.0154, and the probability of experiencing 3 mutations decreasing the set point is 0.01305. In combination therefore the probability of experiencing 5 positive and 3 negative mutations comes out at $p = 0.00154 \times 0.01305 = 0.0000219 - 2$ in 100,000 individuals. Since each mutation is assumed to move the set point by 8 units, these individuals would have a BMI set point at the end of the drift period of $20 + (5 \times 8) - (3 \times 8) = 36$. We will assume that there is no negative implication of a high set point because these individuals are not selected against by the absence of predation, but that set points less then 20 are selected against by differential mortality during disease and starvation periods. Thus a lineage with 1 positive and 2 nega-

tive movements in the set point will be eliminated because the resultant set point is only 12 BMI units. We can combine the probabilities that have resultant set points of $\geqq 20$ to evaluate the resultant drifted distribution (Fig. 3). This distribution shows the expected pattern of variation in BMI in a population after 1.8 million years of absence of selection against high set points. This model predicts that 38% of the population would experience no net movement in their set points and still regulate their body weights at BMIs between 16 and 24 (centered on 20), but that the remaining 62% would have BMIs greater than 24, and 30% would have BMIs greater than 30, and 7.9% with BMIs greater than 36. This expected distribution under drift is actually very similar to the present (2000) distribution of BMI in the United States in which 35.5% of the population has BMI <25, and in the remaining population with BMI >25, 30.5% have BMI greater than 30, and 4.7% BMI >40 (Flegal et al., 2002: NHANES III data; www.cdc.gov/nchs). The important point, however, is not that the model closely

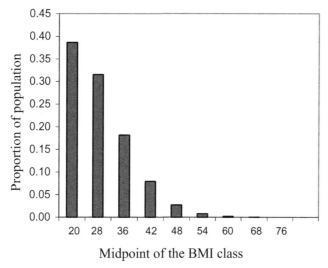

FIGURE 3   The expected distribution of body mass indices in a population after 1.8 million years of drift in the presence of strong selection against reduced set points, but with no selection opposing random mutations causing increased set points to the regulatory system. The model is based on best estimates for the rates of mutation per generation, human generation times, and the time period of 1.8 million years since early hominids dramatically reduced their risks of predation.

predicts the modern distribution of body weights because we do not know how realistic the model assumptions are. If the set point is defined by 20 genes with an effect of 0.2 BMI units, the predicted distribution would have lower numbers of obese, and if it was governed by 80 genes with the same magnitude of effect, it would generate much greater numbers of obese. The key point is that realistic parameterization of the drift model results in a predicted distribution of obesity phenotypes that is not wildly discordant with the current distribution. The "nonadaptive drift model" therefore overcomes the key challenge facing evolutionary scenarios of explaining why only a small proportion of the population gets massively obese (BMI >40) even when the entire population is faced with unlimited access to energy supplies.

VI. TIME TRENDS IN OBESITY PREVALENCE: THE INTERACTION OF PHYSIOLOGY AND SOCIAL FACTORS

The nonadaptive drift model provides a potential explanation for the modern day distribution of body weights. However, it is an inadequate description of the trends in obesity prevalence over time, because it alone cannot explain why trends in obesity have only changed over the past 50 years. Indeed the hypothetical distribution of BMIs in Fig. 3 after 1.8 million years of drift is virtually identical to the distribution predicted if only 1.75 million years of drift had occurred. On one hand this makes the model robust to estimates of the time that social structures evolved, fire and weapons were discovered, and predation eliminated. However, it predicts that the modern distribution of phenotypes (with over 30% of the population obese) should have already been present by the start of the Paleolithic period 50,000 years ago. So some additional factor is needed to reconcile this prediction with the observation that for most commu-

nities the distribution until about 50 years ago conformed pretty much to that anticipated under selection at both ends of the distribution (i.e., averaging a BMI of 20, with few individuals obese: Helmchen and Henderson (2004)).

I suggest that there have been several additional factors operating to historically constrain the actual distribution of body weights from attaining the drifted set points expected from the model. Before discussing these, however, one factor that we need to eliminate as important is changes in life expectancy. Many studies have shown that there is an age-related component to the regulation of body weight. Although the reasons are not yet clear why, the data suggest that our regulated body weights get higher as we get older (DiPietro, 1999; Flegal and Troiano, 2000; Vardi and Pinas-Hamiel, 2000). I developed the drift model predictions (Fig. 3) independent of any considerations of age-related factors. However, if the BMI set point increases with age then changes in life expectancy could in theory have a large impact on the reported BMI distributions. It is generally assumed that life expectancy at birth during most of human history cannot have been lower than about 20 years. Demographic modeling indicates this is a lower limit because the human population would be unable to sustain itself at our maximum rates of fertility if people on average died younger than 20. By 1700, life expectancy in the OECD countries is estimated to have been 30 years (Fogel, 1994). Thus over the entire Paleolithic, Neolithic, and early historical periods most people would have died before they were 30. Life expectancy has increased enormously since the 1700s. On average we live 47 years longer now than we did in 1700 (Fogel, 1994). Modern summaries of BMI distributions of adults therefore pool data over a much broader age spectrum than would be appropriate for societies in the past. Perhaps then the apparent synchrony of the modern distribution with the predictions of the drift model only comes about because of an arte-fact of including much older cohorts in the modern summaries. While this is in theory possible, that data concerning the trends of BMI with age suggest that BMI increases on average only at a rate of about 0.1 units per year. Four decades of extra life conse-quently move the expected distribution upwards by about 4 BMI units (12 kg for a 1.73-m tall individual) (Flegal and Troiano, 2000). Changes in life expectancy over time therefore cannot explain why until 50 years ago people in Western societies were not attaining their drifted set points, whereas more recently they do.

I suggest that there are two separate factors of importance—the level of food supply and the social distribution of it. Before the Neolithic period the most critical factor was probably the level of attainable food supply. Paleolithic individuals proba-bly could not increase their body masses to reach their drifted set points simply because there was insufficient food avail-able for them to do so. At this stage, like most wild animals, each individual or small group would be foraging entirely for their own needs. Things, however, changed in the Neolithic with the advent of agriculture. Subsistence agriculture is not much differ-ent from hunter-gathering—in that each individual grows and harvests food for themselves and/or a small grouping. As yields from agricultural practice improve, however, the numbers of people needed to grow and harvest food as a percentage of the total population starts to decline. It is at this stage that we witness the emergence of much more complex human societies (Diamond, 1997).

These societies are only feasible because it is possible for a subset of individuals to grow and harvest the food for a wider set of individuals to eat. This wider group of individuals then becomes engaged in tasks and activities that would be unfeasible if they spent their whole time growing and harvesting food for themselves—such as building projects with stone, making

pottery, iron, and bronzeware that require the very high temperatures of a kiln and mining ores, and engaging in time-consuming activities such as leisure, organ-ized religion, politics, the arts, and wars. These activities are possible not only if the yields from crops are sufficiently high enough to release some individuals from crop raising, but also if the control of food supply is centralized, so that any food that is produced by one section of society can be distributed to those that do not produce it. This effectively requires monetary systems, most of which have their origins in the shadow of Neolithic agricultural develop-ment. This central control of food supply is important because people can only attain their drifted set points if there is adequate supplies of food for them to do so.

In the Paleolithic period most people could not get access to these resources because there were insufficient resources available to them. After the Neolithic, however, most people could also not get access to unlimited food supplies to attain their drifted set points because of a central societal regulation of access to food. This pattern of access leads to a class-related pattern of variation in body weights. In the lower classes where food supply is restricted people cannot attain their set points, whereas in higher levels of society access to food is effectively unlimited and in these groups attainment of the drifted set point is possible. Consequently, obesity is restricted to the wealthy and powerful. Not all wealthy and powerful people become obese (only those with the genetic predis-position to do so, that is, highly drifted set points), but none of the poorer classes do. Reports of people being obese stretch back at least into early Greek times. In the fifth century B.C., Hippocrates suggested some potential cures for obesity (Procope, 1952). Obviously there would be no need for a cure if nobody suffered from it or its preva-lence was so rare that it was not worth both-ering with. (An interesting point, however, is that even at this stage obesity was not seen as something advantageous or desirable—but something to be "cured." This attitude provides additional evidence against the famine hypothesis—since obesity at this time, when famines were still supposed to be a major selective pressure should have been viewed as advantageous.)

Records from agricultural historians of the levels of food production back to the 1700s support the contention that for the most part people were under socially restricted food supplies that meant they could not attain their drifted set points. In the late 1700s, for example, *per capita* average intake of energy was estimated to be 2060 kcal (8.6 MJ) in Britain, and in France it was 1900 kcal (7.9 MJ). Moreover, it is estimated that 70% of the population of Britain and 90% of the population of France were consuming less than 12 MJ each day. Supplies of food to the lowest 20% of society are considered to have been so low that these people could not effectively participate in the labor force. If only 10% of the population had free access to unlimited energy supplies, then only this proportion of the population would be expected to attain their drifted set points. Obesity prevalence would be expected to be less than 3%. This was the prevalence of obesity in the United States in 1890 (Helmchen and Henderson, 2004). Studies in modern day, third world societies have similar inequalities in nutritional distribution to that experienced in the West before the 1900s (e.g., in India: Sharriff and Mallick (1999)). In these societies obesity is rare, confined to rich upper-class groups, and the majority of the population has BMIs around the low end of the normal range.

It seems that the social conditions concerning food supplies only really started to change to any extent in Western societies after the First World War, in the 1920s. If it was the case that following the First World War greater access to nutritional resources occurred, why did an obesity epidemic not occur then? The answer is that it did. In his classic treatise on basal metabolism published in 1936, Eugene DuBois devoted an entire chapter to the links between obesity and metabolism. His opening statement in that chapter is very revealing, *"Overnutrition is prevalent in our country and obesity is one of the most common diseases. Perhaps it is more discussed by the laity and the public press than any other nutritional disorder."* This statement would not be out of place if he had written it today. These trends, however, were largely reversed when the Western world went back to war in 1939. The modern obesity epidemic, that has developed over the past 50 years, reflects a second wave of obesity as societal access to nutritional resources has again became widespread across all social levels, and the whole of Western society has started to again attain their drifted set points. It has been frequently noted that increases in obesity in other societies coincides with the economic transition of societies from being largely rural to largely urban. Explanations for this trend have largely concerned alterations in levels of physical activity. The evidence supporting the effects of physical activity, however, is relatively weak. It seems plausible that what really alters during these transitions is accessibility to food resources, allowing people to reach their drifted set points.

VII. IMPLICATIONS

The "nonadaptive drift" model presented here provides a very different perspective on the observation that obese people have high leptin levels. One interpretation of these high levels is that humans cannot have a functioning lipostatic regulation system or else they would act on this signal and reduce their body fatness accordingly. Attention has therefore been directed toward understanding why this signal is not being correctly read—the concept of "leptin resistance." However, such an interpretation is rooted in the assumption that the set point of any regulatory system is set at the level of a lean person with a BMI around 20. However, if a person with a BMI of 35 has a set point of their regulatory system set at a BMI of 35, as suggested by the drift model, then the system may be working exactly as it should do. The fat stores of the obese person produces an appropriate leptin signal for the level of adiposity, the brain detects this signal normally, and regulates the food intake (if it is available) accordingly, driving them to eat more food if their BMI falls below 35 and to eat less if it rises greater. Leptin resistance is therefore only a real phenomenon if one considers that the set point is fixed in everyone at a low level. There is no evidence to support this interpretation.

If the suggestion that obese people have functioning lipostatic systems centered on drifted set points is correct, then it is valid to inquire why clinical trials using exogenous leptin have been so unsuccessful (Heymsfield et al., 1999). In fact, clinical trials using leptin have been more successful than is widely acknowledged. For example, the data from the first trial show that extent of mass loss was related to leptin treatment in a dose-dependent manner, as anticipated it should be, and those on the highest dose (0.3 mg/kg/day) on average lost 10 times the amount of weight (and fat) of those on the lowest dose (0.01 mg/ kg/day) (7.1 kg versus 0.7 kg). More recent use of leptin in patients following weight loss to stem postrestriction weight rebound have been very successful (Rosenbaum et al., 2002, 2005). These patterns are exactly what would be predicted if obese subjects have an intact regulatory system, but what differs between somebody with a BMI of 35 and somebody with a BMI of 20 is not that the regulatory system is somehow broken in the person with a BMI of 35, but that it is simply working with a different target (see also Leibel, 1990). The model I have presented here provides an evolutionary framework for understanding why this diversity of targets exists.

The model of drifting set points has several additional implications. The first is that the obesity epidemic is predicted to have a limited extent. Once populations attain the drifted set-point distribution it is predicted that there will be a state of stability and the epidemic will grow no worse. Unfortunately, because we know next to nothing about the genetic basis of the system we cannot predict when that state will be attained. The model I presented in Fig. 3 is consistent with the current phenotype distribution in the United States, but whether this model is realistic or not (in terms of the numbers of genes and magnitudes of their effects) remains to be tested.

The second implication is that dietary interventions alone are doomed to failure. Around 420 B.C. Hippocrates suggested that the cure for obesity was to eat less food. One might have thought that two and a half thousand years of failure using this strategy would have been enough to get the message across—but the main problem is that dietary interventions do work in the short term. This is because one can overcome the regulatory system for a while by cognitive means. We can decide to ignore the signals telling us to eat more food. But we can only do this for a limited time period. Ultimately, our physiology gains the upper hand again, and our body weights increase back to the target level. Food companies are able to profit therefore, selling people the dream that they have a simple dietary intervention system that works, without seriously diminishing their target market by providing something that actually does work in the long term.

The third implication is that, like dietary treatments, drug treatments that interfere with the appetite system (or theoretically the energy expenditure system) will also work as long as the drugs are being taken. Given the manner in which the system works, it is completely unrealistic to expect any obesity treatment that is drug-based and relies on interference with appetite or energy absorption alone to have an impact that extends beyond the time that the drug is being taken. The final implication is that an important key to understanding and potentially discovering new solutions to the obesity problem is to find the molecular basis for the actual set point of the system. Unfortunately, while we can postulate that such a system exists and we know much about the likely peripheral signals that code the level of fat storage, we know virtually nothing about the coding in the brain to which these levels are compared (and few people are even looking). Yet once we know how the set point is encoded, manipulating it will become a rich vein of pharmaceutical discovery. Even then, however, it seems probable that pharmacological interventions will only be successful as long as the drugs are being taken.

Acknowledgments

I am grateful to many people for interesting discussions about the ideas in this chapter and their constructive suggestions. I am particularly grateful to Peter Grant for his complete refusal to believe the arguments and his ability to come up with counter suggestions to everything I come up with. I am also grateful to Ela Krol for her meticulous reading of initial drafts, and her suggestions for many improvements in the chapter and the model.

References

Abdel-Hamid, T. K. (2002). Modeling the dynamics of human energy regulation and its implications for obesity treatment. *Syst. Dynam. Rev.* **18**, 431–471.

Adami, G. F., et al. (1998). Relationship between body mass index and serum leptin concentration. *Diabetes Nutr. Metab.* **11**, 17–19.

Addis, G., Urga, K., and Dikasso, D. (2005). Ethnobotanical study of edible wild plants in some selected districts of Ethiopia. *Hum. Ecol.* **33**, 83–118.

Adriaensen, F., et al. (1998). Stabilizing selection on blue tit fledgling mass in the presence of sparrowhawks. *Proc. R. Soc. Lond. Ser. B Biol. Sci.* **265**, 1011–1016.

Alemu, T., and Lindtjorn, B. (1995). Physical-activity, illness and nutritional-status among adults in a rural Ethiopian community. *Int. J. Epidemiol.* **24**, 977–983.

Baldellou, M., and Henzi, S. P. (1992). Vigilance, predator detection and the presence of supernumerary males in vervet monkey troops. *Anim. Behav.* **43**, 451–461.

Ballor, D. L., and Poehlman, E. T. (1995). A meta analysis of the effects of exercise and/or dietary restriction on resting metabolic-rate. *Eur. J. Appl. Physiol. Occupat. Physiol.* **71**, 535–542.

Barsh, G. S., Farooqi, I. S., and O'Rahilly, S. (2000). Genetics of body-weight regulation. *Nature* **404**, 644–651.

Bobek, B. (1969). Survival, production and turnover of small rodents in a beech forest. *Acta Theriol.* **14**, 191–210.

Boyle, P. C., Storlien, L. H., Harper, A. E., and Keesey, R. E. (1981). Oxygen consumption and locomotor activity during restricted feeding and realimentation. *Am. J. Physiol.* **241**, R392–R397.

Brain, C. K. (1981). The hunters or the hunted? "An Introduction to African Cave Taphonomy." University of Chicago Press, Chicago.

Bribiescas, R. G. (2001). Serum leptin levels and anthropometric correlates in Ache Amerindians of eastern Paraguay. *Am. J. Phys. Anthropol.* **115**, 297–303.

Broberger, C. (2005). Brain regulation of food intake and appetite: molecules and networks. *J. Intern. Med.* **258**, 301–327.

Brodin, A. (2000). Why do hoarding birds gain fat in winter in the wrong way? Suggestions from a dynamic model. *Behav. Ecol.* **11**, 27–39.

Brodin, A. (2001). Mass-dependent predation and metabolic expenditure in wintering birds: is there a trade-off between different forms of predation? *Anim. Behav.* **62**, 993–999.

Brown, M. M., et al. (1992). Silhouettes elicit alarm calls from captive vervet monkeys (*Cercopithecus aethiops*). *J. Comp. Psychol.* **106**, 350–359.

Campbell, B., O'Rourke, M. T., and Lipson, S. F. (2003). Salivary testosterone and body composition among Ariaal males. *Am. J. Hum. Biol.* **15**, 697–708.

Campbell, B. C., and Cajigal, A. (2001). Diabetes: energetics, development and human evolution. *Med. Hypoth.* **57**, 64–67.

Carmichael, A. (1983). Infection, hidden hunger and history. *In* "Hunger and History: The Impact of Changing Food Production and Consumption Patterns on Society" (R. I. Rotberg and T. K. Rabb, eds.), pp. 51–66. Cambridge University Press, Cambridge.

Chakravarthy, M. V., and Booth, F. W. (2004). Eating, exercise, and "thrifty" genotypes: connecting the dots toward an evolutionary understanding of modern chronic diseases. *J. Appl. Physiol.* **96**, 3–10.

Cheney, D. L., and Seyfarth, R. M. (1985). Vervet monkey alarm calls—manipulation through shared information. *Behaviour* **94**, 150–166.

Considine, R. V., et al. (1996). Serum immunoreactive leptin concentrations in normal-weight and obese humans. *N. Engl. J. Med.* **334**, 292–295.

Covas, R., et al. (2002). Stabilizing selection on body mass in the sociable weaver *Philetairus socius*. *Proc. R. Soc. Lond. Ser. B Biol. Sci.* **269**, 1905–1909.

Cresswell, W. (1998). Diurnal and seasonal mass variation in blackbirds *Turdus merula*: consequences for mass-dependent predation risk. *J. Anim. Ecol.* **67**, 78–90.

Crow, J. (1999). The odds of losing at genetic roulette. *Nature* **397**, 293–294.

Cuthill, I. C., et al. (2000). Body mass regulation in response to changes in feeding predictability and overnight energy expenditure. *Behav. Ecol.* **11**, 189–195.

De Castro, J. M. (2004). Genes, the environment and the control of food intake. *Br. J. Nutr.* **92**, S59–S62.

De Castro, J. M., and Plunkett, S. (2002). A general model of intake regulation. *Neurosci. Biobehav. Rev.* **26**, 581–595

Diamond, J. (1997). "Guns, Germs and Steel: The Fates of Human Societies." Norton, New York.

Diamond, J. (2003). The double puzzle of diabetes. *Nature* 599–602.

DiPietro, L. (1999). Physical activity in the prevention of obesity: current evidence and research issues. *Med. Sci. Sports Exer.* **31**(Suppl. S), S542–S546.

Dirks, R. (1993). Famine and disease. *In* "The Cambridge World History of Human Disease" (K. F. Kiple, ed.), pp. 157–163. Cambridge University Press, Cambridge.

Dulloo, A. G. (1997). Human pattern of hyperphagia and feed partitioning during weight recovery after starvation: a theory of autoregulation of body composition. *Proc. Nutr. Soc.* **56**, 25–40.

Dulloo, A. G. (2005). A role for suppressed skeletal muscle thermogenesis in the pathways from weight fluctuations to the insulin resistance syndrome. *Acta Physiol. Scand.* **184**, 295–307.

Dupaquier, J. (1979). L'analyse statistique des crises de mortalitie. *In* "The Great Mortalities" (H. Charbonneau and A. LaRose, eds.), pp. 83–112. Ordina, Liege.

Eaton, S. B., Konner, M., and Shostak, M. (1988). Stoneagers in the fast lane: chronic degenerative diseases in evolutionary perspective. *Am. J. Med.* **84**, 739–749.

Elia, M. (2000). Hunger disease. *Clin. Nutr.* **19**, 379–386.

Erlanson-Albertsson, C. (2005). How palatable food disrupts appetite regulation. *Basic Clin. Pharmacol. Toxicol.* **97**, 61–73.

Even, P. C., and Nicolaidis, S. (1993). Adaptive changes in energy expenditure during mild and severe feed restriction in the rat. *Br. J. Nutr.* **70**, 421–431.

Eyre-Walker, A., and Keightley, P. D. (1999). High genomic deleterious mutation rates in hominids. *Nature* **397**, 344–347.

Flegal, K. M. (2005). Epidemiologic aspects of overweight and obesity in the United States. *Physiol. Behav.* **86**, 599–602.

Flegal, K. M., and Troiano, R. P. (2000). Changes in the distribution of body mass index of adults and children in the US population. *Int. J. Obes.* **24**, 807–818.

Flegal, K. M., et al. (2002). Prevalence and trends in obesity among US adults, 1999–2000. *JAMA* **288**, 1723–1727.

Fogel, R. W. (1994). New findings about trends in life expectation and chronic disease: the implications for health costs and pensions. Selected paper 76. Chicago Business School, Chicago.

Fransson, T., and Weber, T. P. (1997). Migratory fuelling in blackcaps (*Sylvia atricapilla*) under perceived risk of predation. *Behav. Ecol. Sociobiol.* **41**, 75–80.

Friedman, J. M. (2002). The function of leptin in nutrition; weight, and physiology. *Nutr. Rev.* **60**, S1–S14.

Gentle, L. K., and Gosler, A. G. (2001). Fat reserves and perceived predation risk in the great tit, *Parus major*. *Proc. R. Soc. Lond. Ser. B Biol. Sci.* **268**, 487–491.

Girardier, L. (1994). Self-regulation of weight and body-composition in humans—systemic approach using models and simulation. *Arch. Int. Physiol. Biochim. Biophys.* **102**, A23–A35.

Gonzales-Pacheco, D. M., Buss, W. C., Koehler, K. M., Woodside, W. F., and Alpert, S. S. (1993). Energy restriction reduces metabolic rate in adult male Fisher-344 rats. *J. Nutr.* **123**, 90–97.

Haffner, S. M., et al. (1996). Leptin concentrations in relation to overall adiposity and regional body fat distribution in Mexican Americans. *Int. J. Obes.* **20**, 904–908.

Halaas, J. L., et al. (1997). Physiological response to long-term peripheral and central leptin infusion in lean and obese mice. *Proc. Nat. Acad. Sci. U.S.A.* **94**, 8878–8883.

Hambly, C., and Speakman, J. R. (2005). Contribution of different mechanisms to compensation for energy restriction in the mouse. *Obes. Res.* **13**, 1548–1557.

Harris, R. B. S. (1990). Role of set-point theory in regulation of body-weight *FASEB J.* **4**, 3310–3318.

Hart, D., and Sussman, R. W. (2005). "Man the Hunted. Primates, Predators and Human Evolution." Westview Press, Boulder, CO.

Hebebrand, J. (2005). Is obesity heritable? Connection between genetics and overweight. *Ernahrungs-Umschau* **52**, 90.

Helmchen, L. A., and Henderson, R. M. (2004). Changes in the distribution of body mass index of white US men, 1890–2000. *Ann. Hum. Biol.* **31**, 174–181.

Heymsfield, S. B., et al. (1999). Recombinant leptin for weight loss in obese and lean adults—A randomized, controlled, dose-escalation trial. *JAMA* **282**, 1568–1575.

Hill, J. O., Fried, S. K., and DiGirolamo, M. (1984). Effects of fasting and restricted feeding on utilization of ingested energy in rats. *Am. J. Physiol. Regul. Integr. Comp. Physiol.* **248**, R318–R327.

Hill, J. O., Latiff, A., and DiGirolamo, M. (1985). Effects of variable caloric restriction on utilisation of ingested energy in rats. *Am. Physiol. Soc.* **85**, R549–R559.

Hinney, A., et al. (1997). Absence of leptin deficiency mutation in extremely obese German children and adolescents. *Int. J. Obes.* **21**, 1190.

Hirvonen, M. D., and Keesey, R. E. (2000). The regulation of body weight: set-points and obesity. *Handb. Exp. Pharmacol.* **149**, 133–151.

Ho, P. T. (1959). "Studies on the Population of China 1368–1953". Harvard University Press, Cambridge, MA.

Kalra, S. P. (1997). Appetite and body weight regulation: Is it all in the brain? *Neuron* **19**, 227–230.

Keesey, R. E. (1995). A set-point of body weight regulation and its implications for obesity. *In* "Comprehensive Textbook of Eating Disorders and Obesity" (K. D. Brownell and C. G. Fairburn, eds.), pp. 46–50. Guilford Press, New York.

Keesey, R. E., and Corbett, S. W. (1984). Metabolic defence of the body weight set-point. *In* "Eating and Its Disorders" (A. J. Stunkard and E. Stellar, eds.), pp 87–96. Raven Press, New York.

Keesey, R. E., and Corbett, S. W. (1990). Adjustments in daily energy expenditure to caloric restriction and weight loss in adult obese and lean Zucker rats. *Int. J. Obes.* **14**, 1079–1084.

Keesey, R. E., and Hirvonen, M. D. (1997). Body weight set-points: Determination and adjustment. *J. Nutr.* **127**, S1875–S1883.

Kennedy, G. C. (1953). The role of depot fat in the hypothalamic control of food intake in the rat. *Proc. R. Soc. Lond. Biol. Sci.* **140**, 578–592.

Kesteloot, H., et al. (1997). Serum lipid levels in a Pygmy and Bantu population sample from Cameroon. *Nutr. Metab. Cardiovasc. Dis.* **7**, 383–387.

Keys, A. J., et al. (1950). "The Biology of Human Starvation." University of Minnesota Press, Minnesota.

Kirchengast, S. (1998). Weight status of adult !Kung San and Kavango people from northern Namibia. *Ann. Hum. Biol.* **25**, 541–551.

Kullberg, C., Fransson, T., and Jakobsson, S. (1996). Impaired predator evasion in fat blackcaps (*Sylvia atricapilla*). *Proc. R. Soc. Lond. Ser. B Biol. Sci.* **263**, 1671–1675.

Leibel, R. L. (1990). Is obesity due to a heritable difference in set point for adiposity? *West. J. Med.* **153**, 429–431.

Leibel, R. L., Rosenbaum, M., and Hirsch, J. (1995). Changes in energy expenditure resulting from altered body weight. *N. Engl. J. Med.* **332**, 621–628.

Levin, B. E. (2005). Factors promoting and ameliorating the development of obesity. *Physiol. Behav.* **86**, 633–639.

Levin, B. E., and Keesey, R. E. (1998). Defense of differing body weight set-points in diet-induced obese and resistant rats. *Am. J. Physiol.* **274**, R412–R419.

Levitsky, D. A. (2002). Putting behavior back into feeding behavior: a tribute to George Collier. *Appetite* **38**, 143–148.

Lev-Ran, A. (1999). Thrifty genotype: how applicable is it to obesity and type 2 diabetes? *Diabetes Rev.* **7**, 1–22.

Lev-Ran, A. (2001). Human obesity: an evolutionary approach to understanding our bulging waistline. *Diabetes Metab. Res. Rev.* **17**, 347–362.

Lindtjorn B. (1990). Famine in southern Ethiopia, 1985–86: population structure, nutritional state and incidence of death. *Br. Med. J.* **301**, 1123–1127.

Macias, A. E. (2004). Experimental demonstration of human weight homeostasis: implications for understanding obesity. *Br. J. Nutr.* **91**, 479–484.

Maffei, M., et al. (1995). Leptin levels in human and rodent—measurement of plasma leptin and ob RNA in obese and weight-reduced subjects. *Nat. Med.* **1**, 1155–1161.

Masuzaki, H., Ogawa, Y., Hosoda, K., Hiraoka, J., Hanaoka, I., Miyawaki, T., Matsuoka, N., and Nakao, K. (1997). Generation of transgenic mice overexpressing leptin and long-term effects of leptin in the regulation of body weight. *Diabetes* **46**, 990.

Mauer, M. M., Harris, R. B. S., and Bartness, T. J. (2001). The regulation of total body fat: lessons learned from lipectomy studies. *Neurosci. Biobehav. Rev.* **25**, 15–28.

McCance, R. A. (1975). Famines of history and of today. *Proc. Nutr. Soc.* **34**, 161–166.

Mellinkoff, S., Franklin, M., Boyle, D., and Geipell, G. (1956). Relationship between serum amino acid concentration and fluctuation in appetite. *J. Appl. Physiol.* **8**, 535–538.

Mercer, J. G., and Speakman, J. R. (2001). Hypothalamic neuropeptide mechanisms for regulating energy balance: from rodent models to human obesity. *Neurosci. Biobehav. Rev.* **25**, 101–116.

Millward, D. J., and Wijesinghe, D. G. N. G. (1998). Energy partitioning and the regulation of body weight. *Br. J. Nutr.* **79**, 111–113.

Mistry, A. M., Swick, A. G., and Romsos, D. R. (1997). Leptin rapidly lowers food intake and elevates metabolic rates in lean and ob/ob mice. *J. Nutr.* **127**, 2065–2072.

Montague, C. T., et al. (1997). Congenital leptin deficiency is associated with severe early-onset obesity in humans. *Nature* **387**, 903–908.

Morgan, P. J., and Mercer, J. G. (2001). The regulation of body weight: lessons from the seasonal animal. *Proc. Nutr. Soc.* **60**, 127–134.

Munch, I. C., Markussen, N. H., and Oritsland, N. A. (1993). Resting oxygen-consumption in rats during food restriction, starvation and refeeding. *Acta Physiol. Scand.* **148**, 335–340.

Murray, M., et al. (1976). Somali food shelters in the Ogaden famine and their impact on health. *Lancet* **332**, 1283–1285.

Nair, K. S., et al. (1981). Effect of caloric restriction in obese subjects on resting metabolic-rate, protein-turnover, peripheral t4 metabolism and glucose-oxidation. *Clin. Sci.* **61**, 9–10.

Neel, J. V. (1962). Diabetes mellitus a "thrifty" genotype rendered detrimental by "progress"? *Am. J. Hum. Genet.* **14**, 352–353.

Odea, K. (1991). Cardiovascular-disease risk-factors in Australian aborigines. *Clin. Exp. Pharmacol. Physiol.* **18**, 85–88.

Owren, M. J., and Bernacki, R. H. (1988). The acoustic features of vervet monkey alarm calls. *J. Acoust. Soc. Am.* **83**, 1927–1935.

Park, A. J., and Bloom, S. R. (2005). Neuroendocrine control of food intake. *Curr. Opin. Gastroenterol.* **21**, 228–233.

Payne, P. R., and Dugdale, A. E. (1977). A model for the prediction of energy balance and body weight. *Ann. Hum. Biol.* **4**, 525–535.

Peacock, W. L., and Speakman, J. R. (2001). Effect of high-fat diet on body mass and energy balance in the bank vole. *Physiol. Behav.* **74**, 65–70.

Pelleymounter, M. A. (1995). Effects of the obese gene product on body weight regulation in ob/ob mice. *Science* **269**, 540–543.

Pelleymounter, M. A., et al. (1998). Efficacy of exogenous recombinant murine leptin in lean and obese 10- to 12-mo-old female CD-1 mice. *Am. J. Physiol. Regul. Integr. Comp. Physiol.* **44**, R950–R959.

Perusse, L., et al. (1998). Genetic epidemiology and molecular genetics of obesity: results from the Quebec Family Study. *Med. Sci.* **14**, 914–924.

Pickering, T. R., Clarke, R. J., and Moggi-Cecchi, J. (2004). Role of carnivores in the accumulation of the Sterkfontein Member 4 hominid assemblage: a taphonomic reassessment of the complete hominid fossil sample (1936–1999). *Am. J. Phys. Anthropol.* **125**, 1–15.

Platek, S. M., Gallup, G. G., and Fryer, B. D. (2002). The fireside hypothesis: was there differential selection to tolerate air pollution during human evolution? *Med. Hypoth.* **58**, 1–5.

Pravosudov, V. V., and Grubb, T. C. (1998). Management of fat reserves in tufted titmice (*Parus bicolor*): evidence against a trade-off with food hoards. *Behav. Ecol. Sociobiol.* **42**, 57–62.

Pravosudov, V. V., and Lucas, J. R. (2001). Daily patterns of energy storage in food-caching birds under variable daily predation risk: a dynamic state variable model. *Behav. Ecol. Sociobiol.* **50**, 239–250.

Prentice, A. M. (2001). Obesity and its potential mechanistic basis. *Br. Med. Bull.* **60**, 51–67.

Prentice, A. M. (2005a). Early influences on human energy regulation: thrifty genotypes and thrifty phenotypes. *Physiol. Behav.* **86**, 640–645.

Prentice, A. M. (2005b). Starvation in humans: Evolutionary background and contemporary implications. *Mech. Age. Dev.* **126**, 976–981.

Prentice, A. M., Rayco-Solon, P., and Moore, S. E. (2005). Insights from the developing world: thrifty genotypes and thrifty phenotypes. *Proc. Nutr. Soc.* **64**, 153–161.

Procope, J. (1952). "Hippocrates on Diet and Hygiene." Zeno, London.

Ravussin, E. (2002). Cellular sensors of feast and famine. *J. Clin. Invest.* **109**, 1537–1540.

Romsos, D. R., et al. (1996). Intracerebroventricular recombinant leptin decreases food intake and increases metabolic rate in ob/ob mice. *FASEB J.* **10**, 1287.

Rosenbaum, M., et al. (2002). Low dose leptin administration reverses effects of sustained weight-reduction on energy expenditure and circulating concentrations of thyroid hormones. *J. Clin. Endocrinol. Metab.* **87**, 2391–2394.

Rosenbaum, M., et al. (2005). Low-dose leptin reverses skeletal muscle, autonomic, and neuroendocrine adaptations to maintenance of reduced weight. *J. Clin. Invest.* **115**, 3579–3586.

Samec, S., Seydoux, J., and Dulloo, A. G. (1998). Interorgan signaling between adipose tissue metabolism and skeletal muscle uncoupling protein homologs—Is there a role for circulating free fatty acids? *Diabetes* **47**, 1693–1698.

Schmidt-Nielsen, K. (1997). "Animal Physiology: Adaptation and Environment," 5th ed. Cambridge University Press, Cambridge.

Schwarz, J. M., et al. (1997). Effect of diet on leptin: a signal of surplus dietary fat in lean and obese men? *Diabetes* **46**, 942.

Schwartz, J. M., et al. (1999). Model for the regulation of energy balance and adiposity by the central nervous system. *Am. J. Clin. Nutr.* **69**, 584–596.

Shariff, A., and Mallick, A. C. (1999). Dynamics of food intake and nutrition by expenditure class in India. *Econ. Polit. Weekly* **34**, 1790–1800.

Speakman, J. R. (2004). Obesity: the integrated roles of environment and genetics. *J. Nutr.* **134**, 2090–2105S.

Speakman, J. R. (2006a). The genetics of obesity: five fundamental problems with the famine hypothesis. In "Adipose Tissue and Adipokines in Health and Disease" (G. Fantuzzi and T. Mazzone, eds.), (in press). Humana Press.

Speakman, J. R. (2006b). "Thrifty genes" for obesity and the metabolic syndrome: time to call off the search? *Diabetes Vasc. Res.* (in press).

Speakman, J. R., Stubbs, R. J., and Mercer, J. G. (2002). Does body mass play a role in the regulation of body mass? *Proc. Nutr. Soc.* **61**, 473–487.

Speakman, J. R., et al. (2003). Resting and daily energy expenditures of free-living field voles are positively correlated but reflect extrinsic rather than intrinsic effects. *Proc. Nat. Acad. Sci. U.S.A.* **100**, 14057–14062.

Stearns, P. N. (2001). "The Encylopedia of World History," 6th ed. Houghton Mifflin, New York.

Tartaglia, L. A., et al. (1995). Identification and expression cloning of a leptin receptor, OB-R. *Cell* **83**, 1263–1271.

Toole, M. J., and Waldman, R. J. (1988). An analysis of mortality trends among refugee populations in Somalia, Sudan, and Thailand. *Bull. WHO* **66**, 237–247.

Treves, A., and Naughton-Treves, L. (1999). Risk and opportunity for humans coexisting with large carnivores. *J. Hum. Evol.* **36**, 275–282.

van der Veen, I. T. (1999). Effects of predation risk on diurnal mass dynamics and foraging routines of yellowhammers (*Emberiza citrinella*). *Behav. Ecol.* **10**, 545–551.

van der Veen, I. T., and Sivars, L. E. (2000). Causes and consequences of mass loss upon predator encounter: feeding interruption, stress or fit-for-flight? *Funct. Ecol.* **14**, 638–644.

Vardi, P., and Pinhas-Hamiel, O., (2000). The young hunter hypothesis: age-related weight gain—a tribute to the thrifty theories. *Med. Hypoth.* **55**, 521–523.

Venter, J. C., et al. (2001). The sequence of the human genome. *Science* **291**, 1304–1351

Vergoni, A. V., and Bertolini, A. (2000). Role of melanocortins in the central control of feeding. *Eur. J. Pharmacol.* **405**, 25–32.

Walker, A. R. P., Adam, F., and Walker, B. F. (2001). World pandemic of obesity: the situation in Southern African populations. *Public Health* **115**, 368–372.

Watkins, S. C., and Menken, J. (1985). Famines in historical perspective. *Popul. Dev. Rev.* **11**, 647–675.

Weinsier, R. L, Nagy, T. R., Hunter, G. R., Darnell, B. E., Hensrud, D. D., and Weis, H. L. (2000). Do adaptive changes in metabolic rate favor weight regain in weight-reduced individuals? An examination of the set-point theory. *Am. J. Clin. Nutr.* **72**, 1088–1094.

Widmaier, D. (1998). "Why Geese Don't Get Obese: and We Do. How Evolution's Strategies for Survival Affect Our Everyday Lives." Freeman, New York.

Wilkin, T. J., and Voss, L. D. (2004). Metabolic syndrome: maladaptation to a modern world. *J. R. Soc. Med.* **97**, 511–520.

Williams, G., Harrold, J. A., and Cutler, D. J. (2000). The hypothalamus and the regulation of energy homeostasis: lifting the lid on a black box. *Proc. Nutr. Soc.* **59**, 385–396.

Wirtshafter, and Davies (1977). Set points, settling points, and control of body-weight. *Physiol. Behav.* **19**, 75–78.

WHO (1998). Obesity: preventing and managing the global epidemic. Report 894. Geneva, WHO.

Wrigley, E. A., and Schofield, R. (1981). "The Population History of England, 1541–1871." Harvard University Press, Cambridge, MA.

Yun, S., et al. (2006). A comparison of national estimates of obesity prevalence from the behavioral risk factor surveillance system and the national health and nutrition examination survey. *Int. J. Obes.* **30**, 164–170.

Zhang, Y., et al. (1994). Positional cloning of the mouse obese gene and its human homologue. *Nature* **372**, 425–432.

13

Preclinical Developments in Antiobesity Drugs

STEVEN P. VICKERS
Nottingham, RenaSci Consultancy Ltd.
SHARON C. CHEETHAM
RenaSci Consultancy Ltd., Nottingham

I. Introduction
II. CNS Targets for Novel Antiobesity Drugs
    A. Cannabinoid CB_1 Receptor Antagonists
    B. 5-HT_{2C} Receptor Agonists
    C. 5-HT_6 Receptor Antagonists
    D. Melanin-Concentrating Hormone Receptor Antagonists
    E. MC4 Receptor Antagonists
    F. Neuropeptide Y
    G. Other Central Strategies
III. Peripheral Targets for Novel Antiobesity Drugs
    A. Leptin
    B. PYY
    C. Carbonic Anhydrase Inhibitors
    D. 11β-HSD1 Inhibitors
    E. GLP-1, and DPP-IV Inhibitors
    F. Additional Peripheral Targets
IV. Conclusion
    References

The worldwide prevalence of obesity is estimated to exceed 300 million, with more than 30% of U.S. adults being obese. This incidence is despite the increased availability of low calorie and low fat foods, fitness center, and "healthy lifestyle" magazine literature. Accordingly, it is increasingly recognized that drug treatment is required to help address the problem. The clinical utility of current antiobesity treatments is limited by the incidence of cardiovascular-related side effects in the case of sibutramine (Meridia/Reductil), gastrointestinal disturbance in the case of orlistat (Xenical), and poor efficacy in the case of both drugs. Consequently, there is much effort in developing more effective and safer treatments than the antiobesity drugs currently on the market. The regulation of energy balance, or its imbalance in obesity, is the result of a complex multilayered system involving hormones and neurotransmitters which signal to the brain, the gut, the liver, and other organs which integrate this information. As the understanding of these systems increases, so does the number of potential targets for novel drug development. Not surprisingly, preclinical approaches in the development of antiobesity treatments include the targeting of central and peripheral mechanisms. This chapter reviews the preclinical targets that we believe warrant special attention and could lead to the development of safer, more effective, drugs for the treatment of obesity.

I. INTRODUCTION

The worldwide prevalence of obesity is estimated to exceed 300 million, with the greatest incidence being in the United States, where over 30% of adults are classified as obese. Worryingly, estimates suggest that the worldwide incidence of obesity in children was 30–45 million in 2000 (Lobstein et al., 2004). Obesity is a major risk factor in the development of chronic life-threatening diseases such as type II diabetes, coronary heart disease, hypertension, stroke, osteoarthritis, and some cancers (Bonow and Eckel, 2003; http://www.iotf.org/). Accordingly, obesity has become a major public health problem that significantly reduces life expectancy (Fontaine et al., 2003) and poses a considerable economic burden. For example, it has been estimated that obesity in England accounted for 18 million days of sickness absence and 30,000 premature deaths in 1998 (National Audit Office U.K., 2001). In the United States, obesity-related costs in 1995 were estimated to be an astounding US$99.5 billion (Wolf and Colditz, 1998). Not surprisingly, since this rising incidence of obesity and associated costs is occurring despite a marked increase in the availability of low fat foods, fitness centers, and "healthy living" literature in newspapers and magazines, there is considerable inter-est in the research and development of novel, safe, and effective drugs for the treatment of obesity.

Current drugs prescribed for the treatment of obesity are the peripheral lipase inhibitor, orlistat (Xenical), the serotonin (5-hydroxytryptamine (5-HT)) and noradrenaline reuptake inhibitor, sibutramine (Meridia/Reductil), and the generic catecholamine, phentermine (a releaser of noradrenaline and dopamine). On average, chronic treatment with these drugs typically leads to a 5–10% weight loss, a value significantly less than that desired by patients and clinicians. Furthermore, not only do these drugs deliver poor efficacy but, in addition, they possess unfavorable side-effect profiles. Hence, clinical utility is limited by the incidence of increased blood pressure (sibutramine, phentermine), tachycardia (sibutramine), gastrointestinal disturbance (orlistat), and abuse liability (phentermine). Accordingly, there remains a clear, unmet medical need for safe and effective drugs for the treatment of obesity. With global sales of obesity treatments expected to reach more than US$1.5 billion in 2005 (Nicholas Hall & Co), and some analysts predicting that the market size for an efficacious and safe drug could reach US$3.7 billion by 2008, it is not surprising that a number of pharmaceutical companies have well resourced research programs focused on the development of novel drugs for the treatment of obesity.

The regulation of energy balance, or its dysregulation in obesity, is the result of a complex multilayered system involving hormones and neurotransmitters which communicate between, and integrate information from the brain, gut, liver, and other organs. As the understanding of these systems increases, so does the number of potential targets for novel drug development. Not surprisingly, current preclinical approaches in the development of antiobesity treatments include the targeting of central and peripheral mechanisms.

II. CNS TARGETS FOR NOVEL ANTIOBESITY DRUGS

The role of the central nervous system (CNS) in mediating energy balance is long established, with the hypothalamus believed to be of special importance (Horvath, 2005; see also Chapter 4). The most promising molecular strategies that target the CNS are detailed later and are also summarized in Table 1.

A. Cannabinoid CB₁ Receptor Antagonists

A substantial literature exists in support of the potential utility of CB_1 receptor antagonists for the treatment of obesity.

Indeed, the topic has been subject to a large number of reviews (e.g., Black, 2004; Kirkham and Williams, 2004; Kirkham, 2005; Vickers and Kennett, 2005). In a number of species, including man, administration of exogenous and endogenous cannabinoids leads to robust increases in food intake, an effect mediated through activation of the CB_1 receptor (Foltin et al., 1988; Williams et al., 1998; Williams and Kirkham, 1999, 2002; Kirkham and Williams 2001; Kirkham, 2005).

The first potent, selective, and orally available CB_1 receptor antagonist to be discovered was SR141716 (rimonabant, Acomplia; Rinaldi-Carmona et al., 1994). In light of reports that CB_1 receptor activation increases food intake, it is perhaps

TABLE 1   Preclinical Antiobesity Drug Targets: A Selection

Molecular target	Compound	Companies
Cannabinoid CB_1 receptor antagonists	Rimonabant (Acomplia; SR141716)	Sanofi-Aventis
	SR147778	
	SLV-319	Solvay
	V24343	Vernalis
	E-6776	Esteve
$5\text{-}HT_{2C}$ receptor antagonists	APD356	Arena Pharmaceuticals
	V17627	Vernalis
$5\text{-}HT_6$ receptor antagonists		Biovitrum
	SUVN503; SUVN504	Suven Pharmaceuticals
	PRX-07034	Predix
		Esteve
MCH_1 receptor antagonists	T-226296	Takeda
		Schering-Plough
		Merck
		Neurocrine
	AMG-076	Amgen
	856464	GSK
MC_4 receptor agonists	PGE-657022	Proctor & Gamble
Y2/Y4 receptor agonists	TM-30338	7TM Pharma
Y4 receptor agonists	TM-30339	7TM Pharma
Leptin agonists	CBT-1452	Cambridge Biotechnology/ Biovitrum
$11\beta\text{-}HSD1$ inhibitors	BVT-2733; BVT-3498	Biovitrum/Amgen Merck
DPP-IV inhibitors	Vildagliptin (LAF237)	Novartis
	PSN9301	Prosidion Biovitrum/Santhera

unsurprising that rimonabant has been widely reported to reduce both food intake and body weight in lean and obese laboratory animals (Colombo et al., 1998; Ravinet-Trillou et al., 2003; Vickers et al., 2003). These findings have been reinforced by reports that CB_1 receptor knockout mice are resistant to the development of diet-induced obesity (Ravinet-Trillou et al., 2004).

Interestingly, there is increasing evidence from the preclinical literature that the reductions in food intake and, more particularly of body weight, observed after treatment with CB_1 receptor antagonists such as rimonabant, may be mediated via the activation of metabolic processes. For example, rimonabant-treated mice exhibit an increased weight loss in response to a 24-h fast compared to vehicle-treated animals (Ravinet-Trillou et al., 2003). Furthermore, at least some pair-feeding studies have shown that rimonabant-induced weight loss is greater than that observed in pair-fed controls (Ravinet-Trillou et al., 2003). Consistent with the hypothesis that the body weight loss consequent to chronic CB_1 receptor antagonist administration is, at least in part, independent of food intake, it has been reported that rimonabant stimulates adiponectin mRNA expression in adipose tissue by a direct action on adipocytes (Bensaid et al., 2003; Gary-Bobo et al., 2006). Since adiponectin (Acrp30) is an adipokine that promotes free fatty acid oxidation and body weight loss, it has been suggested that this action of rimonabant may account for the action of the drug in reducing body fat mass (Gary-Bobo et al., 2006).

Rimonabant (Sanofi-Aventis) is currently undergoing regulatory evaluation in the United States and in Europe. Results from two major 1- and 2-year trials have been published (Despres et al., 2005; Van Gaal et al., 2005; Pi-Sunyer et al., 2006). The weight-loss effect of rimonabant was similar in each trial, with patients on the clinically effective 20-mg dose losing on average approximately 5 kg versus placebo over 1 year (Despres et al., 2005; Van Gaal et al., 2005) when the more conservative last observation carried forward (LOCF) analysis is applied. Patients administered the drug for a second year maintained the weight loss achieved in year 1, but did not lose substantially more with the prolonged treatment (Pi-Sunyer et al., 2006). In these trials, rimonabant-induced weight loss was coupled with significant drops in waist circumference and blood pressure together with improvements in plasma lipid profiles and moderate reductions in glycemic control (Despres et al., 2005; Van Gaal et al., 2005; Pi-Sunyer et al., 2006). The main reasons stated for discontinuation included both anxiety (Despres et al., 2005; Van Gaal et al., 2005; Pi-Sunyer et al., 2006) and depression (Despres et al., 2005), along with nausea. These adverse events suggest that blockade of endogenous cannabinoid reward pathways can lead to decreased affect in some susceptible individuals, a finding that is in agreement with the preclinical literature (Gobbi et al., 2005). In terms of its clinical profile, rimonabant delivers weight loss that is approximately equivalent to the other approved centrally acting antiobesity drug, sibutramine, but without the potential to increase blood pressure and heart rate. However, this modest weight loss efficacy is unlikely to satisfy the expectation of obese patients.

A number of companies have developed, or are in the process of developing, CB_1 receptor antagonists for the treatment of obesity, and a number of these are illustrated in Table 1. These compounds are in various stages of clinical or preclinical development. Whether other companies can develop CB_1 receptor antagonists with improved efficacy compared to rimonabant, is eagerly awaited.

In light of the proposed action of rimonabant on adiponectin expression, and other studies suggesting that the effects of the compound on body weight—and, more controversially, on food intake—are mediated through peripheral CB_1 receptors (Gomez et al., 2002; Kirkham, 2005; Vickers and Kennett, 2005), one approach likely to be addressed by pharmaceutical companies at the preclinical stage is the discovery of non brain-penetrant CB_1 receptor antagonists. Such compounds should minimize the incidence of CNS side effects, such as anxiety and depression, which have been observed with rimonabant (Despres et al., 2005; Van Gaal, 2005). This strategy may result in the improved efficacy and an optimal side-effect profile of the CB_1 receptor antagonist approach.

B. 5-HT$_{2C}$ Receptor Agonists

A role for the serotonergic system in the regulation of ingestive behavior in a wide range of species, including humans, is well-established. Fourteen 5-HT receptor subtypes are currently described (for review, see Hoyer et al., 2002) and there is an extensive literature on the role of these receptor subtypes on ingestive behavior (for review, see Vickers and Dourish, 2004). Of these receptor subtypes, 5-HT$_{2C}$ receptor agonists and 5-HT$_6$ receptor antagonists are currently receiving the most attention in terms of the preclinical development of antiobesity drugs.

5-HT$_{2C}$ receptor agonists reduce food intake and body weight in a variety of preclinical models (for review, see Bickerdike et al., 1999). In addition, studies using 5-HT$_{2C}$ receptor knockout mice (Vickers et al., 1999) and selective 5-HT$_{2C}$ receptor antagonists (Vickers et al., 2001) have demonstrated that this receptor subtype mediates the hypophagic effect of d-fenfluramine (dexfenfluramine). Since d-fenfluramine was an effective treatment for obesity, although withdrawn in 1997 due to an increased risk of mitral valve damage, many companies have focused on 5-HT$_{2C}$ receptor agonism as an approach to developing safer alternative drug treatments.

A key challenge in the development of 5-HT$_{2C}$ receptor antagonists is to synthesize molecules with selectivity over the closely related 5-HT$_{2A}$ and 5-HT$_{2B}$ receptor subtypes. This is of particular importance since activity at these receptor subtypes has been linked to hallucinogenic activity (5-HT$_{2A}$ receptor) and valvular heart disease (5-HT$_{2B}$). The most advanced 5-HT$_{2C}$ receptor agonist in terms of development is APD356 (Arena Pharmaceuticals), which is currently in Phase IIb clinical trials. This compound reportedly has 15-fold selectivity over the 5-HT$_{2A}$ receptor subtype and 100-fold selectivity over the 5-HT$_{2B}$ receptor subtype (www.arenapharm.com). Initial results suggest patients treated with APD356 lost approximately 3 kg over an 85 day period.

C. 5-HT$_6$ Receptor Antagonists

5-HT$_6$ receptors have also been implicated in the control of food intake and body weight, with a number of companies developing selective 5-HT$_6$ receptor antagonists for the treatment of obesity. Ro-04-6790 was the first centrally active 5-HT$_6$ antagonist to enter the public domain (Sleight et al., 1998). Despite exhibiting only moderate affinity for the human 5-HT$_6$ receptor and modest brain penetration, Ro-04-6790 has been reported to inhibit food intake in a food restriction paradigm and to produce a small reduction in the body weight gain (3.2%) of normal, lean, growing rats (Bentley et al., 1999; Woolley et al., 2001). Central application of antisense to 5-HT$_6$ receptors has also been reported to decrease body weight gain (5.6%) in normal rats (Woolley et al., 2001). Furthermore, 5-HT$_6$ receptor knockout mice are resistant to weight gain when exposed to a high fat diet (Caldirola, 2003).

BVT 5182C, a high affinity 5-HT$_6$ receptor antagonist, decreases food intake and body weight in dietary-induced obese mice and rats due, at least in part, to a reduction in body fat (Caldirola, 2003; Caldirola and Svartengren, 2005). The compounds were not reported to affect locomotor activity or pica behavior, and meal pattern analysis indicated that the reduction in food intake was related to increased intermeal intervals (Caldirola and Svartengren, 2005). Such data suggest that 5-HT$_6$ receptor antagonists affect ingestive behavior by a behaviorally specific mechanism. Similar findings on rat food intake and body weight gain have been reported with SUVN503 and SUVN504 (Pendharkar et al., 2005). These antagonists reportedly have 100-fold selectivity for the 5-HT$_6$ receptor (Pendharkar et al., 2005).

Importantly, considerable caution should be taken when interpreting data obtained using the mouse. Hence, despite significant sequence homology between human, rat, and mouse 5-HT$_6$ receptors, the central distribution and pharmacological profile of the mouse 5-HT$_6$ receptor is significantly different to the rat and human receptor (Hirst et al., 2003; Setola and Roth, 2003). In rat and human brain, 5-HT$_6$ receptors are widely expressed and densely populated in the basal ganglia, whereas the mouse exhibits much lower levels throughout the brain (Hirst et al., 2003). Furthermore, radioligand binding, site-directed mutagenesis, and molecular modeling suggest that the 5-HT$_6$ receptor binding pocket is different in the mouse compared to the rat.

A number of companies have information in the public domain relating to 5-HT$_6$ receptor ligands. For example, Wyeth, Esteve, GSK, Predix Pharmaceuticals, Biovitrum, and others have all disclosed, either on a company website, as a patent application, or as a journal publication, an interest in 5-HT$_6$ receptor antagonists. Indeed, current reports suggest Biovitrum and Predix Pharmaceuticals are in preclinical development with this approach (see the respective company websites). However, 5-HT$_6$ receptor antagonists are also widely reported to exhibit efficacy in animal models of cognition (Woolley et al., 2004) and, accordingly, it is expected that a large number of companies are developing such ligands for the treatment of dementia.

D. Melanin-Concentrating Hormone Receptor Antagonists

Melanin-concentrating hormone plays a major role in the regulation of ingestive behavior and energy balance (Collins and Kym, 2003; Chapter 4). Melanin-concentrating hormone (MCH) is a 19 amino acid cyclic neuropeptide predominantly located in the neurons in the lateral hypothalamus and zona incerta with extensive projections throughout the brain (Skofitsch et al., 1985; Bittencourt et al., 1992). Two MCH receptors (both G-protein coupled) have currently been identified: MCH-1 and MCH-2 receptors. The MCH-1 receptor is widely distributed throughout the brain, including areas such as the nucleus accumbens and various hypothalamic nuclei. Unlike the MCH-1 receptor, MCH-2 receptors are not expressed in rodents but are present in ferrets, dogs, and humans (Tan et al., 2002). Accordingly, this raises obvious issues concerning the use of animal models in drug development.

There is a considerable body of evidence to indicate a role for MCH in the control of body weight. For the reasons detailed before, this is largely restricted to the role of the MCH-1 receptor. Central application of MCH in rats and mice results in an increase in food intake (Qu et al., 1996; Rossi et al., 1999). Furthermore, the mRNA for the MCH precursor protein is overexpressed in the hypothalamus of genetically obese ob/ob mice and expression is increased by fasting in both normal and ob/ob mice (Qu et al., 1996). Mice lacking the gene encoding MCH exhibit reduced body weight and a lean phenotype due to hypophagia and increased metabolic rate (Shimada et al., 1998).

An analysis of the patent literature suggests that a large number of companies are pursuing MCH-1 receptor antagonists for the treatment of obesity, exploring a wide range of diverse structural types (Kowalski and McBriar, 2004, Dyke and Ray, 2005). Several compounds have been reported to have efficacy in rodent models of obesity, some of which suggest that this approach has the potential to deliver greater weight loss in the clinic than do existing drugs (Borowsky et al., 2002; Collins and Kym, 2003; Shearman et al., 2003). At least two MCH-1 receptor antagonists have entered clinical development. Amgen has initiated a randomized, double-blind placebo-controlled Phase I single-ascending-dose study to evaluate the effect of AMG-076 on appetite and satiety (Dyke and Ray, 2005). Similarly, GlaxoSmithKline (GSK) have initiated a clinical trial with 856464. Data from these trials are eagerly awaited.

While there is a sound preclinical rationale that MCH-1 receptor antagonists may have utility for the treatment of obesity, it should be mentioned that MCH-1 receptor knockout mice exhibit osteoporosis (Bohlooly et al., 2004). Furthermore, the wide CNS distribution of the MCH-1 receptor increases the likelihood of selective compounds having effects on signaling pathways other than those involved in ingestive behavior.

E. MC4 Receptor Antagonists

The melanocortins (i.e., adrenocorticotrophin (ACTH) and the α-, β-, and γ-melanocyte-stimulating hormones (MSH)) interact with a family of receptors, MC1–5. Mutant mouse models (e.g., *ob/ob* mice), gene knockout mice, and feeding studies with MC3/4 receptor agonists and antagonists indicate that MC4 receptor agonists may be effective weight-loss agents (Fan et al., 1997; Huszar et al., 1997; Goodfellow and Saunders, 2003). Furthermore, mutations in the gene encoding the MC4 receptor have been found in approximately 4% of morbidly obese people.

Despite the high level of interest in this target, the synthesis of small molecule MC4 receptor agonists has proved to be a major challenge, with very few successes to date (O'Rahilly et al., 2004). That said, PGE-657022, a high affinity MC4 receptor agonist, has been reported to decrease food intake, body weight, and fat mass when administered chronically to obese rodents (R. Sheldon, British Pharmacological Society 3rd Focused Meeting "Obesity: Potential Pharmacological Targets" April, 2004, Buckingham, U.K.). Other companies believed to be focusing on the development of MC4-receptor agonists include Chiron Corp., Amgen, Novartis, Neurocrine, and Merck. Importantly, the potential to produce cardiovascular side effects and stimulate erectile activity may also limit the utility of this approach to the treatment in obesity.

F. Neuropeptide Y

Neuropeptide Y (NPY), a 36 amino acid peptide, is one of the most potent endogenous orexigenic agents yet to be discovered (King, 2005). Six NPY receptors have been identified: Y1, Y2, Y3, Y4, Y5, and Y6. Initially, Y1 and then subsequently Y5 receptors were thought to be of principal importance in the mediation of the feeding effects of NPY. This led to a marked interest in the potential of NPY receptor antagonists as antiobesity agents. Despite the identification of numerous novel Y5 receptor antagonists, their effects on food intake and body weight in rodent models have been disappointing (Turnbull et al., 2002). This is coupled with anecdotal evidence of lack of efficacy in Phase IIa trials testing Y5 receptor antagonists in human volunteers.

In light of such reports, interest has instead focused on Y2 and Y4 receptors. Within the hypothalamus, Y2 receptors are predominantly located presynaptically where they act to inhibit NPY release. Accordingly, stimulation of Y2 receptors, either using synthetic peptide ligands or the gut-derived peptide, PYY_{3-36}, has been reported to reduce food intake (Feletou and Levens, 2005).

Both constitutive (total gene activation in all cells) and conditional (tissue-specific gene activation) Y2 receptor knockout mice have been described (Sainsbury et al., 2002). Interestingly, despite exhibiting increased food intake, mice lacking hypothalamic Y2 receptors also exhibit a significant reduction in body weight (Sainsbury et al., 2002). The phenotype of Y4 knockout mice is similar to that of Y2 mice, exhibiting both reduced body weight and white adipose tissue mass (Sainsbury et al., 2003). Y2/Y4 receptor double knockout mice show marked reductions in adiposity, and are also hypoleptinemic and hypoinsulinemic (Sainsbury et al., 2003). Furthermore, Y2/Y4 knockout mice are resistant to diet-induced obesity, and are characterized by reduced food intake and improved glucose tolerance (Sainsbury et al., 2006). In contrast, Y1 knockout, or Y1/Y2, or Y1/Y4 double receptor knockout mice develop exacerbated diet-induced obesity (Sainsbury et al., 2006).

Companies currently pursuing compounds with affinity for Y2 and/or Y4 receptors include 7TM Pharma. Based on insight into the relative molecular recognition epitopes of PP-fold peptides, such as PYY, 7TM Pharma have reported the discovery of highly selective Y2 and Y4 receptor agonists. Furthermore the same company has discovered potent, dual selective, Y2/Y4 receptor agonists. Chronic administration of these ligands in diet-induced obese mice, has confirmed the potential of these approaches to reduce body weight through the selective reduction of fat mass. TM30338, a dual Y2/Y4 receptor agonist, is currently in Phase I/II clinical trials (personal communication, Christian E. Elling, 7TM Pharma).

G. Other Central Strategies

In addition to the targets detailed earlier, additional approaches that focus on the CNS which warrant at least a brief mention include the more recent decision by GSK to develop the noradrenaline and dopamine reuptake inhibitor, radafaxine, for obesity treatment. This compound is an active metabolite of the antidepressant, bupropion, and has reportedly begun Phase I evaluation. In addition, Neurosearch are developing the triple monoamine reuptake inhibitor, tasofensine (NS2330). This compound was initially evaluated for the treatment of Alzheimer's (AD) and Parkinson's (PD) diseases. However, weight loss, equivalent to that seen with sibutramine and orlistat, was observed in 312 obese patients with AD and PD (12/14 week trials). Preclinical studies in dietary-induced obese rats have also demonstrated that the compound reduces body weight. Depending on the potency of the noradrenaline reuptake component of these two compounds, they are likely to have similar side-effect issues of increased heart rate and blood pressure to sibutramine, which may limit the dose range and hence their efficacy in man. The dopaminergic component of these drugs may also provide an additional challenge in light of the previous history of abuse surrounding the amphetamines.

III. PERIPHERAL TARGETS FOR NOVEL ANTIOBESITY DRUGS

As illustrated earlier, neural pathways in the brain control appetite and influence body weight, and these mechanisms can be exploited for the development of novel drug treatments for obesity. However, the CNS also acts in response to blood-borne factors that signal nutritional status from the periphery. Investigation of these feedback mechanisms, and how they may be defective in obesity, has led to a number of targets being assessed for their utility in the treatment of the condition. These are summarized in Table 1.

A. Leptin

Much attention has been focused on this circulating hormone over the past 20 years. It is generally accepted that leptin signals information to the CNS regarding the size of adipose tissue mass, and the peptide has been shown to attenuate feeding via its actions on central regulatory systems (see Chapter 5). Trials by Amgen Inc. with peripherally administered recombinant leptin or leptin analogs have been disappointing (King, 2005). The lack of efficacy is possibly due to "leptin resistance" whereby high plasma leptin levels, as seen in obese individuals, results in impaired transport of leptin into the CNS. Interest in leptin has been renewed with the discovery by Cambridge Biotechnology (acquired by Biovitrum in 2005) of leptin receptor agonists (e.g., CBT-001452) that are small enough to bypass the leptin transporter system and diffuse into the brain. It remains to be seen if such an approach can deliver efficacy in the clinic.

B. PYY

Although peptide YY (PYY) is a 36 amino acid peptide closely related to neuropeptide Y, it is excreted in the gastrointestinal tract and is, accordingly, a peripheral rather than central signal involved in energy homeostasis. A fragment of this peptide, PYY_{3-36}, has more recently been shown to decrease food intake in animals and man, but not in Y2 receptor knockout mice, indicating that its hypophagic action is mediated by this receptor subtype (Batterham et al., 2002). This peptide is currently in development for obesity treatments by Amylin and Merck/Nastech. However, despite these findings PYY_{3-36} remains a controvertial antiobesity target with several groups being unable to replicate the findings in preclinical studies (Boggiano et al., 2005).

C. Carbonic Anhydrase Inhibitors

Carbonic anhydrase (CA) is physiologically important in catalyzing the reversible hydration reaction of CO_2. It is expressed in a number of isoforms (CA I–XIV) with varying degrees of enzymatic activity. Of these isoenzymes, types V and II have been

proposed as novel targets for antiobesity agents (Antel et al., 2002; Supuran, 2003; Winum et al., 2005).

CA isoenzymes are involved in several important functions relating to respiration and transport of CO_2/bicarbonate between metabolizing tissues and the lungs, pH and CO_2 homeostasis, electrolyte secretion in a variety of tissues and organs, and biosynthetic reactions such as ureagenesis, gluconeogenesis, and lipogenesis (Supuran, 2003). Importantly, two CA isoenzymes are critical to the entire process of lipogenesis: CA V in the mitochondria and CA II in the cytosol (Supuran, 2003). Thus, selective inhibition of CA V and dual inhibition of CA V and CA II should reduce fatty acid synthesis and, therefore, fat deposition. This is supported by the finding that sulfonamides, such as trifluoromethylsulfonamide, a potent nonspecific CA inhibitor has been shown to decrease *de novo* fatty acid synthesis in adipocytes (Lynch et al., 1995).

Interestingly, the clinically effective anticonvulsant drug, topiramate, leads to weight loss in some patients (Gordon and Price, 1999) and the drug's utility in treating obesity has been linked to its ability to inhibit CA V and probably CA II (Antel et al., 2002). Although topiramate produces clinically significant weight loss in obese patients, its development for obesity has now been discontinued by Johnson and Johnson due to unacceptable CNS side effects, including cognitive dysfunction, confusion, and tingling/numbness in limb extremeties (Bray et al., 2003; Astrup et al., 2004; Wilding et al., 2004). It remains to be seen whether CA II and V inhibitors can deliver weight loss in the absence of such side effects.

D. 11β-HSD1 Inhibitors

The enzyme 11b-hydroxysteroid dehydrogenase type 1 (11βHSD-1) converts inactive cortisone into the active glucocorticoid, cortisol. Excessive stimulation of glucocorticoid receptors may be one cause of insulin resistance and other features of the metabolic syndrome, including visceral adiposity (for review see Wang, 2005). For example, mice which overexpress the 11βHSD-1 enzyme selectively in adipose tissue to produce increased levels of glucocorticoids, develop obesity with visceral predominance (Masuzaki et al., 2001). Accordingly, inhibitors of this enzyme may have utility in treating obesity and related disorders.

Companies currently focusing on this approach include Merck and Biovitrum, the latter having a cooperation agreement with U.S. biotech, Amgen. The 11βHSD-1 inhibitor, BVT.2733, has been reported to improve insulin sensitivity in different mouse models of type 2 diabetes (Alberts et al., 2003). The compound BVT.3498 began clinical trials in March 2002 and progressed to Phase II; however, at the time of writing both Amgen and Biovitrum report that this program is in preclinical development on their websites, perhaps suggesting that clinical data failed to meet initial expectations.

E. GLP-1, and DPP-IV Inhibitors

Glucagon-like peptide (GLP-1) is a hormone released in the gut in response to food ingestion and is degraded by the enzyme, dipeptidyl peptidase IV (DPP-IV). Much evidence suggests that GLP-1 is a short-acting endogenous peptide with a role in satiety (Edwards, 2005). However, the utility of this hormone as a treatment for obesity or diabetes is limited by its short half-life (less than 2 min). In response, a number of pharmaceutical companies are focusing on the development of chemical inhibitors of DPP-IV so as to extend the duration of action of endogenous GLP-1. Compounds of note include Novartis' Vildagliptin (LAF 237; Ristic et al., 2005) which is currently in phase III trials for diabetes. Other companies pursuing this area include Biovitrum and Santhera (DPP-IV inhibitors), Prosidion (DPP-IV inhibitors in

Phase II), and Novo Nordisk (GLP-1 derivative such as Liraglutide).

F. Additional Peripheral Targets

Other targets of interest at the present time include ghrelin and the more recently characterized peptide, obestatin (Zhang et al., 2006). Ghrelin is a 28 amino acid peptide that binds to the growth hormone secretagogue receptor (GHSR) and exhibits stimulatory effects on food intake and body weight gain in animal models (Tschop et al., 2000). Ghrelin is derived by posttranslational processing from a prohormone, proghrelin. Interestingly, another peptide, obestatin, is also derived from proghrelin but reportedly acts to reduce food intake and body weight gain through activation of the orphan G-protein coupled receptor, GPR39 (Zhang et al., 2006). Hence, two hormones with opposing actions in the regulation of food intake and body weight are derived from the same ghrelin gene. Confirmation of obestatin's actions are required, but it will be of interest to see if antagonism of the ghrelin GHSR system or agonism of the obestatin GPR39 system offers viable pharmacological approaches to develop antiobesity drugs.

IV. CONCLUSION

Due to the current obesity epidemic, it is not surprising that a large number of preclinical approaches are being evaluated by pharmaceutical companies in order to develop drugs for the treatment of obesity and related disorders. While not an exhaustive review of every conceivable target, we have attempted to highlight not only the most "drugable" and "popular" targets but those that we believe have the highest chance of working on the clinical stage.

One thing that can be seen from this chapter is that numerous factors and pathways are involved in the control of body weight regulation. As a result it remains to be seen whether a single pharmacological approach to the development of antiobesity drugs will ever provide the efficacy that is required by both patients and clinicians.

References

Alberts, P., Nilsson, C., Selen, G., Engblom, L. O., Edling, N. H., Norling, S., Klingstrom, G., Larsson, C., Forsgren, M., Ashkzari, M., Nilsson, C. E., Fiedler, M., Bergqvist, E., Ohman, B., Bjorkstrand, E., and Abrahmsen, L. B. (2003). Selective inhibition of 11 beta-hydroxysteroid dehydrogenase type 1 improves hepatic insulin sensitivity in hyperglycemic mice strains. *Endocrinology* **144**, 4755–4762.

Antel, J., Hebebrand, J., Preuschoff, U., David, S., Sann, H., and Weske, M. (2002). Method for locating compounds which are suitable for the treatment and/or prophylaxis of obesity. WO Patent 0207821.

Arch, J. R. S., Wang, S. J. Y., Birtles, S., Smith, D. M., and Turnbull, A. (2005). Effects of an inhibitor of 11beta-hydroxysteroid dehydrogenase type 1 inhibitor on energy balance and glucose homeostasis in diet-induced obesity. *Diabetologia* 48(Suppl 1), A238, Abstract 650.

Astrup, A., Caterson, I., Zelissen, P., Guy-Grand, B., Carruba, M., Levy, M., Sun, X., and Fitchet, M. (2004). Topiramate: Long-term maintenance of weight-loss induced by a low-calorie diet in obese subjects. *Obes. Res.* **12**, 1658–1669.

Batterham, R. L., Cowley, M. A., Small, C. J., Herzog, H., Cohen, M. A., Dakin, C. L., Wren, A. M., Brynes, A. E., Low, M. J., Ghatei, M. A., Cone, R. D., and Bloom, S. R. (2002). Gut hormone PYY(3–36) physiologically inhibits food intake. *Nature* **418**, 650–654.

Bensaid, M., Gary-Bobo, M., Esclangon, A., Maffrand, J. P., Le Fur, G., Oury-Donat, F., and Soubrié, P. (2003). The cannabinoid CB1 receptor antagonist SR141716 increases Acrp30 mRNA expression in adipose tissue of obese fa/fa rats and in cultured adipocyte cells. *Mol. Pharmacol.* **63**, 908–914.

Bentley, J. C., Marsden, C. A., Sleight, A. J., and Fone, K. C. F. (1999). Effect of the 5-HT6 antagonist, Ro 04-6790 on food consumption in rats trained to a fixed feeding regime. *Br. J. Pharmacol.* **126**, 66P.

Bickerdike, M. J., Vickers, S. P., and Dourish, C. T. (1999). 5-HT2C receptor modulation and the treatment of obesity. *Diabetes Obes. Metab.* **1**, 207–214.

Bittencourt, J. C., Presse, F., Arias, C., Peto, C., Vaughan, J., Nahon, J. L., Vale, W., and Sawchenko, P. E. (2002). The melanin-concentrating hormone system of the rat brain: an immuno- and hybridization histochemical characterization. *J. Comp. Neurol.* **319(2)**, 218–245.

Black, S. C. (2004). Cannabinoid receptor antagonists and obesity. *Curr. Opin. Invest. Drugs* **5**, 389–394.

Boggiano, M. M., Chandler, P. C., Oswald, K. D., Rodgers, R. J., Blundell, J. E., Ishii, Y., Beattie, A. H., Holch, P., Allison, D. B., Schindler, M., Arndt, K., Rudolf, K., Mark, M., Schoelch, C., Joost, H. G., Klaus, S., Thone-Reineke, C., Benoit, S. C., Seeley, R. J., Beck-Sickinger, A. G., Koglin, N., Raun, K., Madsen, K., Wulff, B. S., Stidsen, C. E., Birringer, M., Kreuzer, O. J., Deng, X. Y., Whitcomb, D. C., Halem, H., Taylor, J., Dong, J., Datta, R., Culler, M., Ortmann, S., Castaneda, T. R., and Tschop, M. (2005). PYY3–36 as an anti-obesity drug target. *Obes. Rev.* **6**, 307–322.

Bohlooly, Y. M., Mahlapuu, M., Andersen, H., Astrand, A., Hjorth, S., Svensson, L., Tornell, J., Snaith, M. R., Morgan, D. G., and Ohlsson, C. (2004). Osteoporosis in MCHR1-deficient mice. *Biochem. Biophys. Res. Commun.* **318**, 964–969.

Bonow, R. O., and Eckel, R. H. (2003). Diet, obesity and cardiovascular risk. *N. Engl. J. Med.* **348**, 2057–2058.

Borowsky, B., Durkin, M. M., Ogozalek, K., Marzabadi, M. R., DeLeon, J., Lagu, B., Heurich, R., Lichtblau, H., Shaposhnik, Z., Daniewska, I., Blackburn, T. P., Branchek, T. A., Gerald, C., Vaysse, P. J., and Forray, C. (2002). Antidepressant, anxiolytic and anorectic effects of a melanin-concentrating hormone-1 receptor antagonist. *Nat. Med.* **8**, 825–830.

Bray, G. A., Hollander, P., Klein, S., Kushner, R., Levy, B., Fitchet, M., and Perry, B. H. (2003). A 6-month randomized, placebo-controlled, dose-ranging trial of topiramate for weight loss in obesity. *Obes. Res.* **11**, 722–733.

Caldirola, P. M. (2003). 5-HT$_6$ receptor antagonism, a novel mechanism for the management of obesity. SMi Conf Obesity and Related Disorders, London.

Caldirola, P. M., and Svartengren, J. (2005). Selective serotonin 5-HT$_6$ receptor antagonist(s) for the treatment of obesity. *Neuropsychopharmacology* **30**(Suppl. 1), S54.

Collins, C. A., and Kym, P. R. (2003). Prospects for obesity treatment: MCH receptor antagonists. *Curr. Opin. Invest. Drugs* **4**, 386–394.

Colombo, G., Agabio, R., Diaz, G., Lobina, C., Reali, R., and Gessa, G. L. (1998). Appetite suppression and weight loss after the cannabinoid antagonist SR 141716. *Life Sci.* **63**, 113–117.

Despres, J. P., Golay, A., and Sjostrom, L. (2005). Rimonabant in Obesity-Lipids Study Group. Effects of rimonabant on metabolic risk factors in overweight patients with dyslipidemia. *N. Engl. J. Med.* **353**, 2121–2134.

Dyke, H. J., and Ray, N. C. (2005). Recent developments in the discovery of MCH-1R antagonists for the treatment of obesity—an update. *Expert Opin. Ther. Patents* **15**, 1303–1313.

Edwards, C. M. (2005). The GLP-1 system as a therapeutic target. *Ann. Med.* **37**, 314–322.

Fan, W., Boston, B. A., Kesterson, R. A., Hruby, V. J., and Cone, R. D., (1997). Role of melanocortinergic neurons in feeding and the agouti obesity syndrome. *Nature* **385**, 165–168.

Feletou, M., and Levens, N. R. (2005). Neuropeptide Y2 receptors as drug targets for the central regulation of body weight. *Curr. Opin. Invest. Drugs* **6**, 1002–1011.

Foltin, R. W., Fischman, M. W., and Byrne, M. F. (1988). Effects of smoked marijuana on food intake and body weight of humans living in a residential laboratory. *Appetite* **11**, 1–14.

Fontaine, K. R., Redden, D. T., Wang, C., Westfall, A. O., and Allison, D. B. (2003). Years of life lost due to obesity. *JAMA* **289**, 187–193.

Gary-Bobo, M., Elachouri, G., Scatton, B., Le Fur, G., Oury-Donat, F., and Bensaid, M. (2006). The cannabinoid CB1 receptor antagonist rimonabant (SR141716) inhibits cell proliferation and increases markers of adipocyte maturation in cultured mouse 3T3 F442A preadipocytes. *Mol. Pharmacol.* **69**, 471–478.

Gobbi, G., Bambico, F. R., Mangieri, R., Bortolato, M., Campolongo, P., Solinas, M., Cassano, T., Morgese, M. G., Debonnel, G., Duranti, A., Tontini, A., Tarzia, G., Mor, M., Trezza, V., Goldberg, S. R., Cuomo, V., and Piomelli, D. (2005). Antidepressant-like activity and modulation of brain monoaminergic transmission by blockade of anandamide hydrolysis. *Proc. Natl. Acad. Sci. U.S.A.* **102**, 18620–18625.

Gomez, R., Navarro, M., Ferrer, B., Trigo, J. M., Bilbao, A., Del Arco, I., Cippitelli, A., Nava, F., Piomelli, D., and Rodriguez de Fonseca, F. (2002). A peripheral mechanism for CB1 cannabinoid receptor-dependent modulation of feeding. *J. Neurosci.* **22(21)**, 9612–9617.

Goodfellow, V. S., and Saunders, J. (2003). The melanocortin system and its role in obesity and cachexia. *Curr. Top. Med. Chem.* **3**, 855–883.

Gordon, A., and Price, H. (1999). Mood stabilisation and weight loss with topiramate. *Am. J. Psychiatry* **156**, 968–969.

Hirst, W. D., Abrahamsen, B., Blaney, F. E., Calver, A. R., Aloj, L., Price, G. W., and Medhurst, A. D. (2003). Differences in the central nervous system distribution and pharmacology of the mouse 5-hydroxytryptamine-6 receptor compared with rat and human receptors investigated by radioligand binding, site-directed mutagenesis, and molecular modeling. *Mol. Pharmacol.* **64**, 1295–1308.

Horvath, T. L. (2005). The hardship of obesity: a soft-wired hypothalamus. *Nature Neurosci.* **8**, 561–565.

Hoyer, D., Hannon, J. P., and Martin, G. R. (2002). Molecular, pharmacological and functional diversity of 5-HT receptors. *Pharmacol. Biochem. Behav.* **71**, 533–554.

Huszar, D., Lynch, C. A., Fairchild-Huntress, V., Dunmore, J. H., Fang, Q., Berkemeier, L. R., Gu, W., Kesterson, R. A., Boston, B. A., Cone, R. D., Smith, F. J., Campfield, L. A., Burn, P., and Lee, F. (1997). Targeted disruption of the melanocortin-4 receptor results in obesity in mice. *Cell* **88**, 131–141.

King, P. J. (2005). The hypothalamus and obesity. *Curr. Drug Targets* **6**, 225–240.

Kirkham, T. C. (2005). Endocannabinoids in the regulation of appetite and body weight. *Behav. Pharmacol.* **16**, 297–313.

Kirkham, T. C., and Williams, C. M. (2001). Endogenous cannabinoids and appetite. *Nutr. Res. Rev.* **14**, 65–86.

Kirkham, T. C., and Williams, C. M. (2004). Endocannabinoid receptor antagonists: potential for obesity treatment. *Treat. Endocrinol.* **3**, 345–360.

Kowalski, T. J., and McBriar, M. D. (2004). Therapeutic potential of melanin-concentrating hormone-1 receptor antagonists for the treatment of obesity. *Expert Opin. Invest. Drugs* **13**, 1113–1122.

Lobstein, T., Baur, L., and Uauy, R. (2004). Obesity in children and young people. A crisis in public health. *Obes. Rev.* **5**(Suppl.), 4–104.

Lynch, C. J., Fox, H., Hazen, S. A., Stanley, B. A., Dodgson, S. J., and Lanoue, K. F. (1995). Role of hepatic carbonic anhydrase in de novo lipogenesis. *Biochem. J.* **310**, 197–202.

Masuzaki, H., Paterson, J., Shinyama, H., Morton, N. M., Mullins, J. J., Seckl, J. R., and Flier, J. S. (2001). A transgenic model of visceral obesity and the metabolic syndrome. *Science* **294**, 2166–2170.

National Audit Office, U.K. (2001). Controller and Auditor General. Tackling Obesity in England. The Stationary Office, London.

O'Rahilly, S., Yeo, G. S., and Farooqi, I. S. (2004). Melanocortin receptors weigh in. *Nat. Med.* **10**, 351–352.

Pendharkar, V. V., Vishwakarma, S. L., Patel, A. S., Shrisath, V. S., Kambhampati, S. R., and Nirogi, V. S. (2005). Effect of selective 5-HT$_6$ receptor antagonists on food intake and body weight gain. *Soc. Neurosci. Abstr.* P533.7.

Pi-Sunyer, F., Aronne, L., Heshmati, H., Devin, J., and Rosenstock, J. (2006). Effect of rimonabant, a cannabinoid-1 receptor blocker, on weight and cardiometabolic risk factors in overweight or obese patients. RIO-North America: a randomized controlled trial. *JAMA* **295**, 761–775.

Qu, D., Ludwig, D. S., Gammeltoft, S., Piper, M., and Pelleymounter, M. A. (1996). A role for melanin-concentrating hormone in the central regulation of feeding behaviour. *Nature* **380**, 243–247.

Ravinet-Trillou, C., Arnone, M., Delgorge, C., Gonalons, N., Keane, P., Maffrand, J. P., and Soubrie, P. (2003). Anti-obesity effect of SR141716, a CB1 receptor antagonist, in diet-induced obese mice. *Am. J. Physiol. Regul. Integr. Comp. Physiol.* **284**, R345–R353.

Ravinet-Trillou, C., Delgorge, C., Menet, C., Arnone, M., and Soubrie, P. (2004). CB1 cannabinoid receptor knockout in mice leads to leanness, resistance to diet-induced obesity and enhanced leptin sensitivity. *Int. J. Obes. Relat. Metab. Disord.* **28**, 640–648.

Rinaldi-Carmona, M., Barth, F., Heaulme, M., Shire, D., Calandra, B., Congy, C., Martinez, S., Maruani, J., Neliat, G., and Caput, D. (1994). SR141716A, a potent and selective antagonist of the brain cannabinoid receptor. *FEBS Lett.* **350**, 240–244.

Ristic, S., Byiers, S., Foley, J., and Holmes, D. (2005). Improved glycaemic control with dipeptidyl peptidase-4 inhibition in patients with type 2 diabetes: vildagliptin (LAF237) dose response. *Diabetes Obes. Metab.* **7**, 692–698.

Rossi, M., Beak, S. A., Choi, S. J., Small, C. J., Morgan, D. G., Ghatei, M. A., Smith, D. M., and Bloom, S. R. (1999). Investigation of the feeding effects of melanin concentrating hormone on food intake—action independent of galanin and the melanocortin receptors. *Brain Res.* **846**, 164–170.

Sainsbury, A., Schwarzer, C., Couzens, M., Fetissov, S., Furtinger, S., Jenkins, A., Cox, H. M., Sperk, G., Hokfelt, T., and Herzog, H. (2002). Important role of hypothalamic Y2 receptors in body weight regulation revealed in conditional knockout mice. *Proc. Natl. Acad. Sci. U.S.A.* **99**, 8938–8943.

Sainsbury, A., Baldock, P. A., Schwarzer, C., Ueno, N., Enriquez, R. F., Couzens, M., Inui, A., Herzog, H., and Gardiner, E. M. (2003). Synergistic effects of Y2 and Y4 receptors on adiposity and bone mass revealed in double knockout mice. *Mol. Cell. Biol.* **23**, 5225–5233.

Sainsbury, A., Bergen, H. T., Boey, D., Bamming, D., Cooney, G. J., Lin, S., Couzens, M., Stroth, N., Lee, N. J., Lindner, D., Singewald, N., Karl, T., Duffy, L., Enriquez, R., Slack, K., Sperk, G., and Herzog, H. (2006). Y2Y4 receptor double knockout protects against obesity due to a high-fat diet or Y1 receptor deficiency in mice. *Diabetes.* **55**, 19–26.

Setola, V., and Roth, B. L. (2003). Why mice are neither miniature humans nor small rats: a cautionary tale involving 5-hydroxytryptamine-6 serotonin receptor species variants. *Mol. Pharmacol.* **64**, 1277–1278.

Shearman, L. P., Camacho, R. E., Sloan Stribling, D., Zhou, D., Bednarek, M. A., Hreniuk, D. L., Feighner, S. D., Tan, C. P., Howard, A. D., Van der Ploeg, L. H., MacIntyre, D. E., Hickey, G. J., and Strack, A. M. (2003). Chronic MCH-1 receptor modulation alters appetite, body weight and adiposity in rats. *Eur. J. Pharmacol.* **475**, 37–47.

Shimada, M., Tritos, N. A., Lowell, B. B., Flier, J. S., and Maratos-Flier, E. (1998). Mice lacking melanin-concentrating hormone are hypophagic and lean. *Nature* **396**, 670–674.

Skofitsch, G., Jacobowitz, D. M., and Zamir. N. (1985). Immunohistochemical localization of a melanin concentrating hormone-like peptide in the rat brain. *Brain Res. Bull.* **15**, 635–649.

Sleight, A. J., Boess, F. G., Bos, M., Levet-Trafit, B., Riemer, C., and Bourson, A. (1998). Characterization of Ro 04-6790 and Ro 63-0563: potent and selective antagonists at human and rat 5-HT6 receptors. *Br. J. Pharmacol.* **124**, 556–562.

Supuran, C. T. (2003). Carbonic anhydrase inhibitors in the treatment and prophylaxis of obesity. *Exp. Opin. Ther. Patents* **13**, 1545–1550.

Tan, C. P., Sano, H., Iwaasa, H., Pan, J., Sailer, A. W., Hreniuk, D. L., Feighner, S. D., Palyha, O. C., Pong, S. S., Figueroa, D. J., Austin, C. P., Jiang, M. M., Yu, H., Ito, J., Ito, M., Ito, M., Guan, X. M., MacNeil, D. J., Kanatani, A., Van der Ploeg, L. H., and Howard, A. D. (2002). Melanin-concentrating hormone receptor subtypes 1 and 2: species-specific gene expression. *Genomics* **79**, 785–792.

Tschop, M., Smiley, D. L., and Heiman, M. L. (2000). Ghrelin induces adiposity in rodents. *Nature* **407**, 908–913.

Turnbull, A. V., Ellershaw, L., Masters, D. J., Birtles, S., Boyer, S., Carroll, D., Clarkson, P., Loxham, S. J., McAulay, P., Teague, J. L., Foote, K. M., Pease, J. E., and Block, M. H. (2002). Selective antagonism of the NPY Y5 receptor does not have a major effect on feeding in rats. *Diabetes* **51**, 2441–2449.

Van Gaal, L. F., Rissanen, A. M., Scheen, A. J., Ziegler, O., and Rossner, S. (2005). Effects of the cannabinoid-1 receptor blocker rimonabant on weight reduction and cardiovascular risk factors in overweight patients: 1-year experience from the RIO-Europe study. *Lancet* **365**, 1389–1397.

Vickers, S. P., and Dourish, C. T. (2004). Serotonin receptor ligands and the treatment of obesity. *Curr. Opin. Invest. Drugs* **5**, 377–388.

Vickers, S. P., and Kennett, G. A. (2005). Cannabinoids and the regulation of ingestive behaviour. *Curr. Drug Targets* **6**, 215–233.

Vickers, S. P., Clifton, P. G., Dourish, C. T., and Tecott, L. H. (1999). Reduced satiating effect of *d*-fenfluramine in serotonin 5-HT(2C) receptor mutant mice. *Psychopharmacology (Berlin)* **143**, 309–314.

Vickers, S. P., Dourish, C. T., and Kennett, G. A. (2001). Evidence that hypophagia induced by *d*-fenfluramine and *d*-norfenfluramine in the rat is mediated by 5-HT2C receptors. *Neuropharmacology* **41**, 200–209.

Vickers, S. P., Webster, L. J., Wyatt, A., Dourish, C. T., and Kennett, G. A. (2003). Preferential effects of the cannabinoid CB1 receptor antagonist, SR 141716, on food intake and body weight gain of obese (*fa/fa*) compared to lean Zucker rats. *Psychopharmacology (Berlin)* **167**, 103–111.

Wang, M. (2005). The role of glucocorticoid action in the pathophysiology of the metabolic syndrome. *Nutr. Metab. (Lond.)* **2**, 3.

Wilding, J., Van Gaal, L., Rissanen, A., Vercruysse, F., and Fitchet, M. (2004). A randomized double-blind placebo-controlled study of the long-term efficacy and safety of topiramate in the treatment of obese subjects. *Int. J. Obes.* **28**, 1399–1410.

Williams, C. M., and Kirkham, T. C. (1999). Anandamide induces overeating: mediation by central cannabinoid (CB1) receptors. *Psychopharmacology (berlin)* **143**, 315–317.

Williams, C. M., and Kirkham, T. C. (2002). Reversal of delta 9-THC hyperphagia by SR141716 and naloxone but not dexfenfluramine. *Pharmacol. Biochem. Behav.* **71**, 333–340.

Williams, C. M., Rogers, P. J., and Kirkham, T. C. (1998). Hyperphagia in pre-fed rats following oral Δ9-THC. *Physiol. Behav.* **65**, 343–346.

Winum, J.-Y., Scozzafava, A., Montero, J.-L., and Supuran, C. T. (2005). Sulfamates and their therapeutic potential. *Med. Res. Rev.* **25**(Suppl 2), 186–228.

Wolf, A. M., and Colditz, G. A. (1998). Current estimates of the economic costs of obesity in the United States. *Obes. Res.* **6**, 97–106.

Woolley, M. L., Bentley, J. C., Sleight, A. J., Marsden, C. A., and Fone, K. C. (2001). A role for 5-ht6 receptors in retention of spatial learning in the Morris water maze. *Neuropharmacology* **41**, 210–219.

Woolley, M. L., Marsden, C. A., and Fone, K. C. (2004). 5-ht6 receptors. *Curr. Drug Targets CNS Neurol. Disord.* **3**, 59–79.

Zhang, J. V., Ren, P. G., Avsian-Kretchmer, O., Luo, C. W., Rauch, R., Klein, C., and Hsueh, A. J. (2006). Obestatin, a peptide encoded by the ghrelin gene, opposes ghrelin's effects on food intake. *Science* **310**, 996–999.

Clinical Investigations of Antiobesity Drugs

JOHN WILDING

Clinical Sciences Centre, Department of Medicine, Diabetes, and
Endrocrinology—Clinical Research Group,
University Hospital, Aintree

 I. Introduction
 II. A Brief History of Obesity Pharmacotherapy
 III. Currently Available Drugs Licensed for Obesity Treatment
    A. Phentermine
    B. Diethylpropion Hydrochloride (Amfepranone, Tenuate Dospan)
    C. Sibutramine
    D. Orlistat
    E. Rimonabant
    F. Other Drugs: Topiramate and Zonisamide
    G. Unlicensed Drugs
 IV. Potential New Targets for Obesity Drug Development
    A. Inhibition of Nutrient Absorption
    B. Enhancement of Peripheral Satiety or Adiposity Signals
    C. Drugs That Alter Metabolism/Substrate Utilization
    D. Drugs Acting on CNS Targets That Affect Energy Balance
 V. Preclinical Testing
    A. Target Validation
    B. Acute (Short-term) Testing
    C. Long-term Testing
 VI. Regulatory Requirements for Clinical Development
    A. Phase I Trials
    B. Phase II Trials
    C. Phase I/II Experimental Medicine Studies
    D. Phase III Trials
    E. Phase IV Trials
    References

Obesity has become a global health problem, and the costs of treating its complications are escalating rapidly. Development of new drugs to treat obesity and its complications has become a priority and is a major focus of pharmaceutical research. Drugs for obesity have had a difficult history, with many being withdrawn because of an adverse risk–benefit ratio; the standards for approval by the regulatory authorities are now set at a high level. In the last 20 years, improved understanding of the biology of energy balance has led to a rapid increase in the identification of potential targets for antiobesity drugs. These include pathways of food absorption, peripheral signals of hunger, satiety and energy balance, metabolic pathways, and central systems involved in energy balance regulation. Each of these targets requires careful validation in preclinical models and often detailed analysis of structure–function relationships before potential drugs can be developed. Clinical testing is time-consuming and costly. Use of experimental medicine models to help decision making at the interface of Phase I and II studies is one way of improving the efficiency of this process, but at present large Phase III trials are still required before a product can be approved for clinical use. In the future, emphasis is likely to shift from simple efficacy in terms of weight loss toward measurable clinical outcomes, such as prevention of diabetes and vascular disease; comparative trials and limited registration with careful follow up of outcomes may become necessary for new drugs in this class. Use of drug combinations to improve outcomes in patients with multiple complications and in morbid obesity may also be investigated.

I. INTRODUCTION

The clinical need and demand for safe and effective drugs for the treatment of obesity has never been greater. Rates of obesity are rapidly increasing worldwide, and with this, the prevalence of common obesity-related complications such as diabetes, heart disease, arthritis, and cancer is set to overwhelm the capacity of health care systems to cope, even in the wealthiest countries.

The history of antiobesity treatment includes many failures and false starts, either because effective therapy was found to be unsafe and poorly tolerated, or because initially promising concepts in pre-

clinical models did not result in clinically meaningful weight loss when tested in humans. The requirements for a safe, clinically effective drug are now set at a very high standard in terms of safety and tolerability and at a modest level for efficacy. The costs of developing such agents is high, so it is vital that those seeking to do so are able to make rational decisions about which mechanisms constitute likely targets, are able to test these in appropriate preclinical models, and confirm proof-of-concept in early clinical models that will then enable a clear decision as to whether to proceed to a full-scale clinical development program. Such programs must themselves fulfil the minimal regulatory requirements, but also provide as comprehensive assessment as possible with regard to efficacy and safety in a range of patients and clinical settings as well as evidence of benefit in respect of obesity-related comorbidity.

In this chapter the historical context in terms of currently available drugs, as well as those that have now been withdrawn will be briefly outlined. The importance of early clinical testing, including a description of what is needed at the preclinical stage before proceeding to a proof-of-concept study, will be discussed. Different techniques for preclinical testing will then be described. Regulatory requirements are important for the design of Phase III trials, and these will be described next. To some extent this will determine the design of Phase III studies. It is likely that regulatory requirements will change in the future. Finally, the design of "outcome" and other Phase IV studies will be discussed. This will be illustrated where possible with examples from previous and current clinical development programs.

II. A BRIEF HISTORY OF OBESITY PHARMACOTHERAPY

Early drugs that were used for weight loss included centrally acting sympathomimetic agents, the first being an amphetamine, desoxyephedrine, approved in the United States in 1947; further amphetamine derivatives followed, and included phentermine and diethylpropion, both agents that are still available today (Colman, 2005). While these agents were effective at producing short-term weight loss, CNS and cardiovascular adverse effects (the potential for abuse and pressor effects) were considered significant, and the use of these agents was restricted to short-term use (12 weeks) in the late 1960s, which probably contributed to a decline in their use over the next two decades, and may have stifled research into potentially beneficial or adverse long-term effects. The appearance of agents that had a mode of action based on serotonin began with the licensing of fenfluramine in 1972, and later with its dextroisomer, dexfenfluramine, which was the first drug to demonstrate efficacy over a 12-month study (Guy-Grand et al., 1989). Although the potential for serotonin-releasing agents to cause primary pulmonary hypertension was recognized, this adverse effect is rare (18 per million population, compared to 6 per million in the general population), and in general because of the perceived benefits of weight loss on cardiovascular risk this was felt to be an acceptable risk at the time (Abenhaim et al., 1996). Based on the result of one small study suggesting impressive efficacy for weight loss (Weintraub et al., 1992), the use of combination therapy with fenfluramine and phentermine became widespread in the United States in the 1990s, only to be curtailed by the observation of cardiac valvulopathy in some patients, which led to the voluntary withdrawal of fenfluramine and dexfenfluramine by the manufacturers in 1997 (Connolly et al., 1997) (see later). More recently licensed agents include sibutramine, an inhibitor of *nor*-adrenaline and serotonin reuptake and orlistat, an intestinal lipase inhibitor. The cannabinoid-1 receptor antagonist, rimonabant, has recently been licensed in Europe and is likely to obtain a licence in the near future in the United States.

III. CURRENTLY AVAILABLE DRUGS LICENSED FOR OBESITY TREATMENT

A. Phentermine

Phentermine is a centrally acting drug that is a sympathomimetic. It is thought to suppress appetite, and possibly have a modest thermogenic effect. It has been evaluated in placebo-controlled trials of up to 36 weeks duration as monotherapy and is effective at producing weight loss of up to 12.2 kg (versus 4.8 kg on placebo) over this timescale (Weintraub et al., 1984, 1992). There are few data on effects of phentermine on obesity-related comorbidities such as diabetes, hypertension, or lipids, although blood pressure did fall during one reported study (Weintraub et al., 1992). Phentermine does have a potential to induce dependency, and has significant CNS side effects such as irritability and anxiety, consistent with its central sympathomimetic mode of action. It remains widely available in the United States, in the European Union, and in many other countries. In the United Kingdom it remains a class C controlled drug, and its use is not recommended in the British National Formulary or in obesity guidelines, such as those issued by the Royal College of Physicians, because of concerns about dependence (Royal College of Physicians, 1998), CNS side effects, and the lack of long-term safety and efficacy data compared to other weight-loss drugs.

B. Diethylpropion Hydrochloride (Amfepranone, Tenuate Dospan)

Diethylpropion is a centrally acting anorectic agent. It is a derivative of amphetamine. The usual dosage prescribed for weight loss is 25 mg three times a day. A modified release preparation (75 mg) is also available.

There is little efficacy data available on this centrally acting agent, which has only been tested in short-term trials. Although it is more effective than placebo at producing weight loss, there have been concerns about its use because of its stimulatory effect on the central nervous system, with potential for dependence and pressor effects on the cardiovascular system (Bray 1999).

C. Sibutramine

Sibutramine is a serotonin and nor-adrenaline reuptake inhibitor. It was originally developed as an antidepressant, but was found to be effective for weight loss, but ineffective for depression early in clinical development (Weintraub et al., 1991; Bray et al., 1995). Sibutramine treatment results in weight loss via reduced energy intake (this amounts to approximately 16% reduction at a test lunch), predominantly by increasing satiety (Rolls et al., 1998; Barkelling et al., 2003), and by a small, but measurable effect on energy expenditure, which has been most clearly demonstrated as a greater than expected measured energy expenditure in individuals who have lost weight while taking sibutramine (Hansen et al., 1998). The efficacy of sibutramine has been shown in studies from 3 months to 2 years duration (James et al., 2000). Its effect on weight loss is dose-dependent (Bray et al., 1999), and it meets current criteria for efficacy of a weight-loss drug in both Europe and the United States. A more recent meta-analysis showed that the odds ratio for achieving a 5% weight loss at 12 months was 5.05 for sibutramine versus placebo (Douketis et al., 2005). The multi-center STORM study, demonstrated that sibutramine is also effective for weight maintenance over 2 years, with only slight weight regain during the second year of treatment (James et al., 2000). Weight loss with sibutramine is associated with improvements in some, but not all obesity-related comorbidities. Notably, HDL cho-

lesterol rises by up to 10%, triglycerides fall by about 20%, and diabetes control improves significantly in patients who successfully lose at least 5% of body weight. These benefits are partially offset by small increases in blood pressure (1 mm Hg) and heart rate (3–5 bpm), although analysis based on risk equations suggests that the net effect should be of benefit to patients who lose weight (James et al., 2000; Lauterbach and Evers, 2002). Other adverse effects include dry mouth, insomnia, and constipation, but these are uncommon and rarely severe.

D. Orlistat

Orlistat is unique among drugs used for the treatment of obesity, in that its mode of action is via inhibition of nutrient absorption, rather than effects on appetite or energy expenditure. Orlistat is an inhibitor of intestinal lipases, resulting in inhibition of the absorption of about 30% of dietary fat (Borgstrom, 1988; Guerciolini, 1997). For an individual on a diet containing 90 g of fat per day, this will result in loss of 30 g fat in the stool, equivalent to about 270 kcal/day energy deficit. Over a period of 3 months, this might be expected to result in 3.5 kg greater weight loss than placebo or lifestyle change alone. This is consistent with the changes in body weight that have been seen in clinical trials (Drent et al., 1995). As with all weight-loss drugs studied to date, weight loss tends to reach a plateau after 6–9 months of treatment; at 1 year mean weight loss with orlistat was 10.3 kg versus 6.1 kg with placebo; after 2 years, most of this weight loss was maintained for the second year, with a difference of 3.6 kg being maintained between orlistat and placebo-treated groups (Sjostrom et al., 1998). Orlistat treatment has also been shown to produce benefits in terms of CV risk factors (modest lowering of LDL cholesterol, blood pressure, and glycemia) (Davidson, 1997; Broom et al., 2002). It is also effective in patients with type 2 diabetes treated with a variety of concomitant oral hypoglycemic therapy and insulin (Hollander et al., 1998; Hanefeld and Sachse, 2002).

E. Rimonabant

Rimonabant, a selective antagonist at cannabinoid-1 (CB-1) receptors, has been given a positive opinion from the European regulators, and is now available for prescription at the time of writing. It also has an approvable letter from the U.S. FDA, but it is not available for prescription at the time of writing. In clinical trials involving over 6000 patients, rimonabant has been shown to be effective at reducing body weight in obese subjects with a range of comorbidities, including dyslipidemia and type 2 diabetes. The main side effects seen in clinical trials were mood alterations with depressive symptoms, depressive disorders, anxiety, dizziness, insomnia, nausea, diarrhea, vomiting, and asthenia/fatigue. Its efficacy for weight loss is similar, or slightly superior to existing agents; in nondiabetic patients, the mean weight loss was 6.8 kg at 1 year, compared to 1.4 kg with placebo (last observation carried forward (LOCF) analysis) (Pi-Sunyer et al., 2006a). It is of interest that improvements in HDL cholesterol and triglycerides were greater than would be expected from weight loss alone (Despres et al., 2005). This suggests independent effects of the drug on metabolism, and are supported by a number of preclinical and mechanistic studies that suggest that CB-1 receptors in adipose tissue may be important in regulation of lipid metabolism (Gary-Bobo et al., 2006). Rimonabant is also effective in patients with type 2 diabetes; and has been found to have independent effects on insulin resistance and glycemic control. These findings are reflected in the European CPMP opinion, that supports the use of rimonabant "As an adjunct to diet and exercise for the treatment of obese patients (BMI $\geq$30 kg/m^2), or overweight patients (BMI >27 kg/m^2) with associated risk factor(s), such as type 2 diabetes or dyslipidemia."

F. Other Drugs: Topiramate and Zonisamide

Topiramate and zonisamide are both anticonvulsant drugs that were observed to result in weight loss during treatment of patients with epilepsy. Both have a complex mechanism of action that includes inhibition of sodium ion channels; topiramate is a sulfamated polysaccharide that is also an inhibitor of carbonic anhydrase; this may contribute to some of its adverse effects. However the precise mechanisms by which these drugs cause weight loss is not known. In clinical trials of up to 12 months duration topiramate has been shown to be effective for weight loss in patients with uncomplicated obesity (Wilding et al., 2004). The effect is dose-related, and placebo-subtracted weight loss is about 8% after 12 months treatment at the highest dose used (256 mg/day). However it is not licensed for obesity, as use is limited by CNS side effects including loss of concentration, memory problems, and in a few patients, suicidal ideation. Peripheral effects related to carbonic anhydrase inhibition include parasthesia, lowered serum bicarbonate, and an increased risk of renal stone formation. Zonisamide has been tested in one 6-month study; weight loss was again dose related, but CNS side effects were relatively common (Gadde et al., 2003). Although these agents are not licensed or recommended for treatment of obesity, their use should be considered in obese patients who require anticonvulsant therapy or exceptionally in those with migraine, but only in line with their licensed indications.

G. Unlicensed Drugs

1. Fenfluramine and Dexfenfluramine

These two drugs (dexfenfluramine is the active d-isomer of fenfluramine, which is a racemic mixture), enhance serotonin release from neurons and other tissues. Although effective at producing weight loss, principally by acting on the hypothalamus to reduce appetite (Carvajal et al., 2000), they have now been withdrawn because they were found to produce carcinoid-like valvular heart lesions, resulting in mitral and aortic regurgitation, which in some cases required valve replacement (Connolly et al., 1997). This side effect was particularly common in patients who were also taking phentermine. The fenfluramines have also been associated with primary pulmonary hypertension; although the incidence of this side effect is low, it may only become apparent some years after taking the drug (Brenot et al., 1993; Abenhaim et al., 1996). It is a relentlessly progressive condition, and can result in severe cardiac and respiratory failure that can only be effectively treated by heart–lung transplantation. Neither of these drugs is currently available, both were withdrawn from sale by the manufacturers in 1997 as a result of the reports of cardiac valvular disease.

2. Ephedrine and Caffeine

The effect of ephedrine to promote weight loss is enhanced by caffeine, which has no effect when used alone. This combination has been licensed for short-term use in some countries, and has been shown to produce weight loss in studies of up to 6 months duration, with modest improvements in lipids. Concerns over cardiovascular side effects, principally rises in heart rate and blood pressure and lack of long-term data have limited its use (Astrup et al., 1992).

3. Phenylpropanolamine

This sympathomimetic, commonly used in cold remedies, has been shown in a few short-term studies to produce weight loss (Caffry et al., 1987). However it does not have a license for this indication, and has been withdrawn in the United States following a number of reports of hemorrhagic stroke. It should not therefore be used in the management of obesity.

4. Thyroxine and Diuretics

Thyroxine and diuretics have both been used as aids to weight loss. Both have significant side effects and have no place in the management of obesity.

IV. POTENTIAL NEW TARGETS FOR OBESITY DRUG DEVELOPMENT

There are currently four main areas of interest that provide valid drug targets for treatment of obesity (Fig. 1); these are agents that block or inhibit absorption of nutrients, those that mimic or enhance peripheral satiety or adiposity signals, drugs that alter metabolic rate or substrate utilization, and finally, those acting at CNS targets that result in altered energy intake or energy expenditure; some drugs may have multiple modes of action that include more than one of these broad mechanisms. Some potential targets in each of these groups are discussed later.

A. Inhibition of Nutrient Absorption

Agents in this class include intestinal lipase inhibitors such as orlistat; other potential mechanisms might include agents that interfere with lipid or other nutrient absorption across the intestinal brush-border membrane.

B. Enhancement of Peripheral Satiety or Adiposity Signals

There is no licensed drug in this group at present. However, this seems a promising area for future development. Potential targets include gut hormones such as GLP-1, CCK, peptide YY (3–36), obestatin, and oxyntomodulin. Leptin itself has not lived up to its early promise except in those rare individuals with leptin deficiency, but enhancement of leptin action in the CNS remains a possibility.

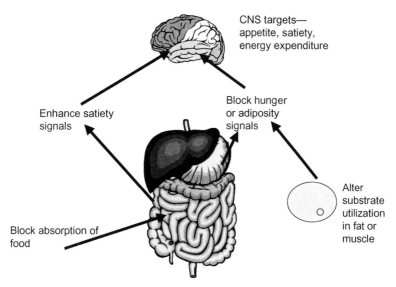

FIGURE 1 Drug targets for obesity treatment. Targets fall into four main groups: those affecting nutrient absorption; those affecting peripheral signals of hunger or satiety; drugs that act on substrate storage or utilization, and CNS targets that influence energy intake and/or energy expenditure.

GLP-1 is a gut hormone that is secreted by intestinal L cells; it is a potent stimulus to insulin secretion, and its main physiological role is thought to be as an incretin hormone (Kreymann et al., 1987). GLP-1 has also been shown to reduce food intake in preclinical studies, an effect that appears to be mediated via receptors in the brainstem and hypothalamus, and is due to enhancement of satiety rather than induction of anorexia or by producing nonspecific illness (Turton et al., 1996). GLP-1 and its analogs also slow gastric emptying, an effect that may also confound measurements of hunger and satiety (O'Halloran et al., 1990). Several human studies (reviewed in Verdich et al., 2001) have shown that infusion of GLP-1 into healthy volunteers, obese subjects, and people with type 2 diabetes may acutely reduce food intake by up to 35%. Exenatide (synthetic exendin-4) is a long-acting GLP-1 analog used for the treatment of type 2 diabetes (Buse et al., 2003, 2004; DeFronzo et al., 2005). The glucose lowering effects of Exenatide mainly occur due to its incretin mimetic effects—stimulation of insulin secretion and suppression of glucagon secretion. Unlike most other drugs used to treat type 2 diabetes exenatide also results in weight loss that may contribute to its antihyperglycemic effect. Weight loss with exenatide is dose related, and appears to be independent of the transient nausea that may be experienced early in treatment. Exendin-4 has been shown to reduce food intake by 19% after intravenous administration to healthy volunteers (0.05 pmol/kg/h) during a test meal; analysis of visual analog questionnaires suggested this was due to a selective affect on eating behavior rather than nausea (Edwards et al., 2001). It is well recognized that weight loss is more difficult to achieve in people with type 2 diabetes and it is unknown whether the effects of exendin seen in healthy volunteers also occur in diabetic subjects after subcutaneous administration of exenatide or whether this effect persists with chronic dosing. GLP-1 also increases body temperature in experimental animals, suggesting a possible effect to increase energy expenditure. Despite this evidence, the use of GLP-1 analogs has not been explored as a specific treatment for obesity.

Oxyntomodulin is a further product produced from the L cells from the pre-proglucagon precursor. It has been shown to decrease food intake in rodents and humans, and appears to act at the same receptor site as GLP-1 (Dakin et al., 2001; Cohen et al., 2003). One short-term study that involved thrice daily subcutaneous administration in obese people has suggested that these effects may result in weight loss (Wynne et al., 2005). The results of longer term trials are awaited.

Cholecystokinin (CCK), may be considered as the archetypal satiety signal; it is released after meals and signals via the vagus nerve to receptors in the brainstem. Administration of CCK to rodents and humans decreases food intake, and blockade of CCK-A receptors peripherally and CCK-B receptors in the CNS increases energy intake (Dourish et al., 1989; Ballinger et al., 1995). CCK receptor agonists have been developed, but have not been demonstrated to be effective in clinical trials to date.

Two stomach-derived hormones, *ghrelin* and *obestatin*, are also worthy of mention. These are produced by differential splicing from the preproghrelin gene. Ghrelin acts via the growth hormone secretagogue receptor in the hypothalamus and is a potent stimulus to food intake; plasma concentrations rise in anticipation of meals and fall after food ingestion (Cummings et al., 2001). However, circulating concentrations are low in obese individuals (English et al., 2002). It remains to be seen whether ghrelin antagonism will reduce food intake in obese subjects. Obestatin is more recently discovered; it does reduce food intake in rodents, but its relevance to humans is currently unknown (Zhang et al., 2005).

Peptide YY (3–36) is released from K cells in the large intestine (Adrian et al., 1985). Its concentrations rise rapidly after ingestion of food, and it is thought to act on the neuropeptide Y Y2 receptors in the hypothalamus, to suppress NPY secretion and thus reduce food intake. PYY concentrations are low in obesity, and PYY infusion decreases food intake in lean and obese subjects (Batterham et al., 2003). Further investigation of its potential to treat obesity is underway.

C. Drugs That Alter Metabolism/ Substrate Utilization

This area has remained one of active interest, although there is no licensed drug that works exclusively via a peripheral metabolic mechanism. Sibutramine probably has some thermogenic activity that contributes to its efficacy. Drugs acting purely as thermogenic agents include β3 adrenoceptor agonists, and enhancers of uncoupling protein action in mitochondria (Clapham, 2004b). Other potential targets in this area include activators of AMP-kinase, drugs that alter fatty acid or triglyceride synthesis and modification of peripheral steroid metabolism (Clapham, 2004a). Some drugs may also have peripheral metabolic effects that are independent of weight loss; examples include the effects of sibutramine and rimonabant on lipid metabolism.

D. Drugs Acting on CNS Targets That Affect Energy Balance

The central nervous system is a rich source of potential targets for the treatment of obesity. Over 50 peptide and classic neurotransmitters have been identified that influence energy balance and most of them have been considered or tested in preclinical models at some point—nevertheless it has proved difficult to translate this into drugs that can proceed into testing in humans. These can be broadly divided into transmitters that decrease energy intake and maybe also increase energy expenditure and those that increase energy intake and/or decrease energy expenditure—the aim is of course to develop agonists of the former and antagonists at the latter. The most promising target in this group at present is the cannabinoid receptor system—the CB1 receptor antagonist rimonabant has completed Phase III trials, and is likely to be available for clinical use in the near future. Many other possible targets are under active investigation; these include neuropeptide Y Y5 receptor antagonists (Marsh et al., 1998; Daniels et al., 2002; Dumont et al., 2004), galanin antagonists (Kyrkouli et al., 1986), ghrelin receptor antagonists or inverse agonists (this is a peripheral hormone that increases food intake via a CNS receptor) (Asakawa et al., 2001; Holst et al., 2003), and melanin-concentrating hormone antagonists (Rossi et al., 1997; Borowsky et al., 2002; Dyke and Ray, 2005). Agonism of receptors that are involved in satiety or reduction of food intake is also possible; classic targets include the serotonin and *nor*-adrenaline systems—this is the target for the licensed reuptake inhibitor sibutramine. Peptide targets include melanocortin-4 receptors, CART, neurotensin, GLP-1, and many others (Wilding et al., 1993b; Turton et al., 1996; Thim et al., 1998; Vaisse et al., 2000).

V. PRECLINICAL TESTING

A detailed description of preclinical models and modes of testing is beyond the scope of this review; a brief description will be provided to help put clinical evaluation into context, and to demonstrate how some of the existing drugs and current targets have been validated in preclinical models.

A. Target Validation

This usually requires a range of studies that demonstrate that the neurotransmitter of interest has an effect on energy balance in both acute and long-term studies—this might involve CNS administration of the molecule in healthy animals (usually rats or mice) in appropriate experimental conditions. In the absence of pharmacological agonists or antagonists, demonstration that an effect is of physiological relevance may

include use of RNA interference or anti-sense to block peptide synthesis or use of potent antibodies to block effects. Demonstration that the molecule is regulated in conditions of altered energy balance, such as overfeeding or food deprivation may also provide supporting data, as may study of genetic overexpression or gene knockout, although the latter may sometimes give conflicting data, unless conditional, tissue-specific knockouts are used. Demonstration that the transmitter has a role in models of obesity, may also be helpful, but should include studies of both dietary and genetic forms of obesity. Examples of such validation studies is given for neuropeptide Y in Table 1; of course not all of these are a prerequisite for selecting a target, but they all provide useful corroborative evidence that a particular target has biological validity.

B. Acute (Short-term) Testing

For peripheral and CNS targets this might include single dose studies looking at effects on food intake, energy expenditure, or (in the case of drugs that work by interfering with nutrient absorption) food absorption. The study design must be appropriate to the mode of action of the drug. Preliminary studies are usually carried out in healthy animals, but studies in both dietary and genetic models of obesity may be more informative. It must be remembered that many receptor systems are different between humans and experimental animals and this may require study within humanized systems where possible.

C. Long-term Testing

Long-term validation studies are likely to include studies over several weeks looking at changes in energy balance in both healthy and obese animal models. This may also allow for preliminary study for metabolic effects, such as insulin resistance, glucose, and lipids. *Toxicology and safety* is a necessary requirement before human

TABLE 1   Some Key Studies Validating NPY Y5 Receptor as a Valid Target for Obesity Treatment
(20 years of research)

Study type	Main findings	References
Discovery of peptide; tissue localization	NPY a major neuropeptide in rat and human CNS	Tatemoto, 1982; Adrian et al., 1983; Allen et al., 1983
I.c.v. injection of native peptide	Central NPY stimulates food intake	Clark et al., 1984
Local injection of native peptide	NPY injected into PVN stimulates food intake; chronic injection causes obesity	Stanley and Leibowitz, 1985; Stanley et al., 1985, 1986
Measurement of endogenous concentrations and mRNA in conditions of altered energy balance (fasting)	Fasting increases hypothalamic NPY synthesis, concentration, and release	Sahv et al., 1988; Kalra et al., 1991; Lambert et al., 1994
Measurement of endogenous concentrations and mRNA in obese animal models	Obese animal models have higher hypothalamic NPY mRNA and concentrations	Williams et al., 1991; Wilding et al., 1993a,b
Antibody blockade	Blockade of NPY reduces food intake in fasted animals	Lambert et al., 1993
Defining the "feeding receptor"	Discovery that Y5 agonists influence feeding	Hwa et al., 1999
Knockout models of NPY gene	NPY gene knockout has normal food intake	Erickson et al., 1996
Knockout models of receptor	Y5 receptor knockout has reduced food intake	Marsh et al., 1998
Selective antagonist reduces food intake	Y5 receptor antagonist effective in animal models	Daniels et al., 2002

testing can begin, but will not be discussed here.

VI. REGULATORY REQUIREMENTS FOR CLINICAL DEVELOPMENT

The clinical development of drugs to aid management of obesity is constrained to some extent by the regulatory requirements for registration of such drugs in the main potential markets of Europe and the United States. These have many similarities, and the principles are similar, but there are important differences that may require slightly different trial design to satisfy both regulatory agencies. The general principles are that the drug must show clinically meaningful weight loss over placebo (this is defined differently in Europe and the United States—see Table 2), beneficial effects should be seen in patients with obesity-related comorbidities such as dyslipidemia, hypertension, and diabetes. Finally, the risk–benefit ratio must be clearly in favor of the drug; this is often the most difficult judgment to make, and recent history with drugs in other classes for chronic disease management (such as COX-2 inhibitors and α-γ PPAR activators for diabetes) suggests that the requirements in this respect are becoming increasingly stringent. The requirements are summarized in Table 2; and can be accessed at the following websites (www.emea.eu.int/pdfs/human/ewp/028196en.pdf and www.fda.gov/cder/guidance/obesity.pdf); however, it should be noted that these are currently under review.

A. Phase I Trials

These "first into man" studies are typically conducted in small numbers of

TABLE 2   Criteria Recommended by Regulatory Agencies in Evaluation of Drugs Used in Weight Control

	EMEA (Europe)	FDA (USA)
Primary end point	Weight loss 10% of baseline weight at 12 months (statistically greater than placebo)	Weight loss Drug to result in 5% greater weight loss than placebo
Secondary end points	CV risk factors (BP, lipids), arthritis, cardiac function, OSA; must be statistically significant and clinically relevant. Studies in diabetes recommended	CV risk factors (BP, lipids), arthritis, cardiac function, OSA; must be statistically significant and clinically relevant
Placebo run-in period	Yes—duration not specified	6 weeks recommended
Subject demographics	BMI >30 or >27 with risk factors	BMI >30 or >27 with risk factors
Other studies	Pharmacokinetics Interactions Adverse events	Pharmacokinetics Interactions Adverse events

healthy male volunteers with the express aim of testing for initial safety, and providing provisional pharmacodynamic and pharmacokinetic data. It would be unusual to test for efficacy in such studies, although it might be possible to obtain preliminary data on effects on hunger and satiety scores in such studies.

B. Phase II Trials

These are the first studies that provide data in the patient population of interest. In obesity studies this has usually involved study of 200 or 300 individuals over a period of about 3 months. It is common to study several doses at this stage of development (Bray et al., 1996). In obesity studies the end point is usually % weight loss after 3 months treatment. This information may be supplemented by measurement of other outcomes, such as changes in body fat, waist circumference, and surrogate markers such as lipids, glucose, measurements of insulin sensitivity, and blood pressure. Monitoring for potential adverse events is important at this stage, and usually involves comprehensive recording of symptoms, signs, clinical chemistry, and hematological findings. Other evaluations, such as psychiatric evaluation and health status (usually by questionnaire) may also

be important, depending on the mode of action of the drug.

C. Phase I/II Experimental Medicine Studies

Given the current high attrition rate of drugs entering traditional Phase II studies, pharmaceutical companies are increasingly looking to use experimental medicine models that might be used to select drugs that are likely to succeed in Phase II studies and proceed to further development, and to help with dose finding, so that a smaller Phase II study is needed. For obesity drugs this might be particularly useful for agents that affect energy intake or energy expenditure. An appropriate set up might include acute measurement of energy intake under controlled laboratory conditions. For a drug that reduced hunger or increased satiety, the most appropriate model would be to administer drug or placebo in the morning after an overnight fast, and then provide the subject first with a fixed energy breakfast with assessment of hunger and satiety scores and later in the day with a buffet lunch, or a more simple meal, provided in sufficient quantity that the subjects can be allowed to eat until they are full. The primary outcome measure is food eaten at the lunchtime meal, with supporting data

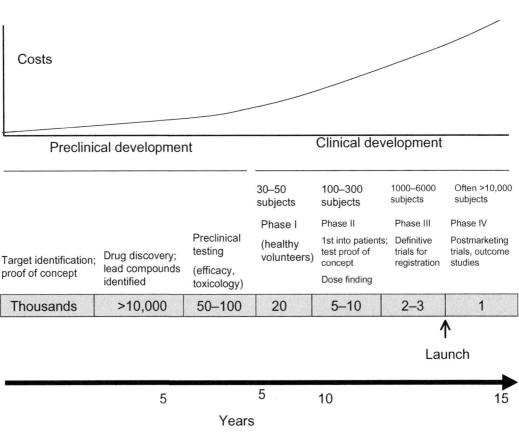

FIGURE 2   Drug development. Stages in drug development: Figures show likely attrition rates at each stage. Development may take 15 years or more.

from the visual analog scores. More sophisticated questionnaires can be used to help distinguish between effects that are truly affecting appetite and satiety, or those that are more akin to aversion, such as nausea. It is possible to use computerized models both to aid collection and analysis of the questionnaire data and to measure real-time food intake using a balance linked to the computer (Fig. 3). Measurements of energy expenditure, using indirect calorimetry can also be made during these studies. It is therefore possible to conduct proof of concept studies in a relatively short timescale and without prolonged patient exposure to the investigational drug. Of course this does not guarantee that a drug will be effective or safe in long-term trials but may allow exclusion of some drugs at an earlier stage and help with dose selection.

D. Phase III Trials

The design of Phase III trials has largely been determined by regulatory requirements as discussed before. These will generally have to be of at least 12 months duration with some studies of 2 years, and include sufficient numbers of patients to accumulate data on efficacy, surrogate disease markers, and effects in specific comorbid conditions (especially diabetes). The whole trial program usually has to include sufficient numbers of individuals to demonstrate an acceptable risk–benefit ratio in terms of safety. This has not been predefined by the regulators, but more recent developmental programs have included at least 2000–4000 patients in total.

There are many controversial areas in the design of Phase III studies in obesity; these

Universal eating monitor

- Balance system linked to PC

- Real-time recording of food intake and appetite scores

FIGURE 3 Computerized eating monitor.

include whether there should be a placebo run-in period, the intensity of ancillary non-pharmacological therapy, the use of active comparators, and evaluation of the drug's effect on weight maintenance as opposed to weight loss. Statistical analysis of the data has also created controversy at times. These may all alter the apparent efficacy of the drugs (Fig. 4).

1. Role of the Placebo Run-in Period

The regulatory authorities currently both support the use of a placebo run-in period, although the duration of this is only specified in the FDA document (6 weeks). This has the advantage that individuals who are poorly compliant with the trial protocol may be excluded, and can also be used to exclude people who are able to lose weight very rapidly with lifestyle change. Some trials report weight loss from the start of the run-in period, for example, those of orlistat (Davidson, 1997), others only from the point of randomization (e.g., rimonabant clinical trials) (Pi-Sunyer et al., 2006b); furthermore, different trials use different run-in periods which can make straightforward comparisons of drugs difficult, unless placebo-subtracted differences in weight from baseline are used.

2. Intensity of Nonpharmacological Treatment

The intensity of additional support provided for patients is also critical. This was

demonstrated more recently for sibutramine, when additional behavior modification therapy was shown to significantly improve the overall effectiveness of the treatment program (Wadden et al., 2005).

3. Trials for Weight Loss or Weight Maintenance?

Most treatments for obesity will result in initial weight loss over the first 6 to 9 months of therapy, and this will be followed by a period of weight stability, or slow weight regain. The effects of drugs on weight maintenance is generally investigated by switching patients to placebo after a period of 6 months or a year on active therapy. Comparisons can then be made between those remaining on the drug and those in whom it has been stopped. In some trial programs, different clinical trials have been conducted to investigate the effects of drugs on weight maintenance after a period of intensive weight loss (e.g., with a very low calorie diet). These provide useful additional information that may be helpful to practitioners seeking to use such an approach in clinical practice (see Figure 4) (Apfelbaum et al., 1999; James et al., 2000; Pi-Sunyer et al., 2006b).

4. Use of Active Comparators

Now that three, moderately effective drugs have approval for use in obesity, the question may be asked as to whether

(a)

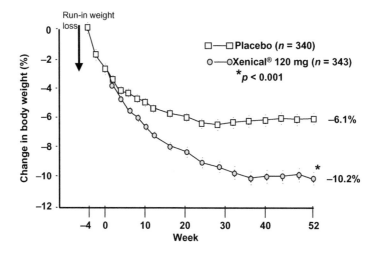

(b)

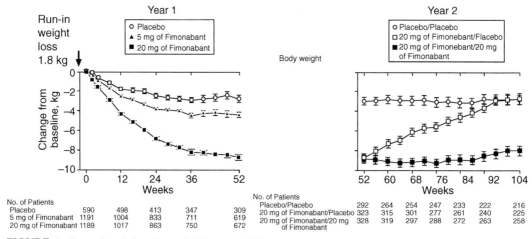

FIGURE 4 Examples of the effects of different trial designs on outcomes: (a) Effect of including run-in period on total weight loss acheived, (b) Active treatment for all to emphasize effect on weight maintenance.

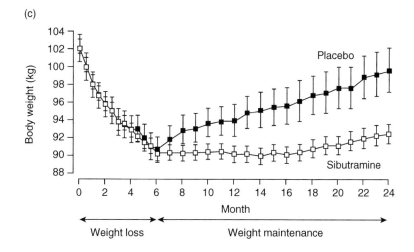

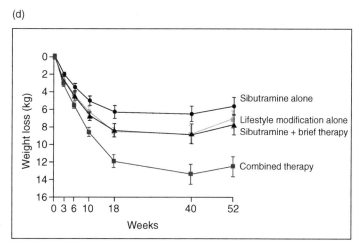

FIGURE 4, *Continued*   (c) Crossover design—run-in data not shown, and (d) How intensity of ancillary therapy can alter results.

placebo-controlled trials will continue to have a place in Phase III programs, particularly for patients with comorbidity, where it might be argued that a placebo-controlled trial is unethical. Under these circumstances, the rules would have to change again, to allow for studies that showed at least equivalence to existing agents.

5. Statistical Issues

In a perfect clinical trial for obesity, there would be no dropouts, and a straightforward comparison of weight loss and other surrogate outcomes between treatment groups would be sufficient. Unfortunately,

particularly in studies of patients with uncomplicated obesity, drop out rates may be high (as much as 40%), and this may have significant effects on the way the trial is analyzed and interpreted. Most trials therefore report results using the "last observation carried forward" (LOCF) analysis. This includes all available data from all patients entered into the trial (intention to treat) and uses the last observation on a particular patient "carried forward" to the end of the study to derive the final data on weight loss. This will tend to underestimate the effect seen in clinical practice for patients who continue to take the drug, especially if there are high dropout rates in nonresponders. It is therefore also important to look at the proportion of patients who will achieve clinically meaningful weight loss (say 5 and 10%) on active therapy.

E. Phase IV Trials

There is now substantial evidence on the effects of existing drugs on weight loss and surrogate outcome measures in patients with obesity-related comorbidity, such as dyslipidemia, hypertension, and diabetes control. Furthermore, the Xendos study, a comparison of orlistat plus diet and lifestyle versus diet and lifestyle alone over 4 years clearly showed benefit of treatment in terms of the risk of development of new-onset diabetes, with an overall 37% relative risk reduction, that was due to a risk reduction of 46% in the 20% of patients with abnormal glucose tolerance at baseline. Patients treated with orlistat also had significant improvements in lipids and blood pressure (Torgerson et al., 2004). However, the key question as to whether obesity treatment with drugs will influence important clinical outcomes such as rates of cardiovascular disease, or death has not yet been answered in a clinical trial. The Sibutramine Cardiovascular OUT-comes (SCOUT) trial has now recruited over 9000 overweight and obese patients with a range of cardiovascular risk factors, including diabetes and a past history of vascular disease, and should help provide an answer to this important question. Similar trials looking at prevention of diabetes and of cardiovascular disease are planned with rimonabant.

Although the focus of trials of obesity treatment has been largely based on prevention and management of cardiovascular disease and diabetes, other obesity-related conditions should not be forgotten, and are potentially important areas for future study. These include arthritis, respiratory disease (especially sleep apnea and asthma), infertility, and cancer. Effects on health status and quality of life are often included in Phase III studies, but have never been fully investigated in the long term. Optimal use of drug therapy, including intermittent use, has been shown to be a promising approach in one study of sibutramine in type 2 diabetes (Redmon et al., 2003), but has never been studied in detail, and may be worth investigating in the future. Finally, the management of most chronic conditions requires the use of different drugs given together to help achieve optimal clinical outcomes; partly because of the problems encountered with the fenfluramine–phentermine combination this has been discouraged by regulators. Nevertheless this is an area that may be worth reexploring as new treatments become available, perhaps initially in more severe cases of obesity. It may then be necessary to test combinations in the context of chronic conditions such as diabetes, and in comparison to surgical treatments that currently represent the "gold-standard" treatment for morbid obesity.

References

Abenhaim, I., Moride, Y., and Brenot, F. (1996). Αππετιτε συππρεσσαντ δρυγσ ανδ τηε ρισκ οφ πριμαρψ πυλμοναρψ ηψπερτενσιον. *N. Engl. J. Med.* **335**, 609–616.

Abenhaim, L., Moride, Y., Brenot, F., Rich, S., Benichou, J., Kurz, X., et al. (1996). Appetite-suppressant drugs and the risk of primary pulmonary hypertension. *N. Engl. J. Med.* **335**, 609–616.

Adrian, T. E., Allen, J. M., Bloom, S. R., Ghatei, M. A., Rosser, M. N., Roberts, G. W., et al. (1983). Neuropeptide Y distribution in human brain. *Nature* **308**, 584–586.

Adrian, T. E., Ferri, G. L., Bacarese-Hamilton, A. J., Fuessl, H. S., Polak, J. M., and Bloom, S. R. (1985). Human distribution and release of a putative new gut hormone, peptide YY. *Gastroenterology* **89**, 1070–1077.

Allen, Y. S., Adrian, T. E., Allen, J. M., Tatemoto, K., Crow, T. J., Bloom, S. R., et al. (1983). Neuropeptide Y distribution in the rat brain. *Science* **221**, 877–879.

Apfelbaum, M., Vague, P., Ziegler, O., Hanotin, C., Thomas, F., and Leutenegger, E. (1999). Long-term maintenance of weight loss after a very-low-calorie diet: a randomized blinded trial of the efficacy and tolerability of sibutramine. *Am. J. Med.* **106**, 179–184.

Asakawa, A., Inui, A., Kaga, T., Yuzuriha, H., Nagata, T., Ueno, N., et al. (2001). Ghrelin is an appetite-stimulatory signal from stomach with structural resemblance to motilin. *Gastroenterology* **120**, 337–345.

Astrup, A., Breum, L., Toubro, S., Hein, P., and Quaade, F. (1992). The effect and safety of an ephedrine caffeine compound compared to ephedrine, caffeine and placebo in obese subjects on an energy restricted diet—a double-blind trial. *Int. J. Obes.* **16**, 269–277.

Ballinger, A., Mcloughlin, L., Medbak, S., and Clark, M. (1995). Cholecystokinin is a satiety hormone in humans at physiological post-prandial plasma concentrations. *Clin. Sci (Colch.)* **89**, 375–381.

Barkeling, B., Elfhag, K., Rooth, P., and Rossner, S. (2003). Short-term effects of sibutramine (Reductil™) on appetite and eating behaviour and the long-term therapeutic outcome. *Int. J. Obes.* **27**, 693–700.

Batterham, R. L., Cohen, M. A., Ellis, S. M., Le Roux, C. W., Withers, D. J., Frost, G. S., et al. (2003). Inhibition of food intake in obese subjects by peptide YY3-36. *N. Engl. J. Med.* **349**, 941–948.

Borgstrom, B. (1988). Mode of action of tetrahydrolipstatin: a derivative of the naturally occurring lipase inhibitor lipstatin. *Biochim. Biophys. Acta* **962**, 308–316.

Borowsky, B., Durkin, M. M., Ogozalek, K., Marzabadi, M. R., DeLeon, J., Heurich, R., et al. (2002). Antidepressant, anxiolytic and anorectic effects of a melanin-concentrating hormone-1 receptor antagonist. *Nat. Med.* **8**, 825–830.

Bray, G. A. (1999). Drug treatment of obesity. *Best Pract. Res. Clin. Endocrinol. Metab.* **13**, 131–148.

Bray, G. A., Ryan, D. H., Gordon, D., Heidingsfelder, S., Macchiavelli, R., and Wilson, K. (1995). Double-blind randomized trial of sibutramine in overweight subjects. *Am. J. Clin. Nutr.* **61**, 912.

Bray, G. A., Ryan, D. H., Gordon, D., Heidingsfelder, S., Cerise, F., and Wilson, K. (1996). Double-blind randomized placebo-controlled trial of sibutramine. *Obes. Res.* **4**, 263–270.

Bray, G. A., Blackburn, G. L., Ferguson, J. M., Greenway, F. L., Jain, A. K., Mendel, C. M., et al. (1999). Sibutramne produces dose-related weight loss. *Obes. Res.* **7**, 189–198.

Brenot, F., Herve, P., Petitpretz, P., Parent, F., Duroux, P., and Simonneau, G. (1993). Primary pulmonary-hypertension and fenfluramine use. *Br. Heart J.* **70**, 537–541.

Broom, I., Wilding, J., Stott, P., and Myers, N. (2002). Randomised trial of the effect of orlistat on body weight and cardiovascular disease risk profile in obese patients: UK multimorbidity study. *Int. J. Clin. Pract.* **56**, 494–499.

Buse, J. B., Henry, R. R., Han, J., Kim, D. D., Fineman, M. S., and Baron, A. D. (2004). Effects of exenatide (exendin-4) on glycemic control over 30 weeks in sulfonylurea-treated patients with type 2 diabetes. *Diabetes Care* **27**, 2628–2635.

Caffry, E. W., Kissileff, H. R., and Thornton, J. C. (1987). Assessment of the effects of phenylpropanolamine on appetite and food intake. *Pharmacol. Biochem. Behav.* **26**, 321–325.

Carvajal, A., del Pozo, J. G., de Diego, I. M., de Castro, A. M. R., and Velasco, A. (2000). Efficacy of fenfluramine and dexfenfluramine in the treatment of obesity: a meta-analysis. *Methods Find Exp. Clin. Pharmacol.* **22**, 285–290.

Clapham, J. C. (2004a). Fat oxidation in obesity: druggable or risky enterprise? *Idrugs* **7**, 238–242.

Clapham, J. C. (2004b). Treating obesity: pharmacology of energy expenditure. *Curr. Drug Targets* **5**, 309–323.

Clark, J. T., Kalra, P. S., Crowley, W. R., and Kalra, S. P. (1984). Neuropeptide Y and pancreatic polypeptide stimulate feeding behavior in rats. *Endocrinology* **115**, 427–429.

Cohen, M. A., Ellis, S. M., Le Roux, C. W., Batterham, R. L., Park, A., Patterson, M., et al. (2003). Oxyntomodulin suppresses appetite and reduces food intake in humans. *J. Clin. Endocrinol. Metab.* **88**, 4696–4701.

Colman, E. (2005). Anorectics on trial: a half century of federal regulation of prescription appetite suppressants. *Ann. Intern. Med.* **143**, 380–385.

Connolly, H. M., Crary, J. L., McGoon, M. D., Hensrud, D. D., Edwards, B. S., Edwards, W. D., et al. (1997). Valvular heart disease associated with fenfluramine-phentermine. *N. Engl. J. Med.* **337**, 581–588.

Cummings, D. E., Purnell, J. Q., Frayo, R. S., Schmidova, K., Wisse, B. E., and Weigle, D. S. (2001). A preprandial rise in plasma ghrelin levels suggests a role in meal initiation in humans. *Diabetes* **50**, 1714–1719.

Dakin, C. L., Gunn, I., Small, C. J., Edwards, C. M. B., Hay, D. L., Smith, D. M., et al. (2001). Oxyntomodulin inhibits food intake in the rat. *Endocrinology* **142**, 4244–4250.

Daniels, A. J., Grizzle, M. K., Wiard, R. P., Matthews, J. E., and Heyer, D. (2002). Food intake inhibition and reduction in body weight gain in lean and obese rodents treated with GW438014A, a potent and selective NPY-Y5 receptor antagonist. *Regul. Pept.* **106**, 47–54.

Davidson, M. (1997). A 2 year, U.S., randomized, controlled study of orlistat, a gastrointestinal lipase inhibitor, for obesity treatment. *JAMA* **96**, 4119.

DeFronzo, R. A., Ratner, R. E., Han, J., Kim, D. D., Fineman, M. S., and Baron, A. D. (2005). Effects of exenatide (exendin-4) on glycemic control and weight over 30 weeks in metformin-treated patients with type 2 diabetes. *Diabetes Care* **28**, 1092–1100.

Despres, J. P., Golay, A., and Sjostrom, L. (2005). Effects of rimonabant on metabolic risk factors in overweight patients with dyslipidemia. *N. Engl. J. Med.* **353**, 2121–2134.

Douketis, J. D., Macie, C., Thabane, L., and Williamson, D. F. (2005). Systematic review of long-term weight loss studies in obese adults: clinical significance and applicability to clinical practice. *Int. J. Obes.* **29**, 1153–1167.

Dourish, C. T., Rycroft, W., and Iversen, S. D. (1989). Postponement of satiety by blockade of cholecystokinin (CCK-B) receptors. *Science* **245**, 1509–1511.

Drent, M. L., Larsson, I., Williamolsson, T., Quaade, F., Czubayko, F., Vonbergmann, K., et al. (1995). Orlistat (Ro-18-0647), a lipase inhibitor, in the treatment of human obesity—a multiple-dose study. *Int. J. Obes.* **19**, 221–226.

Dumont, Y., Thakur, M., Beck-Sickinger, A., Fournier, A., and Quirion, R. (2004). Characterization of a new neuropeptide Y Y5 agonist radioligand: [I-125] [cPP(1-7), NPY(19-23), Ala(31), Aib(32), Gln(34)] hPP. *Neuropeptides* **38**, 163–174.

Dyke, H. J., and Ray, N. C. (2005). Recent developments in the discovery of MCH-1R antagonists for the treatment of obesity—an update. *Expert Opin. Ther. Pats.* **15**, 1303–1313.

Edwards, C. M. B., Stanley, S. A., Davis, R., Brynes, A. E., Frost, G. S., Seal, L. J., et al. (2001). Exendin-4 reduces fasting and postprandial glucose and decreases energy intake in healthy volunteers. *Am. J. Physiol. Endocrinol. Metab.* **281**, E155–E161.

English, P. J., Ghatei, M. A., Malik, I. A., Bloom, S. R., and Wilding, J. P. H. (2002). Food fails to suppress ghrelin levels in obese humans. *J. Clin. Endocrinol. Metab.* **87**, 2984–2987.

Erickson, J. C., Clegg, K. E., and Palmiter, R. D. (1996). Sensitivity to leptin and susceptibility to seizures of mice lacking neuropeptide Y. *Nature* **381**, 415–418.

Gadde, K. M., Franciscy, D. M., Wagner, H. R., and Krishnan, K. R. R. (2003). Zonisamide for weight loss in obese adults—A randomized controlled trial. *JAMA* **289**, 1820–1825.

Gary-Bobo, M., Elachouri, G., Scatton, B., Le Fur, G., Oury-Donat, F., and Bensaid, M. (2006). The cannabinoid CB1 receptor antagonist rimonabant (SR141716) inhibits cell proliferation and increases markers of adipocyte maturation in cultured mouse 3T3 F442A preadipocytes. *Mol. Pharmacol.* **69**, 471–478.

Guerciolini, R. (1997). Mode of action of orlistat. *Int. J. Obes.* **21**, S12–S23.

Guy-Grand, B., Apfelbaum, M., Crepaldi, G., Gries, A., Lefebvre, P., and Turner, P. (1989). International trial of long term dexfenfluramine in obesity. *Lancet* 1142–1145.

Hanefeld, M., and Sachse, G. (2002). The effects of orlistat on body weight and glycaemic control in overweight patients with type 2 diabetes: a randomized, placebo-controlled trial. *Diabetes Obes. Metab.* **4**, 415–423.

Hansen, D. L., Toubro, S., Stock, M. J., Macdonald, I. A., and Astrup, A. (1998). Thermogenic effects of sibutramine in humans. *Am. J. Clin. Nutr.* **68**, 1180–1186.

Hollander, P. A., Elbein, S. C., Hirsch, I. B., Kelley, D., McGill, J., Taylor, T., et al. (1998). Role of orlistat in the treatment of obese patients with type 2 diabetes—A 1-year randomized double-blind study. *Diabetes Care* **21**, 1288–1294.

Holst, B., Cygankiewicz, A., Jensen, T. H., Ankersen, M., and Schwartz, T. W. (2003). High constitutive signaling of the ghrelin receptor—Identification of a potent inverse agonist. *Mol. Endocrinol.* **17**, 2201–2210.

Hwa, J. J., Witten, M. B., Williams, P., Ghibaudi, L., Gao, J., Salisbury, B. G., et al. (1999). Activation of the NPYY5 receptor regulates both feeding and energy expenditure. *Am. J. Physiol. Regul. Integr. Comp. Physiol.* **277**, R1428–R1434.

James, W. P. T., Astrup, A., Finer, N., Hilsted, J., Kopelman, P., Rossner, S., et al. (2000). Effect of sibutramine on weight maintenance after weight loss: a randomised trial. *Lancet* **356**, 2119–2125.

Kalra, S. P., Dube, M. G., Sahu, A., and Phelps, C. P. (1991). Neuropeptide Y secretion increases in the paraventricular nucleus in association with increased appetite for food. *Proc. Natl. Acad. Sci. U.S.A.* **88**, 10931–10935.

Kolterman, O. G., Buse, J. B., Fineman, M. S., Gaines, E., Heintz, S., Bicsak, T. A., et al. (2003). Synthetic exendin-4 (Exenatide) significantly reduces postprandial and fasting plasma glucose in subjects with type 2 diabetes. *J. Clin. Endocrinol. Metab.* **88**, 3082–3089.

Kreymann, B., Williams, G., Ghatei, M. A., and Bloom, S. R. (1987). Glucagon-like peptide-1 7-36: a physiological incretin in man. *Lancet* **2**, 1300–1304.

Kyrkouli, S. E., Stanley, B. G., and Leibowitz, S. F. (1986). Galanin: stimulation of feeding induced by medial hypothalamic injection of this novel peptide. *Eur. J. Pharmacol.* **122**, 159–160.

Lambert, P. D., Wilding, J. P. H., Al-Dokhayel, A. A. M., Bohuon, C., Comoy, E., Gilbey, S. G., et al. (1993). A role for neuropeptide Y, dynorphin and noradrenaline in the central control of food intake after food deprivation. *Endocrinology* **133**, 29–32.

Lambert, P. D., Wilding, J. P., Turton, M. D., Ghatei, M. A., and Bloom, S. R. (1994). Effect of food deprivation and streptozotocin-induced diabetes on hypothalamic neuropeptide Y release as measured by a radioimmunoassay-linked microdialysis procedure. *Brain Res.* **656**, 135–140.

Lauterbach, K., and Evers, T. (2002). Treatment with sibutramine reduces the estimated risk of coronary heart disease: application of the Munster Heart Study Risk Equation (PROCAM). *Am. J. Clin. Nutr.* **75**, 91.

Marsh, D. J., Hollopeter, G., Kafer, K. E., and Palmiter, R. D. (1998). Role of the Y5 neuropeptide Y receptor in feeding and obesity. *Nat. Med.* **4**, 718–721.

O'Halloran, D. J., Nikou, G. C., Kreymann, B., Ghatei, M. A., and Bloom, S. R. (1990). Glucagon-like peptide-1 (7-36)-NH2: a physiological inhibitor of gastric acid secretion in man. *J. Endocrinol.* **126**, 169–173.

Pi-Sunyer, F., Aronne, L. J., Heshmati, H. M., Devin, J., and Rosenstock, J. (2006a). Effect of rimonabant, a cannabinoid-1 receptor blocker, on weight and cardiometabolic risk factors in overweight or obese patients—RIO-North America: a randomized controlled trial. *JAMA* **295**, 761–775.

Pi-Sunyer, F., Aronne, L. J., Heshmati, H. M., Devin, J., and Rosenstock, J. (2006b). Effect of rimonabant, a cannabinoid-1 receptor blocker, on weight and cardiometabolic risk factors in overweight or obese patients—RIO-North America: a randomized controlled trial. *JAMA* **295**, 761–775.

Redmon, J. B., Raatz, S. K., Reck, K. P., Swanson, J. E., Kwong, C. A., Fan, M. S., Thomas, W., and Bantle, J. P. (2003). One-year outcome of a combination of weight loss therapies for subjects with type 2 diabetes: a randomized trial. *Diabetes Care* **26**, 2505–2511.

Rolls, B. J., Shide, D. J., Thorwart, M. L., and Ulbrecht, J. S. (1998). Sibutramine reduces food intake in non-dieting women with obesity. *Obes. Res.* **6**, 1–11.

Rossi, M., Choi, S. J., Oshea, D., Miyoshi, T., Ghatei, M. A., and Bloom, S. R. (1997). Melanin-concentrating hormone acutely stimulates feeding, but chronic administration has no effect on body weight. *Endocrinology* **138**, 351–355.

Royal College of Physicians. (1998). Clinical management of overweight and obese patients, with particular reference to the use of drugs.

Sahu, A., Kalra, P. S., and Kalra, S. P. (1988). Food deprivation and ingestion induce reciprocal changes in neuropeptide Y concentrations in the paraventricular nucleus. *Peptides* **9**, 83–86.

Sjostrom, L., Rissanen, A., Andersen, T., Boldrin, M., Golay, A., Koppeschaar, H. P. F., et al. (1998). Randomised placebo-controlled trial of orlistat for weight loss and prevention of weight regain in obese patients. *Lancet* **352**, 167–172.

Stanley, B. G., and Leibowitz, S. F. (1985). Neuropeptide Y injected in the paraventricular hypothalamus: a powerful stimulant of feeding behavior. *Proc. Natl. Acad. Sci. U.S.A.* **82**, 3940–3943.

Stanley, B. G., Chin, A. S., and Leibowitz, S. F. (1985). Feeding and drinking elicited by central injection of neuropeptide Y: evidence for a hypothalamic site(s) of action. *Brain. Res. Bull.* **14**, 521–524.

Stanley, B. G., Kyrkouli, S. E., Lampert, S., and Leibowitz, S. F. (1986). Neuropeptide Y chronically injected into the hypothalamus: a powerful neurochemical inducer of hyperphagia and obesity. *Peptides* **7**, 1189–1192.

Tatemoto, K. (1982). Neuropeptide Y: complete amino acid sequence of the brain peptide. *Proc. Natl. Acad. Sci. U.S.A.* **79**, 5485–5489.

Thim, L., Kristensen, P., Larsen, P. J., and Wulff, B. S. (1998). CART, a new anorectic peptide. [Review] [10 refs.]. *Int. J. Biochem. Cell Biol.* **30**, 1281–1284.

Torgerson, J. S., Hauptman, J., Boldrin, M. N., and Sjostrom, L. (2004). XENical in the prevention of diabetes in obese subjects (XENDOS) study. *Diabetes Care* **27**, 155–161.

Turton, M. D., O'Shea, D., Gunn, I., Beak, S. A., Edwards, C. M. B., Meeran, K., et al. (1996). A role for glucagon-like peptide-1 in the central control of feeding. *Nature* **379**, 69–72.

Vaisse, C., Clement, K., Durand, E., Hercberg, S., Guy-Grand, B., and Froguel, P. (2000). Melanocortin-4 receptor mutations are a frequent and heterogeneous cause of morbid obesity. *J. Clin. Invest.* **106**, 253–262.

Verdich, C., Flint, A., Gutzwiller, J. P., Naslund, E., Beglinger, C., Hellstrom, P. M., et al. (2001). A meta-analysis of the effect of glucagon-like peptide-1 (7-36) amide on ad libitum energy intake in humans. *J. Clin. Endocrinol. Metab.* **86**, 4382–4389.

Wadden, T. A., Berkowitz, R. I., Womble, L. G., Sarwer, D. B., Phelan, S., Cato, R. K., et al. (2005). Randomized trial of lifestyle modification and pharmacotherapy for obesity. *N. Engl. J. Med.* **353**, 2111–2120.

Weintraub, M., Hasday, J. D., Mushlin, A. I., and Lockwood, D. H. (1984). A double-blind clinical-trial in weight control—use of fenfluramine and phentermine alone and in combination. *Arch. Intern. Med.* **144**, 1143–1148.

Weintraub, M., Rubio, A., Golik, A., Byrne, L., and Scheinbaum, M. L. (1991). Sibutramine in weight control—a dose-ranging, efficacy study. *Clin. Pharmacol. Ther.* **50**, 330–337.

Weintraub, M., Sundaresan, P. R., Madan, M., Schuster, B., Balder, A., Lasagna, L., et al. (1992). Long-term weight control study, 1. (weeks 0 to 34)—the enhancement of behavior-modification, caloric restriction, and exercise by fenfluramine plus phentermine versus placebo. *Clin. Pharmacol. Ther.* **51**, 586–594.

Wilding, J., Van Gaal, L., Rissanen, A., Vercruysse, F., and Fitchet, M. (2004). A randomized double-blind placebo-controlled study of the long-term efficacy and safety of topiramate in the treatment of obese subjects. *Int. J. Obes.* **28**, 1399–1410.

Wilding, J. P. H., Gilbey, S. G., Bailey, C. J., Batt, R. A. L., Williams, G., Ghatei, M. A., et al. (1993a,b). Increased neuropeptide Y messenger RNA and decreased neurotensin messenger RNA in the hypothalamus of the obese (*ob/ob*) mouse. *Endocrinology* **132**, 1939–1944.

Williams, G., Cardoso, H. M., Lee, Y. C., Ball, J. M., Ghatei, M. A., Stock, M. J., et al. (1991). Hypothalamic regulatory peptides in obese and lean Zucker rats. *Clin. Sci.* **80**, 419–426.

Wynne, K., Park, A. J., Small, C. J., Patterson, M., Ellis, S. M., Murphy, K. G., et al. (2005). Subcutaneous oxyntomodulin reduces body weight in overweight and obese subjects—A double-blind, randomized, controlled trial. *Diabetes* **54**, 2390–2395.

Zhang, J. V., Ren, P. G., Avsian-Kretchmer, O., Luo, C. W., Rauch, R., Klein, C., et al. (2005). Obestatin, a peptide encoded by the ghrelin gene, opposes ghrelin's effects on food intake. *Science* **310**, 996–999.

Index

A

Abecarnil, 226
AC187, 113
AcbSh. *See* Nucleus accumbens shell
Action–outcome learning, 276
Adaptive food choice, 273
Addiction
  drug, 205–206
  food, 206–208
  incentive sensitization theory of, 205–206
Adipokines, 113
Adiponectin, 326
Adipose tissue
  peptides from, 113–114
  storage of, 248
Adiposity, 112–113, 308
Adiposity signals
  amylin as, 153
  in animals, 308
  drug targeting of, 343–344
  ghrelin as, 157
  leptin as, 168
  modulation of, 167
  peripheral, 167, 179
Ageusia, 10
Agouti gene-related protein, 121, 177
Agouti-related protein
  description of, 52
  neuropeptide Y and, 69
Agriculture-based societies, 302
Alloesthesia, 7
Allometric sense, 194

Amfepranone. *See* Diethylpropion hydrochloride
AMG-076, 329
AMPA, 30
Amygdala
  basolateral, 220
  central nucleus of, 169
  emotional response to feeding regulated by, 173
  food selection and, 220
  taste-specific neurons in, 10
Amylin
  agonists of, 155
  animal models, 154
  antagonism of, 152–153
  in area postrema, 159–160
  as adiposity signal, 153
  blood-borne, 153
  body weight and, 153–155
  brain regions, 153–154
  central mechanisms of, 153–154
  characteristics of, 144t
  cholecystokinin and, 112, 150t, 154
  description of, 35, 112–113
  diabetes mellitus treated with, 155
  functions of, 152
  insulin and, 152
  meal-ending satiation, 152–153
  obesity treatment and, 154–155

peripheral mechanisms of, 153–154
  physiological dose of, 152
  secretion of, 152
  synthesis of, 152
Amylin receptors, 152–153
Anandamide
  description of, 72, 175
  food reward affected by, 198–199
Anhedonia, 202
Anorexia nervosa. *See also* Eating disorders
  brain reward system dysfunction in, 192
  leptin levels in, 180
Antiobesity drugs
  carbonic anhydrase inhibitors, 331
  CB$_1$ receptor antagonists, 325–327, 344
  description of, 337–338
  development of
    active comparators for, 351
    phase I trials, 346–348
    phase II trials, 347–348
    phase III trials, 348–351
    phase IV trials, 351–352
    placebo run-in period, 350
    regulatory requirements, 346–352
    stages in, 347f
  diethylpropion hydrochloride, 339–340

DPP-IV inhibitors, 325t, 332
11β-HSD1 inhibitors, 325t,
    331–332
ephedrine and caffeine, 342
fenfluramine and
    dexfenfluramine, 327,
    339, 341–342
global sales of, 324
history of, 338–339
5-HT$_6$ receptor antagonists,
    325t, 327–328
5-HT$_{2C}$ receptor agonists,
    325t, 327
leptin agonists, 325t, 330–331
MC$_4$ receptor agonists, 325t,
    329
melanin-concentrating
    hormone receptor
    antagonists, 325t,
    328–329
neuropeptide Y receptor
    agonists, 325t, 329–330
opioid receptor antagonists,
    263
orlistat, 323–324, 340
phentermine, 324, 339
phenylpropanolamine, 342
preclinical testing of,
    344–346
regulatory requirements for
    development of,
    346–352
rimonabant, 3, 326, 340–341
sibutramine, 323–324, 340,
    344, 351–352
targets for
    adiposity signals, 343–344
    central nervous system,
        344
    description of, 342
    ghrelin, 332
    metabolism/substrate
        utilization, 344
    nutrient absorption,
        342–343
    peptide YY, 331
    peripheral, 330–332
    peripheral satiety
        enhancement, 343–344
    validation of, 345
testing of, 344–346
theoretical mechanisms, 317
thyroxine and diuretics, 342

topiramate, 331, 341
zonisamide, 341
Antiobesity treatments
amylin, 154–155
description of, 3
drugs. See Antiobesity
    drugs
gastric mechanoreception
    and, 146
ghrelin, 158
global sales of, 324
history of, 338
melanin-concentrating
    hormone and, 81
neuropeptide Y and, 75
nonpharmacological,
    350–351
palatability and, 263–264
weight loss secondary to,
    351
APD356, 327
Apolipoprotein A IV, 144t
Appetite
hedonic stimulation of, 257
learned, for energy, 277–278
palatability effects on
    control of appetite,
        257–262
    hedonics, 250
    hunger–satiety
        interactions, 258–260
    ingestion measures, 249
    laboratory studies,
        248–250
    manipulated palatability
        studies, 250–252
    neurochemical basis,
        253–255
    short-term appetite,
        250–253
for protein, 284–286
sodium, 283–284
Appetite stimulation, 198
2-Arachidonoyl glycerol, 175
Arc, 52
Arcuate nucleus
ghrelin's action in, 160
neurons
    leptin effects on, 120
    neuropeptide Y synthesis
        by, 69, 115
    proopiomelanocortin
        neurons in, 121

Area postrema
amylin's action in, 159–160
cholecystokinin injections'
    effect in, 108
description of, 104
Associative learning, 274, 276,
    294
Australopithecus afraensis, 311
Aversive responses
description of, 225
morphine effects on, 230
Awareness, and dietary
    learning, 290–293

B
Bank voles, 306–309
Basolateral amygdala, 39–40
Benzodiazepine(s)
food preferences and,
    220–221
ingestive responses affected
    by, 225
palatability modulated by,
    228, 231, 237
taste preferences and,
    225–228, 237
taste reactivity and, 225–228
Benzodiazepine receptors
agonists
    description of, 220
    inverse, 226–227
    partial, 226
antagonists, 226–227
central, 226
β cells, 112
Bicuculline, 47
Binge-eating disorder
characteristics of, 282
description of, 179
food cravings in, 282
obesity and, 263
Bitter taste, 197
Body fat
energy content of, 308
regulation of, 168
Body mass index
during famine, 304
mortality and, 311
nonadaptive drift model
    and, 312–313
set point and, 317
trends in, 315
Body size, 304

Body weight
  amylin's effect on, 153–155
  cholecystokinin effects on, 151
  ghrelin's effect on, 157
  glucagon-like peptide 1 receptor ligands' effect on, 115–116
  insulin therapy effects on, 155
  lateral hypothalamus effects on, 176
  melanocortins and, 118
  neural substrates of, 181
  nonadaptive drift model of, 312–314
  nucleus accumbens shell neurons' role in long-term control of, 38
  regulation of, 168, 309
  set-regulation point for, 309
  in small mammals, 306–307
Bombesin-like peptides
  characteristics of, 144t
  description of, 108–109
Brain. *See also specific anatomy*
  cortical sites in, 194
  hedonic hot spots in, 196–197
  "liking" systems in, 3
  mesocorticolimbic systems, 205
  monosynaptic pathways, 44
  opioid systems, 171–175
  "wanting" systems in, 3
Brain reward systems
  description of, 191–192
  development of, 210–211
  dysfunction of, 192
  in eating, 192, 211
  in eating disorders, 192–193
  for food "liking" and "wanting," 193–201
  for food pleasure, 194–195
  passively distorted, 192
  purpose of, 210–211
  regulatory systems and, 210
  resilience in, 193
Brainstem
  cocaine- and amphetamine-regulated transcript peptide in, 121

feeding circuits, descending peptidergic regulation of
  cocaine- and amphetamine-regulated transcript peptide, 121
  corticotrophin-releasing hormone, 119–120
  description of, 119
  melanin-containing hormone, 122
  neuropeptide Y, 120–121
  orexin, 121–122
  oxytocin, 120
  proopiomelanocortin, 121
  urocortin, 119–120
feeding-specific areas in, 105
ghrelin action in, 111
neuropeptides in, 105
prolactin-releasing peptide production in, 116
taste hedonics processed by, 231
Bretazenil, 226
Bulimia nervosa. *See also* Eating disorders
  binge eating behaviors associated with, 180
  characteristics of, 282
  depression secondary to, 180
  leptin levels in, 180
Bupropion, 330
BVT 2733, 332
BVT 3498, 332
BVT 5182C, 327–328

C
Caffeine, 291, 342
Cannabinoids. *See also* Endocannabinoid(s)
  CB$_1$ receptor antagonists, 325–327, 344
  description of, 198–199
Capsaicin
  ghrelin response affected by, 110
  vagal afferent fibers sensitive to, 103
Carbohydrate intake
  morphine effects on, 223
  neuropeptide Y and, 70–71, 173
  obesity and, 74

β-Carboline drugs, 226
Carbonic anhydrase inhibitors, 331, 341
CB1 cannabinoid receptors
  antagonists, 325–327
  description of, 35, 198
Central amygdaloid nucleus, 41
Central nervous system, 344
Central neuropeptide Y, 53
c-Fos
  cholecystokinin effects on description of, 108, 119
  leptin and, 114
  description of, 103
c-*fos*, 43, 103, 197
CGS 8216, 227
Children
  energy delivery differences, 277
  food preferences in, 275, 294
  obesity in, 324
Chlordiazepoxide, 220–221, 228
Cholecystokinin
  amylin and, 112, 150t, 154
  antagonism of, 148
  body weight affected by, 151
  brain areas of, 107
  capsaicin-sensitive vagal afferent fibers, 103
  central mechanisms of, 149
  c-Fos induced by, 108, 119
  characteristics of, 144t
  conditional taste aversion produced by high levels of, 108
  description of, 36–37, 77, 343
  eating affected by, 146–148
  functions of, 107
  gastric loads and, 150
  glucagon-like peptide 1 and, 150t
  intestinal, 146–152
  leptin and, 113–114, 150t
  nucleus of the solitary tract neurons activated by, 104
  obesity and, 151–152
  pancreatic glucagon and, 150t, 150–151
  in paraventricular hypothalamus, 107

peripheral mechanisms of, 149
physiological dose of, 148
receptors for, 147–148
release of, 103, 107
removal of, 148
satiation, 151, 343
secretion of, 147
signals that interact with, 149–151
synthesis of, 146
Cholecystokinin receptors
CCK$_A$, 107, 148–149, 233, 343
CCK$_B$, 107, 149, 343
Clonazepam, 225
Cocaine- and amphetamine-regulated transcript peptide
brain localization of, 121
description of, 33–34, 52
melanin-concentrating hormone and, 78, 121
nociceptin effects on, 35
in paraventricular thalamic nucleus, 40
Conditional place preference, 170, 174
Conditional taste aversion
description of, 108
disruption of, 220
Conditioned flavor preferences
description of, 232–233
dopamine and, 234–235
"electronic esophagus" preparation, 233
opioids and, 233–234
taste reactivity and, 235–236
types of, 234
Conditioned satiety, 279–280
Conditioned stimulus, 232, 276
Conditioning
flavor-flavor, 232
flavor-nutrient, 232–233
Consummatory behavior, 70
Corticotropin-releasing hormone, 77, 119–120
Cortisol, 285
Cranial nerves, 9
Cravings, 282
Cribriform plate, 13
Cue, 276
Cue-potentiated eating, 280

D
DAMGO
description of, 173, 197
feeding inducing by, 33, 48
food consumption affected by, 223
taste palatability affected by, 230
Decerebrate rats, 100
Desoxyephedrine, 339
Dexfenfluramine, 339, 341–342
2DG, 79
Dietary learning, 290–293
Diet–binge cycles, 207
Diethylpropion hydrochloride, 339–340
Diet-induced obesity, 74
Digestion, vaso-vagal reflexes in, 101
Digestive tract, 8
6,7-Dinitroquinoxaline-2,3-dione, 30
Dipeptidyl peptidase IV inhibitors. *See* DPP-IV inhibitors
Dishabituation, 274
Diuretics, 342
DMCM, 226
Dopamine
conditioned flavor preferences and, 234–235
false hedonic mechanisms, 202
feeding behavior affected by, 254
food preferences and, 223–224
mesolimbic systems, 204
nucleus accumbens release
description of, 223–224
sucrose solution consumption affected by, 231
receptors for, 181–182
suppression of, 204
taste reactivity and, 230–231
Dopamine binding, 181
Dopamine receptors
D$_1$, 234
D$_2$, 202, 224, 255
quinpirole, 224

Dopamine reuptake transporter, 171
Dorsal raphe nuclei, 42
Dorsal vagal complex
area postrema, 104
bombesin-like peptides, 108
description of, 101
humoral interactions with, 107–114
intrinsic peptidergic neurons of
description of, 114
glucagon-like peptides. *See* Glucagon-like peptide 1; Glucagon-like peptide 2
melanocortins, 118–119
neuromedin U, 118
neuropeptide Y. *See* Neuropeptide Y
prolactin-releasing peptide, 116–118
meal-related reflexes, 105
negative feedback mechanisms, 105
neural integration by, 103–105
orexin application, 121–122
thyrotrophin-releasing hormone receptors in, 122
DPP-IV inhibitors, 325t, 332
Drift model, nonadaptive, 312–314, 316
Drive reduction, 209–210
Drug(s). *See* Antiobesity drugs
Drug addiction, 205–206
Drug motivation, 259
Duodenum
cholecystokinin released in, 107
distention of, 103

E
Eating
brain reward systems in, 192
cephalic phase of, feed-forward stimulation in, 102
controls of, 100–101
cue-potentiated, 280
gastric mechanoreceptor signaling effects on, 145

hedonic reactions and, 204, 256
hormonal controls of, 147t
innate influences on, 273–274
sensory-specific satiety effects on, 260
Eating behavior
  drive-reduction theory of, 209–210
  hunger's effects on, 209
  learning influences on, 272, 294
Eating disorders. *See also* Anorexia nervosa; Binge-eating disorder; Bulimia nervosa
  brain reward systems in, 192–193
  incentive salience and, 208
  neural substrates of, 181
  nucleus accumbens shell feeding circuit damage and, 54
  weight management and, 282
"Electronic esophagus," 233
11β-HSD1 inhibitors, 325t, 331–332
Emotion
  food motivation and, 6–7
  regulation of, 194
  research regarding, 7
Emotional state, 7
Endocannabinoid(s). *See also* Cannabinoids
  antagonists, 264
  drugs that affect, 264
  energy regulatory signals and, 175
  feeding behaviors modified by, 255
  functions of, 35
  leptin effects on, 175
  nucleus accumbens shell, 35–36
  opioids and, 255
  orexin and, 77
Endocannabinoid type 1 receptor, 175
Energy balance, 324
Energy density
  child's food preferences and, 272

definition of, 260
habitual snacking and, 283
hunger and, 282
palatability and, 260–261
Energy intake, excess, 248
Energy regulatory signals
  brain opioid systems and, 171–175
  endocannabinoids and, 175
  environmental effects on, 169
  food reward circuitry and, 169–170
  lateral hypothalamus area circuitry and, 176–178
  mesolimbic dopamine circuitry and, 170–171
  negative feedback loop and, 168
Enterostatin, 144t
Ephedrine and caffeine, 342
Estradiol, 150t
Estrogen receptor-α, 149
Evaluative conditioning, 287–288
Exenatide, 343
Extinction, 276

F
Facial nerve, 9
Famine
  description of, 302
  evidence regarding, 303–305, 316
  length of, 305
  mortality during, 304–305
  prevalence of, 304–305
  starvation during, 305
Fasting
  glucocorticoids affected by, 72, 171
  insulin affected by, 72, 172
  leptin affected by, 72, 172
Fatty meal, 106
Feed-forward sensory information
  in cephalic phase of eating, 102
  description of, 100
Feeding
  brainstem areas associated with, 105

control of
  brainstem motor systems involved in, 101
  direct, 101
  indirect, 101
  negative sensory feedback from gut, 102–103
  visceromotor control column in, 101–102
  duodenal distention effects on, 103
  endocannabinoids' role in, 175
  forebrain involvement in, 182
  neuropeptide effects on, 105–106
  pancreatic polypeptide effects on, 111
  psychological modulation of, 167
Feeding behavior
  consummatory phase of, 70
  description of, 68
  dopamine effects on, 254
  leptin effects on, 113
  neuropeptide Y stimulation of, 69–70
  nucleus accumbens shell regulation of
    behavioral specificity of changes, 31–32
    description of, 177
    endocannabinoids, 36
    excitatory and inhibitory amino acid circuits in, 30–31
    feeding circuit effects on, 37–38
    hypothalamus, 49–53
    neuropeptide systems, 32–37
    overview of, 28–29
    ventral pallidum in, 47–49
  opioids role in, 172
  orbitofrontal cortex and, 178
  oxytocin effects on, 120
  regulation of
    brainstem motor systems in, 101
    description of, 100

nucleus accumbens shell
in. *See* Feeding behavior,
nucleus accumbens
shell regulation of
subcortical forebrain regions
involved in, 28
suppression of, 54
Feelings, 7
d-Fenfluramine, 327, 339,
341–342
FG 7142, 226–227
Flavor
energy consumption and,
272
neural encoding of, 256
pleasantness of, 260f
Flavor preferences
acquisition of, 232, 237
characteristics of, 236
conditioned
description of, 232–233
dopamine and, 234–235,
237
"electronic esophagus"
preparation, 233
opioids and, 233–234
taste reactivity and,
235–236
types of, 234
learned
acquisition of, 232
description of, 218
evidence of, 278
naltrexone effects on, 234,
236
in rats, 232
sucrose-conditioned, 234
in young children, 271–272
Flavor-consequence learning
description of, 261, 277
flavor-flavor learning vs.,
286–287
principles of, 289
Flavor-flavor conditioning,
232, 291–292
Flavor-flavor learning
animal studies of, 286–287
as evaluative conditioning,
287–288
description of, 261, 286, 294
dietary behavior and,
289–290
evidence for, 286–287

flavor-consequence learning
vs., 286–287
human studies of, 287–289
Flavor-nutrient conditioning
description of, 232–233, 237
"electronic esophagus"
testing, 233
Flavor-nutrient pairing, 291
Flavor-postingestive
consequences, 291
Flavor-preference learning, 291
Flumazenil, 226–227
Flupentixol, 173
Food
amygdala's role in selection
of, 220
availability of, 218
choice of, 217–218
cortical representations of
identity of, 6
energy density of, 260–261
hedonic functions, 196
palatability of. *See*
Palatability
protein content of, 284–285
sensing of, 8
sensory properties of, 282
Food addiction, 206–208
Food anticipatory behavior
neuropeptide Y's role in, 69
paraventricular thalamic
nucleus' role in, 40
Food choice
adaptive, 273
neuropeptide Y effects on, 70
regulation of, 282–283
Food consumption
benzodiazepine receptor
agonists' effect on, 220,
227
DAMGO effects on, 223
homeostatic processes, 6
opioids' effect on, 221
palatability effects on, 237
Food deprivation
conditional place preference
affected by, 174
melanin-concentrating
hormone levels affected
by, 79
neuropeptide levels in
nucleus accumbens
shell affected by, 37

opioid levels affected by,
172
orexin neurons affected by,
76
Food intake
CB$_1$ receptor antagonist
effects on, 326
cholecystokinin effects on,
36
computational principles of,
6, 20
cortical processing involved
in, 5–6, 20–21
ghrelin effects on, 111
glucagon-like peptide 1
effects on, 343
hedonic component of, 6–7
homeostatic processes
involved in, 6
5-HT$_{2C}$ receptor agonist
effects on, 327
lateral hypothalamus
regulation of, 176
melanin-concentrating
hormone effects on, 79,
177
naltrexone effects on, 255f
neural mechanisms that
regulate, 7
neuropeptides' effect on
description of, 68–69
neuropeptide Y, 71
nucleus accumbens shell
feeding circuit's role in,
55
nutritional requirements of,
7
opioids effect on, 172–173
orexin effects on, 53, 75–76
palatability effects on, 253
prolactin-releasing peptide
effects on, 117
sensory inputs involved in
olfactory system. *See*
Olfactory system
overview of, 8
taste system. *See* Taste
system
somatosensory input in, 16
subcortical forebrain regions
involved in, 28
summary of, 20
swallowing processes, 16

Food "liking"
    GABA effects on, 199
    lateral hypothalamus lesion
        effects on, 200
Food motivation
    drug motivation and, 259
    emotion and, 6–7
Food neophobia, 219–220
Food pleasure, 194–195
Food preferences
    benzodiazepines and,
        220–221
    characteristics of, 236
    in children, 294
    dopamine and, 223–224
    expressing of, 218
    naloxone effects on, 222
    opioids and, 221–223
    pharmacology of, 219
    in rats, 219–220, 232
    studies of, in children, 275
Food restriction, 71
Food rewards
    central nervous system's role
        in, 169
    conditional place preference
        studies, 170
    definition of, 168
    description of, 168
    energy regulatory signals
        and, 169–170
    hedonic value, 209
    interactions with other
        reward systems, 182
    neurotransmitter systems
        involved in, 253
    opioid neurotransmitters in,
        196
    prefrontal cortex's role in,
        180
Food supply, 315–316
Food "wanting," 200
Food-aversion learning, 292
Forebrain
    brainstem connection to, 195
    feeding behavior regulated
        by subcortical regions
        of, 28
    in food reward/motivation,
        178
    limbic hedonic hot spot, in
        nucleus accumbens,
        195–199

Fos
    expression of, 43
    muscimol injections' effect
        on, 50–51
Fos plume mapping technique,
    197
Frontal operculum
    description of, 10
    gustatory area in, 11
β-Funaltrexamine, 228

G
GABA
    description of, 226
    "liking" affected by, 199
    in nucleus accumbens shell,
        30
GABA receptors
    A, 33, 38, 48, 52, 226
    B, 38, 52
    blockers of, 50
    in perifornical lateral
        hypothalamus, 51
GABA transaminase, 30
Galanin, 119
Galanin-like peptide, 77
Gastric distention, 104
Gastric emptying
    corticotrophin-releasing
        hormone effects on, 120
    description of, 145–146
    glucagon-like peptide 1
        effects on, 343
Gastric mechanoreception,
    145–146
Gastrin-releasing peptide,
    108
Genetic predisposition
    alternative models, 306–314
    description of, 303, 306
    models of, 72–74
Gestalt, 286
Ghrelin
    antagonism of, 156–157
    in arcuate nucleus, 160
    as adiposity signal, 157
    body weight control, 157
    characteristics of, 144t
    description of, 80, 110–111
    discovery of, 155
    drug targeting of, 332
    eating control and, 156–157
    functions of, 155–156

knockout mouse model, 156
    mechanisms of action,
        157–158, 344
    obesity treatments and, 158
    orexigenic effect of, 156
    physiological dose of, 156
    secretion of, 156
Ghrelin receptors, 156–158
Glicentin, 109
Glossopharyngeal nerve, 9
Glucagon, 144t
Glucagon-like peptide 1
    aversive stimuli regulated
        by, 116
    brain distribution of, 115
    characteristics of, 144t
    cholecystokinin and, 150t
    definition of, 332, 343
    description of, 77, 80
    Exenatide, 343
    food intake affected by, 343
    gut production of, 109, 115
    inhibitors of, 332
    interoceptive stress signals
        relayed by, 116
    neurons, 115
    orexinergic neurons
        stimulated by, 122
    paraventricular nucleus
        injections of, 116
    receptor, 115
Glucagon-like peptide 2
    brain distribution of, 115
    description of, 109
    gut production of, 109, 115
    neurons, 115
Glucocorticoids
    fasting effects on, 72, 171
    melanin-concentrating
        hormone regulation
        and, 80
    orexin system regulation
        and, 77
GPR39, 332
Growth hormone secretagogue
    receptor, 155, 332
Gut
    cholecystokinin from. See
        Cholecystokinin
    negative sensory feedback
        from, 102–103
    peptides, 144t
Gut hormone, 279

Gut-brain signals
   advantages of, 159
   disadvantages of, 159
   meal size affected by, 160
   summary of, 159

H
Habituation, 274
Haloperidol, 231
Hedonic behavior
   anandamide effects on, 199
   endocannabinoids effect on,
      255
   food intake and, 6
   incentive motivation and, 7
   need states and, 7
   neural basis of, 255–257
   neuroimaging studies of,
      17–20, 18f
   opioid involvement in, 254
   sensory systems and, 21f
   ventral pallidum effects on,
      201
Hedonic systems
   brain areas, 196–197
   brainstem processing of, 231
   neuropeptide Y and, 72
   opioid effects on, 231
Hemispheric specialization, of
      odor processing, 15
High fat diet
   description of, 173
   neural inhibition of learned
      appetite affected by, 294
   palatability of, 218
Hippocrates, 315, 317
Homeostatic processes, 6
Hominids, 311
5-HT, 324
5-HT$_6$ receptor antagonists,
      325t, 327–328
5-HT$_{2C}$ receptor agonists, 325t,
      327
Human genome, 312
Hunger
   eating behavior and, 209
   energy density and, 282
   mesolimbic activation by,
      207
   satiety and, 258–260
Hunter-gatherers, 304
6-Hydroxydopamine, 230
Hyperphagia, 113, 307

Hypocretin, 75
Hypophagia, 108
Hypothalamo-pituitary-
      adrenal system, 120
Hypothalamus
   galanin production in, 119
   ghrelin, 110
   hindbrain and, 70
   lateral. *See* Lateral
      hypothalamus
   medial, 52–53
   mediobasal, 169
   melanin-concentrating
      hormone in, 79–80
   neuropeptide Y in, 69
   nucleus accumbens shell
      projections, 41–42
   paraventricular, 107
   pathways of, 68

I
Implicit learning, 290
Impulsive eating, 180
Incentive motivation, 7
Incentive salience, 205, 208
Incentive sensitization theory
   of drug addiction, 205–206
   of food addiction, 206–208
Infralimbic cortex
   food deprivation effects on,
      39
   systemic processes
      controlled by, 39
Ingestive responses, 225
Instrumental learning, 276
Insulin
   amylin and, 152
   body weight affected by, 155
   description of, 112
   fasting effects on, 72, 171
   hypoglycemia induced by,
      76
   lateral hypothalamus self-
      stimulation affected by,
      176
   striatal dopamine reuptake
      and, 171
Insulin receptors, 112
Interganglionic laminar
      endings, 102–103
Intermuscular arrays, 102–103
Inverse agonists, 226
Islets of Langerhans, 111

J
Jacobson's organ, 13

K
K cells, 344

L
L cells, 343
Lateral hypothalamus
   description of, 49–52, 75
   energy regulatory signals
      and, 176–178
   food intake regulated by,
      176, 199
   food reward and, 169, 199
   food "wanting" affected by,
      200
   glucose sensing neurons in,
      176
   lesions of, 200
   melanin-concentrating
      hormone expression in,
      177
   orexin production by, 121
   prepro-orexin expression in,
      77
   self-stimulation studies, 176
   studies of, 169
Lateral olfactory tract, 13
Lateral orbitofrontal cortex, 19
Learned appetite
   description of, 277–278
   nutrient-specific, 283–286
Learned flavor preferences
   acquisition of, 232
   description of, 218
   evidence of, 278
Learned response, 276
Learned satiety
   description of, 261, 280
   flavor-preference learning
      and, 291
Learned taste aversion
   description of, 220
   inhibition of, 274
Learning
   action–outcome, 276
   associative, 274, 276, 294
   definition of, 272
   description of, 271–272
   dietary, 290–293
   eating behavior influenced
      by, 272

fetal, 284
flavor-consequence
    description of, 261, 277
    flavor-flavor learning vs., 286–287
    principles of, 289
flavor-flavor. *See* Flavor-flavor learning
flavor-postingestive consequences, 291
flavor-preference, 291
food-aversion, 292
habituation, 274
implicit, 290
instrumental, 276
nonassociative, 274, 294
two-stage process of, 276
types of, 274–276
Leptin
    adiposity and, 308
    in anorexia nervosa, 180
    arcuate nucleus neurons affected by, 120
    as adiposity signal, 168
    binge eating disorder and, 179
    in bulimia nervosa, 180
    cholecystokinin and, 113–114, 150t
    clinical trials of, 317
    definition of, 308
    dorsal raphe neuron accumulation of, 42
    drug targeting of, 330–331
    endocannabinoids affected by, 175
    fasting effects on, 72, 171
    feeding behavior affected by, 113
    food intake suppression and, 108
    gastric, 144t
    genetic models of, 175
    melanin-concentrating hormone affected by, 79–80
    night eating syndrome and, 179
    orexin affected by, 76–77
Leptin receptor, 175
Leptin resistance, 316–317
Life expectancy, 314

"Liking"
    brain systems for, 3
    "wanting" and, 203–204
Limbic reward system, 169
Lipostatic regulation system, 307–308, 316
Lithium chloride, 108
L-type enteroendocrine cells, 109

M

Macronutrients, 222–223, 277
Marijuana, 198
MC$_4$ receptor agonists, 325t, 329
Meal(s)
    amylin effects on, 112
    cortisol levels after, 285
    stimuli related to, 102–103
Meal size
    feedback systems, 253
    learned control of, 278–282
    palatability and, 258
    signals that affect, 160
Mechanoreceptors, 145–146
Medial hypothalamus, 52–53, 168
Medial prefrontal cortex, 38–39
Median raphe nuclei, 42
Mediodorsal thalamic nucleus, 47–48
Melanin-concentrating hormone
    antiobesity treatment and, 81, 328–329
    chronic injections of, 79
    cocaine- and amphetamine-regulated transcript peptide and, 78, 121
    degradation of, 79
    description of, 34, 78
    2DG effects on, 79
    in DIO rats, 80
    food intake stimulated by, 79, 177
    glucocorticoids regulation of, 80
    hypothalamic, 79–80
    in lateral hypothalamus, 177
    leptin effects on, 79–80
    neurons, 122, 177
    neuropeptide Y and, 80
    obesity and, 80–81
    ob/ob mouse model, 80

orexigenic effects of, 78–79
    OX1-receptors on neurons of, 78
Melanin-concentrating hormone receptors
    antagonists, 325t, 328–329
    description of, 79
    types of, 328
Melanocortin-4 receptor, 179
Melanocortins, 77, 118–119, 329
Meridia. *See* Sibutramine
Mesocorticolimbic dopamine neurons, 170
Mesolimbic systems
    dopamine, 204
    hunger, 207
    hypothalamic systems and, 210
Metabolic expectancies, 281
Midazolam, 228
Midbrain dopaminergic cell bodies, 169
Minimal circuit element analysis, 101
Morphine
    aversive responses affected by, 230
    carbohydrate intake affected by, 221, 223
    description of, 48
    saccharin intake and, 229
    sweet solution consumption affected by, 228–229
Mortality
    body mass index and, 311
    during famines, 304–305
Motivation
    drive theories of, 7
    drug, 259
    food, 6–7, 259
    incentive, 7
Motivational hedonics, 20
Mucosal terminals, 102–103
Muscimol
    feeding induced by, 33
    intra-basolateral amygdala injections of, 40

N

Naloxone
    conditioned flavor preference suppressed by, 233–234

description of, 172–174
food preferences affected by,
    221–222
salt solution consumption
    affected by, 229
Naltrexone
    description of, 173, 177
    flavor preferences and, 234,
        236
    food intake affected by, 255f
    palatability affected by, 254
    salt solution consumption
        affected by, 229
Nausea-producing agents, 108
Need states
    hedonic behavior and, 7
    types of, 7
Neolithic period, 314
Neophilia, 219
Neophobia, 219–220, 274–275
Neuromedin B, 108
Neuromedin U, 118
Neuromodulators, 68
Neurons
    glucagon-like peptide 1, 115
    glucagon-like peptide 2, 115
    lateral hypothalamus, 49
    melanin-concentrating
        hormone, 122, 177
    mesocorticolimbic
        dopamine, 170
    neuropeptide Y, 53
    nucleus of the solitary tract,
        104
    oxytocin, 120
    proopiomelanocortin,
        118–119
Neuropeptide(s)
    ablation of, using
        transgenesis, 106
    amylin, 35
    cholecystokinin, 36–37
    classification of, 68
    cocaine- and amphetamine-
        regulated transcript
        peptide, 33–34
    definition of, 100
    description of, 32, 68
    endocannabinoids, 35–36
    factors that affect, 105
    feeding affected by, 105–106
    food intake affected by,
        68–69

gut function affected by, 106
melanin-concentrating
    hormone, 34
nociceptin, 35
opioid receptor-like 1
    receptor, 35
opioids, 32–33
orexin. See Orexin
role of, 105–107
summary of, 122
transgenesis-related ablation
    of, 106
Neuropeptide W, 77
Neuropeptide Y
    agouti-related protein and, 69
    antibodies to, 70
    antiobesity treatments and,
        75
    arcuate nucleus neurons, 69
    carbohydrate intake affected
        by, 70–71, 173
    central, 53
    consummatory behavior
        reduced by, 115
    description of, 52, 69, 329
    dietary preferences and,
        70–71
    food choice affected by, 70
    food intake induced by, 71
    food restriction effects on, 71
    functions of, 69
    glucose availability effects
        on, 71–72
    hedonic systems and, 72
    in hypothalamus, 69
    macronutrients and, 70–71
    melanin-concentrating
        hormone and, 80
    murine studies of, 74
    negative energy balance
        activation of, 71
    obesity and, 72–75
    orexigenic effects of, 69–70,
        114
    orexin and, interactions
        between, 76
    in paraventricular nuclei, 71
    Zucker rat studies of, 72–74
Neuropeptide Y receptors
    agonists and antagonists of,
        75
    antiobesity treatment uses,
        329–330

brain regions of, 115
description of, 72, 329
Y2, 329–330
Y4, 330
Y5, 329, 345
Neurotransmitters, 253
Newborns
    sodium appetite in, 284
    sweet taste responses,
        224–225
Night eating syndrome, 179
NMDA receptors, 50
Nociceptin, 35
Nonadaptive drift model,
    312–314, 316
Nonassociative learning, 274,
    294
Nor-binaltorphimine, 228
Nucleus accumbens
    DAMGO administration in,
        223
    dopamine release in
        description of, 223–224
        sucrose solution
            consumption affected
            by, 231
    forebrain limbic hedonic hot
        spot in, 195–199
    glutamate blockade, 199
    hedonic neurotransmitters
        in, 198
    hypothalamus and, 210
    opioid modulation of, 254
    output projections, 199
    subdivisions of, 195
Nucleus accumbens shell
    afferent projections to
        basolateral amygdala,
            39–40
        caudal midbrain, 42
        dorsal raphe nuclei, 42
        hypothalamus, 41–42
        medial prefrontal cortex,
            38–39
        median raphe nuclei, 42
        nucleus of the solitary
            tract, 42
        overview of, 38
        paraventricular thalamic
            nucleus, 40–41
        ventral pallidum, 41, 47–49
    amino acid systems, 30
    AMPA effects on, 30–31

behavioral changes effected by, 43
caudal, 39
connectivity in, 29
description of, 195–196
excitatory amino acid inputs, 30–31, 38–41
fast synaptic activity in, 30
feeding behavior regulation by
  behavioral specificity of changes, 31–32
  description of, 177
  endocannabinoids, 36
  excitatory and inhibitory amino acid circuits in, 30–31
  feeding circuit effects on, 37–38
  hypothalamus, 49–53
  neuropeptide systems, 32–37
  overview of, 28–29
  ventral pallidum in, 47–49
feeding circuit
  description of, 29, 53–54
  downstream components of, 43–53
  food intake and, 54
  homeostatic feeding behavior regulated by, 37–38
  immediate-early gene studies of, 43–44
  pathologic and therapeutic interventions for eating disorders, 53–55
  postnatal damage to, 54
  reentrant projections in, 43
  studies of, 43–45
Fos expression, 43, 45
GABAergic neurons, 45
GABA-mediated feeding in, 30
glutamate levels in, 37
hyperphagia, 52
inhibition of
  brain regions activated by, 45–47
  description of, 51
  medial hypothalamus neurons activated by, 52
  unilateral, 46–47

inhibitory amino acid circuits in, 30–31
inputs to, 28
medial
  chemical inhibition of neurons in, 30
  cortical input to, 39
  hypothalamic projections, 50
monosynaptic pathways, 44
neural activity in, 30
neurons, body weight control affected by, 38
neuropeptide systems
  amylin, 35
  cholecystokinin, 36–37
  cocaine- and amphetamine-regulated transcript peptide, 33–34
  description of, 32
  endocannabinoids, 35–36
  food deprivation effects on, 37
  melanin-concentrating hormone, 34
  nociceptin, 35
  opioid receptor-like 1 receptor, 35
  opioids, 32–33
  orexin, 34–35
neurotransmitters that affect, 30
opioids, 173, 197–198
overview of, 27–28
recurrent collaterals from, 28
Nucleus of the solitary tract
  bombesin-like peptides' effect, 108
  caudal, 42
  c-Fos expression, 104
  cholecystokinin effects on, 104
  cholecystokinin$_A$ receptors in, 107
  description of, 6, 9, 149
  medial subnucleus of, 103–104
  neuromedin U, 118
  neurons, 104
  neuropeptide Y-positive neurons, 114

nucleus accumbens shell projections, 42
orexin effects, 122
preproglucagon neurons in, 116
prolactin-releasing peptide production in, 116
topographic organization of, 103–104

O
Obesity
  binge-eating disorder and, 263
  body mass index and, 304
  in children, 324
  cholecystokinin and, 151–152
  diet-induced, 74
  diseases and conditions associated with, 324, 352
  disordered eating associated with, 179
  dopamine receptors in
    D$_2$ receptor, 202
    description of, 181
  environment that promotes, 168
  externality theory of, 262
  famine hypothesis of, 302
  food overconsumption and, 218, 247–248
  gene–environment interactions, 301–303
  genetic predisposition to
    alternative models, 306–314
    description of, 303, 306
    models of, 72–74
  global prevalence of, 324
  health risks associated with, 168
  historical descriptions of, 315
  inducement of, in laboratory animals, 74
  mediation of, 264
  neuropeptide Y and, 72–75
  overview of, 1
  palatability and, 262–264
  prevalence of
    description of, 302–303
    global, 324

historical description of,
  316
reasons for increase in,
  262–263, 301
statistics, 316, 323–324
time trends in, 314–316
thrifty genotype and, 303
Obesogenic environment, 306
Obestatin, 344
ob/ob mouse model, 80, 308
Ob-Rb, 113
Odors
  detection of, 15
  locations and, 293
  processing of, 15
  taste associations with, 17
Odor–taste integration, 256
Olfaction
  cells involved in, 13
  dissociable encoding of, 16
  neuroimaging studies of,
    15–16
  vision and, 17
Olfactory cortex, 13–15
Olfactory system
  description of, 256
  food evaluations by, 8
  human lesion studies, 14–15
  orexins in, 77
  vomeronasal, 13
1229U91, 77
Operant conditioning, 276
Opioid(s)
  cannabinoids and, 199
  conditioned flavor
    preferences and,
      233–234
  description of, 196
  endocannabinoids and,
    255
  endogenous
    description of, 221
    salt taste palatability
      affected by, 229
  energy state's effect on, 174
  food consumption affected
    by, 221
  food intake affected by, 172
  food preferences and,
    221–223
  food rewards and, 171
  hedonic behavior and, 254
  hedonic hot spots, 198

macronutrient intake
  affected by, 222–223
neural basis of, 254
nucleus accumbens areas
  modulated by, 254
in nucleus accumbens shell
  control of feeding
  behavior, 32–33
palatability and, 221,
  253–254, 263
Opioid antagonists
  conditioned flavor
    preferences affected by,
      233, 237
  description of, 172
  food preferences and,
    222–223, 228
  obesity treated with, 263
Opioid receptor(s)
  kappa, 174
  mu, 39, 174, 196–198, 228
  in ventral striatum, 32
Opioid receptor-like 1 receptor,
  35
Orbitofrontal cortex
  anterior, 17
  damage to, 19
  description of, 5
  efferent connections to, 178
  feeding behaviors and, 178
  food preferences affected by
    lesions in, 20
  in hedonic experiences, 19
  information flow linked to,
    9f
  lateral, 19
  lesions studies in, 19–20
  in motivational hedonics, 20
  in odor–taste integration,
    256
  olfactory pathways, 14
  pain studies, 19
  reward-based activation of,
    19–20
  studies of, 181
Orexigenic neuropeptides, 2
Orexin
  antagonists of, 78
  down-regulation of, 78
  endocannabinoids and, 77
  food intake stimulated by,
    53, 75–76
  forms of, 75

glucocorticoids effect on, 77
hyperphagia, 76
lateral hypothalamus
  production of, 121
leptin effects on, 76–77
neuromodulators that
  interact with, 77
neuropeptide Y and,
  interactions between, 76
nucleus of the solitary tract
  affected by, 122
obesity and, 77–78
in *ob/ob* mouse models,
  77–78
in olfactory system, 77
perifornical lateral
  hypothalamic, 53
regulation of, 76–77
sleep–wake cycle regulation
  by, 75, 121
synthesis of, 75
in Zucker rat models, 78
Orexin peptides, 34–35
Orexin receptors
  activation of, 51
  description of, 34–35
  leptin effects on, 77
  OR1-R, 76
  OR2-R, 76
Orlistat, 323–324, 340
Orosensory stimulation, 260
Otsuka Long-Evans
  Tokushima Fatty rat,
    117
Overconsumption
  obesity and, 218, 247–248
  palatability and, 218, 237
Overweight, 302
Oxyntomodulin, 109, 115, 343
Oxytocin, 120

P
Palatability
  antiobesity treatments and,
    263–264
  appetite affected by
    control of, 257–262
    hedonics, 250
    hunger–satiety
      interactions, 258–260
    ingestion measures, 249
    laboratory studies,
      248–250

manipulated palatability studies, 250–252
neurochemical basis, 253–255
opioids effect, 263
short-term appetite, 250–253
as hedonic expression of homeostatic controls, 257–258
benzodiazepine modulation of, 228, 231, 237
DAMGO effects, 230
definition of, 252
description of, 172, 174, 237
energy density of foods and, 260–261
factors that affect, 261–262
food consumption affected by, 237
food overconsumption and, 218, 237
influences on, 261–262
learned satiety and, 261
meal size and, 258
naloxone effects on, 223
naltrexone effects on, 254
neurochemical basis of, 253–257
obesity and, 262–264
of high fat diet, 218
of saccharin solutions, 229
opioids and, 221, 253–254, 263
oversensitivity to, in obesity, 262–263
positive feedback mechanisms and, 253
responsivity to, 263
salt taste, 229
satiety and, 258–260
sweet taste, 228
taste reactivity measures for assessment of, 227
ventral striatum effect on, 256
Paleolithic period, 314–315
Pancreatic polypeptide, 111–112
Parabrachial nucleus, 107
Paraventricular nucleus
corticotrophin-releasing hormone neurons in, 120

description of, 40–41, 169
energy homeostasis role of, 173
Pars tuberalis, 118
Pavlovian-instrumental transfer, 276
Pelman vs. McDonalds, 292
Peptide YY
characteristics of, 144t
definition of, 331
description of, 104, 109–110, 159, 344
PYY$_{3-36}$, 109, 329, 331
Phase I trials, 346–348
Phase II trials, 347–348
Phase III trials, 348–351
Phase IV trials, 351–352
Phentermine, 324, 339
Phenylpropanolamine, 342
Phenylthiocarbamide, 289
Pimozide, 230–231
Piriform cortex, 15
Pleasure
brain systems for, 194–195
definition of, 193
food, 194–195
hierarchy of, 195
objective aspects of, 193–194
ventral pallidum effects on, 200
Pleasure "liking"
description of, 194–195
opioid effects on, 197–198
Point mutations, 312
Postrestriction hyperphagia, 307, 310
Pramlintide, 155
Prefrontal cortex, 178
Preproglucagon, 115
Prepro-orexin, 77
Primary olfactory cortex, 13–15
Primates, 311
Prolactin-releasing peptide, 116–118
Prolactin-releasing peptide receptor, 116
Proopiomelanocortin
description of, 52
neurons
in arcuate nucleus, 121
brain regions of, 121
description of, 118–119
peptide YY effects on, 110

Protein, appetite for, 284–286
Protein–energy malnutrition, 285
Pyloric cuff technique, 146

Q
Quinpirole, 224

R
Raclopride, 181, 234, 254
Radafaxine, 330
RAMPs, 153, 155
Rats
flavor preferences in, 232
food preferences in, 219–220, 232
salt appetite in, 284
wild, 219
Zucker. *See* Zucker rats
Reductil. *See* Sibutramine
Rewards
brain systems. *See* Brain reward systems
drive reduction and, 209–210
food. *See* Food rewards
hedonic impact of, 7
incentive salience of, 7
motivated behaviors elicited by, 6–7
olfactory processing and, 16
orbitofrontal cortex activation and, 19–20
representations of, 6
"wanting," suppression of, 202
RFamide receptors, 117
Rimonabant, 3, 326, 340–341
Ro15-4513, 227
Ro17-1812, 226

S
Saccharin, 228–229
Salmon calcitonin, 155
Salt taste palatability, 229
Satiety
cholecystokinin and, 343
conditioned, 279–280
hunger and, 258–260
learned, 261, 280, 291
neuropeptide Y effects on, 70
palatability and, 258–260
sensory-specific
definition of, 10, 260, 274

eating cessation and, 260
example of, 275
neuroimaging studies, 17
onset of, 274
Satiety factors, 113
SCH 23390, 234
Sensory processing, 8
Sensory systems
in food intake, 8
hedonic systems and, 21f
multimodal integration of,
16–17
olfaction. *See* Olfaction
smell. *See* Smell sense
taste. *See* Taste(s)
Sensory-specific satiety
definition of, 10, 260, 274
eating cessation and, 260
example of, 275
neuroimaging studies, 17
onset of, 274
Serotonin. *See* 5-HT
Set-point control systems
body weight and, 311–312
criticism of, 309–310
description of, 307–308, 317
drifting set points, 317
evolution and, 308–309, 315
in obese individuals, 310
settling point system vs.,
310
Sibutramine, 323–324, 340, 344,
351–352
Signaling peptides, 105
SLC-1, 79
Smell sense
food evaluations by, 8
olfactory receptors, 13
taste and, cross-modal
integration between, 17
Sodium appetite, 283–284
Somatosensory cortex, 10
SR141716, 325
Starvation, 305
STAT3, 113
"Sublenticular extended
amygdala," 200
Subsistence agriculture, 315
Subthalamic nucleus, 178
Sucrose
abecarnil effects on
consumption of, 226
"liking" of, 200

opioid receptor antagonists'
effect on consumption
of, 228
solutions, dopamine release
in nucleus accumbens
affected by, 231
Swallowing, 16
Sweet solution consumption
morphine effects on, 228–229
naloxone effects on, 228–229
Sweet taste
"disliking" of, 197
FG 7142 effects on, 227
innate preferences, 273
newborn responses, 224–225
opioid drug effects on, 197,
228
palatability, 228, 231
responses to, 224
salt taste and, 229

T
Tasofensine, 330
TASR38, 273
Taste(s)
brain areas receiving
information about, 10
classification of, 224
hedonic value of, 209
motivational states' effect
on, 10
neural correlates of, 12
neuroimaging studies of
description of, 11
motivation-independent
representations, 11–12
reward-dependent
representations, 12–13
odor associations with, 17
odor–taste integration, 256
qualities of, 9
responses to, 224
smell and, cross-modal
integration between, 17
types of, 224.256
Taste aversion
conditional
description of, 108
disruption of, 220
elicitation of, 225
glucagon-like peptide 1 and,
116
learned, 220

Taste buds, 8–9
Taste "liking"
description of, 194, 201
false hedonic mechanisms,
201
individual differences in,
289
Taste preference
benzodiazepines and,
225–228, 237
measurement of, 224–225
Taste reactivity
benzodiazepines and,
225–228
chlordiazepoxide effects on,
228
conditioned flavor
preferences and,
235–236
dopamine and, 230–231
measurement of, 224–225,
227
Taste reactivity tests
description of, 225, 229
haloperidol, 231
Taste system
central pathways of, 9–10
cranial nerves involved in, 9
higher order cortices, 10
human lesion studies, 10–11
primary taste cortex, 10
receptor cells of, 8–9
ventral posterior medial
nucleus of the thalamus
in, 9–10
Tenuate Dospan. *See*
Diethylpropion
hydrochloride
Thalamus, 9
Thrifty genes, 302, 304–305
Thyrotrophin-releasing
hormone, 122
Thyroxine and diuretics, 342
Topiramate, 331, 341
Transgenesis, 106
Treatments. *See* Antiobesity
treatments

U
Umami, 9
Unconditioned stimulus, 232,
276
Urocortin, 119–120

V

Vagus nerve, 9
Vaso-vagal reflexes, 101, 106
Ventral pallidum
  bicuculline effects, 48
  description of, 200
  hedonic impact affected by,
    201
  inputs to, 48
  lateral hypothalamus
    projections, 50
  medial, 54
  monosynaptic projection, 49
  morphine effects, 48
  mu receptors in, 48
  nucleus accumbens shell
    projections, 41
  in nucleus accumbens
    shell-mediated feeding,
    47–49

pleasure enhancement by, 200
posterior, 201
subcommissural, 47–48
Ventral posterior medial
    nucleus of the thalamus,
    9
Ventral striatum, 256
Visceromotor control column,
    101–102
Voluntary stimulus sampling,
    225
Vomeronasal olfactory system,
    13
VR-1, 103

W

"Wanting"
  addiction and, 205–206
  brain systems, 3
  cognitive goals and, 208

cue-triggered, 205, 206
definition of, 203, 205
incentive, 203
lateral hypothalamus effects,
  200
"liking" vs., 203–204
White adipose tissue, 113
World War I, 316

X

Xenical. *See* Orlistat

Z

Zonisamide, 341
Zucker rats
  neuropeptide Y studies in,
    72–74
  orexin studies in, 78
  prolactin-releasing peptide
    studies in, 117

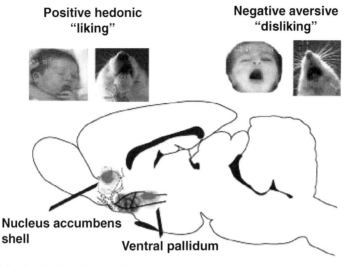

Opioid hedonic hot spots

FIGURE 1 "Liking" reactions and brain hedonic hot spots. Top: Positive hedonic "liking" reactions are elicited by sucrose taste from human infant and adult rat (e.g., rhythmic tongue protrusion). By contrast, negative aversive "disliking" reactions are elicited by bitter quinine taste. Lower: Forebrain hedonic hot spots in limbic structures where μ-opioid activation causes a brighter pleasure gloss to be painted on sweet sensation. Red/yellow shows hot spots in nucleus accumbens and ventral pallidum where opioid microinjections caused the biggest increases in the number of sweet-elicited "liking" reactions. Modified from Peciña and Berridge (2005) and Smith and Berridge (2005).

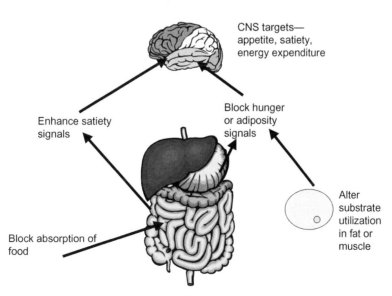

FIGURE 1 Drug Targets for obesity treatment. Targets fall into four main groups: those affecting nutrient absorption; those affecting peripheral signals of hunger or satiety; drugs that act on substrate storage or utilization, and CNS targets that influence energy intake and/or energy expenditure.

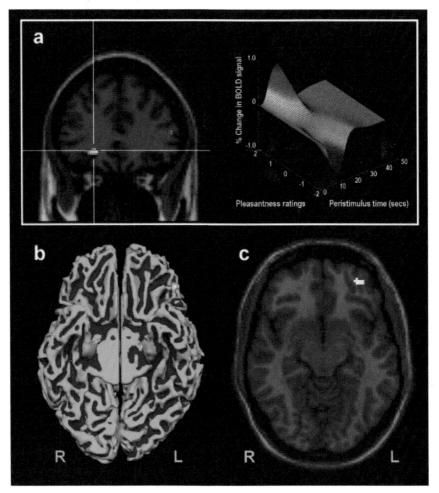

FIGURE 2 Hedonic experience. (a) A neuroimaging study using selective satiation found that mid-anterior parts of the orbitofrontal cortex are correlated with the subjects' subjective pleasantness ratings of the foods throughout the experiment (Kringelbach et al., 2003). On the right is shown a plot of the magnitude of the fitted hemodynamic response from a representative single subject against the subjective pleasantness ratings (on a scale from −2 to +2) and peristimulus time in seconds. (b) Additional evidence for the role of the orbitofrontal cortex in subjective experience comes from another neuroimaging experiment investigating the supra-additive effects of combining the umami tastants monosodium glutamate and inosine monophosphate (De Araujo et al., 2003a). The figure shows the region of mid-anterior orbitofrontal cortex showing synergistic effects (rendered on the ventral surface of human cortical areas with the cerebellum removed). The perceived synergy is unlikely to be expressed in the taste receptors themselves and the activity in the orbitofrontal cortex may thus reflect the subjective enhancement of umami taste which must be closely linked to subjective experience. (c) Adding strawberry odor to a sucrose taste solution makes the combination significantly more pleasant than the sum of each of the individual components. The supralinear effects reflecting the subjective enhancement were found to significantly activate a lateral region of the left anterior orbitofrontal cortex, which is remarkably similar to that found in the other experiments (De Araujo et al., 2003c).